STELLAR ASTROPHYSICS

ASTROPHYSICS AND SPACE SCIENCE LIBRARY

VOLUME 254

STELLAR ASTROPHYSICS

Proceedings of the Pacific Rim Conference held in Hong Kong, 1999

edited by

K.S. CHENG
H.F. CHAU
University of Hong Kong

K.L. CHAN
Hong Kong University of Science and Technology

and

K.C. LEUNG
University of Nebraska

KLUWER ACADEMIC PUBLISHERS
DORDRECHT / BOSTON / LONDON

A C.I.P. Catalogue record for this book is available from the Library of Congress.

ISBN 0-7923-6659-X

Published by Kluwer Academic Publishers,
P.O. Box 17, 3300 AA Dordrecht, The Netherlands.

Sold and distributed in North, Central and South America
by Kluwer Academic Publishers,
101 Philip Drive, Norwell, MA 02061, U.S.A.

In all other countries, sold and distributed
by Kluwer Academic Publishers,
P.O. Box 322, 3300 AH Dordrecht, The Netherlands.

Printed on acid-free paper

Printed in the Netherlands.

SCIENTIFIC ORGANISING COMMITEE:

Co-Chair:

Kwing-Lam Chan (Hong Kong, China)
Kwong-Sang Cheng (Hong Kong, China)
Kam-Ching Leung (U.S.A.)

Members:

Helmut A. Abt (U.S.A.)
Ali Alpar (Turkey)
Bambang Hidayat (Indonesia)
Edward F. Guinan (U.S.A.)
Tomokazu Kogure (Japan)
Wlodek Kluzniak (U.S.A.)
Ti-Pei Li (China)
Acadio Poveda (Mexico)
Sabatino Sofia (U.S.A.)
M.A. Ruderman (U.S.A.)
Gou-Xiang Ai (China)
Trevor Weekes (U.S.A.)
Shuang-Nan Zhang (U.S.A.)

LOCAL ORGANISING COMMITEE:

Kwing-Lam Chan (Hong Kong University of Science and Technology)
Hoi-Fung Chau (The University of Hong Kong)
Kwong-Sang Cheng, Chair (The University of Hong Kong)
Ming-Chung Chu (The Chinese University of Hong Kong)

SPONSORS:

Astronomical Society of China
Astrophysical Society of Hong Kong
K.C. Wong Education Foundation
The University of Hong Kong
The Chinese University of Hong Kong
Hong Kong University of Science and Technology

CONTENTS

Session V - High Energy Phenomena I: Observations

Session VI - High Energy Phenomena II: Theories

Poster Session

PREFACE

The Pacific Rim Conference originally started with one research concentration only - binary star research. The first Conference was held in Beijing, China, 1985, the second one in Seoul and Taejon, South Korea, 1990 and the third one in Chiang Mai, Thailand, 1995. In recent years, the conference series evolved into a much broader area of stellar astrophysics. The first such conference was held in Hong Kong in 1997. Kwong-Sang Cheng, a.k.a. one of the three Musketeers, documented the "accidental" development in writing in the Proceedings of the1997 Pacific Rim Conference on Stellar Astrophysics (Volume 138 of the ASP Conference Series)! The meeting at Hong Kong University of Science and Technology covered three major topics: binary stars, compact stars and solar type stars. The conference was extremely successful. There was a general feeling among the participants that the conference on stellar astrophysics provided a good means to share ideas between such closely related disciplines. Unfortunately after the very successful meeting at HKST, Kwing L. Chan (another Musketeer) thought that he had already served and would not like to chair for another LOC for at least five years! After a few drinks at one of the watering holes in Wan Chai district of Hong Kong, Kwong-Sang Cheng was in very high spirit and volunteered to taking on the responsibility of hosting the 5th Pacific Rim Conference at Hong Kong University in 1999.

The meeting at HKU was most successful and we saw only smiling faces. All the participants seemed to find that the integration of three research areas in one conference very informative and helpful for their own research. It is now generally recognized that Hong Kong is a place for good food and for excellent scientific conferences!

Needless to say, to have a good conference we have to have good people to serve on the Scientific Organizing Committee. We are very grateful to all our members of the SOC and to those who took their duties seriously in chairing the scientific sessions. We are very fortunate to have dedicated LOC members for running an extremely successful and faultless meeting. We would like to express our special thanks to the following sponsors of the conference: Astrophysical Society of Hong Kong, The Chinese University of Hong Kong, K.C. Wong Education Foundation, University of Hong Kong, Hong Kong University of Science and Technology. Finally we would like to thank our conference assistants Miss Anisia Tang and Mr. S.F. Ko for their dedication to the conference administration work and clerical work in the proceeding.

We are looking forward in seeing you at the next Rim Conference.

Kam-Ching Leung
Chairman of SOC

LIST OF PARTICIPANTS

Abt, Helmut A.	Kitt Peak National Observatory
Allen, Christine	Instituto de Astronomia
Becker, Werner	Max-Planck-Institut f"ur extraterr. Physik
Belvedere, Gaetano	University of Catania
Bombaci, Ignazio	University of Pisa
Brandenburg, Axel	University of Newcastle
Chan, David	Chinese University of Hong Kong
Chan, Kin-Wing	University of Tokyo
Chan, Kwing Lam	Hong Kong University of Science and Technology
Chan, Sze Ting	University of Hong Kong
Chang, Rui-xiang	Shanghai Observatory
Chau, Hoi Fung	University of Hong Kong
Chen, Da-ming	Shanghai Observatory
Chen, Li	Shanghai Observatory
Cheng, K.S.	University of Hong Kong
Cheung, Oswald	
Chu, Ming Chung	Chinese University of Hong Kong
Dai, Zigao	Nanjing University
Deng, Licai	Beijing Astronomical Observatory
Ding, Shixue	Huazhong Normal Uinversity
Edalati Sharbaf, Mohammad Taghi	University of Mashad
Epstein, Richard	Los Alamos National Laboratory
Fan, Gailing	University of Hong Kong
Fan, Junhui	Guangzhou Normal University
Gil, Janusz	Astronomy Centre, Poland
Gimenez, Alvaro	LAEFF
Goderya, Shaukat	Illinois State University
Gong, B.P.	University of Hong Kong
Harko, T	University of Hong Kong
Hoard, Donald W.	Cerro Tololo Inter-American Observatory
Hon, Kam Yuen Dennis	Hong Kong University of Science and Technology
Ikeda, Yuji	Tohoku University
Jiang, Xiaojun	Beijing Astronomical Observatory
Ko, ShuFung	University of Hong Kong
Kogure, Tomokazu	Bisei Astronomical Observatory

Kong, Albert	Oxford University
Lai, Dong	Cornell University
Lai, Kelvin	University of Hong Kong
Leung, Chun Sing	University of Hong Kong
Leung, Kam Ching	University of Nebraska
Li, Tipei	Institute of High Energy Physics
Liang, Yanchun	Beijing Astronomical Observatory
Lie, HoYin	University of Hong Kong
Link, Bennett	Montana State University
Liu, Qingyao	Yunnan Observatory
Lu, Hui Qing	Shanghai University
Lu, Tan	Nanjing University
Lu, Ye	Beijing Astronomical Observatory
Ludwig, Hans-Guenter	Hong Kong University of Science and Technology
Mak, Man.Kwong	Institute of Vocational Education, HK
Manchester, Richard	Australia Telescope National Facility
Ng, Kin Wang	Institute of Astronomy and Astrophysics,Taipei
Poveda, Arcadio	Instituto de Astronomía
Qiu, HongMei	Beijing Astronomical Observatory
Robinson, Frank	Hong Kong University of Science and Technology
Ruediger, Guenther	Astrophysikalisches Institut Potsdam, Germany
Savonije, Gertjan	University of Amsterdam
Shaw, J. Scott	University of Georgia
Shibazaki, Noriaki	Rikkyo University
Shih, I Chun	Oxford University
Sion, Edward	Villanova University
Siu, Lai Wa	University of Hong Kong
Soonthornthum, Boonraksar	Chiang Mai University
Steffen, Mattias	Astrophysical Institute Potsdam, Germany
Suen, Wai Mo	Washington University
Tamura, Shin'ichi	Tohoku University
Tang, Anisia	University of Hong Kong
Tang, KingLoy	University of Hong Kong
Tang, WingShun	University of Hong Kong
Tian, Kaiping	Shanghai Astronomical Observatory
Tsuruta, Sachiko	Montana State University
van der Hucht, Karel A.	Space Research Organization Netherlands
Wachter, Stefanie	Cerro Tololo Interamerican Observatory
Wang, Ding-Xiong	Huazhong University of Science and Technology
Wang, Jian Min	Beijing Normal University
Wang, Jun-Jie	Beijing Astronomical Observatory
Wang, Xiao Feng	Beijing Normal University

Wei, Daming	Purple Mountain Observatory
Wen, Linqing	Massachusetts Institute of Technology
Wong, Samuel S.M.	University of Toronto
Wu, Xinji	Peking University
Wu, Xuebing	Beijing Astronomical Observatory
Wu, Xuejun	Nanjing Normal Univerisity
Xu, Chongming	Nanjing Normal University
Xu, Ren Xin	Peking University
Yan, KoonKiu	University of Hong Kong
Yang, Lantian	Huazhong Normal Uinversity
Yang, Pibo	Hua Zhong Normal University
Yang, Yulan	Yunnan Observatory
Ying, Xiao	Shanghai Observatory
Yu, Wenfei	Institute of High Energy Physics
Yu, Zhiyao	Shanghai Astronomical Observatory
Yuan, Feng	Nanjing University
Zdziarski, Andrzej	N. Copernicus Astronomical Center
Zhang, Li	University of Hong Kong
Zhang, Shuang Nan	University of Alabama in Huntsville
Zhang, Yijun	University of Hong Kong
Zhang, Yuan-zhong	Institute of Theoretical Physics

THE DYNAMO EFFECT IN STARS

Axel Brandenburg
NORDITA, Blegdamsvej 17, DK-2100 Copenhagen Ø, Denmark; and
Department of Mathematics, University of Newcastle upon Tyne, NE1 7RU, UK
brandenb@nordita.dk

Abstract Stars with outer convection zones are all magnetically active. They possess magnetic fields that have a strong large scale component, which can sometimes show cyclic reversals, like in the sun. Over the past thirty years mean-field dynamo theory has been used to explain structure and evolution of those large scale fields. The main ingredients of this theory are the alpha-effect and turbulent diffusion, but the physical nature of these effects has shifted from a purely hydrodynamical origin to more magnetically controlled scenarios, where thermal buoyancy, for example, is replaced by magnetic buoyancy and other magnetically driven instabilities.

Keywords: dynamos, stellar cycles, magnetohydrodynamics

1. INTRODUCTION

Stellar chromospheric and coronal activity is usually explained by some kind of dynamo process, which converts kinetic energy into magnetic energy. For example, for turbulent convection at sufficiently high magnetic Reynolds numbers (small enough magnetic diffusivity) small scale magnetic fields are produced (Meneguzzi & Pouquet 1989, Nordlund et al. 1992, Brandenburg et al. 1996). However, many stars show cyclic behavior. Explaining such behavior requires some extra ingredients, such as rotation, shear, and vertical density stratification (e.g. Moffatt 1978). Those extra ingredients tend to give the flow some swirl and make it helical–just like cyclones. The shear from *differential* rotation tends to align the field with the toroidal direction, converting poloidal magnetic field into toroidal. To close the cycle, poloidal magnetic field is generated from toroidal via cyclonic convection. This effect is usually referred to as the alpha-effect.

K.S. Cheng et al. (eds.), Stellar Astrophysics, 1–8.

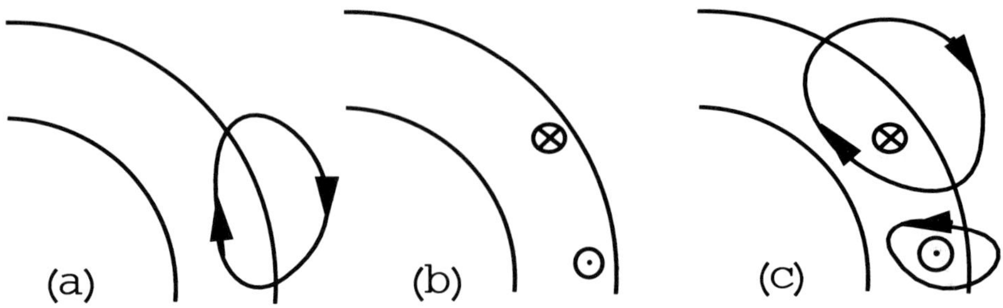

Figure 1 A poloidal magnetic field loop is being sheared out by differential rotation causing toroidal fields. The α-effect produces new poloidal field loops, but shifted polewards.

In the standard picture, rotation twists a rising flux loop such that an extra *turbulent* electric field is induced (the turbulent electromotive force $\boldsymbol{\mathcal{E}}$), which has a component perpendicular to the mean magnetic field, i.e. $\boldsymbol{\mathcal{E}} = \alpha\langle\boldsymbol{B}\rangle$.

The α parameter gives rise to exponentially growing solutions of the induction equation provided the magnitude of α is large enough (large enough dynamo number). If this α-effect is supplemented by differential rotation (the Ω-effect), one talks about an $\alpha - \Omega$ dynamo (e.g., Parker 1979, Krause & Rädler 1980). Figure 1 gives a qualitative explanation of why this α-effect can lead to dynamo waves propagating in the toroidal direction. Consider first some pre-existing poloidal magnetic field loop (Fig. 1a). In lower latitudes the deeper regions of the sun spin slower, giving rise to a toroidal field as shown in Fig. 1b. This toroidal field induces a current and the α-effect produces a new magnetic field parallel to it (Fig. 1c).

Comparing Figs. 1a and 1c we notice the emergence of a new loop near the equator with *opposite* orientation relative to the original loop in Fig. 1a. The larger loop in 1c, however, has the same orientation as the loop in Fig. 1a, which therefore appears to have migrated away from the equator. (Three further applications of shear and α-effect bring the situation back full circle to the configuration in Fig. 1a.) The field migration seen in this model is of course in the "wrong" direction, because in the sun the sunspot belts migrate equatorward. This is known as the dynamo dilemma (Parker 1987). In the early days of dynamo theory, before helioseismology told us otherwise, one used to believe that the sun spun faster in deeper layers than at the surface. In that case the dynamo waves would go in the right direction. There are some indications supporting this possibility in the case of accretion discs (Brandenburg &

Donner 1997), but it is not clear that similar circumstances apply to the case of stars. An opposite angular velocity gradient would indeed be consistent with the observation that very young sunspots have faster proper motion than older spots. (There is at present no good explanation for this property of sunspot proper motions. However, helioseismology is now beginning to resolve a negative radial gradient of the angular velocity in low latitudes and near the surface, see also Chan's article in these proceedings.)

If the sign of α was for some reason reversed (negative in the northern hemisphere) then this would turn the dynamo wave into the right direction. Another possibility that has been discussed already by Parker (1987), and more recently by Durney (1996) and Choudhuri, Schüssler, & Dikpati (1995) is to invoke meridional circulation to turn the dynamo wave around. The most recent model of that type is by Dikpati & Charbonneau (1999), where a more realistic profile of differential rotation has been adopted.

2. MAGNETICALLY DRIVEN α-EFFECTS

Over the past few years there have been several suggestions that the dynamo effect should actually increase with field strength. In the paper of Brandenburg, Saar & Turpin (1998) this was just a hypothesis that appeared plausible in view of other simulated dynamos that share the property of becoming more and more effective as the magnetic field strength increases. One possible and straightforward explanation would be that α may *not* be driven by thermal buoyancy, but by *magnetic* buoyancy. This idea goes back to Schmitt (1985), who was the first to derive in detail the α-effect resulting from such a system. Recent simulations have been presented by Brandenburg & Schmitt (1998), and model calculations have been carried out by Moss, Shukurov & Sokoloff (1999). The stronger the magnetic field, the more the flux tubes are evacuated (total pressure = magnetic pressure + gas pressure) and the more buoyant they are. It may therefore not be so implausible that α could indeed increase with increasing field strength.

If α really does increase with field strength we need some other mechanism for saturation of the dynamo. This could be again magnetic buoyancy: once the magnetic buoyancy effect exceeds a certain value it would no longer lead to field growth, because the generated flux would be removed too quickly from the dynamo-active region. In the case of the magnetorotational instability, which is primarily relevant to *accretion discs*, the growth would cease once the Alfvén speed becomes so large that the travel distance of an Alfvén wave within one orbit becomes comparable to some relevant scale of the disc (e.g. the disc height in the case of a vertical field). This would effectively limit the mean field strength. This system provides an important example of a magnetically driven α-effect (Brandenburg et al. 1995, Brandenburg & Donner 1997). Here the turbulence is driven by the magnetorotational or Balbus-Hawley (1991) instability.

3. SIMULATIONS

Brandenburg, Nordlund & Stein (1999) have simulated a convective dynamo with imposed shear trying to capture both the effects of latitudinal differential rotation in the convection zone proper and vertical shear at the bottom of the convection zone. In that simulation the total magnetic energy, $\langle \boldsymbol{B}^2 \rangle$, as well as the energy in the mean magnetic field, $\langle \boldsymbol{B} \rangle^2$, increase exponentially until saturation is reached. The mean field shows unsteady behavior without real cycles and field reversals. However this is strongly related to geometrical effects and boundary conditions, because the large scale field extends over the scale of the box making global effects important. Furthermore, the energy of the mean field to the total magnetic energy, $f = \langle \boldsymbol{B} \rangle^2 / \langle \boldsymbol{B}^2 \rangle$, which is a measure of the filling factor, also increase with time. Thus, again, the large scale field becomes better defined (relative to the fluctuations) once it reaches appreciable field strength. Those results are encouraging in that they confirm the observational finding that the solar magnetic field shows a great deal of coherence even though it is basically of turbulent origin.

In the case of local turbulence simulations of accretion disc dynamos (Brandenburg et al. 1995) we found that the mean magnetic field (averaged azimuthally and over some radius interval) shows spatio-temporal coherence as evidenced by a "butterfly-type" diagram of the mean toroidal field as a function of time and height above and below the midplane of the disc. This result is however markedly dependent on boundary conditions. If one adopts perfectly conducting boundary conditions instead of vacuum boundary conditions one finds steady dipole-type solutions instead of oscillatory quadrupole-type solutions (Brandenburg 1998). It is remarkable, however, that the same change of behavior is reproduced by an $\alpha - \Omega$ dynamo model. In that sense simulations and $\alpha - \Omega$ model show an important similarity.

There is another point that needs to be emphasized. While simulations such as the accretion disc simulations show fairly well-defined large scale fields, they also display an extremely "noisy" behavior for the turbulent electromotive force and hence the α-effect. Although it has been shown that in the presence of shear and turbulent diffusion,

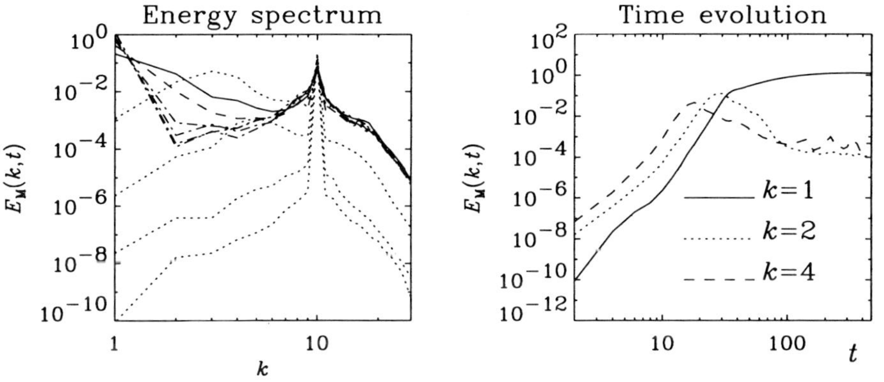

Figure 2 Spectral magnetic energy, $E_{\rm M}(k,t)$, as a function of wavenumber k for different times: dotted lines are for early times ($t = 2, 4, 10, 20$), the solid and dashed lines are for $t = 40$ and 60, respectively, and the dotted-dashed lines are for later times ($t = 80, 100, 200, 400$).

noisy α-effects are quite capable of producing mean fields that are not very noisy (Vishniac & Brandenburg 1997), it remains still somewhat of a mystery as to how such a noisy α-effect can have anything to do with a fairly well-behaved large scale magnetic field as seen in the simulations.

4. THE INVERSE MAGNETIC CASCADE

It is actually very difficult to verify that it is really the α-effect that is responsible for the large scale field generation. From the seminal papers of Frisch et al. (1975) and Pouquet et al. (1976) it is clear that the amplification of large scale fields can be explained by an inverse cascade of magnetic helicity. This effect too is rather difficult to isolate in simulations of astrophysical turbulence. However, under somewhat more idealized conditions, for example when magnetic energy is injected at high wave numbers, one clearly sees how the magnetic energy increases at large scales; see Fig. 2. (For more details of those calculations see Brandenburg 1999.)

A somewhat different situation is encountered in the absence of any forcing where some initial magnetic field can only decay. However, if initially most of the magnetic energy is in the small scales, there is the possibility that magnetic helicity and thereby also magnetic energy is transferred to large scales. This is exactly what happens (Fig. 3), provided there is initially some net helicity. The inset of Fig. 3 shows that in the absence of initial net helicity the field at large scales remains

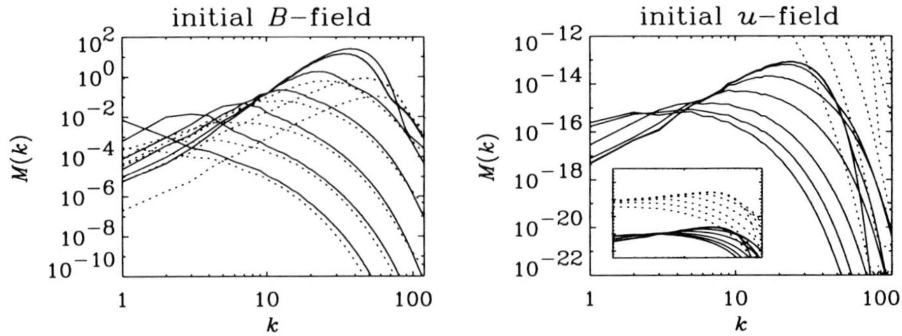

Figure 3 Power spectra of magnetic energy (solid lines) and kinetic energy (dotted lines) in a decay run. The left hand panel is for a case where the flow is only driven by an initial helical magnetic field. In the right hand panel the field is weak and governed by strong decaying fluid turbulence. The inset shows both velocity and magnetic spectra in the same plot.

unchanged, until diffusion kicks in and destroys the remaining field at very late times.

If the magnetic field has the possibility to tap energy also from the large scale velocity the situation is somewhat different and a large scale magnetic field can also be driven without net helicity. In that case even without any helicity (kinetic or magnetic helicity) the large scale field can increase. In astrophysical settings there is usually large scale shear from which energy can be tapped. This was indeed the case in the simulations discussed in the previous section.

5. CONCLUSIONS

While dynamo theory in its present form is in principle able to reproduce basic behavior of solar and stellar magnetic fields and cycles, there are a number of problems of theoretical and practical nature, as well as a number of new hypotheses that could resolve some of these problems.

The main theoretical problem is related to the functional dependence between the electromotive force and the mean magnetic field. Comparison with simulations suggests tentatively that α may work preferentially at the largest possible scale. If that is true one could solve the (practical) problem of explaining the increase of stellar cycle frequencies with increasing inverse Rossby number by assuming that the α-effect increases with field strength (anti-quenching).

Another rather practical problem concerns the shape of the solar butterfly diagram. Theoretically one would expect that the dynamo wave should migrate poleward; see the pioneering simulations of Gilman (1983) and Glatzmaier (1985). In order to explain the observed equatorward migration one would either need to have a negative α in the northern hemisphere (some simulations do predict this, but it is not clear that this applies really to the solar regime), or one might be able to explain the migration directly by invoking a suitable meridional circulation. Recent work by G. Rüdiger and collaborators (private communication) suggests that this is indeed a viable possibility (see also Dikpati & Charbonneau 1999). This was first suggested by Parker (1987) and Durney (1996), and confirmed by a model calculation by Choudhuri et al. (1995), but it seemed to be a rather special case given that meridional circulation usually tends to make oscillatory models stationary (Rädler 1986).

In any case, the theoretical foundations of $\alpha - \Omega$ are sufficiently shaky that one may consider a realistic high-resolution simulation of stellar dynamos as absolutely crucial before one can try to use $\alpha - \Omega$ type dynamos with real predictive power.

Acknowledgements

This work has been supported in part by the British Council (JRS 98/39).

References

Balbus, S. A., & Hawley, J. F. 1991, ApJ 376, 214

Brandenburg, A. 1998, in Theory of Black Hole Accretion Discs, ed. M. A. Abramowicz et al. (Cambridge University Press), p.61

Brandenburg, A. 1999, in Magnetic Helicity, ed. A. A. Pevtsov (American Geophysical Union, Florida), (in press)

Brandenburg, A., & Donner, K. J. 1997, MNRAS 288, L29

Brandenburg, A., & Schmitt, D. 1998, A&A 338, L55

Brandenburg, A., Nordlund, Å., & Stein, R. F. 1999, A&A (to be submitted)

Brandenburg, A., Saar, S. H., & Turpin, C. R. 1998, ApJ (Letters) 498, L51

Brandenburg, A., Nordlund, Å., Stein, R. F., & Torkelsson, U. 1995, ApJ 446, 741

Brandenburg, A., Jennings, R. L., Nordlund, Å., Rieutord, M., Stein, R. F., & Tuominen, I. 1996, J. Fluid Mech. 306, 325

Choudhuri, A. R., Schüssler, M., & Dikpati, M. 1995, A&A 303, L29

Dikpati, M. & Charbonneau, P. 1999, ApJ 518, 508

Durney, B.R. 1996, Solar Physics 166, 231

Frisch, U., Pouquet, A., Léorat, J., Mazure, A. 1975, J. Fluid Mech. 68, 769
Gilman, P. A. 1983, ApJ Suppl. 53, 243
Glatzmaier, G. A. 1985, ApJ 291, 300
Krause, F., & Rädler, K.-H. 1980, Mean-Field Magnetohydrodynamics and Dynamo Theory (Akademie-Verlag, Berlin; also Pergamon Press, Oxford)
Meneguzzi, M., Pouquet, A. 1989, J. Fluid Mech. 205, 297
Moffatt, H. K. 1978, Magnetic Field Generation in Electrically Conducting Fluids (Cambridge University Press, Cambridge)
Moss D., Shukurov A., Sokoloff D. 1999, A&A 343, 120
Nordlund, Å., Brandenburg, A., Jennings, R. L., Rieutord, M., Ruokolainen, J., Stein, R. F., & Tuominen, I. 1992, ApJ 392, 647
Parker, E. N. 1979, Cosmical Magnetic Fields (Clarendon Press, Oxford)
Parker, E. N. 1987, Solar Phys. 110, 11
Pouquet, A., Frisch, U., Léorat, J. 1976, J. Fluid Mech. 77, 321
Rädler, K.-H. 1986, Astr. Nachr. 307, 89
Schmitt, D. 1985, Dynamowirkung magnetostrophischer Wellen (PhD thesis, University of Göttingen)
Vishniac, E. T., & Brandenburg, A. 1997, ApJ 475, 263

DIFFERENTIAL ROTATION, MERIDIONAL FLOW AND A HIGH-PRANDTL NUMBER SOLAR/STELLAR DYNAMO

Günther Rüdiger

Astrophysikalisches Institut Potsdam, An der Sternwarte 16, D-14482 Potsdam, Germany

Abstract Meridional flow and rotation law are computed for convection zones of solar-type stars. The solar differential rotation known by helioseismology is reproduced with the hydrodynamic theory of the Lambda-effect. The meridional flow in the convection zone is of order 5–10 m/s and returns to the equator at its base. The present-day models only yield one cell in latitude in each hemisphere. For the theory of the polar branch a double-cell flow pattern would be helpful. For eddy diffusivities of 10^{11} cm^2/s the meridional flow adopts magnetic Reynolds numbers of order $10^{2...3}$ so that drastic consequences for the solar dynamo can be expected.

We find indeed a solar-type butterfly diagram even for positive alpha-effect (in the northern hemisphere). Moreover, the resulting cycle times as well as the obtained phase relation of the radial and the toroidal magnetic field components are reflecting the observations in a rather good approach. It might thus be true that in the light of the presented model calculations the dynamo theory of solar/stellar activity must strongly be modified.

Keywords: differential rotation, meridional flow, butterfly diagram

1. INTRODUCTION

Our understanding of the stellar activity bases on the stellar dynamo theory. One of its basic inputs is the internal rotation law. A dynamo for rigid rotation strongly differs from a dynamo with differential rotation. The axisymmetric large-scale magnetic fields of the Sun and solar-type stars are only understandable with differential rotation. The only star with known internal rotation is the Sun. If we did not know the internal solar rotation from helioseismology we had only speculations about

K.S. Cheng et al. (eds.), Stellar Astrophysics, 9–16.

the solar dynamo. Helioseismology, however, gives us three great mysteries apart from the well-known surface phenomenon of the equatorial acceleration of $\delta\Omega \simeq 0.06$ per day.

The first observation is that we have super-rotation beneath the equator and sub-rotation beneath the poles. We have thus cylinder-like isorotation contours at the equatorial region and disk-like isorotation contours in the polar regions. The second observation concerns the very fast transition beneath the convection zone towards rigid rotation along about 10% of the solar radius. It is the so-called 'tachocline problem'.

The third observation is the rigid rotation of the solar core with practically the same rotation rate as that of the surface mid-latitudes. The core already knows – after only 4.6 Gyrs – that there was a substantial spin-down of the surface during this time. It is certainly not possible to reproduce such a strong core spin-down only with the angular momentum transport of the microscopic viscosity of (say) 100 cm^2/s due to the radiative non-turbulent material in the solar interiour.

The latter problem is unclear sofar. If there was a slight shear flow in the radiative core we can believe on Balbus-Hawley instability to work there. It seems to be hard to find an angular momentum transporting instability for the observed rigid rotation. In the present paper we are only discussing the large-scale flow pattern in the solar convection zone.

2. DIFFERENTIAL ROTATION

In close correspondence to dynamo theory, we develop the theory of differential rotation in a mean-field formulation starting from the conservation law of angular momentum, i.e.

$$\frac{\partial}{\partial t}\left(\rho r^2 \sin^2\theta\Omega\right) + \frac{\partial}{\partial x_i}\left(\rho\left(r^2\sin^2\theta\Omega u_i^{\mathrm{m}} + r\sin\theta Q_{i\phi}\right)\right) = 0, \qquad (1)$$

where the Reynolds stress derived from the correlation tensor

$$Q_{ij} = \langle u_i'(\mathbf{x},t)u_j'(\mathbf{x},t)\rangle$$

corresponds to the EMF in the mean-field electrodynamics.

The correlation tensor involves both dissipation ('eddy viscosity') as well as 'induction' (Λ-effect): $Q_{ij} = \Lambda_{ijk}\Omega_k - \mathcal{N}_{ijkl}\Omega_{k,l}$ (cf. Rüdiger 1989). Both effects are represented by tensors and must be computed carefully. All coefficients are found to be strongly dependent on the Coriolis number $\Omega^* = 2\tau_{\mathrm{corr}}\,\Omega$ with τ_{corr} as the correlation time, i.e. the convective turnover time. Moreover, the most important terms of the Λ-effect correspond to higher orders of the Coriolis number. The Coriolis number exceeds unity almost everywhere in the convection zone except

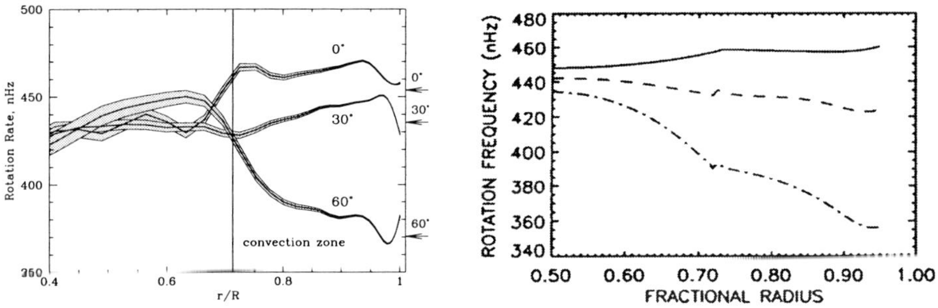

Figure 1 LEFT: The internal rotation of the Sun after the inversion of SOHO data (Kosovichev *et al.* , 1997). RIGHT: Theory of the solar internal rotation by Kitchatinov and Rüdiger (1995). The rotation frequency is given for the equator (solid), mid-latitudes (dashed) and poles (dashed-dotted).

the surface layers. That is true for all stars – all main-sequence stars are rapid rotators in this sense. *Theories linear in Ω are not appropriate for stellar activity physics.* The α-effect in the dynamo theory will thus never run with $\cos\theta$ in its first power.

The solution of (9) is given in Küker *et al.* (1993) and Kitchatinov and Rüdiger (1995) using a mixing-length model by Stix and Skaley (1990). With a given mixing-length ratio ($\alpha_{\rm MLT} = 5/3$), we find the correct equatorial acceleration of about 0.06 per day. There is a clear radial sub-rotation ($\partial\Omega/\partial r < 0$) below the poles, while in mid-latitudes and below the equator, the rotation is basically rigid (right panel in Fig. 1). In this way the bottom value of the pole-equator difference is reduced and the resulting profiles of the internal angular velocity are close to the observed ones. Fig. 2 presents the results of an extension of the theory to a sample of main-sequence stellar models given in Kitchatinov and Rüdiger (1999). We find that for one and the same spectral type the approximation $\delta\Omega \simeq 0.06$ per day should be not too rough. Recent observations of AB Dor and PZ Tel seem to confirm this surprising result. Also the differences for the different spectral classes G and K are only small.

Also interesting is the meridional flow within the convection zone. It always flows equatorwards at the bottom of the convection zone with amplitudes between 5 m/s and 10 m/s, the latter for rapid rotation (Fig. 3). We shall see that such flows are highly important for a dynamo model in case that the eddy diffusivity basically remains below the often used $10^{12\dots13}$ cm^2/s. Only in that case the magnetic Reynolds number of the flow becomes high enough in order to influence the dynamo regime.

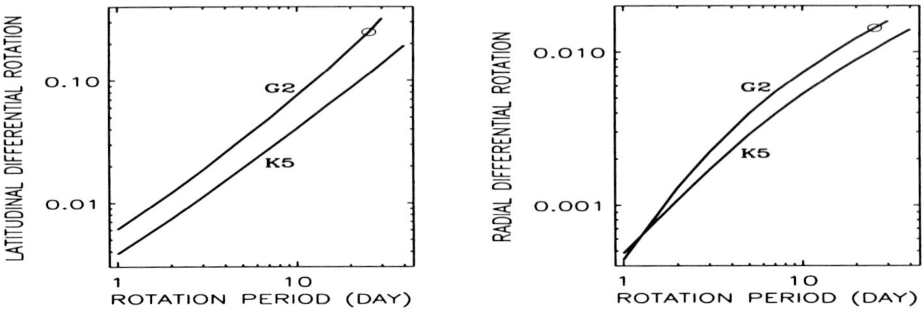

Figure 2 The rotation differences $\delta\Omega/\Omega$ for both a solar model and a K5 main-sequence star vs. basic rotation period after Kitchatinov and Rüdiger (1999). LEFT: normalized equator-pole difference of surface rotation, RIGHT: radial rotation differences.

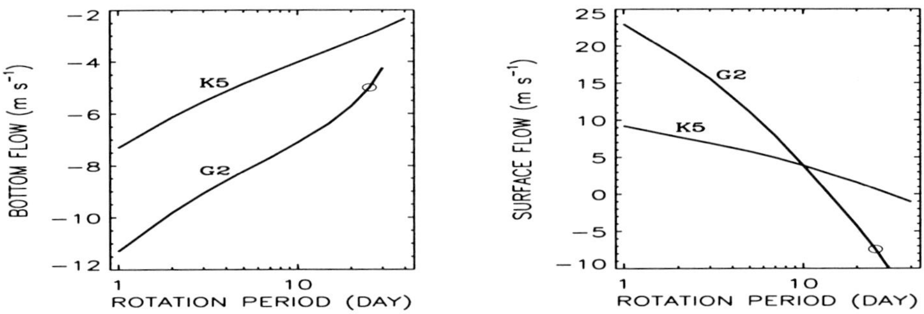

Figure 3 The meridional circulation at mid-latitudes (45°) at the bottom (LEFT) and the top (RIGHT) of the convection zone for both a solar model and a K5 main-sequence star as a function of the basic rotation period (Kitchatinov and Rüdiger, 1999). The circle represents the solar case. The flow at the bottom is equatorwards.

3. DISTRIBUTED DYNAMOS WITH MERIDIONAL CIRCULATION

In the following the dynamo equations are solved with the inclusion of the induction by meridional circulation. In order to produce a solar-type butterfly diagram with $\mathbf{u} = 0$ the α-effect must be taken as negative in the northern hemisphere (Parker, 1987). Our rotation law is super-rotation and uniform in latitude and a one-cell flow pattern is adopted with prescribed amplitude. The $u^{\rm m}$ approximates the latitudinal drift close to the convection zone bottom, the circulation is counterclockwise (in the first quadrant) for positive $u^{\rm m}$. The latter forms a poleward flow at the surface in agreement with observations. For too strong meridional circulation ($u^{\rm m} \geq 50$ m/s) the dynamo stops operation. The same happens already for $u^{\rm m} \approx -10$ m/s if the flow at the bottom is opposite to the magnetic drift wave.

However, it can be hard to explain that the convection zone itself is producing the *negative* α-values used in the model. More easy to understand are positive values. Of course, without circulation the butterfly diagram becomes then 'wrong', with a poleward migration of the toroidal magnetic belts. In order to change the situation basically, the meridional flow must be fast enough, or, more exactly speaking, its magnetic Reynolds number must be high enough. This is only possible with low magnetic eddy diffusivity. For the eddy diffusivity here we used 10^{11} cm^2/s taken from the sunspot decay observations hence for the magnetic Reynolds number $\text{Rm} = u^{\rm m}R/\eta_{\rm T}$ follows 100–1000 (see also Brandenburg, 1993). In their models Choudhuri *et al.* (1995) apply Reynolds numbers of order 500 while Dikpati and Charbonneau (1999) even take $\text{Rm} = 1400$. In Table 1 the (half-)cycle periods in years are given for dynamos with positive-α-effect in the northern hemisphere. Bold numbers are indicating all the dynamos with a solar-type butterfly diagram. We only find them for fast flow and low diffusivity. Slow flow and higher diffusivity yields poleward migration of the activity belts. Note also that the cycle period for slow flow runs proportionate to $1/\eta_{\rm T}$ but it does not for fast flow!

In Fig. 4 four models are presented for one and the same turbulence electromotive force but for various amplitudes of the meridional circulation proceeding equatorwards at the bottom of the convection zone. The first model has zero amplitude. Its butterfly diagram has the wrong shape as it works with positive α. The $B_r B_\phi$ is positive, i.e. also wrong (cf. Stix, 1976). The latter is changed to negative sign for a meridional flow with 3 m/s. Still the butterfly diagram is wrong. It starts to change its overall shape for 4 m/s where the migration of activity belts stops.

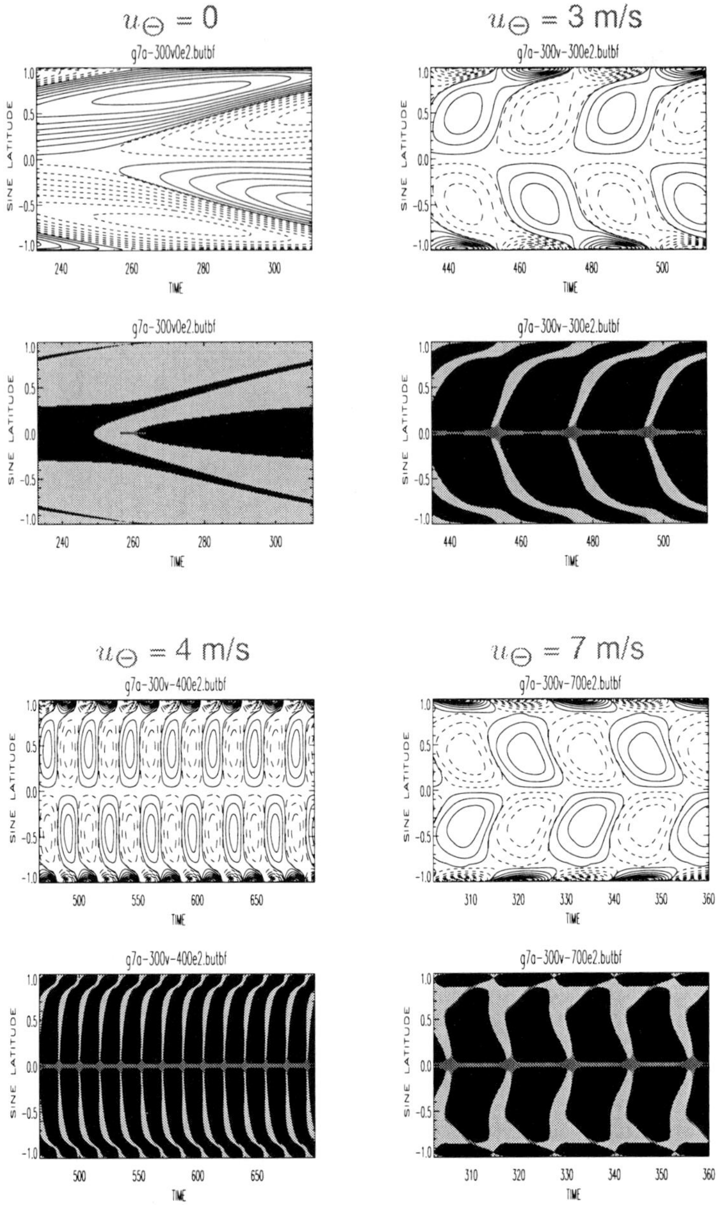

Figure 4 Dynamo models with positive α-effect (3 m/s), with eddy diffusivity of $1 \cdot 10^{11}$ cm^2/s and various bottom values of the meridional flow. Each model is represented by its butterfly diagram (top) and the phase relation of the radial and the toroidal magnetic field components close to the surface. Black color means negative $B_r B_\phi$.

A meridional flow of 7 m/s produces a solar-type butterfly diagram with the same α-effect and turbulent diffusivity as before. This model also exhibits the observed negative sign of $B_r B_\phi$.

We have thus demonstrated how strong for low dissipation the role of the meridional flow is. It might easily be true that the solar/stellar dynamo story must be rewritten in the light of the presented new results. To this end we are in particular interested to find the bifurcation diagrams for the stability of the kinematic dynamo model with meridional flow included. It is not a trivial work to investigate the convergency of the solutions for smaller and smaller magnetic dissipation (Bonanno *et al.* 2000).

References

A. Brandenburg. Simulating the solar dynamo. *The cosmic dynamo*, F. Krause and K.-H. Rädler and G. Rüdiger (eds.):111, 1993.

A. R. Choudhuri and M. Schüssler and M. Dikpati. The solar dynamo with meridional circulation. *Astron. and Astrophys.*, 303:29, 1995.

M. Dikpati and P. Charbonneau. A Babcock-Leighton flux transport dynamo with solar-like differential rotation. *Astrophys. J.*, 518:508, 1999.

L. L. Kitchatinov and G. Rüdiger. Differential rotation in solar-type stars: revisiting the Taylor number puzzle. *Astron. and Astrophys.*, 299:446, 1995.

L. L. Kitchatinov and G. Rüdiger. Differential rotation models for late-type dwarfs and giants. *Astron. and Astrophys.*, 344:911, 1999.

A. G. Kosovichev and J. Schou and P. H. Scherrer et al. Structure and rotation of the solar interior: Initial results from the MDI medium-l program. *Solar Phys.*, 170:43, 1997.

M. Küker and G. Rüdiger and L. L. Kitchatinov. An alpha-omega model of the solar differential rotation. *Astron. and Astrophys.*, 279:1, 1993.

M. Küker and R. Arlt and G. Rüdiger. The Maunder minimum as due to magnetic Λ-quenching. *Astron. and Astrophys.*, 343:977, 1999.

M. Stix. Differential rotation and the solar dynamo. *Astron. Astrophys.*, 47:243, 1976.

M. Stix and D. Skaley. The equation of state and the frequencies of solar p modes. *Astron. and Astrophys.*, 232:234, 1990.

Table 1 Cycle periods in years of solar-type-butterfly dynamos (bold) with positive α-effect ($= 3$ m/s) and meridional flow of u^{m} at the bottom of the convection zone, the eddy diffusivity given in the headline. Signature '-' indicates polwards migration, 'dec' means decaying fields

u^{m} [m/s]	$1 \cdot 10^{11}$ cm^2/s	$2 \cdot 10^{11}$ cm^2/s	$5 \cdot 10^{11}$ cm^2/s
1	-120	-70	-36
3	**60**	-45	-24
5	**45**	**40**	dec
7	**31**	**26**	dec
11	**20**	**20**	dec

GENERATION OF THE ANGULAR VELOCITY SHEAR NEAR THE SOLAR SURFACE

Kwing L. Chan
Mathematics Department
The Hong Kong University of Science & Technology
Hong Kong
maklchan@ust.hk

Abstract Helioseismology has shown that there are two shear layers in the angular velocity distribution of the sun. The principal one is the one between the bottom of the convection zone and the radiative interior. This layer manifests the transition from an almost uniform rotation in the interior to the substantial differential rotation in the convection zone. The second shear layer is right beneath the surface of the sun, where the angular velocity changes by a few percent over a distance of $\sim 5\% R_\odot$.

We have recently made a parametric study of rotating convection using a large number of numerical experiments specified with different energy fluxes, rotation rates, and inclination of the rotation vector. Based on the results, we propose that the sharp change of angular velocity near the surface of the sun can be explained by the behavior of the rotating turbulence and the stratification (not by simple angular momentum conservation). This mechanism is compatible with a radially independent distribution of angular velocity in the deeper part of the convection zone.

Keywords: solar convection,differential rotation, turbulence, Reynolds stress

1. INTRODUCTION

Helioseismology has provided us with much information about the interior condition of the sun. In particular, the non-cylindrical distribution of angular velocity has received more confirmation in recent years (e.g. Thompson et al. 1996, Korzennik et al. 1996). Two shear layers exist at the upper and lower boundaries of the convection zone. The major one is of course the one at the lower radiation/convection transition layer. If one can explain this shear layer, the problem of differential rotation can

K.S. Cheng et al. (eds.), Stellar Astrophysics, 17–24.

be considered solved. Based on our numerical study of rotating convection, we believe that turbulent Reynolds stress alone (e.g. Küker et al. 1993) cannot solve this problem. However, the upper boundary shear can be explained by the behavior of the meridional-vertical component of the turbulence Reynolds stress in a rather simple manner.

2. THE NUMERICAL EXPERIMENTS

We solve the Navier Stokes equations of an ideal gas in f-boxes in which the rotation vector is constant. The box can be viewed as representing a small piece of a spherical shell located at a specific colatitude (polar angle) θ. The shell rotates around the polar axis from west to east. The horizontal and vertical coordinates of the box are x, y, and z, respectively. In terms of the shell geometry, the coordinates x and y are along the north-to-south and west-to-east directions, respectively, and z is radially outward. The rotation vector Ω lies in the second quadrant of the xz-plane.

At the bottom of the box, a constant energy flux F is injected, and at the top the temperature is maintained constant. At both of these boundaries, the vertical velocity is zero while the horizontal velocities satisfy stress-free boundary conditions. At the four side boundaries, all variables are treated as periodic. Aside from the rotation, the situation is similar to those describe in Chan and Sofia (1996). More specific description of the rotating cases can be found in a previous paper (Chan 1999). Here, we concentrate on discussing the behavior of the mean zonal (azimuthal; y direction) flow $\overline{v_y}$. A shear in this flow corresponds to a shear in the angular velocity.

The overline is used to denote combined averaging over time and both horizontal directions, i.e.

$$\overline{q} = \frac{\int\int\int q dx dy dt}{\int\int\int dx dy dt} \tag{1}$$

The fluctuating part of an arbitrary quantity q is defined as: $q' = q - \overline{q}$. The symbol q'' is used to denote the root-mean-square (rms) fluctuation of q , i.e. $q'' = \overline{q'^2}^{1/2}$. The covariance of two quantities q_1 and q_2 is $\overline{q_1 q_2}$, and the correlation coefficient is defined to be

$$C[q_1, q_2] = \frac{\overline{q_1 q_2}}{\overline{q_1^2}^{1/2}\overline{q_2^2}^{1/2}}. \tag{2}$$

Numerically, the averaging over time has to be taken over a period more than 100 turnover time as the convergence to a statistical mean is generally very slow.

All quantities are expressed in terms of units that make the total depth of the box, the initial temperature, density, and pressure at the top all equal to 1. As a consequence, the unit for velocity is the isothermal sound speed.

3. BEHAVIOR OF THE MEAN ZONAL FLOW

Examples of the dependence of $\overline{v_y}$ on Ω are shown in Figure 1. The cases have the same flux (= 0.03125) and the same θ (45°). The dotted, dashed, dot-dashed, and triple dot-dashed curves are for cases with $\Omega =$ 0.125, 0.25, 0.5, and 1, respectively; the solid lines are for the non-rotating situation.

For small Ω (corresponding to Rossby number $Ro > 1$), the distribution of $\overline{v_y}$ is shear-like, with larger negative values at the top and smaller positive values at the bottom. For large Ω ($Ro < 1$), the profile of $\overline{v_y}$ develops a dip structure near the top of the convection zone. In the interior, the mean velocity turns small. The width of the dip decreases with increasing Ω.

The dependence of $\overline{v_y}$ on θ is illustrated in Figure 2. The cases have the same flux as before and Ω is fixed at 1/2. The solid, dotted, dashed, dot-dashed, and triple dot-dashed curves represent cases with θ at 0°, 22.5°, 45°, 67.5°, and 78.75°, respectively. The magnitude of the dip generally increases toward the equator.

4. DIFFERENT RESOLUTIONS AND ASPECT RATIOS

To test whether behavior of the mean zonal flow is robust, we make a number of numerical experiments with various configurations and compare the results. First, we double the resolution. Second, we double the horizontal dimensions of the box (widths in x and y directions). Since both variations are computationally expensive, we only perform calculations for one fixed combination of parameters, namely $(F, \Omega, \theta) = (0.03125, 0.5, 67.5°)$. The comparison is shown in Figure 4, which illustrates that the characteristics of $\overline{v_y}$ is quite robust.

5. MEAN HORIZONTAL MOMENTUM BALANCE

In our model, since variables are periodic in the horizontal directions, the averaged horizontal momentum equations have simple forms:

$$-\partial_z \overline{\rho v_z v_x} + 2\Omega \cos\theta \, \overline{\rho v_y} + \partial_z \overline{\Sigma_{zx}} = 0, \tag{3}$$

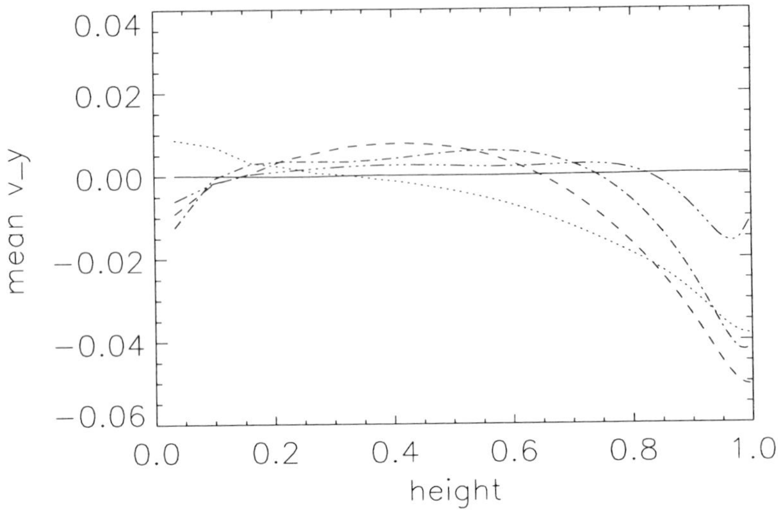

Figure 1 Vertical profiles of $\overline{v_y}$ for cases with fixed F (= 0.03125) and θ (= 45°). The solid, dotted, dashed, dot-dashed, and triple dot-dashed curves are for Ω = 0.125, 0.25, 0.5, and 1, respectively.

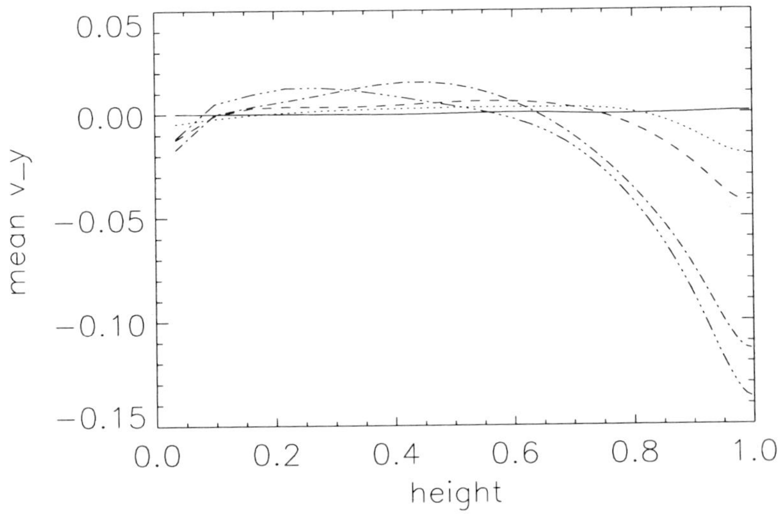

Figure 2 Vertical profiles of $\overline{v_y}$ for cases with fixed F (= 0.03125) and Ω (= 0.5). The dotted, dashed, dot-dashed, and triple dot-dashed curves are for θ = 0°, 22.5°, 45°, 67.5°, and 78.75°, respectively.

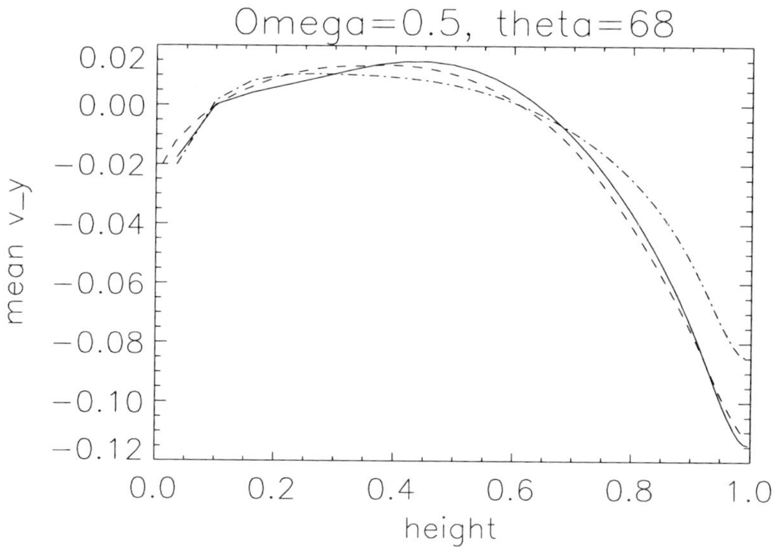

Figure 3 Vertical profiles of $\overline{v_y}$ for cases with same F (= 0.03125), Ω (= 0.5), and θ (= 45°), but different box sizes and resolutions. The solid curve represents a standard case with a 35 × 35 × 39 mesh and an aspect ratio 1.5 (width/depth). The dashed curve is for a high-resolution case with a 79 × 70 × 80 mesh (same aspect ratio). The dot-dashed curve is for a case with doubled widths (aspect ratio = 3, 68 × 68 × 39 mesh).

$$-\partial_z \overline{\rho v_z v_y} - 2\Omega \cos\theta \, \overline{\rho v_x} + \partial_z \overline{\Sigma_{zy}} \quad = \quad 0. \tag{4}$$

First, note that the mean zonal momentum appears in the meridional momentum equation rather than the zonal momentum equation. Second, the numerical study have provided us with two relevant pieces of information: (i) the mean viscous stress $\overline{\Sigma}$ is small in magnitude compared to the other terms. (ii) the approximations $\overline{\rho v_y} \approx \overline{\rho}\ \overline{v_y}$ and $\overline{\rho v_z v_x} \approx \overline{\rho} \overline{v_z' v_x'} = \overline{\rho} C[v_z', v_x'] v_z'' v_x''$ are very accurate. Then, the mean zonal velocity can be related to the vertical-meridional component of the turbulence Reynolds stress in a very simple way:

$$\overline{v_y} \simeq \frac{1}{f_c \, \overline{\rho}} \frac{\partial}{\partial z} (\overline{\rho} C[v_z', v_x'] v_z'' v_x''). \tag{5}$$

The z-dependence of the correlation coefficient $C[v_z', v_x']$ thus have a direct bearing on the mean zonal flow.

Examples of $C[v_z', v_x']$ distributions for cases with fixed F (= 0.03125) and θ (= 45°) but different rotation rates are shown in Figure 5. The symbols of the curves, representing the values of different Ω, are the same as those used in Figure 1. Examples with the same F and fixed

Ω (= 0.5) but different latitudes are shown in Figure 3. The symbols for the different θs are the same as those of Figure 2. In all the computed cases, the values of $C[v'_z, v'_x]$ never exceed the range ± 0.35.

$C[v'_z, v'_x]$ generally has a positive peak just below the top of the convection zone and (sometimes) a negative valley in the lower part. In the cases with $Ro > 1$, it is everywhere positive, for all the mid-latitudes. In the cases with $Ro < 1$, a negative region at the bottom first appears in low-θ cases, and its extent increases with the rotation rate.

The positive peak of $C[v'_z, v'_x]$ near the top of the convection zone is responsible for producing the negative dip of $\overline{v_y}$ there. First, note that the Reynolds stress component inside the differentiation is positive from the peak of $C[v'_z, v'_x]$ to the top of the convection zone. The magnitude of this component decreases with height and approaches zero towards the top (as v''_z must drop there). Thus the sign of the derivative is negative, making $\overline{v_y}$ also negative in this region. Second, note that the length scale for the variation of $C[v'_z, v'_x]$ is large (on the order of the total depth in the numerical model) compared to those of other factors that compose the Reynolds term (the length scales are the local scale heights). The variation of the magnitude of $\overline{v_y}$ therefore mainly follows the part $(1/\overline{\rho})\partial_z(\overline{\rho} v''_z v''_x)$. According to the approximate relations R1 and R21 given in Table 2 of Chan & Sofia (1989),

$$v''_x \sim v''_z \sim (1/\overline{\rho})^{1/3} \tag{6}$$

where '$\sim$' means 'approximately proportional to'. Therefore, the factor inside the differentiation is $\sim \overline{\rho}^{1/3}$, and $(1/\overline{\rho})\partial_z(\overline{\rho} v''_z v''_x) \sim 1/(\overline{\rho}^{2/3} H_\rho)$ where H_ρ is the density scale height $1/|\partial_z ln(\overline{\rho})|$. Inside the convection zone, the stratification is almost adiabatic so that $\overline{\rho}^{2/3} \sim \overline{T}$ and $H_\rho \sim \overline{T}$. The resulting estimate for the height dependence of the magnitude of the mean zonal velocity is thus

$$\overline{v_y} \sim 1/H_\rho^2 \sim 1/\overline{T}^2 \tag{7}$$

which sharply increases towards the top. This is the cause of the negative dip in $\overline{v_y}$ (as well as the low magnitude in the deeper region). This simple process, we believe, is responsible for generating the upper boundary shear in the solar angular velocity distribution.

In the deeper layers, as $\overline{T}$ becomes larger, $\overline{v_y}$ gets smaller and smaller. This is consistent with the lack of radial shear in angular velocity in the central region of the solar convection zone. For this to be an explanation, it is necessary that a number of other conditions on latitudinal transport of angular momentum are satisfied. The current f-box experiments are inadequate for addressing such issues.

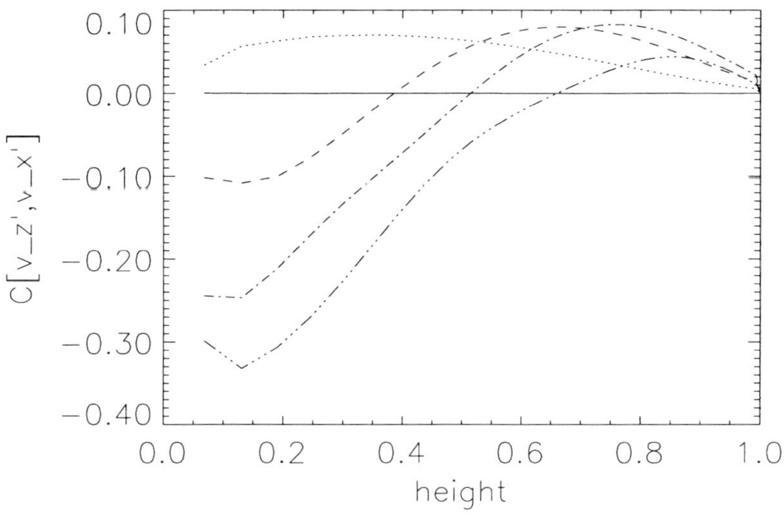

Figure 4 Vertical profiles of $C[v_z', v_x']$ for cases with fixed F (= 0.03125) and θ (= 45°). The dotted, dashed, dot-dashed, and triple dot-dashed curves are for Ω = 0.125, 0.25, 0.5, and 1, respectively.

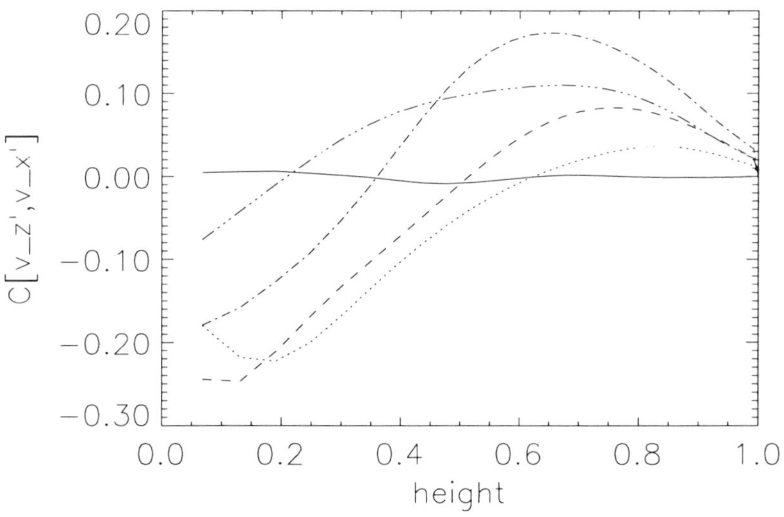

Figure 5 Vertical profiles of $C[v_z', v_x']$ for cases with fixed F (= 0.03125) and Ω (= 0.5). The solid, dotted, dashed, dot-dashed, and triple dot-dashed curves are for θ = 0°, 22.5°, 45°, 67.5°, and 78.75°, respectively.

Acknowledgements

This research is supported by the Research Grant Council of Hong Kong.

References

Chan, K.L., & Sofia, S. 1989, ApJ, 336, 1022
Chan, K.L., & Sofia, S. 1996, ApJ, 466, 372
Chan, K.L. 1999, in *Theory and Tests of Convection in Stellar Structure*, eds. A. Gimenes, E.F. Guinan, & B. Montesinos, ASP Conference Series, Vol. 173, p. 249
Korzennik, S.G. et al. 1997, in *IAU Symp. No. 181: Sounding Solar and Stellar Interiors*, eds. J. Provost & F.-X. Schmider, p. 211, Kluwer Academic Publishers
Küker, M., Rüdiger, G., & Kitchatinov, L.L. 1993, A&A, 279, L1
Thompson, M.J. et al. 1996, Science, 272, 1300

2D NUMERICAL SIMULATION OF STELLAR CONVECTION

An Overview

Matthias Steffen *

Astrophysikalisches Institut Potsdam, An der Sternwarte 16, D-14482 Potsdam, Germany

msteffen@aip.de

Abstract The dynamics and thermal structure of the surface layers of stars with outer convection zones can be studied in some detail by means of numerical simulations of time-dependent compressible convection. In an effort to investigate the properties of "stellar granulation" as a function of spectral type, we have carried out elaborate 2-dimensional radiation hydrodynamics calculations of surface convection for a variety of stellar parameters. The main features of these simulations are reviewed, with particular reference to standard mixing length models.

1. INTRODUCTION

Convection is a universal feature. Essentially all types of stars have either a convective core, a convective envelope, or both. Low-mass stars are fully convective, giants may accommodate several distinct convective shells. In the case of the Sun, the energy transport in the inner parts is entirely due to radiation, while in the outer 28.7% (in radius) it is primarily due to large-scale convective currents. At the surface, the solar granulation is the visible imprint of gas flows in the outermost layers of the convection zone.

The role of stellar convection is far-reaching: Convective energy transport determines the internal temperature structure of a star and its radius (which decreases with increasing convective efficiency), and hence controls the star's global properties. Convective regions are chemically

*Institut für Theoretische Phsyik und Astrophysik, Universität Kiel, D-24098 Kiel, Germany

K.S. Cheng et al. (eds.), Stellar Astrophysics, 25–36.

completely mixed, and overshooting convective flows lead to partial mixing of the adjacent radiative layers. "Overshoot" and similar mixing processes which are not confined to the convectively unstable regions are thought to be responsible for the existence of carbon stars, carbon-rich white dwarfs, and for the destruction of lithium in solar-type stars. Convective motions and concomitant temperature fluctuations exert a direct influence on stellar spectra, causing small but practically relevant changes in wavelength position, shape and strength of spectral lines. The stochastic convective motions can excite stellar oscillations (like in the Sun) and are a source of acoustic energy, contributing to the heating of stellar chromospheres. Finally, convection and rotation are necessary preconditions for the operation of the magnetic dynamo mechanism, and hence for stellar activity.

Unfortunately, a closed analytical theory of stellar convection is lacking due to the complexity of the underlying hydrodynamical problem. So far, stellar structure models still rely on a phenomenological approximation, the so-called *mixing-length theory* (MLT).

2. HYDRODYNAMICAL MODELING

The differential equations governing the physics of stellar convection are well-known but difficult to solve: the conservation equations of hydrodynamics, coupled with the equations of radiative energy transfer. While their highly non-linear and non-local character precludes an adequate analytical treatment, the numerical integration of the system of partial differential equations is now feasible. This approach constitutes an increasingly powerful method to study in detail the time-dependent properties of a radiating, partially ionized fluid under stellar conditions. Using modern supercomputers, it is possible to perform 3D numerical simulations of stellar (surface) convection with realistic background physics. But even 2D convection models require substantial amounts of computer time, particularly for solving the radiative transfer problem.

Just like "classical" stellar atmospheres, the hydrodynamical models are characterized by effective temperature, $T_{\rm eff}$, surface gravity, g, and chemical composition of the plasma. But in contrast to the mixing-length models, there is no longer any free parameter to adjust the efficiency of the convective energy transport. Based on first principles, radiation hydrodynamics (RHD) simulations provide physically consistent *ab initio* models of stellar convection which can serve to check the validity of MLT.

3. STATE-OF-THE-ART RESULTS

In the following, we present some results of elaborate 2-dimensional RHD simulations of solar and stellar surface convection. The models comprise small sections near the stellar surface, typically extending over 10 pressure scale height in vertical direction, including the photosphere, the thermal boundary layer near optical depth $\tau_{\rm Ross} = 1$, and parts of the subphotospheric layers. For the hotter stars of type A and early F, the surface convection zone(s) are shallow enough to be entirely fitted into the computational box. For the cooler, solar-like stars, only the uppermost layers of the deep convection zones can be included into the model, requiring an open lower boundary.

These simulations are designed to resolve the "stellar granulation". The effect of the smaller scales, which cannot be resolved numerically, is modeled by means of a subgrid scale viscosity (so-called Large Eddy Simulation approach). Spatial scales larger than the computational box are ignored. We employ a realistic equation of state, accounting for ionization of H, HeI, HeII and H_2 molecule formation. In order to avoid problematic approximations (like the diffusion or Eddington approximation), we solve the non-local radiative transfer problem along a large number ($\approx 10\,000$) of rays in a number of wavelength bands, with realistic opacities including the influence of spectral lines. For further details see Ludwig et al. (1994) and Freytag et al. (1996).

3.1 SOLAR-TYPE SURFACE CONVECTION

A grid of about 60 simulation runs of solar-type surface convection, covering the range $4\,300$ K $\leq T_{\rm eff} \leq 7\,100$ K; $2.54 \leq \log g \leq 4.74$ (solar metallicity), is presently available. In general, the hydrodynamical simulations exhibit *unsteady convective flows* and superimposed *oscillations.*

General properties. The basic features of the numerical convection models are illustrated in Fig. 1, showing a representative snapshot from a well-relaxed simulation of the solar granulation. In the convectively unstable, subphotospheric layers, fast ($v \lesssim c_s$) narrow downdrafts (also called plumes or jets) stand out as the most prominent feature. They are embedded in broad ascending regions (the granules) where velocities are significantly lower. We find enormous temperature differences between the cool downflows and the hot upflows ($\Delta T_{\rm max} \approx 5\,000$ K at $z \approx -100$ km in the solar case). Near optical depth $\tau = 1$, efficient radiative surface cooling produces a very thin thermal boundary layer over the ascending parts of the flow, where the temperature drops sharply with height and the gas density exhibits a local inversion.

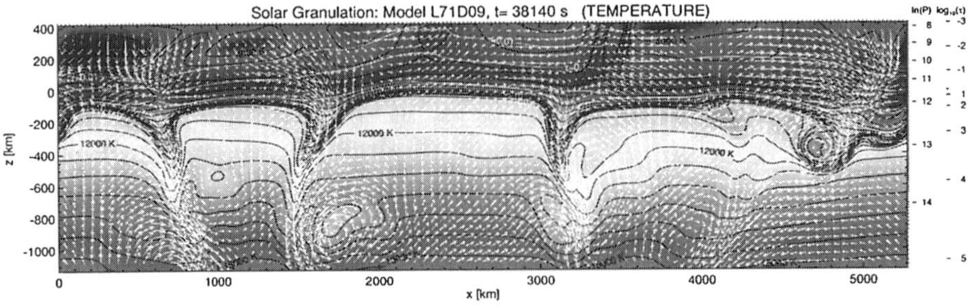

Figure 1 Representative snapshot from a 2-dimensional numerical simulation of solar surface convection after 38 140 s of simulated time. This model was computed on a Cartesian grid with 210x106 mesh points (tick marks along upper and right side), with periodic lateral boundary conditions ($L = 5\,250$ km). "Open" boundary conditions at the bottom and top of the computational domain are designed to minimize artificial distortions of the flow. The velocity field is represented by pseudo streamlines, indicating the displacement of a test particle over 20 sec (maximum velocity is 10.1 km/s at this moment); the temperature structure is outlined by temperature contours in steps of 500 K. Geometrical height $z = 0$ (scale at left) corresponds to $\tau_{\mathrm{Ross}} \approx 1$; scales at right refer to the horizontally averaged gas pressure [dyn cm^{-2}] and Rosseland optical depth τ_{Ross}.

Although convectively stable according to the Schwarzschild criterion, the photosphere ($\tau \lesssim 1$) is by no means static. Convective flows overshooting into the stable layers from below are decelerated here and deflected sideways. This can sometimes result in transonic horizontal streams which lead to the formation of shocks in the vicinity of strong downdrafts (see vertical fronts in Fig. 1 near $x = 3\,400$ km, $z \gtrsim 200$ km). Oscillations excited by the stochastic convective motions contribute to the photospheric velocity field as well. Interestingly, the oscillation periods lie in the range 150 to 500 s for the solar simulation, in close agreements with the observed 5 minute oscillations. Controlled by the balance between dynamical cooling (due to adiabatic expansion) and radiative heating (in the spectral lines), the photospheric temperature stratification is not at all plane-parallel. At least for the Sun, the resulting average temperature in the photosphere is slightly cooler than in radiative equilibrium.

A typical sequence of events seen in the time-dependent simulations is the formation of new downdrafts, their horizontal advection, and finally their merging with another downdraft, as illustrated in Fig. 2. We found that merging of downdrafts is a major source of acoustic energy flux. Fig. 2 also demonstrates the strong correlation between vertical convective velocity and emergent continuum intensity. Downdrafts are clearly seen as narrow dark regions in intensity. On the other hand,

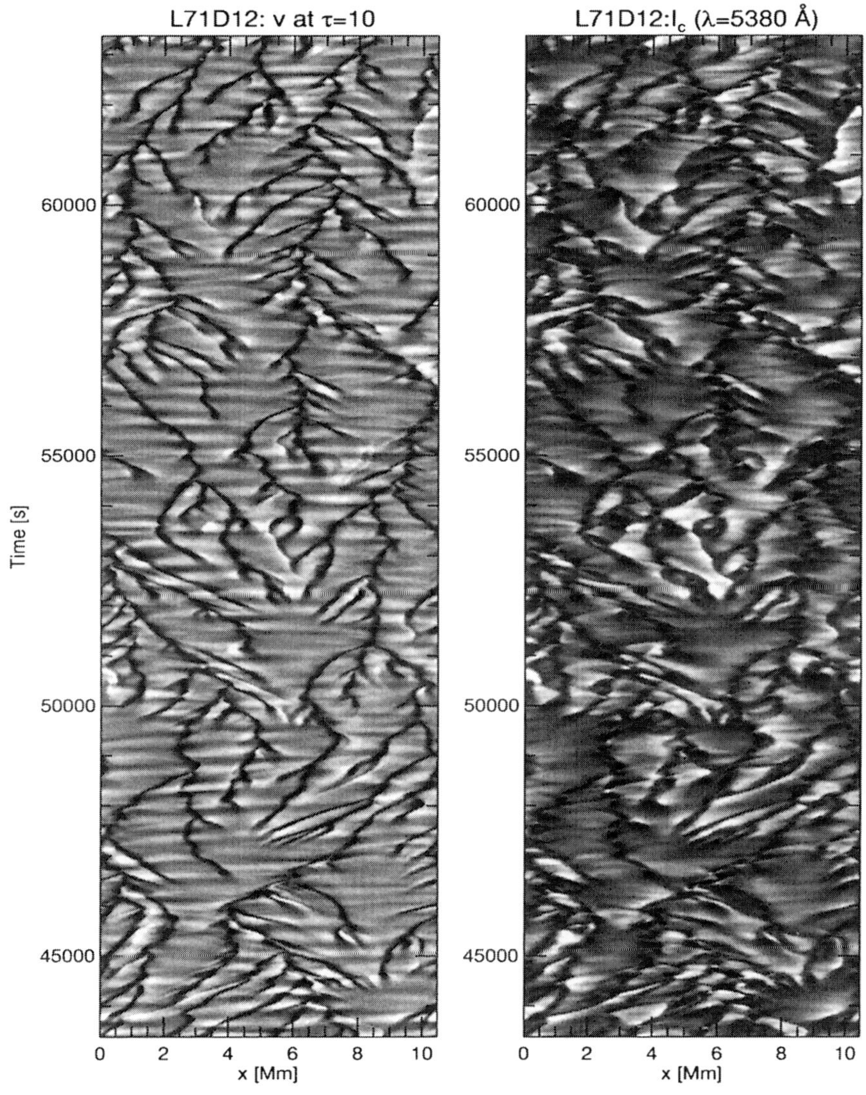

Figure 2 Grey-scale plot of vertical velocity at $\tau_{\rm Ross} \approx 10$ (left) and emergent continuum intensity at $\lambda\,5380$ Å (right) as a function of horizontal position and time. Dark shades indicate downward velocities and low intensities, respectively. Note that the signature of superimposed "5 min oscillations" is clearly seen in both maps. Data were taken from a solar simulation that is similar to the one shown in Fig. 1 but has twice the horizontal size (420x106 grid).

the correlation is not perfect: in the ascending regions there is more structure in the intensity data than in velocity. Small "granules" tend to be brighter than large ones, which often show the highest intensity at their edges. Indeed, the power spectra of velocity and intensity differ significantly; the p-modes stand out much more clearly in velocity power.

Comparison with MLT. The numerical simulations contain a wealth of information and may be analyzed in various ways. For comparison with mixing-length theory, we have averaged the quantities of interest horizontally (over planes of constant geometrical height or over surfaces of constant optical depth) and over time. Overshoot into the photosphere is substantial according to the hydrodynamical models (Fig. 3). Due to the local nature of MLT, however, it is suppressed in the mixing-length models. But even in the unstable layers it is impossible to adjust the mixing-length parameter α such as to reproduce the average 2D or 3D velocity field in the framework of MLT. The situation is similar for the thermal structure $T(\tau)$ (not shown). The difference between the 2D and the 3D simulation results is probably related to the higher acoustic activity of the 2D models (Ludwig & Nordlund, this volume).

The turbulent pressure P_{turb} amounts to a significant fraction of the gas pressure in the solar atmosphere (bottom panel of Fig. 3) and hence affects the geometric scale of the surface layers. Due to the lack of overshoot and the uncertainty in the proper choice of α, mixing-length models fail to account for the correct distribution of the turbulent pressure, often even assuming $P_{\text{turb}} = 0$.

Spectral line formation. It is not surprising that the photospheric velocity field and temperature inhomogeneities due to convective overshoot affect the formation of spectral lines. We have computed LTE synthetic profiles for a set of weak ($W_\lambda < 10$ mÅ) fictitious lines ($\lambda =$ 5500 Å) of different elements and ions, based on an ensemble of representative snapshots from the 2D solar simulation. For each snapshot, we compared the equivalent width of the horizontally averaged line profile, $W_{\text{2D}} = \langle W_\lambda(\text{RHD})\rangle$, with the equivalent width of the same line computed from the averaged (on τ-surfaces) RHD model, $W_{\text{1D}} = W_\lambda(\langle\text{RHD}\rangle)$. Averaged over all snapshots, the ratio $W_{\text{1D}}/W_{\text{2D}}$ gives the correction factor that has to be applied to standard 1D abundance determinations in order to correct for the influence of photospheric inhomogeneities. The current results for a variety of different lines are summarized in Fig. 4. According to this investigation, lines of neutral minority species originating from the ground state are most affected. In the most extreme example, the standard analysis based on 0 eV TiI lines would overestimate the titanium abundance by a factor of 2. The effects may be even larger for F-type stars!

Calibration of Mixing-Length Theory. As we have seen, the mean dynamical and thermal structure of the superadiabatic stellar surface layers is systematically different in RHD and MLT models, respectively. In this sense, there is no way to establish a unique calibration of MLT by means of hydrodynamical simulations. However, it is possible to match particular properties by adjusting α. For stellar evolution, the key quantity is the entropy jump, Δs^*, from the surface to the interior. Based on our grid of solar-type RHD models, we have calibrated MLT by matching the quantity Δs^*, which can be "measured" from the simulations. This calibration yields the correct adiabat and depth of the convective envelope, and hence the appropriate $\alpha^*(T_{\rm eff}, \log g)$ for stellar evolution calculations. To our knowledge, it is the first theoretical prediction of how the mixing-length parameter varies across (parts of) the Hertzsprung-Russell diagram (for details see Ludwig, Freytag & Steffen 1999).

For the Sun, the result is $\alpha^*(\text{2D RHD}) \approx 1.61 \pm 0.05$ and $\alpha^*(\text{3D RHD}) \approx 1.68 \pm 0.05$, in excellent agreement with the helioseismic value of $\alpha^*(\text{Helios.}) \approx 1.71 \pm 0.02$. In the covered range of stellar parameters, we find that α^* depends primarily on $T_{\rm eff}$ and varies systematically between about 1.3 for F dwarfs and 1.8 for K subgiants. Certainly, the fact that α^* varies only moderately with stellar type is one of the reasons for the surprising success of MLT over the past 40 years.

3.2 CONVECTION IN A-TYPE STARS

Our grid of hydrodynamical models of A and early F main sequence stars comprises about 30 simulation runs in the range $7\,200\ \text{K} \leq T_{\rm eff} \leq 9\,500\ \text{K}$; $\log g = 4.44$. These type of stars exhibit two distinct convection zones: the one at the surface is driven by the combined first ionization of hydrogen and helium, the deeper one is related to the second ionization of helium. In the example shown in Fig. 5, the Schwarzschild-unstable regions are separated by a stable radiative layer of about two pressure scale heights. Nevertheless, the hydrodynamical simulations indicate that the two convection zones are effectively connected by vigorous currents and that the stable buffer layer is completely mixed, in contrast to MLT predictions.

A closer inspection of the simulation data showed that the velocity field due to overshooting convective motions declines exponentially with distance from the Schwarzschild boundary (in the example of Fig. 5, the velocity scale height below the HeII convection zone is $H_{\rm v} \approx 0.35\, H_{\rm p}$). We found that mixing due to overshoot can be described as a diffusion process, with a diffusion coefficient $D \sim v_{\rm rms}^2$. For a slightly hotter star

($T_{\rm eff} = 7943$ K, $\log g = 4.34$) we have shown that the convectively mixed mass is underestimated by a factor of 10 if overshoot is ignored (Freytag, Ludwig & Steffen 1996).

In stellar structure models, overshoot is usually implemented by specifying the distance ($d_{\rm over} = \delta\, H_{\rm p}$) by which the convection zone proper is extended due to "penetration" into the stable layers. This empirical approach ignores the hydrodynamical results indicating that overshoot is characterized by an exponential velocity field and (partial) diffusive mixing.

4. CONCLUSIONS

Radiation hydrodynamics simulations of stellar surface convection have now reached a level of sophistication far beyond idealized numerical experiments. They are the key for a better understanding of the thermal structure and dynamics of stellar convection zones, including overshoot and its role for mixing.

The comparison with "classical" mixing-length models reveals quantitative and qualitative differences. In the simulations, the dynamics of convection is dominated by fast, cool, narrow downdrafts, which form coherent structures extending over many pressure scale heights. A pronounced up/down asymmetry is a general feature of the numerical models (2D as well as 3D), seen for all types of stars investigated so far. Obviously, this result is in stark contrast with the symmetric picture of MLT where "bubbles" are assumed to travel for about one pressure scale height before dissolving. Since the "jets" are driven by radiative cooling at the stellar surface, convection is an extremely non-local process. The properties of the surface determine the dynamics and structure of the whole convection zone. Certainly, the assumption of locality is a major problem with MLT.

Hydrodynamical model atmospheres can be used to study the formation of spectral lines in an inhomogeneous medium. For the Sun, we found that "granulation corrections" for spectroscopic abundance determinations can amount to -0.3 dex for the most temperature sensitive lines. These corrections are much larger than the well-known NLTE corrections that have been investigated so far. Work is under way to check whether these findings have notable consequences for the currently adopted chemical composition of the Sun and other stars.

While is seems hopeless to successfully model the structure of the superadiabatic surface layers within the framework of MLT, it is possible to calibrate MLT through hydrodynamical simulations for application to stellar evolution. Based on 2D radiation hydrodynamics, our present calibration ultimately needs to be verified by 3D simulations. In the meantime, we are working to extend the calibration to metal-poor stars, down to [M/H]=-2.

The hydrodynamical simulations for A-type stars (and White Dwarfs) demonstrate that the velocity field due to overshooting convective motions declines exponentially with distance from the Schwarzschild boundary and leads to diffusive mixing. Although this result is not directly applicable to strongly adiabatic conditions, we believe that the exponential depth-dependence of the diffusion coefficient is a general feature of overshoot and may have important implications for mixing and nucleosynthesis in stellar interiors.

Acknowledgements

The author is grateful to H.-G. Ludwig for making available some of his 3D simulation results and for computing MLT reference models. B. Freytag computed the A-star model and helped with processing the simulation data.

References

Böhm-Vitense, E. (1958), Z. Astrophys. 46, 108

Freytag, B., Ludwig, H.-G., Steffen, M. (1996), Astron. Astrophys. 313, 497

Ludwig, H.-G., Jordan, S., Steffen, M. (1994), Astron. Astrophys. 284, 105

Ludwig, H.-G., Freytag, B., Steffen, M. (1999), Astron. Astrophys. 346, 111

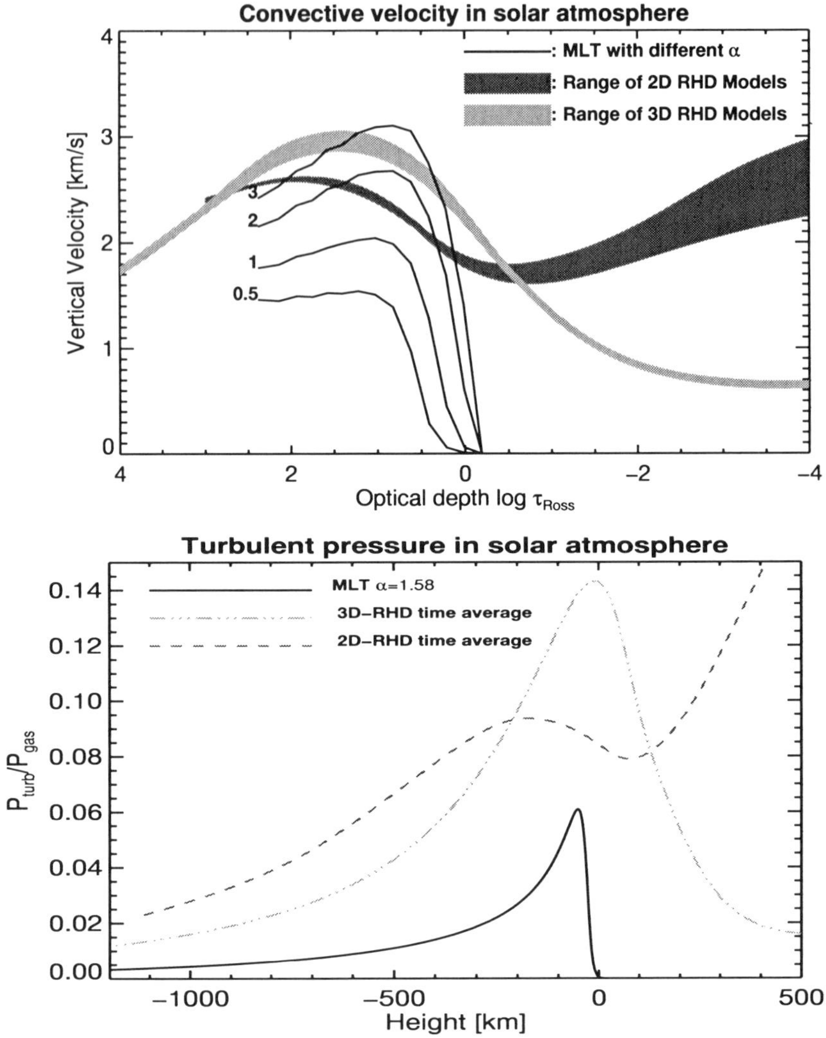

Figure 3 **Top:** Vertical convective velocity v as a function of optical depth $\tau_{\rm Ross}$ as obtained from standard MLT ($\alpha = 0.5, 1, 2, 3$) compared with $v_{\rm rms}$ derived from 2D and 3D numerical simulations by averaging over planes of constant $\tau_{\rm Ross}$ and over time. **Bottom:** Ratio of turbulent pressure ρv^2 to gas pressure P as a function of geometrical height as obtained from MLT for $\alpha = 1.58$ (adequate for evolution calculations) compared with the results obtained from 2D and 3D numerical simulations by averaging ρv^2 and P over horizontal planes and over time. Note that a self-consistent MLT solution does not exist. Rather, we obtained the plotted results *a posteriori* from a standard model computed without the effect of turbulent pressure according to the formalism by Böhm-Vitense (1958), using a non-grey $T(\tau)$ relation that is consistent with the numerical simulations. The 3D simulation results are courtesy of H.-G. Ludwig (see Ludwig & Nordlund, this volume).

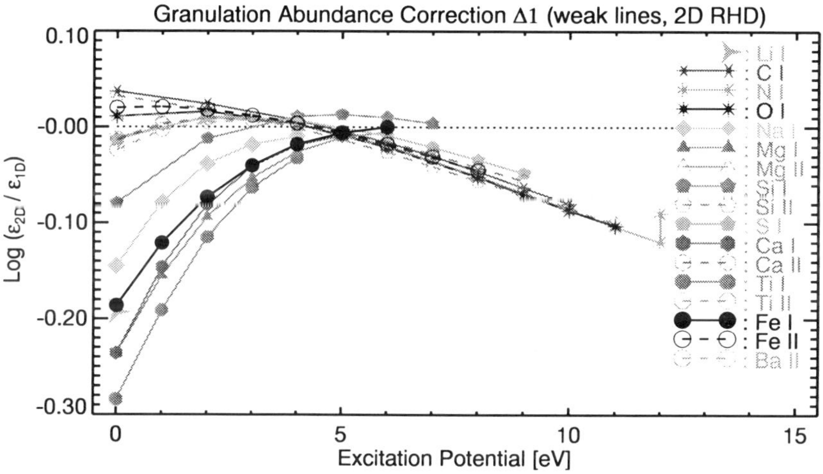

Figure 4 Logarithmic abundance correction to be applied to standard spectroscopic determinations of solar chemical abundances in order to compensate for the effects of photospheric temperature inhomogeneities. These "granulation corrections" depend in a systematic way on the excitation potential of the spectral line's lower level and on the ionization potential of the atom or ion being considered. Corrections are valid for weak lines at $\lambda 5\,500$ Å. The NI 12 eV line was also computed for $\lambda 10\,000$ Å (upper symbol). In general, 5 eV lines are "safe".

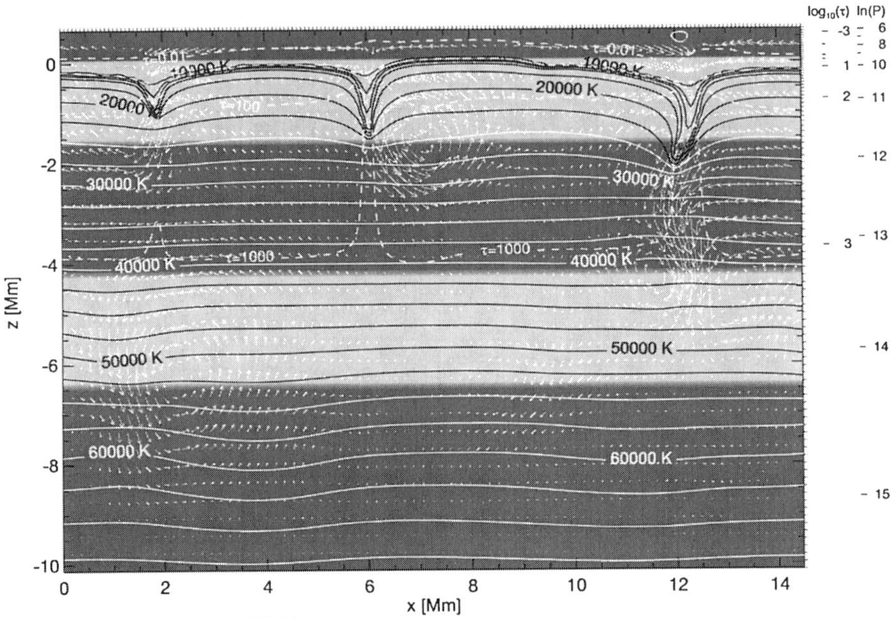

Figure 5 Snapshot from a 2D simulation of convection in an A8 main sequence star ($T_{\rm eff} = 7\,600$ K, $\log g = 4.4$ [cgs units]) with 2 distinct convection zones. The unstable regions ($\mathrm{d}\overline{s}/\mathrm{d}z < 0$, where $\overline{s}$ is the horizontally averaged specific entropy) are shaded light gray. The velocity field is represented by pseudo streamlines, indicating the displacement of a test particle over 40 sec (maximum velocity is 19.0 km/s at this moment); the temperature structure is outlined by temperature contours in steps of 2500 K. Dashed lines trace levels of constant $\tau = 0.01, 1, 100, 1000$. Located in the lower radiative zone, a closed lower boundary is appropriate. Grid size 182x95.

THE ATMOSPHERIC DYNAMICS IN 2D AND 3D SIMULATIONS OF STELLAR SURFACE CONVECTION

Hans-Günter Ludwig[1] and Åke Nordlund[2]
[1]*ENS-CRAL, 46 allée d'Italie, 69364 Lyon Cedex 07, France,*
[2]*Astronomical Observatory, Juliane Maries Vej 30, 2100 Copenhagen Ø, Denmark*
[1]hludwig@ens-lyon.fr,
[2]aake@astro.ku.dk

1. INTRODUCTION

A pressing problem in our understanding of the structure and evolution of late-type stars is the lack of a reliable theory of convection. Convective flows transport mass, momentum, and energy, thus influencing the internal distribution of energy, chemical species, and angular momentum. They drive the stellar magnetic and oscillatory activity. Although the underlying physical principles are well known, the nonlinear and nonlocal character of the equations describing a radiating, partially ionized fluid has prevented the development of a closed analytical theory.

However, the situation is improving since the advent of powerful computers which allow numerical solutions of the governing radiation-hydrodynamics (RHD) equations. During the last two decades computer simulations have dramatically improved our principle understanding of stellar convection. But due to the large computational demands RHD models have not yet become a standard ingredient in models of stellar structure, as would be desirable considering the importance of the process. In order to cut computational costs RHD models restricted to two spatial dimensions (2D) have been constructed (e.g. Gadun et al. 1999, Ludwig et al. 1999). The 2D approach allows an extensive exploration of the parameter space, but it is unclear whether in 3D the findings would be significantly altered.

In this paper we report on an effort to identify and quantify differences between 2D and 3D models by looking at a small set of representative models which are computed on a strictly differential basis. I.e. beside the dimensionality the input physics (opacity, equation of state) and

K.S. Cheng et al. (eds.), Stellar Astrophysics, 37–44.

Table 1 List of model stars. The effective temperature $T_{\rm eff}$, the gravitational acceleration $\log g$, the simulated time for 2D and 3D models, whether the 3D time-sequence includes the transition from a 2D to a 3D flow pattern (marked with "x"), and the geometrical height and width of the computational box are given. Note the significant variation in $T_{\rm eff}$ and $\log g$. In all 3D models 125x125x82 (x-y-z direction) mesh points were used, with 125x82 (x-z direction) mesh points in 2D models. All models have the same (solar) chemical composition.

Spectral type	$T_{\rm eff}$ [K]	$\log g$ [cm/s^2]	2D [h]	3D [h]	2D→3D	height [Mm]	width [Mm]
G V (Sun)	5600	4.44	47.3	1.33		3.09	6.00
K V	4600	4.44	10.7	2.37	x	2.36	5.21
F V	6600	4.44	11.8	1.80	x	4.50	6.76
K IV	4500	2.94	710	30.8	x	94.0	141

numerical properties (resolution, boundary conditions, numerical viscosity) are exactly the same. They are local models representing a small volume at the surface of stars with a convective envelope. As an important property they include the upper part of the convective envelope as well as the optically thin stellar atmosphere. They are — like classical model atmospheres — characterized by effective temperature, gravitational acceleration, and chemical composition. The main feature which they capture are the convective cells, the energetically dominant entities in the flow.

Work is in progress and in the following we give mostly a qualitative interpretation of our findings concentrating on the *atmospheric dynamics*. Interestingly, it appears that the *characteristic horizontal scale* of the flow field plays a central role here. Understanding the differences between 2D and 3D models provides us with new insight into stellar surface convection as such.

2. THE MODELS SET

Table 1 summarizes the properties of the set of models. The models were computed with the RHD code of Nordlund & Stein (1990, for a description). Note that in Tab. 1 the duration of the runs is given in hours. In terms of sound-crossing or turn-over time the most extensive runs were performed for the G V and K IV star.

3. 2D TO 3D TRANSITION

For a given stellar parameter set usually a 2D run was performed until we obtained a relaxed model, i.e. a model whose statistical properties do not undergo long-term changes any more. The corresponding 3D run was set up by expanding the 2D model along the symmetry direction. Since the models use periodic conditions at the lateral boundaries this model is a possible 3D configuration and would keep its symmetry even within a 3D model run. To break the symmetry a random disturbance was added to the x-velocity component (typically 5% of the velocity amplitude). Figure 1 illustrates in the case of the K IV model star the decay of the 2D convective roles into a 3D configuration consisting of convective cellular structures. The emergent intensity in the vertical direction from the 3D computational domain is shown, corresponding closely to the impression an observer would obtain when looking at the disk center of the star.

The main points are: i) The 2D pattern is indeed heavily unstable towards symmetry breaking disturbances. The decay time of the 2D pattern is compatible with what one would expect for a convective instability acting on the Brunt-Väisälä time scale (for this model about 12 min, time difference between the individual frames 15 min, sound-crossing time through the computational volume about 150 min). ii) As the third frame of the sequence shows the convective instability grows fastest on small spatial scales (the ripples along the vertical direction), smaller than the typical 2D or 3D scales. iii) The resulting 3D structures are smaller than the 2D ones (this is perhaps non-obvious from Fig. 1 but is substantiated by Fig. 2).

Point iii) came as a surprise and we consider the understanding of this feature as a key issue for interpreting the different properties of 2D and 3D models. The surprise came about from the following argument: mass conservation suggests that in 2D it is harder to over turn material that was advected upwards and has cooled on the surface. In 2D Cartesian geometry there is only one horizontal direction in which material can be transported while in 3D two directions are available. At a fixed horizontal scale and vertical mass flux divergence (connected with the density scale height) higher horizontal velocities and pressure fluctuations would be necessary for the sidewise transport of material in 2D. From this argument it was plausible to expect that a 2D system would react in a way to reduce the horizontal scales with respect to a 3D one.

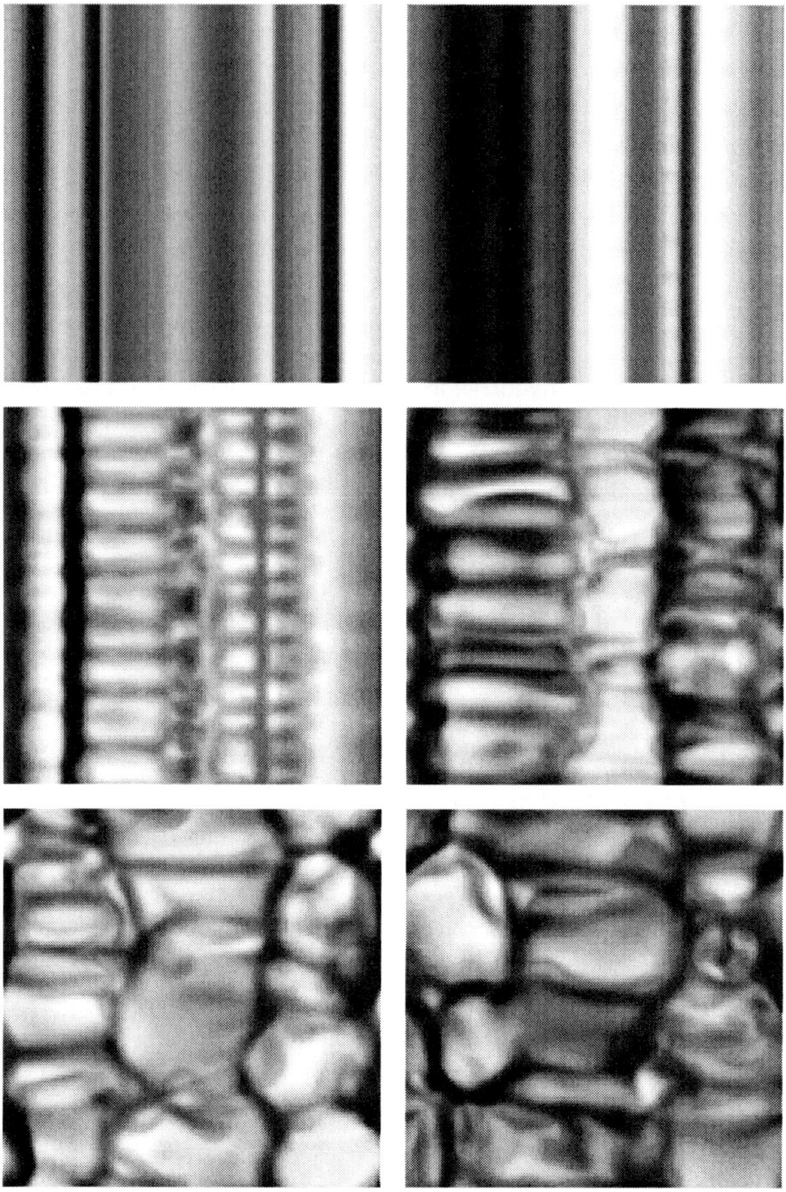

Figure 1 Decay of an initially 2D into a 3D flow pattern for the K IV subgiant. The six frames show from upper-left to lower-right the temporal evolution of the (frequency integrated) emergent intensity in a 3D simulation. The symmetry of the 2D pattern was broken by adding a random velocity disturbance to the horizontal velocity. The time interval between the frames amounts to 15 min, the size of the computational domain is 141x141x94 Mm^3.

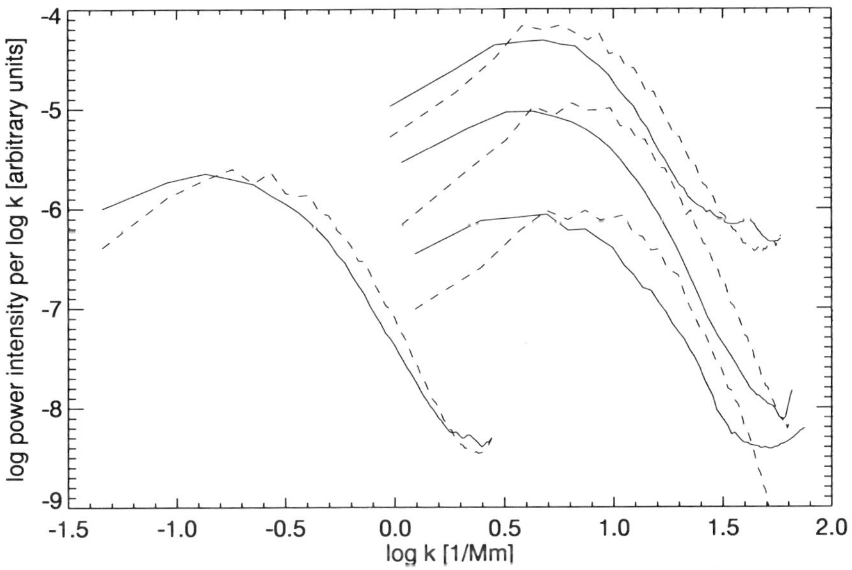

Figure 2 Spatial power spectra of the emergent intensity. Each spectrum represents a long term temporal average. 2D models are depicted with solid, 3D ones with dashed lines. The curves at the left belong to the K IV subgiant, the groups on the right to the dwarf models with F to K spectral class from top to bottom.

4. HORIZONTAL SCALES AND ATMOSPHERIC DYNAMICS

To quantify this further power spectra of intensity images (such as the ones in Fig. 1) were computed and averaged over time for relaxed model sequences. They are shown in Fig. 2. The figure confirms the visual impression from Fig. 1 that generally the spatial scales are smaller in 3D than in 2D. As a side point the figure also shows the change of the intensity contrast of the convective pattern at the stellar surface (basically the integral of the power spectrum). This reflects the change in the horizontal temperature differences at the surface, which are in turn related to the overall convective heat transport. The figure demonstrates that the relative scaling of the intensity contrast between 2D and 3D models is rather similar, and that in this respect the K IV star is quite similar to the Sun.

As discussed previously one expects that the larger spatial scales in 2D are associated with larger horizontal pressure fluctuations. Figure 3 demonstrates that the relative horizontal pressure fluctuations are indeed systematically larger in the 2D case. This refers in particular to the surface region where the overturning of the upflowing material takes place. The most extreme ratio between 2D and 3D fluctuations (about a

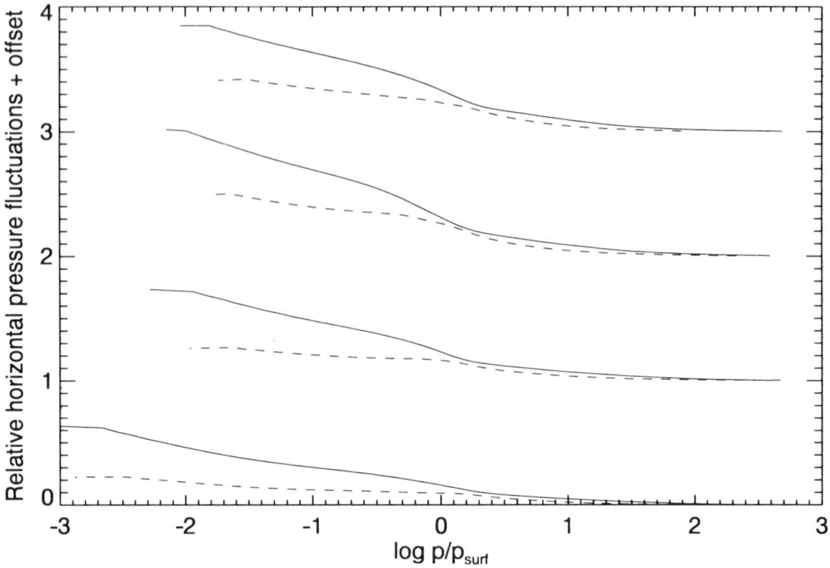

Figure 3 Relative horizontal pressure fluctuations versus pressure. The pressure for all models is given in units of their surface pressure, i.e. the pressure at optical depth unity. The curves represent long term horizontal averages. Solid lines represent 2D models, dashed lines 3D models with the K IV, F V, G V, K V star from top to bottom. For clarity the curves for the individual stars were shifted by 3, 2, 1, and 0 respectively.

factor of 2) is found for the K V star despite that its horizontal temperature fluctuations are smallest. This clearly indicates that the pressure fluctuations are dominated by dynamical effects and do not just reflect effects due to temperature variations.

Another interesting aspect is the increase of the pressure fluctuations with height in the atmospheric layers. If the pressure field in the atmosphere was governed only by the constraint to provide the forces necessary to overturn the upflowing material one would expect the pressure fluctuations to decrease with height. Their increase hints towards another process, and indeed we interpret it as imprint of wave motions.

Figure 4 shows the root-mean-square vertical velocity for the four reference models. Again, a systematic pattern appears: i) The maximum of the vertical velocity amplitude in the overturning region is smaller in 2D. ii) The decline of the velocity is more gradual in 2D. iii) The atmospheric velocities in the 2D models do not drop as far as in the 3D models. iv) The 2D models show a distinct increase of the vertical velocity amplitude towards the higher atmospheric layers, apparently missing in the 3D models. We use the expression "apparently missing" since 3D models that extend into higher atmospheric layers than studied

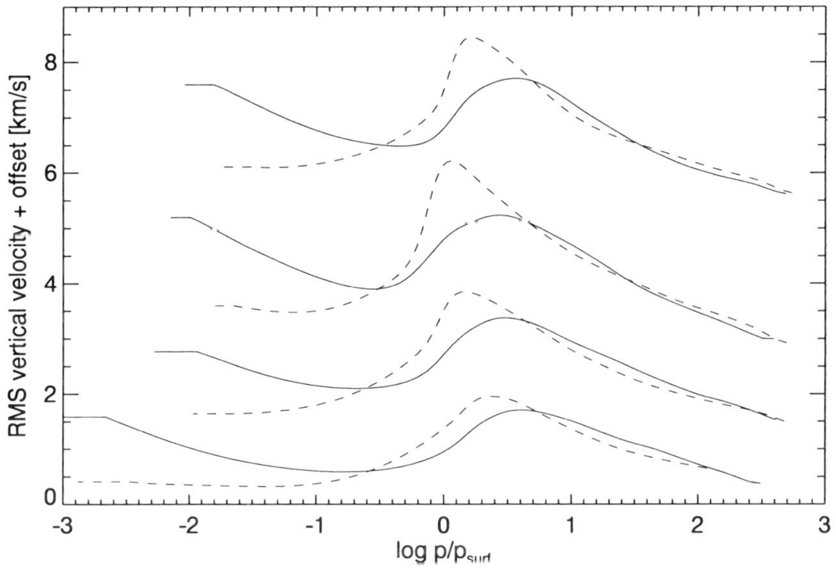

Figure 4 Root-mean-square of the vertical velocity versus pressure. The pressure for all models is given in units of their surface pressure, i.e. the pressure at optical depth unity. The curves represent long term horizontal averages. Solid lines represent 2D models, dashed lines 3D models with the K IV, F V, G V, K V star from top to bottom. For clarity the curves for the individual stars were shifted by 5, 2, 1, and 0 km/s respectively.

here also show a rapid increase of the vertical velocity amplitude. From this perspective one should rather say that in the 2D models the increase just happens in deeper atmospheric layers.

How can the simulation results described above be put into a coherent picture? As discussed before simple mass conservation arguments fail to predict their gross behaviour. We believe that the horizontal scales are a consequence of a competition between the creation and destruction of downflows: the convective instability produces new downflows while the downflows disappear by merging. While the convective instability works with similar efficiency in 2D and 3D, the destruction of downflows is a more efficient process in 2D. If two downflows approach each other in 2D their merging is rapid and destroys a large downflowing area. In 3D, downflows cannot move as freely since they are part of a network, meaning that they are to some extend anchored in the surroundings. In 3D the merging of downdrafts therefore happens more gradually.

If one accepts at face value the existence of larger scales in 2D, than their higher horizontal pressure fluctuations are a natural consequence. The differences in the variation of vertical velocity with depth are a result of the tendency to mitigate the scale effect by spreading the overturning region over a larger vertical layer. The differences in the velocity maxima can be understood along similar lines, and from the so called "buoyancy-breaking", the effect that high horizontal pressure fluctuations not only accelerate horizontal motions but also decelerate vertical ones.

The merging of downflows is accompanied by strong pressure disturbances which are strong sources of sound waves. We have verified that the velocity field in the higher atmosphere is dominated by upward traveling acoustic waves. The higher merging efficiency in 2D models leads to a stronger production of sound waves and ultimately a higher amplitude of the velocity in the upper atmospheric layers. In 3D models these waves are also present but due to their smaller energies they have to travel into higher (lower density) regions before their velocity amplitudes become significant.

5. CONCLUSIONS

We have seen that there are systematic differences in the atmospheric dynamics between 2D and 3D RHD models which are connected to the characteristic horizontal scale of the convective cells. This scale is ultimately connected with the topology of the flow. While we have not discussed the thermal properties of the models here it is already clear that the significant different atmospheric velocities must lead to differences in the shapes of synthetic spectral lines computed from 2D or 3D models. Work is under way to quantify these differences (Asplund et al. 2000). Work is also under way to analyze the thermal properties of the models, in particular with respect to the efficiency of the convective energy transport (Ludwig & Nordlund, 2000). This quantity is related to the mixing-length parameter which is an important input parameter in standard stellar structure models.

References

Asplund M., Ludwig H.-G., Nordlund Å., Stein R.F., 2000, in preparation

Gadun A.S., Solanki S.K., Johannesson J., 1999, A&A 350, 1018

Ludwig H.-G., Freytag F., Steffen M., 1999, A&A 346, 111

Ludwig H.-G., Nordlund Å., 2000, in preparation

Nordlund Å., Stein R.F., 1990, Comp. Phys. Com. 59, 119

ROTATING TURBULENT COMPRESSIBLE CONVECTION IN A SHELL

F. J. Robinson

Abstract By performing a large eddy simulation of a section of the convection zone, we find that differential rotation in the solar convection zone is shaped by the thermal structure (specifically the latitudinal entropy gradient), rather than the Reynolds stress or meridional circulation, as suggested in nowadays semi-analytical formulations.

Keywords: Differential rotation, compressible convection, turbulence

1. INTRODUCTION

The purpose of this paper is to explain the observed differential rotation in the solar convection zone (SCZ). Using helioseismology, 1, the isorotation surfaces in the convection zone (outer 28% by radius of the sun) were found to be cone-like (aligned radially), in disagreement with most numerical simulations, which tended to produce cylindrical isorotation contours parallel to the rotation axis. Specifically, observations have revealed that: the angular velocity at the equator first increases then gently decreases with depth; at mid-latitudes is almost constant along radial lines; at high latitudes increases with depth; and on the surface of the sun increases from the poles (35 day period) to the equator (25 day period).

2. OVERALL SETUP

The computational domain is a sector of a spherical shell spanning 30° degrees above and below the equator, 60° in longitude, and a depth of 0.28 from the top (scaled by the radius of the shell). Meridional, zonal and radial directions are labeled θ (increasing southwards), ϕ (increasing eastwards) and r (increasing upwards), respectively. The boundaries are: insulating, stress free and impenetrable in latitude; periodic in longitude; conducting stress free top, with a constant vertical heat flux, f_b,

K.S. Cheng et al. (eds.), Stellar Astrophysics, 45–50.

flowing through the stress free base. Keeping the physics as simple as possible: the fluid is an ideal gas ($\gamma = 5/3$); the initial distribution is a neutrally stable polytrope spanning 5 pressure scale heights; gravity obeys the inverse square law; and the Prandtl number is one. To provide a natural boundary condition, a convectively stable layer is placed at the top of the shell occupying the upper 5 % of the domain. All variables are made dimensionless by scaling with an appropriate function of pressure, temperature and density at the top, e.g. velocity is scaled by the isothermal sound speed.

Using a code developed by 2, the fully compressible Navier Stokes equations are solved in spherical coordinates. An implicit scheme relaxes the fluid to a self consistent thermal equilibrium. This is followed by a second order explicit method which gathers the statistics of the time averaged state. The integration is over 500 turn over times, and requires about 1.5 million time steps. The resolution is 70 $\times$ 70 $\times$ 39, in the latitudinal, longitudinal and radial directions respectively. For a single processor on the ORIGIN2000, the CPU time per integration step is about 3 seconds. Using automatic parallelisation on 4 processors, the speed up factor is about 3.

The effect of rotation on the turbulent fluid can be measured by computing a turbulent inverse Rossby number or Coriolis number Co $= \Omega_o d/v''$, where $v'' = \sqrt{\overline{v^2} - \overline{v}^2}$ is the turbulent velocity computed from the statistically averaged mean flow. The overbar represents a longitudinal and temporal average. Ω_o and d are the rotation rate of the external reference frame and the depth of the convection zone, respectively. If Co is greater than one, then rotation will have a significant effect on the turbulent flow. In the present simulation Co is about 6.5, and the Reynolds number, $Re = v''d/\nu$, is around 500.

3. RESULTING MEAN FLOW

The time and longitude averaged angular velocity, relative to the rotating frame of reference, $\overline{\omega} = \Omega - \Omega_o = \overline{v_\phi}/r\sin\theta$, is presented in figure 1. In a qualitative sense, the shape of the isocontours resemble the helioseismology observations in two distinctive ways. Firstly, an initial increase then decrease in angular velocity from the top inwards at the equator is found, as shown by the two closed circular rotation contours. Secondly, away from the equator towards mid-latitudes, the contours are about half way between the cylindrical contours seen in most earlier global simulations, e.g. 3 and the cone-like shape seen in the sun. There is also good agreement over the range of depth and latitude in common with the latest large eddy simulations by 4.

To gain a more qualitative comparison with the solar case, $\overline{\omega}$ is plotted against depth (rescaled onto the unit interval), figure 2. Co-latitudes (90°-latitude) of $90°, 85°, 79°$ and $67.5°$, are denoted by solid, long-dashed, triple-dot dashed and dot-dashed curves, respectively. In the present model $\Omega_o \approx 3$, which means the shell spins about 3 times in the time it takes a fluid element to traverse the depth of the convection zone. At the top of the shell $\overline{\omega}$ drops by about 0.2 between the equator (solid line) and co-latitude of 67.5° (dot-dashed line). This means a 7% variation in rotation rate over 22.5°, or extrapolating, a pole that spins 28% faster than the equator. These results agree well with observations which suggest that the surface rotation rate of the sun varies between 25 days at the equator to 35 days at the poles, i.e. about 30% faster at the equator than at the pole.

4. WHAT PRODUCES THE DIFFERENTIAL ROTATION?

After averaging over longitude and time, the meridional momentum equation becomes:

$$\frac{1}{r^2}\frac{\partial}{\partial r}(\overline{\rho v_\theta v_r} r^2) + \frac{1}{r\sin\theta}\frac{\partial}{\partial\theta}(\sin\theta\overline{\rho v_\theta^2}) + \frac{1}{r}(\overline{\rho v_\theta v_r} - \cot\theta\overline{\rho v_\phi^2}) + \frac{1}{r}\frac{\partial\overline{p}}{\partial\theta} - 2\Omega\overline{\rho v_\phi}\cos\theta + S_\theta = 0 \qquad (1)$$

Measured at a co-latitude of 78°, figure 3 shows the relative sizes of terms in (1) computed from the averaged flow. From left to right terms are labeled by pluses, stars, diamonds, triangles, boxes and crosses, respectively. As Re is about 500 the viscous stress terms represented by S_θ are, the equator excluded, much smaller than other terms and not included in the momentum balance. Clearly the latitudinal pressure gradient (boxes) and the Coriolis force (crosses) dominate the momentum balance, while the Reynolds stress terms are comparatively small.

If the advection and viscous terms can be ignored, then the momentum equation reduces to:

$$\frac{\nabla p}{\rho} = \mathbf{g} - 2\mathbf{\Omega}_o \times \mathbf{v}. \qquad (2)$$

The curl of the left hand side of (2) produces

$$-\frac{1}{\rho^2}(\nabla p \times \nabla\rho) = \nabla T \times \nabla S \qquad (3)$$

where $S = c_v \ln(p/\rho^\gamma)$. Assuming the convection zone to be approximately adiabatic ($\partial S/\partial r \approx 0$) and axisymmetric ($\partial/\partial\phi \approx 0$), only the ϕ

Figure 1 Angular velocity contours

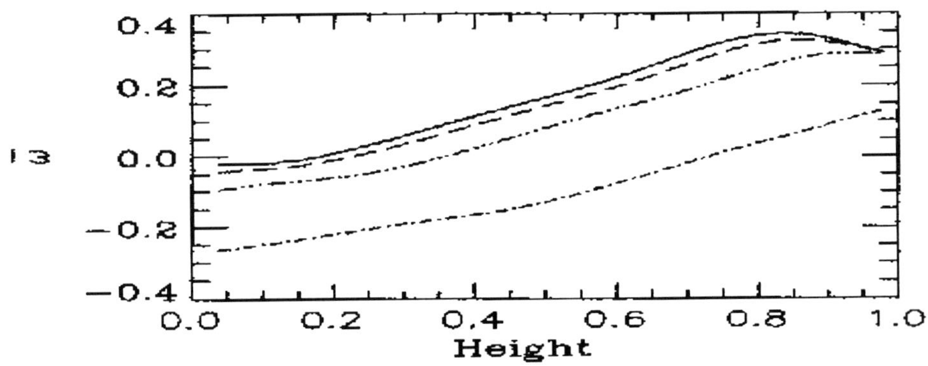

Figure 2 Angular velocity versus height

component of (3) is nonzero. Taking the curl of (2), one obtains

$$2\Omega_o\left(r\cos\theta\frac{\partial v_\phi}{\partial r}-\sin\theta\frac{\partial v_\phi}{\partial\theta}\right)+\frac{\partial T}{\partial r}\frac{\partial S}{\partial\theta}=0 \tag{4}$$

Note as **g** can be written as a potential, the buoyancy term disappears.

Computed from the averaged flow, figure 4 shows $\overline{\partial T/\partial r\cdot\partial S/\partial\theta}$ and $2\Omega_o(r\cos\theta\overline{\partial v_\phi/\partial r}-\sin\theta\overline{\partial v_\phi/\partial\theta})$, denoted by boxes and crosses respectively. Co-latitudes of 90°, 82° and 72° are represented by progressively larger terms. Clearly they are in approximate balance, reinstating that it is indeed the thermal structure that controls the shape of the differential rotation profile, rather than the meridional circulation or Reynolds stresses. These play only a minor role in shaping the differential rotation. A similar conclusion was reached in the framework of certain mean field models of the convection zone e.g 5.

5. CONCLUSION

By performing a large eddy simulation in a section of a spherical shell we have found a rotation profile which is in approximate agreement with observations from helioseismology. Previous analytical models of rotating convection rely on the Reynolds stress and/or meridional circulation to drive the differential rotation. In contrast, we find these to be of only minor importance when compared to the role played by the pressure-density distribution.

Acknowledgements

I would like to thank K.L.Chan and A. Brandenburg for many helpful discussions.

References

Harvey, J., and The GONG Instrument Team. 1995 *ASP Conf. Ser., Vol.* **76**

Chan, K. L. and Wolff, C. L., 1982, *J. of Comp. Phys.* **47,** 109

Glatzmaier, G.A. 1987, *The Internal Solar Angular Velocity, ed. B.R. Durney and S. Sofia (Dordrecht: Reidel), 263*

Elliott, J. R. et al 1998 *Turbulent solar convection and its coupling with rotation Structure and Dynamics of the Interior of the Sun and Sun-like Stars, SOHO 6/GONG 98 Workshop Abstract, June 1-4, 1998, Boston, Massachusetts*

Durney, B.R. 1999 *Apj.* **511,** 945-957

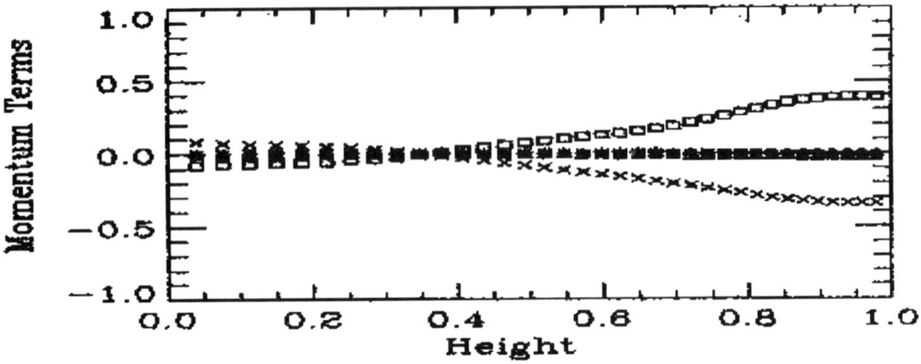

Figure 3 Meridional momentum balance at a co-latitude of 78°.

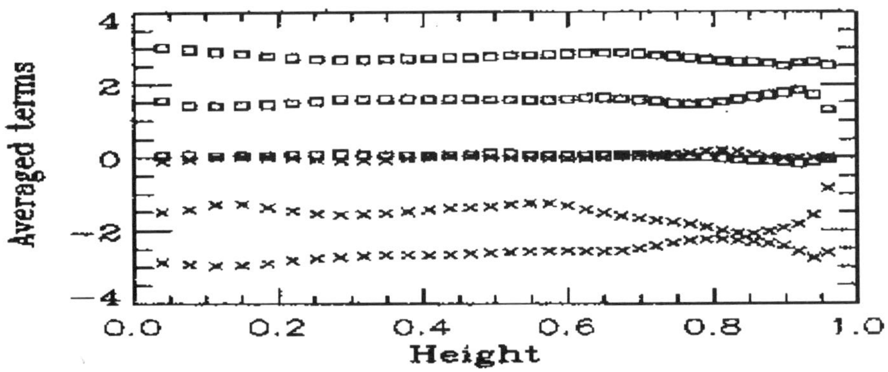

Figure 4 $\overline{\partial T/\partial r \cdot \partial S/\partial \theta}$ and $2\Omega_o(r\cos\theta\overline{\partial v_\phi/\partial r} - \sin\theta\overline{\partial v_\phi/\partial \theta})$, denoted by boxes and crosses respectively.

NUCLEAR ASTROPHYSICS WITH RADIOACTIVE BEAMS

S.S.M. Wong

Department of Physics, University of Toronto, Toronto, Canada M5S 1A7

Abstract The evolution of a star depends to a large extent on the reaction rates between its nuclear constituents. A large fraction of these involves unstable nuclei produced in the environment of high temperature and density. Radioactive beams make it possible to study such reactions in the laboratory where traditional methods fail. A brief review of some of the experiments that are underway at the present and proposed for the near future is given to illustrate some of the possibilities.

The relative abundance of elements we observe today in stars and in intergalactic gas clouds is the result of nuclear reactions taking place during the first few minutes after the big bang and the long aeons of stellar evolution that is still going on today. A good understanding of these reactions is therefore integral to astrophysics. Unlike laboratories on earth, the stellar environment is not an equilibrium and many of the reactions involve nuclei that are unstable and cannot be made into targets in the laboratory. The use of a beam of radioactive particles in measurements overcomes many of the limitations of dealing only with stable nuclei. A brief review is given here on a few examples of the crucial rates that can be measured to good accuracies using the newly commissioned radioactive ion beam facilities.

1. RADIOACTIVE BEAMS

Many reactions of astrophysical interest involve unstable nuclei. Short-lived species created as the result of one reaction may become the participants in subsequent reactions. The most famous example is perhaps the triple-α reaction in which ^{8}Be (half-life $\tau_{1/2} = 7 \times 10^{-17}$ s), created by the chance coalesce of two ^{4}He, captures another α-particle to form ^{12}C. Even though the relative probability of finding ^{8}Be nuclei in a typical helium-burning star is only around one in a billion, it is possible to create enough ^{12}C for nucleosynthesis to continue to heavier nuclei.

K.S. Cheng et al. (eds.), Stellar Astrophysics, 51–59.

The example clearly illustrates the importance of reactions involving radioactive nuclei in nuclear astrophysics and radioactive beams are the new tool for such studies.

It is relatively easy to make radioactive nuclei in the laboratory. As a first approximation, stable nuclei are made of equal numbers of protons and neutrons. However, because of Coulomb repulsion between protons, heavier nuclei with larger numbers of protons require an excess of neutron number N over the number of protons Z to stay stable. In fact, the neutron excess $N - Z$ increases with nuclear mass A $(= N + Z)$. For example, ^{40}Ca has $N = Z = 20$ whereas ^{208}Pb has $Z = 82$ and $N = 126$. If sufficient energy is injected into a heavy nucleus to overcome the binding between its constituents, the nucleus will split up into two or more fragments. Since each one of the daughter nuclei is lighter than its parent, there are more neutrons than required to make it stable and radioactive nuclei are born.

The environment where these nuclei are produced is usually quite "dirty," as the process is not selective and many different types of nuclei are created at the same time. In addition, large numbers of γ-rays and electrons accompany the reaction. Using electromagnetic fields, we can accelerate the charged particles produced and, once they are moving with some velocities, we can make a selection based on their charge to mass ratios. Apart from many extremely difficult technical issues, a beam of unstable ions is produced in this way and directed towards experimental areas.

There are several limitations to this technique. In the first place, the nuclei produced tend to be on the neutron-rich side. This does not pose too sever a problem from a nuclear astrophysics point of view, as neutron-rich nuclei dominate in r-process through which many heavy elements are created. Furthermore, it complements fairly well with heavyion reactions, the other recent advance in nuclear physics, in which proton-rich nuclei are favored. In the second place, very short-lived nuclei are excluded, as it takes some time for a beam to be extracted and transported to the target area. For example, a beam of ^{8}Be is impossible, as the 10^{-16}s lifetime does not allow it even to move more than a few tens of nanometers before it decays into two α-particles. However, as we shall see next through examples, many useful information can be obtained from the new facilities.

2. RADIATIVE PROTON CAPTURE ON ^{7}BE

Let us start with one that is close to home, the cross section for ^{7}Be$(p, \gamma)^8$B reaction that is intimately related to the solar neutrino problem. The ^{7}Be comes from the following chain of reactions in proton burning:

$$p + p \to d + e^+ + \nu_e \qquad d + p \to {}^3\text{He} + \gamma \qquad {}^3\text{He} + {}^4\text{He} \to {}^7\text{Be} + \gamma$$

Figure 1 shows a plot of the expected neutrino flux as a function of neutrino energy from various reactions involved in proton-burning under solar conditions. At the high energy end, for example above the 0.8 MeV threshold of $\nu_e + {}^{37}\text{Cl} \to {}^{37}\text{Ar} + e^-$ reaction, the flux is dominated by neutrinos from β-decay of ^{8}B

$$^8\text{B} \to {}^8\text{Be} + e^+ + \nu_e$$

It is more than an order of magnitude higher than those from *hep* process, involving $^3\text{He}(p, e^+\nu_e)\,^4\text{He}$ reaction. As a result, more than 80% of the neutrinos observed at Homestake and Kamioka comes from ^{8}B produced by the $^7\text{Be}(p, \gamma)^8\text{B}$ reaction.

The "adopted" value1 for the astrophysical S-factor at zero energy for the $^7\text{Be}(p, \gamma)^8\text{B}$ reaction is

$$S_{17}(0) = 22.4 \pm 2.1\text{eV} \cdot \text{b}$$

the value for energy-weighted cross section under conditions existing in solar interior. However, in a more recent measurement,2 the value turns out to be $18.5 \pm 1.7\text{eV} \cdot \text{b}$, indicating that there may still be some question on the real rate of this important reaction. If the value is much lower, it means that fewer ^{8}B are actually produced than what went into standard solar model calculations and there may not be as large a solar neutrino problem. On the other hand, if the value is higher, we have an even bigger discrepancy than what we are aware of so far.

The reaction $^7\text{Be}(p, \gamma)^8\text{B}$ has a positive Q-value of 137 keV. Since the average thermal energy for a proton in the solar core is less than 2 keV ($T = 16 \times 10^6$ K), only the rates in the 0 to 20 keV range is of any importance. Direct measurements of proton scattering cross section on ^{7}Be is difficult because of its short half-life of 53 days. The experiment is usually carried out by first creating ^{7}Be through a (p, n) reaction on ^{7}Li and observe subsequent (p, γ) reactions on the radioactive nucleus created. It is difficult to achieve good accuracies in such a setup especially at low energies where strong Coulomb repulsion between the proton and the ^{7}Be nucleus makes the cross section extremely small comparing with background.

With radioactive beams, the measurements may be done in two different ways. The first is to carry out the scattering with ^{7}Be as the projectile and hydrogen as the target. At TRIUMF, an experiment has already been approved to carry out the $p(^7\text{Be}, \gamma)^8\text{B}$ reaction with a hydrogen gas target and count the ^{8}B produced through its α- and β-decay products. Similar plans are also known to exist in other laboratories.

The second way is to use the inverse reaction by scattering (virtual) γ-rays from ^{8}B. Instead of real γ-rays, a beam of radioactive ^{8}B ($\tau_{1/2} =$ 0.774 s) is directed towards a target made of high-Z nuclei, such as ^{208}Pb. The virtual photon produced by the collision of ^{8}B with ^{208}Pb takes the place of γ-ray in the reaction. This type of experiments is known as Coulomb dissociation, since ^{8}B is dissociated in the Coulomb field of ^{208}Pb into proton and ^{7}Be. Some results have already been reported in 1994.3 The major difficulty here is that it requires a knowledge of each one of the electromagnetic multipoles that contributes to the dissociation process in order to relate the results to (p, γ) cross section and this requires other studies.

3. S-FACTOR FOR $^{12}\text{C}(\alpha, \gamma)\,^{16}\text{O}$ REACTION

After exhausting most of its hydrogen, a star goes into its red-giant stage and begins to convert the helium accumulated to heavier elements. One of the important parameters here is the ^{16}O to ^{12}C ratio at the end of helium burning, as it controls the relative abundance of heavier elements if and when the star ends its life in a supernova explosion.4

Because of the short lifetime of ^{8}Be, the cross section for α-capture to form ^{12}C cannot be measured in the laboratory. However, the energy level structure of ^{12}C, shown on the left of Fig. 2, is sufficiently simple that the value can be inferred from other measurements to an uncertainty of about 20%. This is not true for the $^{12}\text{C}(\alpha, \gamma)^{16}\text{O}$ reaction, the dominant mode for consuming ^{12}C and for creating ^{16}O. As a result, the S-factor for $^{12}\text{C}(\alpha, \gamma)^{16}\text{O}$ reaction at the typical helium burning temperature of 300 keV becomes the major source of uncertainty in estimating the equilibrium ratio of ^{16}O to ^{12}C.

The low-lying level structure of ^{16}O is shown in the right hand panel of Fig. 2. Since the Q-value is 7.16 MeV, α-capture goes mainly to the 1^- state ($E1$) at 7.12 MeV and 2^+ state ($E2$) at 6.92 MeV. Both states are *subthreshold,* as they lie below the threshold for capture. Direct measurements of $^{12}\text{C}(\alpha, \gamma)^{16}\text{O}$ in the laboratory, on the other hand, requires α-particles with incident energies on the order of MeV so as to overcome Coulomb repulsion. At such energies, the process is dominated instead by the 1^- state at 9.60 MeV and 2^+ state at 8.87 MeV. This results in uncertainties for the astrophysical S-factor at 300 keV of approximately 200%.

Recently, we carried out the "inverse reaction" of α-decay of ^{16}O nucleus from its excited states.5 The ^{16}O are obtained from β-decay of ^{16}N ($\tau_{1/2} = 7$ s) coming from a radioactive beam at TRIUMF. The decay goes mainly to the 1^- states of ^{16}O. By detecting the α-particles emitted in conjunction with the β-decay products, we were able to measure the reaction cross section for the 1^--contributions of $^{12}C(\alpha,\gamma)^{16}O$ to the level of 20% uncertainty, giving a value for the $E1$-part of the astrophysical S-factor

$$S_{E1}(300) = 79 \pm 21 \text{ keV barn}$$

We are engaged in finding ways to populate selectively 2^+ states around the threshold from decays of radioactive nuclei nearby so that the $E2$-component of the S-factor can also be known to the same accuracy. However, before we can even do that, we must explore the nuclear structure of unstable nuclei in mass 16 to 19 region that is largely unknown. This program is underway at the moment using the radioactive beam facility at TRIUMF.

4. HOT CNO CYCLE AND *RP*-PROCESS

Explosive hydrogen burning is one of the important processes to produce heavy elements. Under conditions of high temperature ($T \sim 0.1$ to 1×10^9 K) and density ($\rho \sim 10^3$ to 10^6 g cm^{-3}) that exist, for example, in novas and supernovas, nuclear reactions can take place at extremely short time scales.6 This is in contrast to hydrostatic hydrogen burning in which the "normal" carbon-nitrogen-oxygen (CNO) cycle, shown as the central circle in Fig. 3, is limited by the rate of energy production by the slow intervening β-decays.

For example, in the chain of reactions represented by the circle,

$$^{12}C(p,\gamma)^{13}N(\beta^+)^{13}C(p,\gamma)^{14}N(p,\gamma)^{15}O(\beta^+)^{15}N(p,\alpha)^{12}C$$

the rate is dictated by the β-decay half-life of almost 10 minutes for ^{13}N. With higher proton density and kinetic energy, the much faster $^{13}N(p,\gamma)^{14}O$ reaction can take precedent, resulting in the following chain of reactions:

$$^{12}C(p,\gamma)^{13}N(p,\gamma)^{14}O(\beta^+)^{14}N(p,\gamma)^{15}O(\beta^+)^{15}N(p,\alpha)^{12}C$$

as represented by the semi-circle on the right in Fig. 3. This is a part of the hot CNO (HCNO) cycle that starts around $T \sim 2 \times 10^8$ K. The timing is now controlled by ^{14}O ($\tau_{1/2} = 70$ s) and ^{15}O ($\tau_{1/2} = 120$ s) β-decay rates. With the higher rates, there may be enough energy produced to fuel novas as well providing an explanation for the observed higher abundances of ^{14,15}N, the decay products of ^{14,15}O β-decay, in the material ejected (see, e.g., 7).

At even higher temperatures, the rate for energy production increases further through α-particle induced processes, such as

$$^{14}\text{O}(\alpha,p)^{17}\text{F}(p,\gamma)^{18}\text{Ne}(\beta^+)^{18}\text{F}(p,\alpha)^{15}\text{O}$$

The slowest link in this chain is the half-life of around 1 s for ^{18}Ne β-decay, almost two orders of magnitude faster than that for ^{14}O. Under such conditions, it is possible to break out of the HCNO cycle by proton and α-particle reactions to create elements beyond neon. In fact, if there are sufficient amounts of Ne, Na, or Mg, *rp*-processes, rapid processes involving protons and α-particles, can make the energy production rate sufficient high to lead to x-ray bursts and creating elements up to and beyond the iron group.

Many of the reactions taking place in the HCNO and *rp*-processes involve proton-rich nuclei that are unstable. As a result, the rates are poorly known in general. In particular, the following reactions are considered to be pivotal and become part of the central interests in the first experiments to be carried out using radioactive beam facilities.

^{13}N$(p,\gamma)^{14}$O reaction. This is the starting point for hot CNO cycle and reaction rates up to 100 keV ($T \sim 10^9$ K) are important in understanding the conditions leading to break-out to *rp*-process as well. Since ^{13}N is unstable ($\tau_{1/2}$ = 9.93 min.), it cannot be made into a target. Just as in the case for ^{7}Be$(p,\gamma)^8$B reaction, the measurements may be carried out by scattering a radioactive beam of ^{13}N on a target containing hydrogen (e.g., CH_2). The cross section may be deduced from the γ-rays emitted. This is a relatively easy measurement to analyze as the capture is dominated by the $(J^\pi, T) = (1^-, 1)$ first excited state at 5.173 MeV in ^{14}O. The measured results8 from Louvain-la-Neuve has already revised the astrophysical rate connected with it by a factor of two from calculations based on older experiments. However, the uncertainty is still high from direct capture contributions and further experiments are underway both at Louvain-la-Neuve and TRIUMF.

^{19}Ne$(p,\gamma)^{20}$Na reaction. At sufficiently high temperature and density, this reaction can take the path of nucleosynthesis out of the hot CNO cycle. The cross section is dominated by the resonance at 0.45 MeV in the ^{19}Ne + p system. However, the structure of the resonance is not well known and the energy production can be substantially different depending on, for example, whether it has $J^\pi = 1^+$ or 3^+. Both nuclei are unstable, with $\tau_{1/2} = 17$ s for ^{19}Ne and 448 ms for ^{20}Na. Measurements with a radioactive beam of ^{19}Ne is feasible and is under active considerations at several places.

^{18}Ne$(\alpha,p)^{21}$Na reaction. This is another path to break out from HCNO. The resonances of interest is now in ^{22}Mg, i.e., $\alpha+$ ^{18}Ne. The nuclear structure is more complicated than in the previous example. However, its contribution to energy production in stars may be more significant. Laboratory measurements may have to wait till some of the easier ones are carried out first.

5. BIG-BANG NUCLEOSYNTHESIS

The success in accounting for the primordial abundances of light elements, d, ^{3}He, ^{4}He, and ^{7}Li, is one of the strong evidences for big-bang models of cosmology. Since we are dealing here with an environment that neutron and proton densities must be the same, the instability of ^{8}Be essentially stops any accumulation of heavier elements until stars are formed at much later times. The nuclear reactions involved in creating these light elements are between stable nuclei. In spite of some uncertainties in their cross sections at the low energies of interest, there does not seem to be any concerns on the predictions based on their values.

On the other hand, if there are local inequality in the densities of neutrons and protons, as suggested by several alternatives to standard big bang models, ^{6}Li, ^{9}Be and heavier nuclei can also be created. The inhomogeneity allows local regions rich in neutrons and, as a result, reactions involving unstable nuclei around ^{8}Be become important in primordial nucleosynthesis. To understand the processes involved, reactions such as ^{8}Li$(\alpha,n)^{11}$B and ^{8}Li$(n,\gamma)^{9}$Li must be measured in the laboratory at the appropriate low energies. The presence of primordial ^{9}Be is the telltale signature of such asymmetric big-bang models and active searches are underway to find its abundance. An understanding of these reactions are essential in clarifying many of the fundamental questions in big-bang cosmology and studies involving radioactive beams are crucial here.

6. CONCLUSION

In addition to the interests described above, we have also the fact that most of the heavier nuclei are created by rapid neutron capture (r-process), slow neutron capture (s-process), and proton capture (p-process). There are about a hundred reactions in these processes that only estimates of their rates are available. Most of these involve neutron-rich unstable nuclei and radioactive ion beams can make significant contributions. Besides nuclear astrophysics, knowledge of these and other

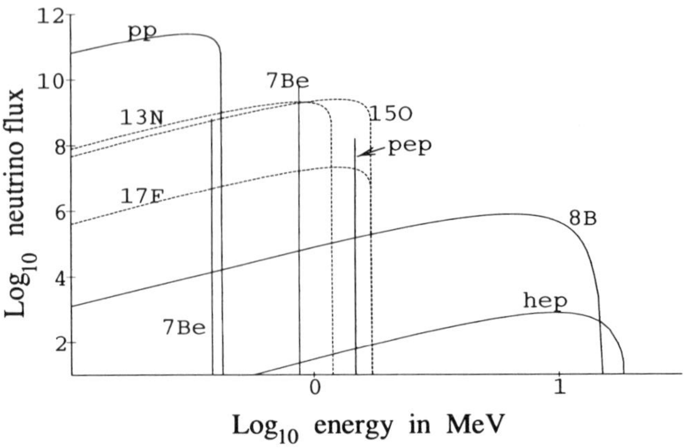

Figure 1 Log-log plot of the expected solar neutrino flux.

unstable nuclei can provide insight to nuclear physics that studies on stable nuclei alone cannot provide. Indeed, radioactive ion beams are going to be an important corner stone for both nuclear physics and nuclear astrophysics in the 21st century.

References

Bahcall, J.N., Krastev, P.I., and Smirnov, A. Yu., 1996, Phys. Rev. D58 096016.

Hammache, F., et al., 1998, Phys. Rev. Lett. 80, 928.

Motobayashi, T., et al., 1994, Phys. Rev. Lett. 73, 2680.

Weaver, T.A., and Woosley, S.E. 1993, Phys. Rep., 227, 65.

Azuma, R.E., et al., 1994, Phys. Rev. C50 1194.

Champagne, A.E., and Wiescher, M. 1992, Annu. Rev. Part. Sci., 42, 39.

Sion, E.M., this proceeding.

Delbar, Th., et al., Phys. Rev. C48, 1993, 3088; Phys. Rev. C56, 1144.

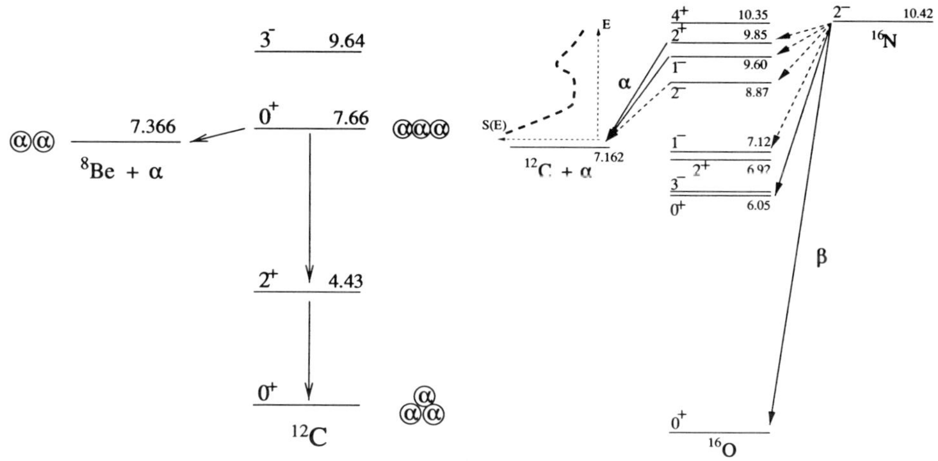

Figure 2 Left, low-lying energy level structure of ^{12}C and the triple-α process. Right, low-lying energy level structure of ^{16}O and the ^{12}C$(\alpha,\gamma)^{16}$O reaction.

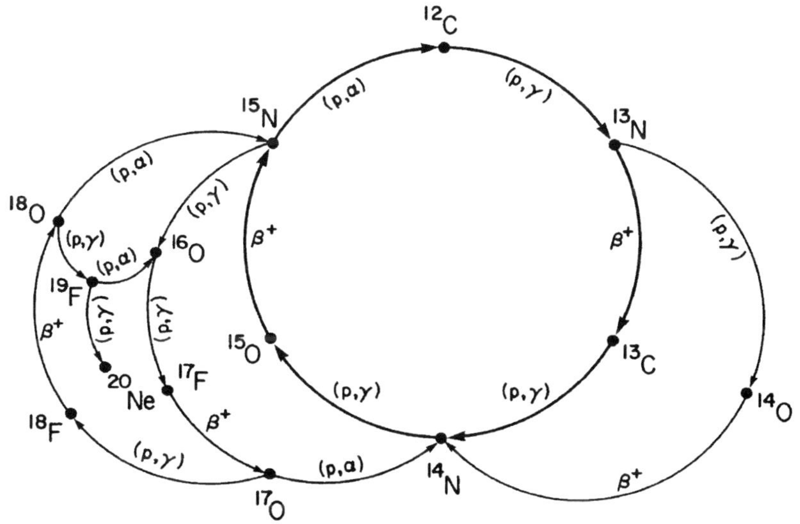

Figure 3 Nuclear reactions involved in CNO cycles.

PULSARS AT PARKES

R. N. Manchester
Australia Telescope National Facility, CSIRO
P.O. Box 76, Epping NSW 1710, Australia; email:rmanches@atnf.csiro.au

Abstract The 64-m Parkes radio telescope of the Australia Telescope National Facility has a long history of observing and finding pulsars. These observations include precision timing, polarization studies and using pulsars to investigate the interstellar medium. Many surveys for pulsars have been undertaken with the Parkes telescope, and with the very successful Multibeam surveys now under way, it has found more pulsars than all other telescopes combined. In this review, I will highlight some recent results in the areas of precision timing, interstellar medium studies and pulsar surveys.

Keywords: pulsars, timing, interstellar medium, surveys

1. INTRODUCTION

Since their discovery more than 30 years ago, pulsars have proved to be a fascinating and productive area of astrophysical research. They vary on all timescales from microseconds to decades and are often unpredictable. Young pulsars sometimes have sudden 'glitches' in their pulse period, and studies of these and the subsequent period variations lead to insights into the interior structure of neutron stars (e.g. Ruderman, Zhu and Chen, 1998). Observations of subpulse and micropulse structure, subpulse drifting, and pulse shape and polarization studies all give information on the pulse emission mechanism (e.g. Rankin, 1986; Lyne and Manchester, 1988).

In contrast, other properties are remarkably predictable, notably the basic pulse periodicity. Especially for millisecond pulsars, this great stability of the pulse period, coupled with the fact that these pulsars are often found in binary systems, has lead to a variety of important applications, including fundamental tests of gravitational theories (Taylor and Weisberg, 1989) and the first detection of extra-solar planets (Wolszczan and Frail, 1992) and studies of stellar and binary evolution. Pulsars are

K.S. Cheng et al. (eds.), Stellar Astrophysics, 61–70.

also great probes of the interstellar medium. The pulsed nature of their emission allows the measurement of signal dispersion and hence estimates of the interstellar free electron density (Taylor and Cordes, 1993). Pulsars generally have significant linear polarization and hence can be used to probe the Galactic magnetic field (Manchester, 1974).

Of course, all of these studies depend on the existence of a catalogue of known pulsars with measurements of their basic properties (e.g. Taylor, Manchester and Lyne, 1993; http://pulsar.princeton.edu/pulsar/). Searches for pulsars are therefore an important aspect of pulsar research. There are many different subclasses of pulsars, e.g., young pulsars, pulsars with drifting subpulses, millisecond pulsars, binary pulsars, pulsars in globular clusters, etc., and many studies depend on having a reasonable sample of each type.

The Parkes radio telescope and the people who use it have been active in all of these areas of research. In this review, I will describe some recent work in three areas: pulse timing, interstellar medium studies, and pulsar surveys.

2. PULSE TIMING

Pulsar periods are remarkably stable, in some cases having a stability rivalling that of the best terrestrial atomic clocks (Kaspi, Taylor & Ryba, 1994). These very precise periods are determined by measuring a series of pulse *times of arrival* (TOAs) at the telescope. A pulse TOA is normally obtained by synchronously adding a few thousand pulses to form a mean pulse profile which is then cross-correlated with a standard profile to give the TOA. Such measurements are repeated at intervals which range from hours to many years. These TOAs are then corrected to refer to the Solar System barycentre, thereby removing the effects of the Earth's motion. A corresponding set of *predicted* arrival times are then computed using the best available parameters for the pulsar and subtracted from the observed TOAs to form a set of timing *residuals*.

The time variations of these residuals give information about offsets of the actual pulsar parameters from those assumed in computing the predicted TOAs. These offsets or corrections can be extracted by performing a multi-parameter least-squares fit of the expected signatures for the various terms. Especially for millisecond pulsars, the parameters are often determined to extraordinary precision. For example, from 3.5 years of timing data at Parkes, Toscano et al. (1999a) obtain a position for PSR J1744−1134 with a precision of 0.5 mas in declination and even better in right ascension. The pulse period is determined with an uncertainty of about one part in 10^{14}. With positions as accurate as this and a few years of timing data, it is usually possible to measure a proper motion and hence derive a space velocity for the pulsar (Toscano et al., 1999b). In a few cases, even the annual parallax can be measured. For PSR J1744−1134 a parallax of 2.8 ± 0.2 mas was obtained, giving a final r.m.s. timing residual of approximately 0.5μs.

In the quest for high precision timing, systematic errors are often more important than random errors. These errors come in many forms, for example, pulse shape changes, uncorrected interstellar delays, calibration errors and clock errors. Calibration errors are particularly significant for PSR J0437−4715, a very strong millisecond pulsar ($P = 5.75$ ms) in a binary orbit of period 5.74 days. This pulsar has an orthogonal polarisation flip right at the peak of the main pulse (Navarro et al., 1997). Consequently any error in the relative amplitude of the two received polarisations is immediately translated into a TOA error with a scale factor of 100 ns for each percentage point calibration error. Errors of several per cent can easily arise from instrumental polarisation in the feed and receiver system. Accurate timing of pulsars such as PSR J0437−4715 therefore requires accurate polarimetry (Britton et al., 1999). Fig. 1 shows timing residuals for PSR J0437−4715 illustrating these effects. Clear systematic daily variations are evident in the residuals plotted in the top part of the figure. Application of precise corrections for instrumental polarisation (accurate at the 0.2% level) removes most of these systematic effects, as shown by the residuals in the middle of the figure. The rms residual over the three days has been reduced from about 0.5 μs to about 0.25 μs.

An interesting analysis by Britton (1999) has shown that the effects of polarisation calibration errors on pulse timing can be avoided completely if the timing is based on the *unpolarised* component of the pulse. The lower group of residuals in Fig. 1 are derived in this way. This method is clearly much simpler than doing a complete polarisation calibration, but has the restriction that it is not very useful for pulsars with very high degrees of polarisation.

Other interesting results to come from recent Parkes pulse timing measurements include the detection by Wang et al. (1999) of 30 glitches in 40 mostly young pulsars. These include the largest known glitch in PSR J1614−5047 ($\Delta\nu/\nu \approx 6.5\times10^{-6}$) and 12 glitches in PSR J1341−6220, making this the most frequently glitching pulsar known.

Analysis of individual pulse data by Young, Manchester and Johnston (1999) showed that the true period of PSR J2144−3933 is 8.5 s, a factor of three greater than previously thought. This gives this pulsar by far the longest known period and puts it below most death lines on the $P-B$ diagram. It therefore calls into question the assumptions, particularly

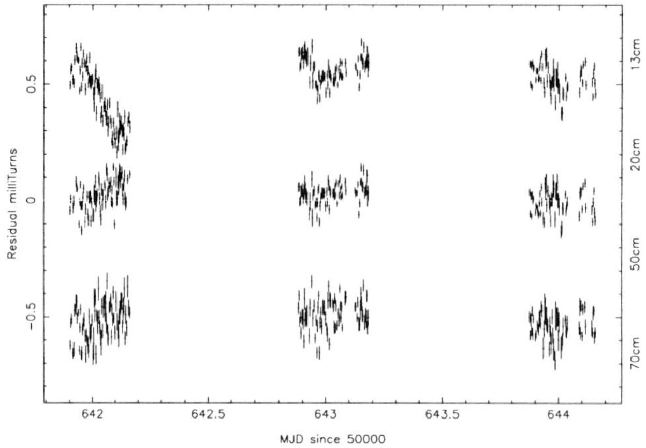

Figure 1 Timing residuals for PSR J0437−4715 derived from data obtained at the Parkes radio telescope at 1400 MHz on three days in July 1997 using the S2 baseband recording system, which has a bandwidth of 16 MHz. Each TOA is based on an observation time of 240 s. In the upper part of the figure, residuals for TOAs obtained using the normal calibration procedure, based on a pulsed noise signal injected into the feed, are shown. The middle section shows residuals after applying corrections for the instrumental polarisation resulting from gain errors and cross-talk between the orthogonal feeds. The lower section shows residuals based on the unpolarised part of the pulsed signal. The vertical offsets between the three sections of the plot are arbitrary. (After Britton et al. 1999)

about pair production in polar gaps, that are implicit in models for the pulse emission mechanism (e.g. Cheng and Ruderman, 1980). This pulsar is also interesting in having the narrowest mean pulse profile (as a fraction of the pulse period) of any pulsar. The half-intensity pulse width is just 0.84 degrees of longitude. This implies that we may be able to detect only a small fraction of such pulsars in the Galaxy.

3. INTERSTELLAR MEDIUM STUDIES

The centre of our Galaxy passes nearly overhead at Parkes and so the telescope is ideally situated for studies of the interstellar medium. In this review I just mention two recent studies, of interstellar scintillation by Johnston, Nicastro and Koribalski (1998), and of Faraday rotation by Han, Manchester and Qiao (1999).

Johnston et al. (1998) obtained scintillation parameters for 49 pulsars, including 13 millisecond pulsars at frequencies between 436 MHz and 22.4 GHz. Transverse velocities derived from these measurements are in good agreement with independently determined velocities, implying that the scattering electrons have a similar scale height to that of the dispersing electrons.

Han et al. (1999) obtained rotation measures for 63 pulsars, allowing an investigation of the structure of the Galactic magnetic field. The results were mostly consistent with the presence of a large-scale antisymmetric field pattern in the Galactic halo. They indicate the presence of several reversals of field direction in the Galactic disc, probably associated with spiral arms. These results have implications for dynamo models for the generation of the Galactic magnetic field.

4. RECENT PULSAR SEARCHES

Parkes has a long history of success in searching for pulsars – as of a couple of years ago, nearly half of the known pulsars had been discovered or confirmed at Parkes. Recent pulsar searches have greatly increased this fraction. These searches are of two main types: searches of globular clusters and searches at low Galactic latitudes. The great success of these recent searches flows from the commissioning of a new and very sensitive 13-beam receiver operating in the 20 cm band (Staveley-Smith et al., 1996).

Although using only the centre beam, this receiver together with a new data acquisition system has provided a very significant increase in number of detectable pulsars in the globular cluster 47 Tucanae (Camilo et al., 2000). A total of nine previously unknown pulsars, all members of binary systems, have been detected. This brings the total number of pulsars known in the cluster to 20 (Table 1). Furthermore, the previously known pulsars are detected far more frequently than with earlier systems, leading to coherent timing solutions and improved estimates of flux densities for most of them.

The full power of the multibeam receiver is exploited in area searches in which data from the 13 beams are simultaneously recorded, leading to an increase in the survey speed of more than an order of magnitude. This requires a very large filterbank, which was constructed at the University of Manchester, Jodrell Bank. There are two searches currently under way using this system. The Parkes multibeam survey is an international collaboration involving groups from the University of Manchester (UK), Massachusetts Institute of Technology (USA), the University of Bologna (Italy) and the ATNF. This survey is covering a 10°-wide strip along the Galactic plane from longitude 260° to 50° (Lyne et al., 1999). The second survey is being carried out by a group from the Swinburne University of Technology (Edwards, 2000) and covers the adjoining strips $5° < |b| < 10°$ over the same longitude range. This survey has a shorter observation time per point (4 min versus 35 min) and a shorter sampling interval (0.125 ms versus 0.250 ms). It is essentially complete and has been very

Table 1 Pulsars in 47 Tucanae

Pulsar	Period (ms)	DM (cm^{-3} pc)	w_{50} (%)	Flux density (mJy)
		Previously Known		
47 Tuc C	5.756780	24.6	13	0.36(4)
47 Tuc D	5.357573	24.7	10	0.22(3)
47 Tuc E	3.536329*	24.3	25	0.21(3)
47 Tuc F	2.623579	24.4	14	0.15(2)
47 Tuc G	4.040379	24.4	13	0.05(2)
47 Tuc H	3.210341*	24.4	13	0.09(2)
47 Tuc I	3.484992*	24.5	14	0.09(1)
47 Tuc J	2.100634*	24.6	15	0.54(6)
47 Tuc L	4.346168	24.4	15	0.04(1)
47 Tuc M	3.676643	24.4	25	0.07(2)
47 Tuc N	3.053954	24.6	11	0.03(1)
		Newly Discovered		
47 Tuc O	2.643343*	24.4	12	0.10(1)
47 Tuc P	3.643021*	24.3	15	–
47 Tuc Q	4.033181*	24.3	11	0.05(2)
47 Tuc R	3.480463*	24.4	15	–
47 Tuc S	2.830*	24.4	9	–
47 Tuc T	7.589*	24.6	17	–
47 Tuc U	4.342827*	24.3	8	0.06(1)
47 Tuc V	4.810*	24.1	7	–
47 Tuc W	2.352344*	24.3	17	–

* Member of a binary system

successful in finding 58 previously unknown pulsars, of which eight are millisecond pulsars.

The excellent performance of the receiver and the long observation time make the Parkes multibeam survey very sensitive, with a limiting flux density for long-period pulsars of 0.15 mJy, about seven times more sensitive that the best previous survey of this type (Johnston et al., 1992). The survey is at present about 50% complete and has already found more than 400 previously unknown pulsars. When complete, it will come close to doubling the number of known pulsars. This large sample of pulsars provides much material for studies of pulsar emission physics, high-energy associations and studies of the interstellar medium. However, much of the interest and motivation for such searches comes from the expectation that a few highly interesting pulsars will be discovered. This is proving to be the case.

Several pulsars with very high rates of spin-down and hence implied surface magnetic fields have been discovered. Two of these, PSR J1119−6127 and PSR J1814−1744, have $B_s > 4 \times 10^{13}$ G (Camilo et al., 1999). PSR J1119−6127 has a period of 0.407 s and is very young, with a characteristic age of only 1600 years, and may have an associated supernova remnant. PSR J1814−1744 has the relatively long period of 3.97 s; this places it very close to the so-called anomalous X-ray pulsars (AXPs) in $P - \dot{P}$ space (Fig. 2). No AXP has been detected at radio wavelengths, suggesting that more than just period and magnetic field strength distinguishs AXPs and radio pulsars.

A total of eight binary pulsars have been discovered so far. Several of these are especially interesting. PSR 1811−1736, with a pulse period of 104 ms, is in a highly eccentric binary orbit of period 18.8 days. The minimum companion mass is large, about 0.7 $M_\odot$, and the system very likely consists of two neutron-stars, only the fourth or fifth such system known in the disk of the Galaxy (Lyne et al., 1999).

PSR J1740−3052 has an even more massive companion, with a minimum mass of 11 $M_\odot$. The companion must be either a massive normal star or a black hole. Unfortunately, it lies very close to the Galactic Centre where obscuration is large. Infrared observations have revealed a possible companion star at the pulsar position and we are currently investigating if that could be the pulsar's binary companion.

Finally, the most recent discovery is a pulsar with period 0.393 s in a mildly eccentric orbit of period only 4.8 hours. This system has minimum companion mass of close to 1.0 $M_\odot$, so it is another good candidate for a double-neutron-star system.

5. CONCLUSION

In this short report, I hope I have given some feeling for the pulsar research being carried out using the Parkes telescope. This covers a wide range of pulsar studies, but there is no doubt that the area where Parkes is pre-eminent is as a discovery instrument. It is, by a wide margin, the most successful telescope in the world at discovering pulsars. This is largely a result of the combination of good location, large size and superb receiver technology. The present surveys have and will provide much material for continuing studies of these fascinating objects.

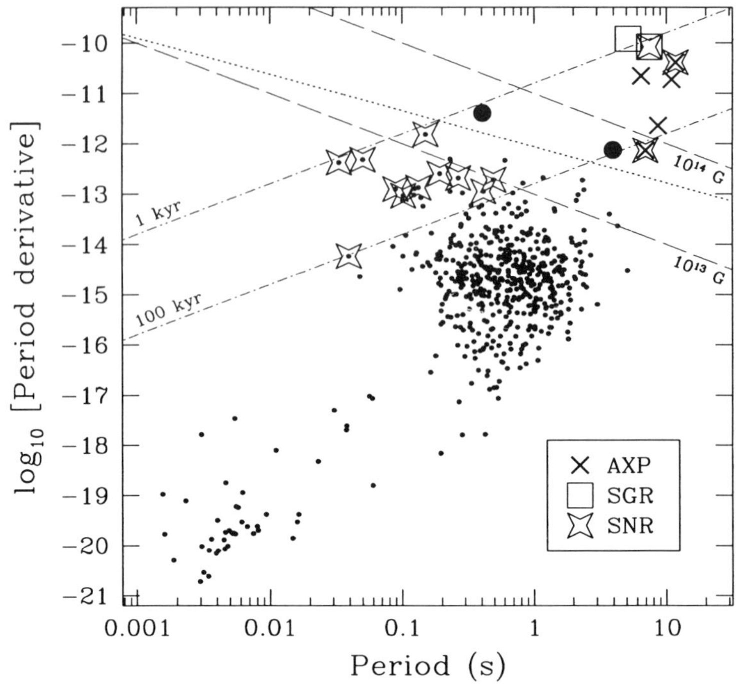

Figure 2 Plot of pulsar period derivative versus pulse period for radio pulsars, anomalous X-ray pulsars (AXPs) and soft gamma-ray repeaters (SGRs) (Camilo et al. 1999). Pulsars associated with supernova remnants (SNRs) are also marked. The dashed lines are lines of constant surface dipole magnetic field, and the dot-dashed lines are lines of constant characteristic age. The dotted line defines a boundary above which radio emission should not be seen in the model of Baring and Harding (1998).

Acknowledgements

I thank my colleagues for their help and inspiration in the work described here, and especially the staff of the Parkes Observatory for their unfailing support. The Parkes radio telescope is part of the Australia Telescope which is funded by the Commonwealth of Australia for operation as a National Facility managed by CSIRO.

References

Baring, M. G. and Harding, A. K. (1998). Radio-quiet pulsars with ultra-strong magnetic fields. *Astrophys. J.*, 507:L55–L58.

Britton, M. C. (2000). Polarimetry and the Lorentz group. *Astrophys. J.* In press.

Britton, M. C., van Straten, W., Bailes, M., Toscano, M., and Manchester, R. N. (1999). Methods of high precision timing and polarimetry. *Astrophys. J.* Submitted.

Camilo, F., Lorimer, D. R., Freire, P., Lyne, A. G., and Manchester, R. N. (2000). Observations of 20 millisecond pulsars in 47 Tucanae at 20 cm. *Astrophys. J.* In press.

Camilo, F. M., Kaspi, V. M., Lyne, A. G., Manchester, R. N., Bell, J. F., D'Amico, N., McKay, N. P. F., and Crawford, F. (1999). Discovery of two high-magnetic-field radio pulsars. *Nature*. Submitted.

Cheng, A. F. and Ruderman, M. A. (1980). Particle acceleration and radio emission above pulsar polar caps. *Astrophys. J.*, 235:576–586.

Edwards, R. T. (2000). Discovery of eight recycled pulsars - the Swinburne intermediate latitude pulsar survey. In Kramer, M. and Wex, N., editors, *Pulsars: Problems and Progress, IAU Colloquium 177*, San Francisco. Astronomical Society of the Pacific.

Han, J. L., Manchester, R. N., and Qiao, G. J. (1999). Pulsar rotation measures and the magnetic structure of our galaxy. *Mon. Not. R. astr. Soc.*, 306:371–380.

Johnston, S., Lyne, A. G., Manchester, R. N., Kniffen, D. A., D'Amico, N., Lim, J., and Ashworth, M. (1992). A high frequency survey of the southern Galactic plane for pulsars. *Mon. Not. R. astr. Soc.*, 255:401–411.

Johnston, S., Nicastro, L., and Koribalski, B. (1998). Scintillation parameters for 49 pulsars. *Mon. Not. R. astr. Soc.*, 297:108–116.

Kaspi, V. M., Taylor, J. H., and Ryba, M. (1994). High-precision timing of millisecond pulsars. III. Long-term monitoring of PSRs B1855+09 and B1937+21. *Astrophys. J.*, 428:713–728.

Lyne, A. G., Camilo, F. M., Manchester, R. N., Bell, J. F., Kaspi, V. M., D'Amico, N., McKay, N. P. F., Crawford, F., Morris, D. J., Sheppard, D. C., and Stairs, I. H. (1999). The Parkes multibeam survey: PSR J1811−1736 – a pulsar in a highly eccentric binary system. *Mon. Not. R. astr. Soc.* In press.

Lyne, A. G. and Manchester, R. N. (1988). The shape of pulsar radio beams. *Mon. Not. R. astr. Soc.*, 234:477–508.

Manchester, R. N. (1974). Structure of the local Galactic magnetic field. *Astrophys. J.*, 188:637–644.

Navarro, J., Manchester, R. N., Sandhu, J. S., Kulkarni, S. R., and Bailes, M. (1997). Mean pulse shape and polarization of PSR J0437−4715. *Astrophys. J.*, 486:1019–1025.

Rankin, J. M. (1986). Toward an empirical theory of pulsar emission. III. Mode changing ,drifting subpulses and pulse nulling. *Astrophys. J.*, 301:901–922.

Ruderman, M., Zhu, T., and Chen, K. (1998). Neutron star magnetic field evolution, crust movement, and glitches. *Astrophys. J.*, 492:267–280.

Staveley-Smith, L., Wilson, W. E., Bird, T. S., Disney, M. J., Ekers, R. D., Freeman, K. C., Haynes, R. F., Sinclair, M. W., Vaile, R. A., Webster, R. L., and Wright, A. E. (1996). The Parkes 21 cm multibeam receiver. *Proc. Astr. Soc. Aust.*, 13:243–248.

Taylor, J. H. and Cordes, J. M. (1993). Pulsar distances and the Galactic distribution of free electrons. *Astrophys. J.*, 411:674–684.

Taylor, J. H., Manchester, R. N., and Lyne, A. G. (1993). Catalog of 558 pulsars. *Astrophys. J. Supp. Series*, 88:529–568.

Taylor, J. H. and Weisberg, J. M. (1989). Further experimental tests of relativistic gravity using the binary pulsar PSR 1913+16. *Astrophys. J.*, 345:434–450.

Toscano, M., Britton, M. C., Manchester, R. N., Bailes, M., Sandhu, J. S., Kulkarni, S. R., and Anderson, S. B. (1999a). Parallax of PSR J1744-1134 and the local interstellar medium. *Astrophys. J.*, 523:L171–L175.

Toscano, M., Sandhu, J. S., Bailes, M., Manchester, R. N., Britton, M. C., Kulkarni, S. R., Anderson, S. B., and Stappers, B. W. (1999b). Millisecond pulsar velocities. *Mon. Not. R. astr. Soc.*, 307:925–933.

Wang, N., Manchester, R. N., Pace, R., Bailes, M., Kaspi, V. M., Stappers, B. W., and Lyne, A. G. (1999). Glitches in southern pulsars. *Mon. Not. R. astr. Soc.* Submitted.

Wolszczan, A. and Frail, D. A. (1992). A planetary system around the millisecond pulsar PSR 1257+12. *Nature*, 355:145–147.

Young, M. D., Manchester, R. N., and Johnston, S. (1999). A radio pulsar with an 8.5-second period that challenges emission models. *Nature*, 400:848–849.

SPARK MODEL FOR DRIFTING SUBPULSES IN PULSARS

J.A. Gil & M. Sendyk

J. Kepler Astronomical Center, Lubuska 2, 65-265 Zielona Gora

jag@astro.ca.wsp.zgora.pl

Abstract We explore the modified Ruderman & Sutherland (1975) spark model in order to explain and reproduce patterns of drifting subpulses in pulsars. We demonstrate that if the observing geometry is determined, these patterns are fully explained by values of pulsar period and slow-down rate. As the best example we present the case of PSR B0943+10, for which were able to reconstruct detailed map of sparks corresponding to the observed pattern of drifting subpulses.

Keywords: pulsars: general – radio emission – drifting subpulses

Introduction

Drifting subpulses seem to provide an important clues concerning pulsar radiation mechanism. In PSR B0809+74, one of the most prototypical drifter, the intensity is almost sinusoidaly modulated along a drift-band, consisting of a succession of regularly drifting subpulses. Moreover, swings of the polarization position angle seem to be frozen into drifting subpulses. In PSR B0809+74, the position angle at a given longitude rotates by about 90 deg through each 11 pulsar period cycle (Taylor et al. (1971)). In other words, drifting subpulses often demonstrate a complex phase-intensity-position angle correlation along drift bands. We believe that this is a strong indication of the angular beaming nature of subpulse modulation, rather than just temporal intensity variations.

Ruderman & Sutherland (1975; RS henceforth) first proposed that the observed subpulse drift phenomenon is caused by the familiar $\mathbf{E}\times\mathbf{B}$ drift occuring in the pulsar polar gap plasma. However, sparks proposed in the RS model as primary sources of the subpulse-associated plasma columns have been criticized, mainly due to short dynamical time scales

K.S. Cheng et al. (eds.), Stellar Astrophysics, 71–80.

($\sim$ 10 μsec) as compared with subpulse time-scales ($\sim$ 1 msec). To account for the subpulse emission they would have to reappear at approximately the same place for a time long relative to their life times. No clear explanation for such kind of a surface memory was proposed by RS. In their model the spark discharges performed a rapid movement towards the magnetic pole, disqualifying the model. In this paper we propose a modification of the RS model, which avoids embarassing problems appearing in the original version of the polar gap model.

New ingredients of our modified RS model include: (i) estimation of the characteristic spark dimension, and (ii) quasi-central spark anchored to the local pole of quasi-axially symetric surface magnetic pole (thus our model is no longer of a hollow-cone type), and (iii) taking into account "non-vacuum" densities of dynamically developing spark plasma (see **??** for details). The critical assumption we need to adopt is an existence of a local magnetic pole (different from the global dipole) at or near the polar cap. This is necessary for two reasons: (a) a radius of curvature about that of the neutron star radius is required to drive and exponentiate spark discharges, and (b) the spark anchored to the local pole implies a quasi-annular pattern of sparks leading to nested cone structure of mean pulsar beam (e.g. (Rankin (1993); Gil et al. (1993); Mitra & Deshpande (1999))). We also assume that the planes of the magnetic fields line planes tend to converge at the local pole, introducing some degree of axial symmetry. As we demonstrate in this paper, a model based on these assumptions reproduces the characteristics of pulsar radiation quite well, especially the patterns of drifting subpulses.

We follow the RS idea that the high potential drop above the polar cap is discharged by a number of localized spark filaments. As RS first suggested, when a sparking discharge begins at some point on the polar cap, the gap potential drop rapidly falls there below the $\gamma - B$ pair formation threshold, which should inhibit another discharge within a distance of about h (Eq. 1). Thus, the polar spark restricts degrees of freedom of other sparks to circumferential motion around the pole, as they cannot approach the polar one within a screening distance h (Eq. 1). The existence of the polar spark is therefore crucial for our model.

Polar gap sparking discharge

As described by RS, the polar gap discharges through a number of localized sparks, separated from each other by a distance approximately equal to the gap height

$$h = 5 \cdot 10^3 B_{12}^{-4/7} \mathcal{R}_6^{2/7} P^{3/7} \quad \text{cm}, \tag{1}$$

where P is the pulsar period in seconds, $B_{12} = B_s/10^{12}$ G is the actual surface magnetic field with a radius of curvature $\mathcal{R}_6 = \mathcal{R}/10^6$ cm. Each spark develops exponentially until the plasma density ρ reaches a value close to the corotational Goldreich-Julian density ρ_{GJ}, screening the gap locally. From the width of subpulses interpreted as radio emission from plasma columns flowing along dipolar field lines connected to sparks, one can estimate the fraction $(D/r_p)^2$ of the polar cap area filled by a spark being in the range 10^{-1} to 10^{-2} (Cheng & Ruderman 1977), where D is a characteristic spark dimension. We can form a dimensionless parameter

$$r_p/h = 5 \cdot 10^4 \mathcal{R}_6^{-2/7} \dot{P}^{2/7} P^{-9/14} , \tag{2}$$

where h is the gap height described by the Eq. (1) with $B_{12} = 3.2 \cdot 10^7 (P \cdot \dot{P})^{1/2}$, and $r_p \approx 10^4 \cdot P^{-1/2}$ cm is the polar cap radius. For a typical pulsar with $P \approx 1s$, $\dot{P} = 10^{-15}$ and $\mathcal{R}_6^{2/7} \approx 0.5$ this equation gives $h_{RS} \approx 0.2\ r_p = 2 \cdot 10^3$ cm. On the other hand $< D/r_p >= \sqrt{0.05} \approx 0.2$ or $D \sim 0.2\ r_p$. Then, by comparison one can conclude that $D \approx h_{RS}$. This means that the characteristic dimension of the subpulse associated plasma column projected onto the polar cap along dipolar field lines is approximately equal to the RS polar gap height (Eq. 1).

The spark plasma is subject to slow "perpendicular" $\mathbf{E} \times \mathbf{B}$ drift (RS) across the planes of field lines with a velocity

$$v_\perp = \mathcal{F} \cdot c \cdot \Delta E / B_s, \tag{3}$$

where

$$\Delta E = \frac{2\pi}{cP} \frac{B_s h}{r_p(1-s)}. \tag{4}$$

Thus

$$v_\perp = \frac{\mathcal{F}}{1-s} \frac{2\pi}{P} \frac{h^2}{r_p}, \tag{5}$$

where $1 < s \leq 0$ is the mapping parameters describing the locus of field lines on the polar cap ($s = 1$ at the polar cap edge) and $\mathcal{F}$ is the filling factor corresponding to the spark termination stage at which plasma density produced by spark approaches the Goldreich-Julian value (corresponding to generation of pulsar radio emission),

$$\mathcal{F} \approx 0.1 \mathcal{R}_6^{-5/21} B_{12}^{1/7} P^{1/7} << 1 \tag{6}$$

(see (Gil & Sendyk (2000)) for details).

Let us consider an avalanche discharge which by chance occured close to the local pole of the surface magnetic field (which is quite likely to happen given a large number of energetic γ-quanta penetrating the gap region). Due to a putative quasi-axial symmetry at the pole (Gil & Sendyk 2000), this discharge will very soon anchore itself to the polar area. Other sparks will perform more or less organized $\mathbf{E}\times\mathbf{B}$ drift motion around the polar one. The spark associated plasma column completes one rotation in a time interval

$$\hat{P}_3 = 2\pi s r_p/v_\perp = (P/\mathcal{F})\cdot a^2 s(1-s), \qquad (7)$$

where $a = r_p/h$ (Eq. 9), and $s = s_{out}$ is the mapping parameter corresponding to the outermost radiation cone.

Polar cap structure and pulsar beams

As argued in (Gil & Sendyk (2000)), the actual pulsar polar cap should be populated by a number of sparks with a characteristic dimension $\mathcal{D} \approx h$, separated from each other also by about h, where h is the actual gap height (Eq. 1). One spark should be anchored at or near the local surface magnetic pole, and other sparks should perform more or less ordered, circumferencial $\mathbf{E}\times\mathbf{B}$ drift motion around the polar spark. In this sense our model differs markedly from the RS hollow-cone version in which sparks occupy only an outermost parts of the polar cap. Since the polar spark prevents the motion of other sparks towards the pole, the drifting sparks should form on the average a multi-ring structure centered on the polar spark, although rings are not expected to be perfectly circular (we only assume a quasi-axial symmetry of field line planes). It is natural to assume that plasma supplied by sparks to the magnetosphere above the polar gap will eventually give rise to the subpulse emission at altitudes of a few percent of the light cylinder radius $R_{LC} = cP/2\pi$ (Kijak & Gil 1997; Kijak & Gil 1998). The subpulses associated with the polar spark should be longitude-stationary and constitute the so called "core component" in the mean pulsar profile, while other subpulses may demonstrate an organized drift from one pulse to another and contribute to the so called "conal components" in the mean pulsar profile (see (Rankin (1983); Rankin (1990)) for the distinction between core and conal components).

The annular ring structure of the polar cap will result in a specific organization of the mean pulsar emission pattern, consisting of a narrow core beam surrounded by a number of nested conal beams. Simulations of pulsar emission within the framework of the above model seem to reproduce the observational properties of both single pulses and mean profiles quite well (Gil et al. 1995; Gil & Krawczyk 1996; Gil & Krawczyk

1997). Such a nested cone structure of pulsar beams have bean suggested by many authors (Rankin 1993; Gil et al. 1993). Most recently, (Mitra & Deshpande (1999)) have confirmed this conal structure identifying clearly up to three nested cones within the overall pulsar beam.

When a spark is modelled with a gaussian of half-power beam width about h (Eq. 1), the number of nested cones surrounding the core beam can be roughly determined by

$$n = Int\left(\frac{a-1}{2}\right), \tag{8}$$

where we introduced a *complexity parameter*

$$a = \frac{r_p}{h} \approx 5 \cdot \dot{P}_{-15}^{2/7} P^{-9/14}, \tag{9}$$

(Eq. 2; see (Gil & Sendyk (2000)) for details of approximation) which describes a complexity of pulsar emission pattern, where $\dot{P}_{-15} = \dot{P}/10^{-15}$. The number of sparks occupying the polar cap area is roughly a^2.

We will simulate the single pulse emission pattern corresponding to a number n of nested cones (Eqs. 8 and 9) using a technique introduced by (Gil (1984)) and developed by (Gil et al. (1995)) and (Gil & Krawczyk (1997)). Each gaussian spark has HPBW diameter of about h (Eq. 1), which is also a typical distance between adjacent sparks (RS). To determine the radio emission geometry we will use empirically determined emission altitudes r_{em} given by (Kijak & Gil (1997)) and (Kijak & Gil (1998)) in the form

$$r_6 = 55 \cdot \nu_{GHz}^{-0.21} \tau_6^{-0.1} P^{0.33} , \tag{10}$$

where $r_6 = r_{em}/10^6$ cm, ν_{GHz} is the frequency in GHz and τ_6 is the timing age in million years.

PSR B0943+10

Recently, (Deshpande & Rankin (1999)) have analysed in unprecedented detail an exquisively stable drifting-subpulse pattern of PSR B0943+10. This is an interesting pulsar showing a secondary drift-bands related to the even-odd modulation caused by the fast subpulse drift close to Nyquist frequency. (Deshpande & Rankin (1999)) determined the observing geometry corresponding to a peripheral sightline grazing the outer beam, in which they were able to identify 20 rotating subbeams producing the drift pattern. Deshande & Rankin (1999) have found this picture consistent with the system of sparks on the polar cap (RS), with dimensions determined by the gap height. In our model, the complexity parameter of this pulsar ($P = 1.098$ s, $\dot{P}_{-15} = 3.52$) $a = 6.73$, suggesting $n = 3$ nested cones around the core beam (Eqs. 8 and 9). The computer simulation shows 7, 14 and 20 spark-associated subpulse beams corresponding to 1st, 2nd and 3rd cone, respectively (Fig. 1). Given the grazing sightline geometry, an observer scans the outermost cone constisting of $N = 20$ sub-beams circulating around the magnetic axis with a linear velocity determined by Eq. 5 and completing one full rotation in the time interval described by Eq. 7. The locus of the outermost cone on the polar cap corresponds to $s_{out} = 0.875$. Thus, if we adopt for a filling factor $\mathcal{F} = 0.13$ (Eq. 6; which for $P = 1.098$ and $B_{12} = 1.99$ means $\mathcal{R}_6 = 0.5$) then for $a = r_p/h = 6.7$, Eq. 9 gives $\hat{P}_3 = 37 \cdot P$. This is exactly the time interval obtained by the sofisticated analysis of (Deshpande & Rankin (1999)). It is worth mentioning that the "vacuum" value of $\hat{P}_3$ given by RS is about $10 \cdot P$ (since they use $\mathcal{F} = 1$ and $s = 0.5$). In fact, $\hat{P}_3/P \approx 5.6 \cdot B_{12}/P^2$ in their model, which for $P \approx 1$ s and $B_{12} \approx 2$ gives $\hat{P}_3 \approx 10P$. Thus, the analysis of drifting subpulses in PSR B0943+10 performed by (Deshpande & Rankin (1999)) is perfectly consistent with our modified RS model of the $\mathbf{E} \times \mathbf{B}$ drift within pulsar polar gap region.

To simulate single pulses of PSR B0943+10 at frequency 0.43 GHz (to be compared with the Arecibo data presented in Fig. 2b), let us consider a locus of the outermost cone containing 20 sparks at a distance $d = s_{out} \cdot r_p = 0.875 \cdot 140m \approx 125$ m from the magnetic pole. The line-of-sight grazes this cone in such a way that the observer can detect up to three spark-associated subpulses in a single pulse.

For $P = 1.098$ s and $\tau_6 = 4.9$ we have emission altitude at 430 MHz $r_6 \approx 60$. Since PSR B0943+10 has never been detected above 600 MHz, we can reasonable assume that the opening angle of the beam of this pulsar at 430 MHz is about the impact angle. Thus $|\beta| \lesssim \rho(0.43) = 1^\circ.24\, r_6^{1/2}(0.43) \cdot P^{-1/2}$, which for $r_6 \approx 60$ (Eq. 10) gives $|\beta| \lesssim 9^\circ$. On the other hand, from the polarization measurements we have $\sin\alpha / \sin\beta = 3$ ((Deshpande & Rankin (1999))) and thus $\alpha \approx \arcsin(3 \cdot \sin 9^\circ) = 28^\circ$.

Assuming $\alpha = 28^\circ$ and $|\beta| = 9^\circ$ as an initial values we will now perform calculations of the radiation emitted tangent to the dipolar field lines at altitude $r_{em}(0.43) = 60 \cdot R = 6 \cdot 10^7$ cm, corresponding to a system of sparks presented in Fig. 1. Each spark circulates uniformly around the central polar spark (shadowed) at a rate of about 9.7 degrees per pulsar period ($360^\circ/37$). The resulting single pulse emission pattern is presented in Fig. 2a, in comparison with the 430 MHz observed pattern presented by (Deshpande & Rankin (1999)) in their Fig. 1, and reproduced here in Fig. 2b. We have obtained a perfect match by fine tuning the geometrical parameters, which finally turned out to be $\alpha = 26°$ and $\beta = -8°$. One can clearly see eight secondary drift bands corresponding to apparent even-odd modulation, in the sequence on 90 individual pulses. If one ignores flux irregularities, this is exactly the sequence presented by (Deshpande & Rankin (1999)). The subpulses along secondary drift bands belong to every other pulse. The primary drift bands are not visible because subpulses drift very fast across the profile from one pulse to another. Less than two pulsar periods ($P_3 = 1.87\ P$/cycle) are needed for the subpulse to reappear at the same longitudinal position, producing an apparent even-odd modulation effect (secondary drift bands). This effect is so specific that it gives full solution (drift pattern and α, β values) in the simulation/fit procedure. We can check for self-consistency that $N = \hat{P}_3/P_3 = 37/1.87 = 20$ (RS), as implied by computer simulations (Fig. 1).

Summary

Recent analysis of drifting subpulses in PSR B0943+10 presented by (Deshpande & Rankin (1999)) revealed in this pulsar 20 spark-associated subpulse beams rotating around the magnetic axis with the speed 9.7 deg/P, thus copmpleting one full cycle in about 37 pulsar periods P. We have applied our modified RS model to this pulsar and fully reproduced all features observed in its single pulses. Both the fundamental modulation periodicity $\hat{P}_3 = 37 \cdot P$ and the number of 20 sparks involved, were self-consistently determined using only P and $\dot{P}$ values. A specific even-odd modulation observed in this pulsar allowed us to determine precisely the observing geometry.

We have simulated the single pulse emission patterns of a number of B0943+10 by performing a number of subsequent steps: (i) determination of a number of nested cones and a number of sparks associated with the outermost cone (Eqs. 8 and 9), (ii) estimation of the mapping parameter s corresponding to the locus of the outermost cone, (iii) estimation of the approximate values of the inclination α and impact β angles using spectral and polarization information, (iv) determination of the fundamental periodicity $\hat{P}_3$ using Eq. 7, (v) estimation the emission altitude from the empirical Eq. 10, (vi) calculations of the single pulse sequence corresponding to a system of drifting sparks (e.g. Fig. 1), (vii) matching the calculated ond observed patterns by fine tuning the values of α and β, and if possible, testing self-consistency of the model by calculationg the polarization angle curve.

We believe that this paper provides strong support for the non-stationary sparking discharge of the pulsar potential drop above the polar cap. It has been first suggested by (Usov (1987)) and then fully explored by (Asséo & Melikidze (1998)), that the non-stationary sparks can lead to the two-stream like plasma instabilities, relatively close to the pulsar surface. In the paper by (Melikidze et al. (2000)) it is argued that a non-linear evolution of Langmuir plasma waves generated by the spark-associated two-stream instabilities leads to formation of charged solitons, which generate a coherent curvature radiation with properties of the observed pulsar radio emission.

Acknowledgements

This paper is supported in part by the KBN Grant 2 P03D 015 12 of the Polish State Committee for Scientific Research. We thank A.A. Deshpande, J.M. Rankin for providing us data with drifting subpulses.

References

Asséo E., Melikidze G., 1998, MNRAS, 301, 59
Cheng A.F., Ruderman M.A., 1977, ApJ, 214, 598
Deshpande A.A., Rankin J.M., 1999, ApJ, 524,1008
Gil J., 1984, A&A 131, 67
Gil J., Kijak J., Seiradakis J.H.,1993, A&A, 272, 268
Gil J., Kijak J., Maron O. and Sendyk M., 1995, A&A, 301, 177
Gil J., Krawczyk A., 1996, MNRAS, 280, 143
Gil J., Krawczyk A., 1997, MNRAS, 285, 561
Gil J., Sendyk M., 2000, ApJ, in press
Goldreich P., Julian H., 1969, ApJ, 157, 869
Kijak J., Gil J., 1997, MNRAS, 288, 631
Kijak, J., Gil, J., 1998, MNRAS, 299, 855
Melikidze G.I., Gil J., Pataraya A., 2000, ApJ,
Mitra D., Deshpande A.A., 1999, A&A, 346, 906
Rankin J.M., 1983, ApJ, 274, 33
Rankin J.M., 1990, ApJ, 352, 314
Rankin J.M., 1993, ApJ, 405, 285
Ruderman M.A., Sutherland, P.G., 1975, ApJ, 196, 51 (RS)
Taylor, J.H., Huquenin, G.R., & Hirsch, R.M., 1971, Astrophys. Lett. 9, 205
Usov, V.V., 1987, ApJ, 320, 333

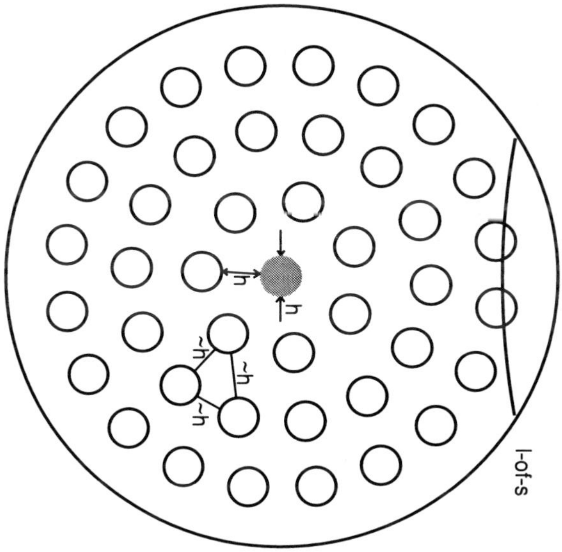

Figure 1 The illustration of instantaneous arrangement of sparks on the polar cap of PSR B0943+10. The shadowed spark is anchored to the local pole of the surface magnetic pole while other sparks perform slow $\mathbf{E} \times \mathbf{B}$ drift around it. The observer's orbit cuts through the outermost ring of 20 equaly spaced sparks at the impact angle $\beta = -8^{\circ}$ from the pole (for comparison see the map of 20 subpulse beams in Deshpande and Rankin 1999). Both the HPBW and a distance between adjacent sparks is $h = r_p/a \sim 17$ m.

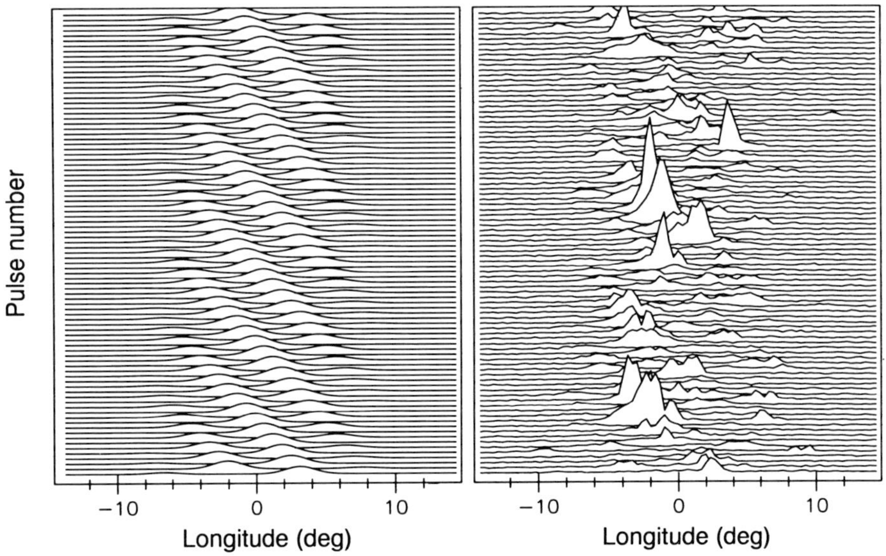

Figure 2 **a**. Simulated subpulse drift patter for PSR B0943+10 (left-hand side) and **b**. observed pattern (right-hand side) after Deshpande & Rankin (1999).

THE OBSERVATION OF PULSARS AT URUMQI ASTRONOMICAL OBSERVATORY

Xinji Wu[1], Na Wang[1,2], Jin Zhang[2], Ali Esamdin[2], R.N. Manchester[3], K.S.Cheng[4], A.G.Lyne[5], Aili Yusup[2],Hongbo Zhang[2], Liansen Kang[6], Shenzen Jin[6]

[1] *Chinese Academic of Science-Peking University joint Beijing Astrophysical Center and Department of Geophysics, Peking University, Beijing ,100871,*

[2] *Urumqi Astronomical Observatory, National Astronomical Observatory, CAS, Urumqi 830011, China,*

[3] *Australia Telescope National Facility, CSIRO, P.O. Box 76, Epping NSW 1710, Australia,*

[4] *Department of Physics, Hong Kong University, Hong Kong, China,*

[5] *University of Manchester, Nuffield Radio Astronomy Laboratories, Jodrell Bank, Macclesfield,Chesire,Sk11 9dl.UK,*

[6] *Beijing Astronomical Observatory, National Astronomical Observatory, CAS, Beijing, 100012, China*

Abstract The pulsar's observation is being developed using the 25m radio telescope at Urumqi Astronomical Observatory (UAO). A pulsar timing system has been set up. It uses a dual polarization receiver operating at 18cm and a filterbank receiver. Timing properties of about 100 pulsars will be monitored with this system. The single channel system at 92 cm works since 1996 for the program of mean pulse profile and long term variation of intensity about 18 pulsars. The observation of spectrum properties for few strong pulsars at 92cm, 18cm, 13cm, 6cm, and 3.6cm is in progress.

1. INTRODUCTION

Observational study of pulsars is scientifically important. The discovery of pulsars and binary pulsars were accorded two separate Nobel prizes, making it a hot topic in research. The observations of timing and intensity of emission give the basic parameters and properties of emission region and structure of neutron star.

K.S. Cheng et al. (eds.), Stellar Astrophysics, 81–87.

The first observation of pulsars in China was started in 1991 using a 15m radio telescope of Beijing Astronomical Observatory and achieved preliminary success at 232 MHz (Wu et al. 1999, Kang et al. 1993). But the work was discontinued because of low sensitivity and external interference. The observations of pulsars was started using the 25m radio telescope at 327MHz at 1996 (Wu et al. 1997). UAO is locate at Nansan of Urumqi of Xinjiang province, which is located in Northwest of China and the geometrical center of Asia. The 25m radio telescope constructed in 1993, which is one of largest radio telescope in China and one of VLBI station of international net. The radio telescope have 6 receivers on 92, 18, 13, 6, 3.6 and 1.3cm. The receivers have cool pre-amplifier on 1.3cm and 6cm only. The electromagnetic interference is rare, which is adequate for the weak radio source of pulsars. The observations of the intensity variation were monitored Science 1998 at 92 cm using a single frequency band for some strongly and nearly pulsars.

We plan to do the timing observation of pulsars at the end of 1996. The international collaboration project "The timing study of pulsars on 18cm band by use of the 25m radio telescope at Urumqi" was formatted. R.N. Manchester, A. G. Lyne and K.S. Cheng join this project with great enthusiasm. Just like Manchester's words: "I certainly am happy to support the pulsar timing project at Urumqi as much as I can", he do the lot of key works. Manchester visited UAO in October 1997 outline the pulsar timing system and May 1999 for testing of the system at UAO using the existing 18cm receiver and observations of pulsars. Wang Na and Aili-Yusup visited ATNF for learnt the techniques of timing analysis and to construct the down-converter system. Lyne also expressed willingness to develop de-dispersion equipment. K.S.Cheng is very interested in this pulsar's timing program, he joining in keeps observation combined with theory and gave the first fund of this project for construct the equipment. This project would improve the research and observation of pulsars in China to international level.

2. THE TIMING OBSERVATION OF PULSARS AT 1500MHZ

The Parkes multi-beam pulsar survey at 1400MHz is expected to find at least 630 pulsars (Manchester 1999, Lyne 1998). The new discovery pulsars are usually rather distant and in or near the Galactic plane. Therefore, they often have large dispersion measures. Many of the pulsars found by these surveys will be young and have significant period irregularities including glitches. Study of these irregularities gives a unique probe into the interior structure of neutron stars. It requires frequent observations which are difficult or impossible with National Facility telescopes such as Parkes.

The scientifically most valuable project that Urumqi could undertake is to make frequent timing observations. The program of timing include that to discover the glitches of young pulsars, to measure significant random period variations of pulsar, to measure parameters of the period and study pulsar's evolution, proper motion and the binary pulsars etc. Considering the effects of Galactic background emission, interstellar scattering, dispersion smearing and the spectrum of pulsars, it was decided to use the 18cm band for the pulsar timing observations. Other higher frequency bands are not suitable for pulsar observing because the flux is low and our system sensitivity is not enough.

It is expected that our telescope will be available for pulsar timing for about two days every week, and so frequent observations will be possible, suitable for monitoring period irregularities such as glitches. That means the work will be science worthwhile at world standard if we can reach sensitivity of 1 mJy in the 18cm band. In order to carry on the project, the 18cm band receiver of the 25m radio telescope at Urumqi must be necessarily re-equipped.

For the first stage, the following equipmentes were constructed: (1) The dual-channel room-temperature preamplifiers with a system noise of approximately 100 K for two polarizations; (2) Down-converter system, which was developed at ATNF of Australia; (3) Filterbank system which has 128 channels for each polarization, the bandwidth of each channel is 2.5MHz. This equipment was developed at Jodrell Bank of UK.

Signals from each channel are then one-bit digitized and recorded by a data acquisition system based on a PC operating under Windows NT. Time and frequency reference signals are obtained from the Observatory H-maser and GPS monitoring system. The online code is a Windows program written in C++. A control card (National Instruments PCI-DIO-96) is used to set the digitizer sampling interval and to control the time synchronization, and a high-speed parallel interface (PCI-DIO-32HS) is used for data transfer. A GPIB card is used for time transfer. The data are unpacked and folded synchronously with the pulsar period by the online program. Folded data are saved to disk at every sub-integration time, typically 120 seconds. A sequence counter and a time flag are checked every sample and on each 5 second respectively. For data analysis we use TREDUCE (supported by Swinburne University of Technology and ATNF) and TEMPO (supported by Princeton University and ATNF) operating under Linux.

First observations were made with the system in mid-1999. So far we have detected more than 80 pulsars, with the lowest mean flux density about 4mJy. Now the radio telescope at Urumqi have the power of undertaking timing study on relatively strong pulsars. The project aims to build a more sensitive dual-polarization cryogenic receiver with system temperature of about 30K so that pulsars with a mean flux density as low as 1 mJy can be observed, which is the task of next stage. The most recent published parameters of many pulsars are up to 20 years old. It is worthwhile to reobserve these pulsars to improve their parameters. About 100 pulsars should be detectable using our current system. The timing properties and proper motions of these pulsars will be investigated.

3. THE OBSERVATIONS AT 92 CM (327MHZ)

World-wide observation of pulsars are mostly above 400MHz, and are few below 327MHz. The flux density at low frequency is higher than that of high frequency and the long term variation of the flux is strong, so the monitor program of pulsar intensity at 327 MHz is good program for our small radio telescope.

The telescope has a system temperature of 130°K and a single bandwidth of 1MHz or 2 MHz at 327MHz. Pulsar data were acquired with a sampling interval of 2 ms or 1 ms. The signals were one-bit digitized using voltage-to-frequency converters and integrating counters, the data stored on disk for off-line analysis. The integrated pulse profiles were formed for each data by folding the data.

The first observational result of mean pulsar profile and flux density for four pulsars was published. For the PSR1133+16, the mean pulse profile is double components at 327 MHz, the normal profile is that the peak value of left component is higher than that of right component, which was observed. But the abnormal profile also recorded, the left component was lower than the right one (Wu et al.1997). Helfand et al.(1975) have put forth some statistical evidence for mode switching in this pulsar and Rankin (1986) has classified it as a candidate of mode change.

The monitor program of pulsar intensity runs whenever the telescope is not being used for the VLBI observations in May 1999. We observe a fixed schedule of 18 pulsars each day, with integration times depending on the strength of the source. Typically, a total of 20-30 minutes is spent observing a given pulsar in order to obtain a higher signal-to-noise ratio.

An variation event of PSR B0809+74 was discovered during the daily monitor of pulsars at 327MHz during a period of 11 hours was recorded.

On 13 July 1999, when we observed the PSR B0809+74 at 4.32(UT), the flux density of this pulsar had largely increased. Subsequently, we continually observed this source till 15.55(UT). A total of 11 data were gotten, each data of a single observation was 33 minutes. The peak value of mean pulse changing from 8.1 to 1.9Jy in a period of 11 hours, show large-amplitude fluctuation (as much as by a factor 4.3). Calibration of the observing system was carried out a regular intervals during every 2 or 3 observations using the continuum source 3c123. The data of PSR B0809+74 obtained on Jul.3,4,9,12,14 and other days in May and April show that the average of peak values of mean pulse was about 1.9 Jy, which is near the value given by Pulsar Catalog (Taylor et al.1993).

The observed temporal variation of pulsar flux densities at radio frequencies can be attributed to either intrinsic luminosity fluctuations or propagation effects along the line of sight, or to some combination of both. (Wu et al. 1995, Stinebring et al.1990, Kaspi et al 1992).

We consider the propagation effects at first. Flux variation of pulsars on short time have been identified as bring due to diffractive interstellar scintillation (DISS) (Scheuer 1968). DISS, characterized by pulsar intensity fluctuation on time scales of minutes. The time scale is dependent upon the distance to the pulsar, the observing frequency and the pulsar's transverse velocity (Stinebring et al. 1990). Some components of the long-term modulation as being due to propagation effects, which has been known as refractive interstellar scintillation (RISS). In RISS, flux variation by electron density irregularities and causes broad-band pulsar intensity fluctuation on time scale of days to year (Cordes et al. 1983).

The time scale of this fluctuation of event of PSR0809+74 during 13 July 1999 is much longer than the time scales of DISS and shorter than the time scales of RISS. The time scale of variation more likely is due to an unknown intrinsic fluctuation of PSR B0809+74 combined with diffractive effect. There also have another possibility in this case that flux arise in a short time because of focusing of the scattered radiation by unusual quickly variation of electron density of interstellar medium in the line of sight. This implies that the apparent some modulation is of unknown origin.

4. THE MULTI-FREQUENCY OBSERVATION OF PSR 0329+54

PSR0329+54 is a strong pulsar with S_{1400}=200mJy. there are some results of spectra of PSR0329+54 with the spectrum index α=2.0 given by Sieber (1973),α=1.6 given by Lorimer et al.(1995),and α=0.7 given by Malofeev (1999). The variations of the flux density of pulsars make it difficult to determine the spectrum. The simultaneous observations at 5 frequencies is impossible. Average of many independent observations is good method. The results of observation of PSR0329+54 at 5 frequencies were listed in table 1. The observation frequencies are the 327Mhz, 1500MHz, 2300MHz, 4800MHz and 8400MHz. The receivers of single channel were used. The system was calibrated with reference to the source 3C123.

Table 1. the hardware details for the experiment and the observational parameters

f(MHz)	327	1500	2300	4800	8400
obs. date	29/04/99	18/05/99	12/03/99	08/06/99	12/03/99
BM(MHz)	1	30	300	320	400
T_{sys}(K)	130	85	130	35	50
$t_{obs}(s)$	1200	1800	2000	2000	2000
$S_{peak}(Jy)$	177.6	13.7	10.2	1.4	0.6
$W_{50}(deg)$	3.9	3.2	2.6	2.4	1.7
$W_{sep.}(deg)$	24.0	23.2	21.2	18.5	
$S_{3c123}(Jy)$	128	41.4	31.7	15.6	9.3

The apparent width of the center component (W_{50}) decrease with the increasing of frequency, from 327MHz to 8400MHz. the separation of the peak values between the two outer components (W_{sep}) also decrease with the increasing of frequency, from 327MHz to 4800MHz. At 8400MHz, the outer components disappear.

In Table 1, the S_{peak} is the peak value of mean pulse of profile, which varies with frequency. After calculation of energy for each mean pulse profile of different frequencies, the spectrum index was obtained. This is a spectrum of the power law. The value of α is 1.72 for the 5 frequencies, which is nearly with the previous results is 2.0 and 1.6 given by Sieber and Lorimer, but there are big difference with the 0.7 given by Malofeev.

In the present spectrum, there are two straight segments, the index α is 1.59 in the range between 327MHz to 2300MHz, and 2.45 in the range between 2300MHz to 8400MHz,which is different with previous results. Certainly, to obtain reliable spectra need the simultaneous observation at different frequencies, but it is impossible for five frequencies. Our observation times of 13cm and 3.6cm are very nearly at same day, so the spectrum index at high frequencies band is reliable.

Acknowledgements

This work is supported by NSF of China, the Climbing Program from the state Science and Technology Commission of China, and by RFDP Of China. The filter-bank and digitiser were constructed at the University of Manchester, Jodrell Bank Observatory, and we thank Prof. Andrew Lyne for his support of this project. The down-converter system was constructed at the Australia Telescope National Facility and we thank them for their assistance.

References

Cordes J.M., Weisberg J.M., Boriakoff V., 1983,APJ,268,370

Helfand D.J., Manchester R.N., Taylor J.H., 1975 ApJ.,198,661

Lorimer D.R.,Yates J.A.,Lyne A.G., et al., 1995,MNRAS,273,411

Lyne A.G., May, 1998, Report at Beijing Astronomical Observatory.

Kang L.S., Huang M. H., Jin S. Z., et.al. Publications of BAO, 1993,Vol.19, p. 81

Kaspi V.M., Stinebring D.R., 1992, APJ, 329, 530

Malofeev V.M. et al, Astron. Rep., 1999, (in press)

Manchester R.N., May, 1999, Report at Peking University

Rankin J.M., 1986 ApJ,301,901

Scheuer P.A.G.,1968, Nature, 218, 920

Stinebring D.R., Cordes J.J., 1990, APJ, 352, 207

Sieber, W., 1973, Astron. Astrophys. 28, 237

Taylor J.H.,Manchester R.N.,Lyne A.G.,1993, APJs, 88, 529

Wu X.J., Kang L.S., Jin S.Z., Li S.Q. Wang N. 1997, Acta Astrophysica Sinica Vol.17 No.1, 37,
Chin.Astron. Astrophys 1997,vol.21.2, 170

Wu X.J.,Zhang J., Wang N., Progress of Astronomy,1999,Vol.17, No.3, P.236

Wu Xinji, Abraham C.-L. Chian, 1995, APJ, 443,261

THERMAL X-RAY PULSES RESULTING FROM PULSAR GLITCHES

Anisia P.S. Tang and K.S. Cheng

Abstract The non-spherical symmetric transport equations and exact thermal evolution model are used to calculate the transient thermal response to the pulsar's glitches. The three ways of energy release originated from glitches, namely the 'shell', 'ring' and 'spot' cases are compared. Their luminosity profiles for various sets of parameters are computed. The 'ring' case is always the middle one in terms of the response time, the response duration and the intensity of the response. This is not unexpected as the volume in which glitch energy is released in the 'ring' case is between those of the 'shell' case and the 'spot' case. The effects of initial core temperature before glitch, the density at which the glitch occurs, the amount of energy released in a glitch and the equations of state on the luminosity profiles are also compared.

The X-ray light curves resulting from the thermal response to the glitches are calculated. Only the 'spot' case needs to be considered because other two cases do not produce modulative X-rays. The relativistic light bending effect and the rotational effect on the photons being emitted in a finite region are considered. Various sets of parameters result in different evolution patterns of light curves. For a given glitch model, we find that this modulated thermal X-ray radiation resulting from glitches is a good method to determine the equations of state for pulsars.

Keywords: dense matter-stars: evolution-star: interiors-stars: neutron-stars: X-rays

1. INTRODUCTION

Glitches are believed to be mainly starquake-driven or superfluid-driven. In our work, only non-accreting pulsars are considered. Glitches are simulated by sudden deposition of thermal energy in the inner crust.

As a pulsar spins down due to dipole radiation, centrifugal force on the crust decreases and gravity pulls the crust towards a less oblate shape. Since pulsars are believed to be neutron stars with a solid crust, the stellar configuration deviating from its equilibrium one induces stress in the

K.S. Cheng et al. (eds.), Stellar Astrophysics, 89–94.

crust until the maximum yield strength is reached. Sudden relaxation of this stress brings the stellar shape to equilibrium. This is the starquake case. It causes localized heating in a small area where the crust cracks and gives the 'spot' case. It can occur anywhere in the crust. Another effect of pulsar spin-down is that the vortices become unpin from the lattice, they corotate with the local superfluid and scatter with the nearby lattices. There are two effects. One is superfluid angular momentum being transferred to the crust and the crust spins up. The other is frictional heating being produced and causes the energy dissipation locally. Since the vortices are parallel to the rotational axis and ω between superfluid and crustal lattices is maximum around rotational equator, so energy deposits in a ring-like structure around the rotational equator. This 'ring' case can occur in the inner crust where n-superfluid and crustal lattices coexist. The spherical symmetric 'shell' case is not realistic. The three cases: 'spot', 'ring' and 'shell' cases are compared. The modulation of the X-ray profile due to gravitational lensing and rotational effects for the 'spot' case is calculated.

2. COMPARISON OF 'SHELL', 'RING' AND 'SPOT' CASES

The non-spherical symmetric general relativistic energy balance and thermal transport equations according to (Cheng, Li, & Suen 1998) are used to study the thermal evolution of the pulsar after a glitch. A finite difference method is used to solve the general relativistic transport and energy balance equations for each of the three cases. A two dimensional grid can be used as azimuthal symmetry can be assumed. The grid is divided into 100×90 cells. Besides the amount of energy deposition in form of heat, other Physics inputs which include equations of state (EOS), composition of the star and initial core temperature are important in determining the luminosity profiles.

Different EOS are used with TOV equations to calculate the structure of the pulsar. In this paper, the major EOS used is UT ((Largaris & Pandharipande 1981); (Friedman & Pandharipande 1981); (Wiringa, Fiks, & Fabrocini 1988)). BPS ((Baym, Pethick, & Sutherland 1971)) and PPS ((Pandharipande, Pines, & Smith 1976)) are used for comparison. Composition of the star is obtained from (Lattimer et al. 1985). Scheme of calculating heat capacities follows (Chong & Cheng 1994) and (Cheng, Li, & Suen 1998).

Thermal conductivity is fitted according to (Itoh et al. 1984) in solid phase and (Itoh et al. 1983) in liquid phase. Initial core temperature is taken to be 10^7 K or 10^8 K. In this temperature range, the most important neutrino emission process will be electron bremsstrahlung in the crust. Surface temperature T_s is calculated according to (Gudmundsson, Pethick, & Epstein 1983). The evolution of heat pulse for the 'ring' case is shown in Figure 1. It is similar to the 'shell' case given in (Chong & Cheng 1994) and the 'spot' case which is given in (Li 1998).

Holding all other Physics inputs constant, the 'ring' case is always in the middle in terms of the response time, the response duration and the intensity of the response as seen in Figure 2.

3. THERMAL X-RAY MODULATION

Schematic illustration of gravitational lensing effect and definition of all the angles following the scheme of (Pechenick, Ftaclas, & Cohen 1983) is shown in Figure 3.

Gravitational lensing effect produces a larger diffraction for a softer EOS. With consideration of the rotational effect and the finite area effect, i.e. the energy is actually emitted from a finite region, the results of the three effects for various EOSs are plotted in Figure 4.

The total energy flux of the pulsar is also calculated in order to compare with the actual data obtained. An example of the expected light curve is given in Figure 5.

4. CONCLUSION AND OUTLOOK

In the above calculations, the interstellar absorption, the magnetospheric effects and the possibility of other diffusion mechanisms besides conducton in the pulsar have been neglected. The magnetic field effect will be considered in (Tang & Cheng 1999). The structure of the pulsar is assumed to be inert to the thermal change of the pulsar. The thermal X-ray emitted from the surface is assumed to be blackbody radiation alone. In comparing with the observed light curves, measuring thermal X-ray radiation resulting from glitches is a good method to determine the equations of state for pulsars.

References

Anderson, P. 1975, University of Cambridge preprint TCM, 1975/5
Baym, G., Pethick, C., & Sutherland, P. 1971, ApJ, 170, 299
Cheng, K.S., Li, Y., & Suen, W.M. 1998, ApJ, 499, L45
Chong, N., & Cheng, K.S. 1994, ApJ, 425, 210
Friedman, B. & Pandharipande, V.R. 1981, Nucl. Phys. A, 361, 502
Gudmundsson, E.H., Pethick,C.J., & Epstein, R.I. 1983, ApJ, 272, 286
Itoh, N., Kohyama,Y., Matsumoto, N., & Seki, M. 1984, ApJ, 285, 758
Itoh, N., Mitake, Y., Iyetomi, H., & Ichimaru, S. 1983, ApJ, 273, 774

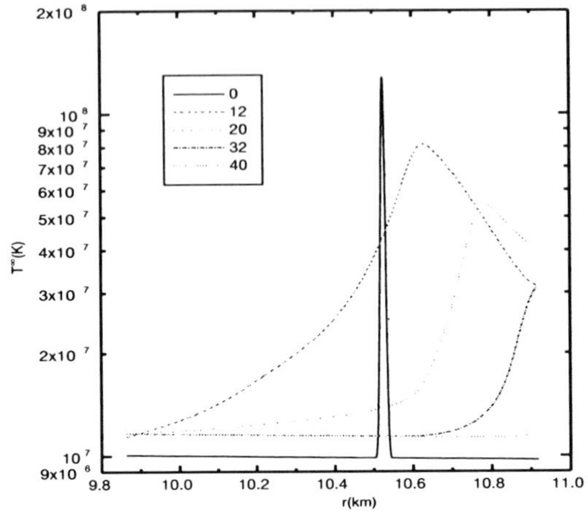

Figure 1 The redshifted temperature profile of a UT-star with $T_{core} = 10^7\ K$, $\Delta E = 10^{42}\ erg$ at $\rho_{glitch} = 10^{13}\ g/cm^3$ for the 'ring' case with respect to the radius from the centre of the star.

Largaris, I.E. & Pandharipande, V.R. 1981, Nucl. Phys. A, 359, 349

Lattimer, J.M., Pethick, C.J., Ravenhall, D.G., & Lamb, D.Q. 1985, Nucl. Phys. A, 432, 646

Li, Y. 1998, M.Phil. Thesis, HKU

Pandharipande, V.R., Pines, D., & Smith, R.A. 1976, ApJ, 208, 550

Pechenick, K.R., Ftaclas,C., & Cohen, J.M. 1983, ApJ, 274, 846

Tang, P.S. & Cheng, K.S. 1999, ApJ, to be published

Wiringa, R.B., Fiks, V., & Fabrocini, A. 1988, Phys. Rev. C, 38,1010

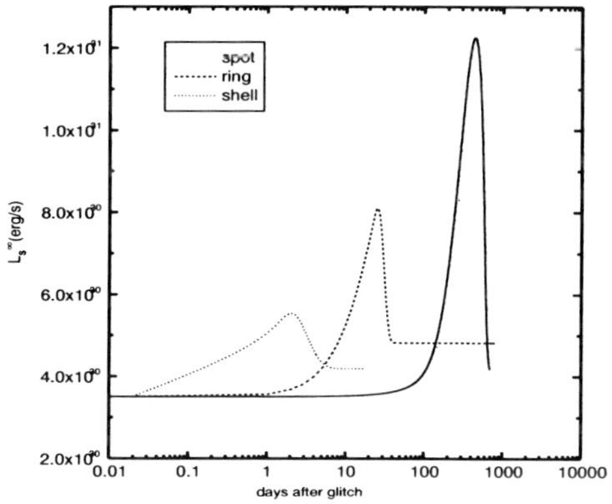

Figure 2 The luminosity profiles of a UT-star with $T_{core} = 10^7\ K$, $\triangle E = 10^{42}\ erg$, $\rho_{glitch} = 10^{13}\ g/cm^3$.

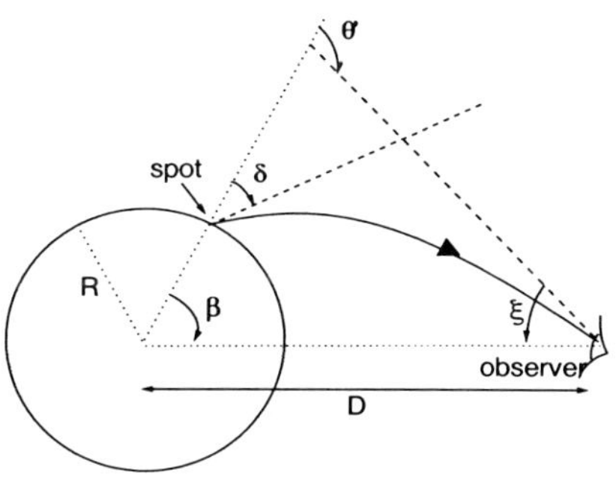

Figure 3 Schematic illustration of gravitational lensing effect.

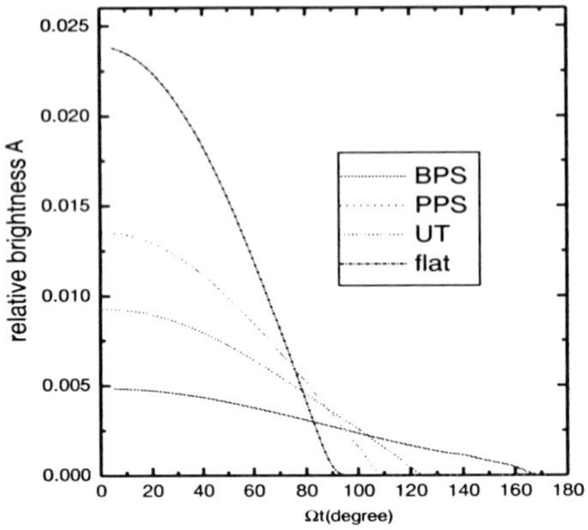

Figure 4 The angular radius of the emission cone is $\alpha = 5^o$. The gravitational lensing effect, the rotational effect and the finite area effect have all been considered.

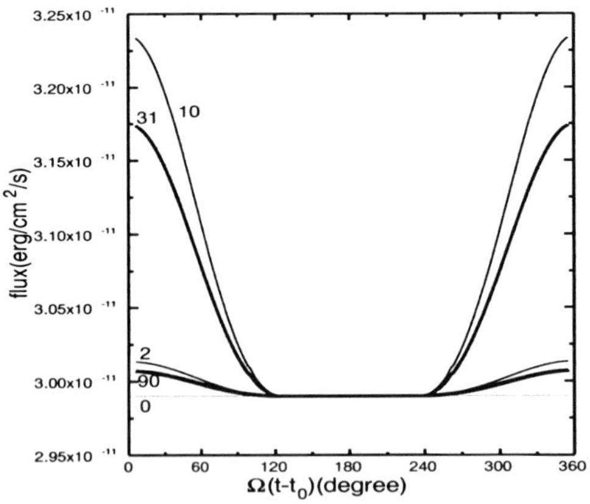

Figure 5 The evolution of total energy flux of a UT-star with $T_{core} = 10^7\ K$ and $T_{core} = 10^8\ K$, $\Delta E = 10^{42}\ erg$, $\rho_{glitch} \sim 10^{13}\ g/cm^3$. The numbers are the days after glitch. ($f_F = 9.0100$ for $T_{core} = 10^7\ K$ and $f_F = 0.0776$ for $T_{core} = 10^8\ K$)

STARQUAKE-INDUCED GLITCHES IN PULSARS

Richard I. Epstein[1] and Bennett Link[2]
[1]*Los Alamos National Laboratory,*
[2]*Montana State University*
Los Alamos National Laboratory
[1]epstein@lanl.gov, [2]blink@dante.physics.montana.edu

Abstract The neutron star crust is rigid material floating on a neutron-proton liquid core. As the star's spin rate slows, the changing stellar shape stresses the crust and causes fractures. These *starquakes* may trigger pulsar glitches as well as the jumps in spin-down rate that are observed to persist after some glitches. Earlier studies found that starquakes in spinning-down neutron stars push matter toward the magnetic poles, causing temporary misalignment of the star's spin and angular momentum. After the star relaxes to a new equilibrium orientation, the magnetic poles are closer to the equator, and the magnetic braking torque is increased. The magnitude and sign of the predicted torque changes are in agreement with the observed persistent spin-down offsets. Here we examine the relaxation processes by which the new equilibrium orientation is reached. We find that the neutron superfluid in the inner crust slows as the star's spin realigns with the angular momentum, causing the crust to spin more rapidly. For plausible parameters the time scale and the magnitude of the crust's spin up agree with the giant glitches in the Vela and other pulsars.

1. INTRODUCTION

Stresses in the crust of a neutron star could produce starquakes that affect the star's spin evolution and generate high-energy emission. As the star's spin rate increases or decreases, changes in the equilibrium shape of the star and the differential rotation between the crust and the interior neutron superfluid generate stress (1). In "magnetars", decay of the superstrong field ($B \gtrsim 10^{14}$ G) could break the crust and drive episodes of intense gamma-ray emission (2). Recent studies showed that starquakes can change the magnetic spin-down torque acting on the

K.S. Cheng et al. (eds.), Stellar Astrophysics, 95–103.

star (3; 4). Starquakes in slowing neutron stars drive matter toward the magnetic poles, distort the star's shape, and excite precession. As the precession damps, the star relaxes to a new rotational state with the magnetic poles closer to the equator. The new magnetic orientation enhances the braking torque on the star and may provide an explanation for the observed increases in the spin-down torque following glitches in the Crab pulsar, PSR1830-08 and PSR0355+54.

Here we investigate the physical processes that allow the star to relax to its post-starquake equilibrium. The most important processes are the coupling between the liquid core and solid crust and the creep of neutron superfluid vortices in the inner crust of the star. We find that the changes produced by large starquakes can trigger catastrophic unpinning of neutron superfluid vortex lines in the star's inner crust (5). As vortices move, the inner crust superfluid rapidly settles to a state of lower angular momentum, while exerting a spin-up torque on the crust. Our estimates show that starquake-triggered events may explain giant pulsar glitches as well as the persistent spin-down offsets. The next section summarizes earlier work on starquakes and spin-down offsets, and the following one describes the post-starquake spin relaxation and glitches.

2. STARQUAKES AND SPIN-DOWN OFFSETS

The crust of a spinning neutron star is oblate with an equatorial bulge. The moment of inertia of the bulge is $I_{eb} \sim I_{\rm crust} R^3\Omega^2/(2GM) \sim 4 \times 10^{-5} I_{\rm crust}\Omega_2^2$ where $R \sim 10^6$ cm is the stellar radius, $\Omega = 100\Omega_2$ rad s^{-1} is the star's spin frequency, and $I_{\rm crust}$ is the characteristic moment of inertia of the crust. The crust contains about 1% of the star's total moment of inertia $I_{\rm total}$. As the star spins down, the equatorial circumference shrinks and the polar radius grows. Because the crust is solid, strain develops as the star's shape changes. As sketched in Figure 1, the strain in the crust is relieved along starquake faults that form at an angle to the star's equator (3; 4). Matter slides along these faults to higher latitudes, and magnetic stresses favor those faults that direct matter along field lines toward the magnetic poles.

An important result of the earlier studies is that starquakes shift the stellar matter asymmetrically, creating excess moment of inertia δI about an axis different from any of the pre-starquake principal axes. This distortion changes the orientation of the principal inertial axis by an angle $\Delta\alpha \sim \delta I/I_{eb} \sim 2.5 \times 10^4 \Omega_2^{-2} \delta I/I_{\rm crust} = 2.5 \times 10^{-3}\Omega_2^{-2}\delta_{-7}$, where $\delta I/I_{\rm crust} \equiv 10^{-7}\delta_{-7}$ (3). The *distortion parameter* δ_{-7} characterizes the size of the starquake.

When the principal axis of inertia of the crust is not aligned with the star's angular momentum, the star precesses and wobbles. Eventually the star relaxes to a new equilibrium in which the axis and angular momentum are again realigned, and the magnetic pole is shifted by the angle $\Delta\alpha$ closer to the rotational equator. In some mechanisms for pulsar spin down, such as the magnetic dipole braking model, this angular shift increases the torque on the star, producing a long-lasting increase in the spin-down rate. The persistent spin-down offsets observed in the Crab pulsar can be explained by this mechanism if $\Delta\alpha \sim 10^{-3}$ (6) corresponding to $\delta_{-7} \simeq 1.6$ ($\Omega_2 \simeq 2$ for the Crab pulsar).

3. THE ORIGIN OF GLITCHES

Starquake-induced asymmetry in the stellar crust excites precession. If the neutron star crust behaved as an isolated rigid body, it would precess or wobble at a frequency $\Omega_w \simeq (I_{eb}/I_{\rm crust})\Omega$ and the angle between the angular velocity and the angular momentum would be $\sim (\Delta\alpha)^2$. The spin behavior of a realistic neutron star is more complicated than this for several reasons. First, the pinning of superfluid vortices in the inner crust acts to stabilize the spin of the crust. Second, the crust and core of the star are not strongly coupled on the precession time scale. Third, the tilting of the crust accelerates vortex creep.

Pinned superfluid. In the inner crust of the star the neutron superfluid vortex lines may pin to the nuclei in the solid crust. The rotation of the superfluid is determined by the location of the vortex lines, and, as long as the vortex lines remained pinned, the superfluid velocity field cannot change. The gyroscopic action of the pinned superfluid works with the equatorial bulge to further stabilize the star (7). The moment of inertia $I_{\rm pinned}$ of the pinned superfluid is comparable to that of the crust and much larger than the moment of inertia of the equatorial bulge; $I_{\rm pinned} \sim 0.01 I_{\rm total} \gg I_{eb}$. The pinned superfluid decreases the equilibrium tilt of the star by a factor $I_{eb}/I_{\rm pinned}$ to an angle $\Delta\alpha_p \sim \delta I/I_{\rm pinned} \sim 10^{-5}\delta_{-7} \ll \Delta\alpha$. This is the tilt angle of the star with completely pinned superfluid *after the precession has damped* (the tilt immediately after the starquake is $\sim [\Delta\alpha_p]^2 \sim 10^{-10}\delta_{-7}^2$). The precession frequency is proportional to $I_{\rm pinned}$ and inversely proportional to moment of inertia coupled to the crust. If the spin of the core and the crust are tightly coupled, the star's precession frequency is $\Omega_w = (I_{\rm pinned}/I_{\rm total})\Omega \simeq \Omega_2$ rad s^{-1}.

Coupling between the crust and the core. Changes in the crust's motion are communicated to the core by MHD-like waves. If the protons in the neutron star core form a type II superconductor, as expected, the magnetic field is confined to thin tubes of flux $\pi\hbar c/e \sim 2 \times 10^{-7}$ G cm^{-2}, with characteristic dimensions $\Lambda \sim 50$ fm and field strengths $B_\phi \sim 10^{15}$ G. Signals travel from the crust to the core on an Alfvèn time $t_{\rm couple} \sim (4\pi\rho)^{1/2}R/(BB_\phi)^{1/2} \sim 4B_{12}^{-1/2}$ s, for an average field of $B = 10^{12}B_{12}$ G and a density of $\rho \sim 10^{15}$ g cm^{-3} (8). Because of the imperfect coupling between the crust and the core, the precession is damped on a time scale (9)

$$t_{\rm damp} \simeq \frac{\Omega}{\Omega_w} t_{\rm couple} \sim \frac{I_{\rm total}}{I_{\rm pinned}} t_{\rm couple} \sim 400\, B_{12}^{-1/2}\ {\rm s}. \qquad (1)$$

The energy of the regular motion of precession is converted into irregular fluid motions in the core. The irregular motions of the superfluid neutrons and superconducting protons are then converted into thermal energy by processes such as electron scattering from vortex lines (10) or flux tube-vortex line interactions (11). Since the core and crust are not strongly coupled on the precession time scale, the effective moment of inertia of the precessing material is less, and the precession frequency may be higher than we have used. The damping time scale would be correspondingly reduced.

After the precession has damped, the star's crust has tilted by $\Delta\alpha_p$. As the pinned vortex lines move to align with the crust's rotation vector, the tilt angle grows until it equals $\Delta\alpha$. We now examine the processes by which the vortex lines move and the effects of this motion on the star's spin behavior.

Motion of vortex lines. In a steady state, the angular velocity of the superfluid closely follows that of the solid crust. As the superfluid rotation slows, the vortex lines move radially with a steady velocity $v_{ss} \simeq -R\dot{\Omega}/(2\Omega) = R/(4t_{\rm age})$, where the spin-down age of the pulsar is $t_{\rm age} \equiv \Omega/(2\dot{\Omega})$ (14). For the Crab pulsar $v_{ss} \simeq 5 \times 10^{-6}$ cm s^{-1} and for the Vela pulsar $v_{ss} \simeq 7 \times 10^{-7}$ cm s^{-1}. If the vortex lines move at v_{ss} following a starquake, they would align with the new direction of the crust's principal axis in a time $t_{\rm align} \sim \Delta\alpha R/v_{ss} \sim 10^{-2}\delta_{-7}\Omega_2^{-2}t_{\rm age}$, years for the Crab pulsar and decades for the Vela pulsar. The vortex velocity can be greatly enhanced by the perturbations produced by the tilt of the spin axis.[1] After the star's spin axis tilts relative to the pre-starquake orientation, some parts of the star are further from the new spin axis; this shift can be as large as $\Delta\alpha_p R$. For these regions, the linear velocity from the star's spin increases by $\Delta\alpha_p R\Omega$. The superfluid velocity, on the other hand, changes much less. If the neutron vortex

[1]Heat produced during a starquake can also increase the vortex velocity (12).

lines in the star vortex were completely pinned, the superfluid velocity in the star's frame would remain unchanged. Even though the superfluid in the core is not pinned, the vortex lines cannot move through the crust, so the crust superfluid velocity remains fixed. The velocity lag v_δ between the superfluid and the solid crust thus changes by as much as $\Delta v_\delta \sim \Delta\alpha_p R\Omega \simeq 10^3\delta_{-7}$ cm s^{-1} in parts of the star; see Figure 2. Velocity differences of this magnitude may have dramatic effects on the vortex velocity, large enough to produce the crustal spin-ups associated with glitches.

Pinned vortex lines in the inner crust can creep outward through thermal activation of vortex segments from their pinning barrier; a line segment unpins from one configuration and migrates to new sites where it repins (13; 14). The general form of the creep velocity is $v_{\rm creep} \propto \exp[-A/kT]$, where A is the minimum activation energy for a segment of the vortex line to unpin (15). The activation energy depends sensitively on the lag velocity v_δ. If the vortex lines are moving at their steady state rate v_{ss} and the lag velocity suddenly jumps by Δv_δ, the new creep rate is

$$v_{\rm creep} = v_{ss} \exp\left[-\frac{dA}{dv_\delta}\frac{\Delta v_\delta}{kT}\right]. \tag{2}$$

The activation energy of a vortex line is characterized by the maximum pinning force $F_{\rm pin}$ between a nucleus and a vortex line, the range of the pinning force $r_{\rm pin}$ and the distance between pinning sites $\ell_{\rm pin}$. The maximum lag velocity between the superfluid and the crust that the pinning forces can support is $v_{\rm max} = F_{\rm pin}/\rho_s\kappa\ell_{\rm pin}$, where $\kappa = \pi\hbar/(m_n) = 2.0 \times 10^{-3}$ cm^2 s^{-1} and ρ_s is the mass density of the neutron superfluid; $\rho_s \sim 10^{14}$ g cm^{-3} is the characteristic density of much of the inner crust superfluid. The activation energy required for a vortex line to unpin depends on the stiffness of the vortex line; if the line is flexible, it can unpin from one nucleus at a time, whereas if it is stiff, the line must unpin from many nuclei simultaneously (15). The parameter $\tau \simeq 0.4\rho_s\kappa^2 r_{\rm pin}/(F_{\rm pin}\ell_{\rm pin})$ characterizes the vortex line's stiffness (15); for the conditions in much of the crusts of the Crab and Vela pulsars $\tau > 1$ and vortex lines move by unpinning from many sites simultaneously. In this limit the appropriate expression for the activation energy is $A \simeq 6.8F_{\rm pin}r_{\rm pin}\tau^{1/2}(1 - v_\delta/v_{\rm max})^{5/4}$, and it's derivative is

$$\frac{dA}{dv_\delta} \simeq -4.8\frac{F_{\rm pin}r_{\rm pin}\tau^{1/2}}{v_{\rm max}} \simeq -3.0\left(\frac{\rho_s^2 r_{\rm pin}^3\kappa^3}{v_{\rm max}}\right)^{1/2}. \tag{3}$$

In obtaining this equation we set $(1 - v_\delta/v_{\rm max})^{1/4} \sim 0.6$, which is a characteristic value for a variety of stellar models (14).

We can use the observations of the Vela pulsar to estimate $v_{\rm max}$. This pulsar exhibited a string of 12 nearly evenly spaced glitches separated with an average interval of $t_{\rm int} = 2.3$ years (16). The regularity of these glitches suggests that the inner crust superfluid remains pinned between glitches until the velocity lag between the solid crust and the superfluid approaches $v_{\rm max}$. With this interpretation, a lower limit to the critical lag velocity is $v_{\rm max} \gtrsim |\dot{\Omega}_{\rm Vela}| t_{\rm int} R = 7.0 \times 10^3$ cm s^{-1}. The value of $v_{\rm max}$ exceeds this limit because the inner crust superfluid might not relax to zero lag velocity at each glitch. For our estimates we use $v_{\rm max} = 10^5 v_5$ cm s^{-1}. For $v_5 \simeq 1$ the pinning force is $F_{\rm pin} \rho_s \kappa \ell_{\rm pin} v_{\rm max} \simeq 13\,(\ell_{\rm pin}/100{\rm fm})$ keV fm^{-1} at a superfluid density of $\rho_s = 10^{14}$ g cm^{-3}. This pinning force is much smaller than that obtained by recent microscopic calculations of vortex-nuclear interactions (17) assuming a perfect crystal, but it is larger than the average pining force estimated for an amorphous crust (18).

Taking $\rho_s = 10^{14}$ g cm^{-3} and interaction distance $r_{\rm pin} = 10$ fm (19), the change in the activation energy in parts of the crust after a starquake is

$$\frac{dA}{dv_\delta}\Delta v_\delta \sim -0.053 v_5^{-1/2} \left(\frac{\Delta v_\delta}{\rm cm\,s^{-1}}\right) {\rm keV} \sim -52\,\delta_{-7}\Omega_2 v_5^{-1/2}\ {\rm keV}. \quad (4)$$

The time scale for the post-starquake relaxation of the vortex lines is $t_{\rm relax} \sim \Delta\alpha R/v_{\rm creep}$; for kT in keV, this gives

$$t_{\rm relax} \sim \frac{\delta_{-7} t_{\rm age}}{100\Omega_2^2} \exp\left[-\frac{52\,\delta_{-7}\Omega_2}{v_5^{1/2} kT}\right]. \quad (5)$$

The vortex relaxation time is very sensitive to the magnitude of the quake-induced shape change. For example, taking $v_5 = 1$, we find that for the Crab pulsar ($kT \sim 20$ keV, $t_{\rm age} \sim 10^3$ yr, $B_{12} \sim 4$) the relaxation time is less than the damping time if the distortion parameter is $\delta_{-7} \gtrsim 8.8$, but it is more than $1000 t_{\rm damp}$ if $\delta_{-7} \lesssim 4.3$. The corresponding values for the distortion parameter for the Vela pulsar ($kT \sim 8$ keV, $t_{\rm age} \sim 10^4$ yr, $B_{12} \sim 8.8$) are $\delta_{-7} \gtrsim 1.9$ and $\delta_{-7} \lesssim 1.2$. For each pulsar, there is a critical value for the distortion parameter δ_{-7}. Starquakes that produce distortions above this threshold trigger rapid vortex motions while smaller events generate only gradual changes.

Spin jumps. In the large events, the rapid outward motion of the vortex lines produces a slowing of the superfluid and a corresponding spin up of the crust. In the regions of the inner crust where Δv_δ is positive the vortex lines rapidly creep a distance $\Delta\alpha R \sim 2.5 \times 10^{-3} \Omega_2^{-2} \delta_{-7} R$. The superfluid in the affected regions of the inner crust will now spin more slowly by corresponding amount: $\Delta\Omega_s/\Omega \sim 2.5 \times 10^{-3}\Omega_2^{-2}\delta_{-7}$. The angular momentum lost by superfluid is imparted to the crust and the core which is strongly coupled to it (10), giving $I_{\rm total}\Delta\Omega_{\rm crust} = -I_{\rm unpinned}\Delta\Omega_s$, where $I_{\rm unpinned}$ is the moment of inertia of the region of rapid vortex creep. Since only 1/2 of the crust has $\Delta v_\delta > 0$, we have $I_{\rm unpinned} \lesssim 0.5 I_{\rm crust}$ and

$$\frac{\Delta\Omega_{\rm crust}}{\Omega} \sim -\frac{I_{\rm crust}}{2I_{\rm total}}\frac{\Delta\Omega_s}{\Omega} \sim 1.3 \times 10^{-5}\frac{\delta_{-7}}{\Omega_2^2}. \tag{6}$$

The rapid creep of vortex lines following a large starquake could explain the giant glitches with $\Delta\Omega_{\rm crust}/\Omega \sim 2 \times 10^{-6}$ observed in the Vela and other pulsars.

In the above estimate we assume that the lag between the superfluid and the crust is large enough to supply the needed angular momentum to the crust. As the vortex lines creep through a distance $\sim \Delta\alpha R$ the local superfluid slows by $\sim \Delta\alpha R\Omega$. If the pre-starquake lag velocity v_δ were less than this value, the vortex creep would stop before the vortex lines move the full distance $\sim \Delta\alpha R$. A necessary condition for the above estimate to be valid is that $v_\delta \gtrsim \Delta\alpha\Omega R$. Since $v_{\rm max} > v_\delta$ we have $v_5 \gtrsim 2.5\delta_{-7}\Omega_2^{-1}$. The limiting factor in the size of a starquake-induced pulsar glitch may be the pre-glitch lag velocity. The reason for the large difference between the size of the Crab glitches and the giant glitches of the Vela pulsar may be that in the Crab pulsar, with its higher internal temperature, vortex creep between glitches limits the build up of a sufficiently large v_δ.

4. SUMMARY

Starquakes tilt the principal axis of inertia of a neutron star. As the star relaxes to its new equilibrium orientation, neutron superfluid vortex lines migrate outward, spinning up the rest of the star. The spin ups from large quakes may explain the giant glitches observed in isolated pulsars. The change in the direction of the magnetic axis may increase the spin-down torque, as observed in the Crab and other pulsars. Detailed calculations of the post-starquake relaxation for both large and smaller events may yield distinctive timing signatures to compare with observations.

Acknowledgements

This work was performed under the auspices of the U. S. Department of Energy, and was supported by IGPP at LANL, NASA EPSCoR Grant 291748, NASA ATP

Grants NAG 53688 and NAG 52863, and by USDOE Grant DOE/DE-FG02-87ER-40317.

References

Ruderman, M., ApJ, **203**, 213 (1976).

Thompson, C. & Duncan, R. C., ApJ, **473**, 322 (1996); Thompson, C. & Blaes, O., Phys. Rev. D, 57, 3219 (1998).

Link, B., Franco, L. M., & Epstein, R. I., ApJ, **508**, 838-843 (1998).

Franco, L. M., Link, B., & Epstein, R. I., preprint, astro-ph/9911105 (1999).

Anderson, P. W., & Itoh, N., Nature, **256**, 25 (1975).

Link, B., Epstein, R. I. & Baym, G., ApJ, **390**, L21 (1992); Link, B. & Epstein, R. I., ApJ, **478**, L91 (1997).

Shaham, J., ApJ, **214**, 251 (1977); Sedrakian, A., Wasserman, I., and Cordes, J. M., ApJ**524**, 341 (1999).

Abney, M., Epstein, R. I. & Olinto, A. V., ApJ, **466**, L91 (1996); Mendell, G., MNRAS, **296**, 903 (1998).

Bondi, H, & Gold, T., MNRAS, **115**, 41 (1955).

Alpar, M. A. and Sauls, J. A., ApJ, **327**, 723 (1988).

Ruderman, M., Zhu, T., and Chen. K., ApJ, **492**, 267 (1998).

Link, B. & Epstein, R. I., ApJ, **457**, 844 (1996).

Alpar, M. A., Anderson, P. W., Pines, D., & Shaham, J., ApJ, **276**, 325 (1984).

Link, B., Epstein, R. I. & Baym, G., ApJ, **403**, 285 (1993)

Link, B. & Epstein, R. I., ApJ, **373**, 592 (1991)

Link, B., Epstein, R. I., & Lattimer, J. M., Phys. Rev. Lett., **83**, 3362 (1999).

Epstein, R. I. & Baym, G., ApJ, **328**, 680 (1988); Pizzochero, P.M., Viverit, L. & Broglia, R. A. Phys. Rev. Lett., **79**, 3347 (1997).

Jones, P. B., MNRAS, **306**, 327 (1999).

Negele, J. W. & Vautherin, D., Nuc.Phys, **A207**, 298 (1973); DeBlasio, F. V. & Elgarøy, Ø. Phys. Rev. Lett., **82**, 1815, (1999).

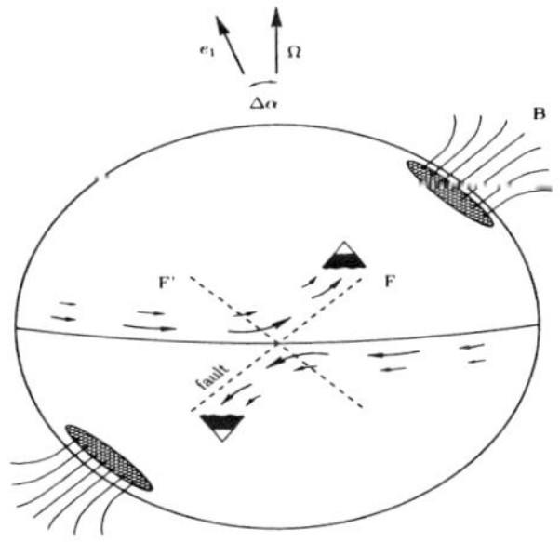

Figure 1 Starquakes relieve the stress that builds as the a neutron star's spin slows. Matter can slide to higher latitudes along faults F or F'. Magnetic stresses favor faults such as F that move matter toward the magnetic poles. The accumulated matter, shown as snow-capped peaks, shift the principal axis of inertia by an angle $\Delta\alpha$ relative to the star.

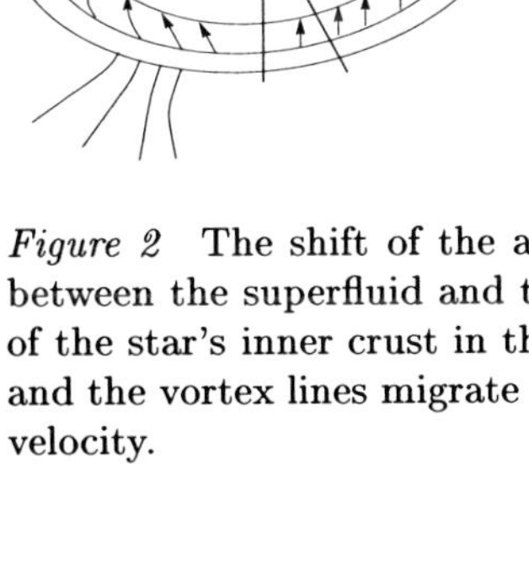

Figure 2 The shift of the angular velocity from Ω to Ω' increases the velocity lag between the superfluid and the solid crust in the upper right and lower left sections of the star's inner crust in this sketch. In these sections the vortex creep accelerates and the vortex lines migrate outward and align with the new direction of the angular velocity.

THERMAL EVOLUTION OF NEUTRON STARS

Sachiko Tsuruta

Abstract We report on the current status of neutron star thermal evolution theories. Recent observational data are compared with theoretical models, and the implications are discussed.

1. INTRODUCTION

The first set of discoveries of Galactic X-ray sources date back to the early 1960's, and neutron star (NS) radiation was proposed as a possible explanation. The first detailed cooling calculations (Tsuruta 1964) showed that a NS can be hot enough to be detectable as an X-ray source for about a million years. However, the X-rays observed in those early years turned out to originate not from the stellar surface, and we had to wait for nearly twenty more years until the launch of the *Einstein* Observatory which gave the first hope for detecting stellar radiation. Nomoto and Tsuruta (1981, 1986) first carried out careful comparison of NS cooling theory with the *Einstein* data, but the temperatures obtained by *Einstein* were only the upper limits. The more recent ROSAT and ASCA observations have proved to be a serious 'turning point' for the detectability of thermal radiation directly from NSs (see, e.g., Tsuruta 1998, hereafter T98). Consequently, it has become possible to compare the observed temperatures, not just upper limits, with NS thermal evolution theories. This prospect has been further enhanced by the recent possible detections of NS thermal radiation in the lower optical to EUV bands, e.g., by HST and EUVE(e.g., Pavlov, Stringfellow, and Cordova 1996, hereafter PSC96).

Recently, Tsuruta (T98) gave a comprehensive review of the current status of NS thermal evolution theories, temperature observations, and the implications of comparison of these observations with theories. The results are suggestive of the need for some fast cooling mechanisms(*1), but the tentative conclusions reached are yet to be definitive - due mainly

K.S. Cheng et al. (eds.), Stellar Astrophysics, 105–115.

to various theoretical uncertainties, such as the effects of the 2D heat flows under strong magnetic fields and atmospheric compositions(*2), as well as lack of numbers of definitive detections and still large error bars associated with the observed data points.

On the other hand, *the prospect for the immediate future appears excellent for this particular area of research. This is mainly due to the availability of numerous data to be expected from the next generation X-ray satellite missions*, e.g., the *Chandra* and XMM already successfully launched in July 1999 and December 1999, respectively. Also, significant progress is expected soon in the theories of NS thermal evolution and atmospheres (T98; Pavlov and Zavlin 2000).

2. OBSERVATIONAL BACKGROUND

By now nearly 30 rotation-powered pulsars have been detected as X-ray sources by ROSAT and ASCA (e.g., Becker's contribution to this volume; Ögelman 1995, Becker and Trümper 1997). Many of these sources appear to be either non-thermal in origin or no spectral information is available. For these sources the spectra had been only assumed and hence only the upper limits to the stellar temperature could be obtained. However, some of these sources revealed a clear thermal component in the softest band of $kT < 0.1$ keV. These pulsars are PSRs 0833-45 (Vela), 0656+14, 0630+18 (Geminga), and 1055-52. With the high spectroscopy of ROSAT's PSPC and long exposure times combined with the data in the lower energy windows (optical to EUV), there is by now some possibility that thermal radiation is detected from somewhere on the stellar surface of at least three more pulsars (PSC96, T98), PSRs 1929+10, PSR 0950+08, and PSR J0437-4715. What is most remarkable is that, due to the better spectral resolution of PSPC, for the first time, some spectral information has become available. The ROSAT results suggest that the observed soft X-ray spectra of the four strongest of these sources, the Vela pulsar, 0656+14, Geminga and 1055-52, consist of at least two components, a definitely thermal softer component and a harder component whose nature was unknown due to ROSAT's soft window. The latest observational evidence, from the combined analyses of both ROSAT and ASCA, now indicate that the harder component is either thermal, non-thermal, or possibly a combination of both. A natural explanation is that depending on the relative strengths of different components, in different sources different components could be resolved. A convincing interpretation is that the softest component comes from the whole stellar surface, while the harder component is from hot spots, somewhere in the magnetosphere, or possibly a combination of both(T98).

Most isolated pulsars have been discovered first in the radio band. Although the optical radiation has been observed in several isolated pulsars the detected radiation was definitely of a non-thermal origin. For some of the middle-aged to older and relatively nearby pulsars, however, there may be a better chance for identifying the low energy tail of the stellar thermal radiation. According to PSC96 it is possible that the thermal-like stellar component was identified by HST in the optical-UV radiation from PSRs 0656+14, 1929+10, and 0950+08.

Recently three more interesting isolated NSs, RX J0822-4300 (Zavlin, Trümper, and Pavlov 1999), RX J0720.4-3125 (Haberl et al. 1997) and RX J185636-3754 (Walter, Wolk, and Neuhäuser 1996) have been found. Among these sources RX J0822-4300 in Pap A appears to be especially interesting when we are to compare cooling theories with observation, because both period and spin-down rate are measured. The spin-down age is 8 x 10^3 year, only a little younger than the Vela pulsar, but the nature of radiation appears to be quite different, because it is predominantly thermal-like with blackbody temperatures of $\sim$ (1 - 5) x 10^6 K, substantially higher than the Vela pulsar, while the magnetic field, of 3.4 x 10^{12} Gauss, is as strong as other typical pulsars. This example may indicate the inhomogeneity of the nature of NSs.

Another recent development in the related fields is the discovery of 'anomalous X-ray pulsars' (AXP) and soft γ ray repeaters (SGR). Both are often classified as 'magnetars', NSs with ultrastrong magnetic fields, of the order of 10^{14} to 10^{16} Gauss (e.g., see Thompson 2000). Many of them are associated with supernova remnants. Many magnetars exhibit thermal components. If the observed radiation is blackbody luminosity L is $\sim$ $10^{33.5}$ - $10^{35.5}$ ergs s^{-1} while the radius of the emitting area is $\sim$ 0.5 - 9 km. They are all pulsed in X-rays with the pulsed fraction of 10% to 70%, with the spin down ages of about 10^3 - 10^5 years (Chakrabarty et al. 2000).

Most recently a point source was detected near the center of Cas A (Tananbaum et al. 1999), which was further confirmed by the analysis of ROSAT data on this supernova remnant (Aschenbach et al. 1999). If it is a NS the radiation observed by *Chandra* is most likely coming from a localized region on the stellar surface, of radius $<\sim$ 1 km (Pavlov et al. 2000; Chakrabarty et al. 2000).

3. THERMAL EVOLUTION MODELS

After a supernova explosion a newly formed NS first cools via various neutrino production mechanisms before the surface photon radiation takes over(e.g., see Tsuruta 1979). With standard cooling(*1) the transition from the neutrino to photon cooling takes place at about a million years(e.g., Nomoto and Tsuruta 1987, hereafter NT87), while for various non-standard cooling scenarios(*1) it is much faster - 10 to 1000 years if the constituent core particles are in the normal state(e.g., see Umeda et al. 1994, hereafter U94; Umeda, Tsuruta and Nomoto 1994, hereafter UTN94). However, if these particles are in a superfluid state the fast cooling can be greatly suppressed (see next section). The one-dimensional (1D) fully general relativistic equations which govern the structure and evolution of a spherically symmetric, non-rotating NS are found, e.g., in NT87. Fig. 1 shows typical standard cooling curves obtained by solving these equations with the exact evolutionary method(*3).

4. EFFECTS OF SUPERFLUIDITY AND HEATING

The most important effect of the presence of superfluid particles is the suppression of neutrino emissivity. This effect is relatively small for standard cooling, but can be significant for non-standard cooling. Depending on the superfluid models (whether the superfluid energy gap is large or small), cooling in various intermediate stages between the standard and nonsuperfluid non-standard case is possible. A NS can be heated in various ways (see, e.g., T98). The most widely studied is the internal frictional heating (see., e.g., Cheng et al. 1992). An example of these effects is illustrated in Fig. 2. Although the direct URCA cooling is adopted here different non-standard scenarios qualitatively produce similar cooling curves(U94, UTN94).

5. EFFECTS OF MAGNETIC FIELDS

Until recently it has been considered that the effects of magnetic fields on the overall NS cooling behavior, typically expressed as the time evolution of total geometrically averaged photon luminosity, is negligible (e.g., Hernquist 1984, Tsuruta 1986). Therefore, in most calculations when our major purpose was to obtain the overall averaged thermal evolution behavior, the effects of magnetic fields on conductivities were generally not included (NT87, U94, UTN94). However, this general conclusion is based on the results from 1D calculations. Under quantized magnetic fields the electron conductivity perpendicular to the magnetic field direction is significantly reduced, while parallel to the field it becomes higher than the zero field case. For an isothermal stellar core, therefore, the polar surface temperature will be higher than the equatorial temperature, and hence the heat flow from the hotter to the cooler region may become important. Unless this heat flow in the θ direction (where θ is the angle from the magnetic polar to the equatorial direction) is properly taken into account, we may not realistically calculate thermal evolution under the quantized magnetic field. The problem then becomes two-dimensional (2D), in the r - θ space. Therefore, recently we constructed the first 2D 'exact' evolutionary code (*3), and carried out sample calculations. Our results are summarized in Fig. 3. We see that a magnetic cooling curve obtained with a 2D method (solid curve) deviates seriously from a 1D curve (dashed curve) after $\sim 10^5$ years. The problem with this 2D method, however, is that although in principle it works it is not practical - in the sense that it takes too long computer time. More seriously it breaks down after about $\sim 10^6$ years. Therefore, we are now currently constructing a more efficient 2D code. Very recently we investigated, semi-analytically, the delay of t_{crit}, the turning point for the sharp decline, and find that it will increase by up to a factor of ~ 10 when the heat flow in the tangential direction is properly included (Teter, Rilett and Tsuruta 2000).

We also calculated the light curve models by integrating the flux from the atmosphere over half the hemisphere facing the observer, taking into account the effects of gravitational bending, surface temperature modulation, flux beaming, etc. (see, e.g., T98). To compare with the observed light curves further modulation by the circumstellar plasmas will be significant in some cases (e.g., Halpern and Ruderman 1993). We expect to carry out such modulation calculations in collaboration with the groups by K.S. Cheng and M.A. Ruderman.

The magnetic fields adopted in our calculations so far have been mostly in the order of 10^{12} Gauss. However, ultra-strong fields of $>> 10^{13}$ Gauss is now required for magnetars. Recently, a few authors extended the conductivity calculations to the ultrastrong regime of $10^{14} - 10^{16}$ Gauss (e.g., Heyl and Hernquist 1998a,b, hereafter HH98). HH98 calculated NS standard cooling and find that their results are marginally consistent with AXP temperatures. They adopted the isothermal method (*3), but some of the magnetars are relatively young and the isothermal equilibrium may be yet to be reached, especially for stiff EOS. Also the surface temperature modulation is yet to be calculated in such ultra-strong field regime. Therefore, we are, in collaboration with Hernquist's group, currently calculating cooling curves for magnetars by adopting the exact evolutionary method. The work is expected to extend to the 2D and temperature modulation calculations.

6. COMPARISON WITH OBSERVATION

Various observed temperature points from ROSAT and ASCA are compared with thermal evolution curves in figures 1 - 3(*4). We note the following properties: (1) When we take into account various uncertainties, e.g., in distance, age, and interstellar column density, we may conclude that most data points are still consistent with the standard cooling scenario if the 1D method is adopted. (2) The Vela data may be too low. (3) When the more realistic 2D method is adopted (the solid curve in Fig. 3), some other data points - e.g., for PSRs 1929+10, 0950+08 and Geminga, may also be too low for standard cooling. This conclusion, however, is still tentative because our current 2D code breaks down at about 10^6 years. However, our improved code which calculates solutions beyond this age should clarify this uncertainty.

In Fig. 2 the 1 $M_\odot$ star (the uppermost dash-chained curve) cools with standard cooling, while the lowest long dash-chained curve shows the direct URCA cooling of a 2 $M_\odot$ nonsuperfluid star. It demonstrates that as the mass changes from 1 to 2 $M_\odot$ the cooling curves can lie anywhere from the standard to the nonsuperfluid non-standard cases. It also illustrates that the sources with apparently too low observed temperatures for the standard scenario can be very naturally explained if these pulsars are indeed cooling with a non-standard scenario with suitable superfluid effects when the mass changes in a narrow range of $\sim$ 1.3 to 1.8 $M_\odot$. One point (PSR 1055-52) is above these curves, but note that these curves are obtained with the 1D method. When the more realistic 2D method is adopted t_{crit} will increase by up to a factor of $\sim$ 10 and hence this apparent discrepancy disappears. Since t_{crit} will be delayed the sources near 10^6 - 10^7 years will become especially important. The detection temperatures (not just upper limits) for these sources will determine whether heating will be necessary, and if so how strong it must be. The temperature determination for the middle aged sources will be also very important because if some of these pulsars are as cool as the Vela pulsar we may need non-standard cooling.

The atmospheric composition is very important, for if it is hydrogen or helium most of the data points as shown in our figures will be significantly lowered, and we definitely will need a non-standard thermal evolution scenario. The atmospheric composition will probably be determined in the near future(*2). On the other hand the pulsars with temperatures consistent with the standard cooling are interpreted simply as being a little less massive than those with lower temperatures.

Our studies may also importantly constrain the EOS and hence the stellar radius. The reason is that the thermal evolution curves are seriously affected by the EOS. Although we have chosen typical intermediate EOS in figures 2 and 3 the curves will be higher (lower) for stiffer (softer) EOS (e.g., see Tsuruta 1986). As we vary various parameters such as heating strength, mass and degree of superfluid suppression within acceptable ranges, for EOS of given stiffness, we may find the data points which disagree with particular EOS. For instance, if none of the softest EOS can agree with a certain data points while stiffer EOS agree, soft EOS must be excluded and then the radius should be $>\sim$ 8 km.

Another important consequence of these comparisons is that it can seriously constrain the composition of the stellar interior. The non-standard cooling by itself is generally too fast to be consistent with most of the observed temperature detection points. However, if the core particles are in the superfluid state the fast cooling is significantly suppressed. Fig. 2 demonstrates that a non-standard cooling with suitable superfluid effects agree well with existing observed temperature data. However, recent studies show that it is hard for the particles to be in a superfluid state for some of the non-standard scenarios (e.g., the direct URCA with nucleons and the scenarios involving kaons)(T98). Then, only pion cooling and possibly the direct URCA with hyperons may remain as the viable options.

RX J0822-4300 is only a little younger than the Vela pulsar, but luminosity is higher, consistent with the standard cooling (Zavlin, Trümper and Pavlov 1999). No pulsation is detected from RX J0720.4-3125 but if their age estimate is valid the observed temperature will be consistent with standard cooling (Walter, Wolk and Neuhäuser 1996). Pulsation is found from RX J185636-3754 but not the spin-down rate, and hence we get only the lower limit to the age which is consistent with standard cooling (Wang et al. 1999). However, the temperature will be too high if the star is older.

The current upper limit to the cooling NS component of Cas A point source is marginally consistent with standard cooling(Umeda et al. 2000), but it could be lower. It is important to measure the actual temperature of this component by future observations by, e.g., *Chandra* and XMM, for if it turns out to be $\sim 10^{33}$ ergs s^{-2} or less that will be extremely interesting because then non-standard cooling will definitely become necessary (see Umeda et al. 2000).

7. CONCLUSION

In summary, the apparently low temperatures observed for some of the pulsars by the ROSAT data are naturally consistent with the current cooling theories, if a non-standard cooling mechanism is in operation and if the core particles are in a superfluid state. While different non-standard scenarios produce qualitatively similar cooling curves, from various theoretical grounds it will be possible to distinguish among these alternatives (T98). The most natural conclusion, recently tentatively reached in Tsuruta's review article (T98), is that stars somewhat less massive than 1.4 $M_{\odot}$ should cool with the standard scenario while those somewhat heavier should cool with a non-standard scenario with the superfluid effects. Whether the non-standard processes can operate or not, and whether the matter is in a superfluid state or not, critically depend on the details in physics of superdense matter beyond the nuclear density. The implication is that the discrepancy between the stellar temperatures predicted by the standard cooling theories and the lower observed temperatures would have a tremendous impact, on theories of not only thermal evolution but also on some of fundamental problems involving high density matter. While the current status is still inconclusive, the combined effort, both observational and theoretical, which will heavily involve the second generation space satellite missions, such as *Chandra* and XMM, will most likely offer more definitive conclusions.

Acknowledgements

We thank the organizers of this conference for hospitality and participants of the conference, especially K.S. Cheng and A. Tang, for useful discussions. Our thanks are due also to M. Teter for helping preparation of the figures. This work is supported in part by NASA grant NAG5-2557.

FOOTNOTES

(*1) Cooling due to various more straightforward neutrino mechanisms, such as the modified URCA, neutrino bremsstrahlung and plasmon neutrino processes, is sometimes called 'standard cooling', while extremely fast cooling caused by some other more unconventional mechanisms, such as the direct URCA and processes involving pions, quarks and kaons, are called 'non-standard cooling'. For convenience, we shall adopt this terminology.

(*2) Romani (1987) first constructed a hydrogen atmospheric code with no magnetic fields. By applying it to ROSAT data he found that the observed surface temperature becomes substantially lower than for blackbody. Pavlov et al. (1995) found that this effect is reduced when the magnetic field is included, although still serious. Further studies have shown that the effect of heavy elements, such as Fe, is to further reduce this effect, to the extent that the temperature deduced for Fe atmospheres is essentially the same as for blackbody (Rajopal, Romani and

Miller 1997). However, we probably can still distinguish between these two cases with sufficiently high spectroscopy because the spectrum from Fe atmospheres will contain many lines (e.g., Pavlov and Zavlin 2000).
(*3) In the isothermal method the hydrostatic and thermodynamic portions of the basic stellar structure-evolution equations are decoupled. We conduct the so-called 'envelope calculations' to obtain the surface temperature vs core temperature relation, by carrying out the integrations from the surface to the core boundary where the temperature gradient vanishes (e.g., Tsuruta 1979). In the 'exact' evolutionary method, on the other hand, the radial integrations of the coupled basic stellar structure-evolution equations are carried out simultaneously, from the stellar center to the surface(see, e.g., NT87).
(*4) The data for figures 1 and 3 are from ROSAT and ASCA taken from Becker 1994, except the Vela pulsar data which are taken from Page, Shibanov and Zavlin 1996. The downward arrow next to PSR 1929+10 at $\sim 5 \times 10^6$ year is PSR 0950+08. The date shown in Fig. 2 are somewhat older, taken from Ögelman 1995.

References

Aschenbach, B., et. al. 1999, IAUC No. 7249

Becker, W. 1994, Ph. D. Thesis, Universitat München

Becker, W., & Trümper, J. 1997, A & A, 326, 682

Chakrabarty, D., Pivovaroff, M.J., Hernquist, L.E., Heyl, J.S. & Narayan, R. 2000, ApJ, submitted (astro-ph/0001026)

Cheng, K.S., Chau, W.Y., Zhang, J.L., & Chau, H.F. 1992, ApJ, 396, 135

Haberl, F., et. al. 1997, A & A, 326, 662

Halpern, J.P., and Ruderman, M. 1993, ApJ, 415, 286

Hernquist, L. 1984, ApJ Suppl., 56, 325

Heyl, J., & Hernquist, L. 1998a, MNRAS, 298, L17; 1998b, MNRAS, 300, 599 (HH98)

Nomoto, K., & Tsuruta, S. 1981, ApJ, 250, L19; 1986, ApJ, 305, L197

Nomoto, K., & Tsuruta, S. 1987, ApJ, 312, 711 (NT87)

Ögelman 1995, Lives of the Neutron Stars, eds. M.A. Alpar & J. van Paradijs (Kluwer), 101

Page, D., Shibanov, Yu.A., & Zavlin, V.E., 1996, in Röntgenstrahlung from the Universe, ed. H.-H. Zimmerman, et. al., MPE Reports, 263, 173

Pavlov, G.G., Shibanov, Yu.A., Zavlin, V.E., & Meyer, R.D. 1995, Lives of the Neutron Stars, eds. M.A. Alpar and J. van Paradijs (Kluwer), 71

Pavlov, G.G., Stringfellow, G.S., & Cordova, F.A. 1996, ApJ, 467, 10 (PSC96)
Pavlov, G.G., Zavlin, V.E., Aschenbach, B., & Trümper, J.E. 2000, ApJ, in press (astro-ph/9912024)
Pavlov, G.G., & Zavlin, V.E. 2000, Proc. IAU Symp. 195, eds. P. Martens and S. Tsuruta (PASP), in press
Rajopal, M., Romani, R., & Miller, M.C. 1997, ApJ, 479, 347
Romani, R.W. 1987, ApJ, 313, 718.
Tananbaum, H., et. al. 1999, IAUC No. 7246
Teter, M., Rilett, D., & Tsuruta, S. 2000, Proc. IAU Symp. 195, eds. P. Martens and S. Tsuruta (PASP), in press
Thompson, C, 2000, Proc. IAU Symp. 195, eds. P. Martens and S. Tsuruta (PASP), in press
Tsuruta, S. 1964, Ph.D. Thesis, Columbia University
Tsuruta, S. 1979, Phys. Reports, 56, 237
Tsuruta, S. 1986, Comments Ap., 11, 151
Tsuruta, S. 1998, Physics Reports, 292, pp 1 - 130 (T98)
Umeda, H., Nomoto, K., Tsuruta, S., Muto, T., & Tatsumi, T. 1994, ApJ, 431, 309 (U94).
Umeda, H., Tsuruta, S., & Nomoto, K. 1994, ApJ, 433, 256 (UTN94).
Umeda, H., Nomoto, K., Tsuruta, S., & Mineshige, S. 2000, submitted to ApJ (astro-ph/9910113)
Walter, F.M., Wolk, S.J., & Neuhäuser, R. 1996, Nature, 379, 233
Wang, J.C.L., Link, B., Van Riper, K., Arnaud, K.A., & Miralles, J.A. 1999, A & A, 345, 869
Zavlin,V.E., Trümper, J., & Pavlov, G.G. 1999, ApJ, 525, 959

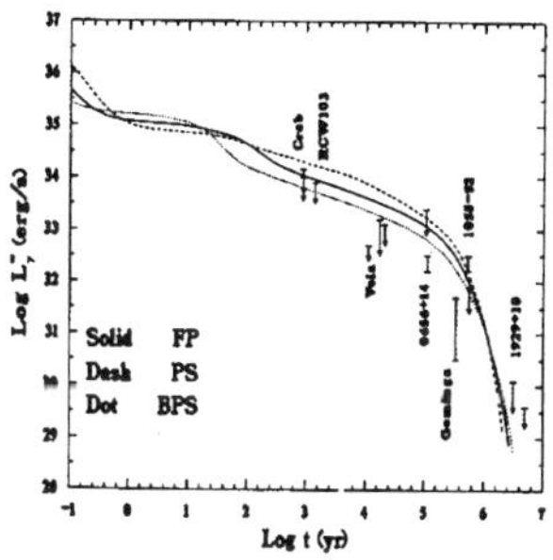

Figure 1 Various observational data(*4) are compared with standard cooling for stars with proper mass $M_A = 1.4M_\odot$, the medium FP (solid), stiff PS(dashed) and soft BPS(dotted) equation of state (EOS), and the T72 superfluid model. Photon luminosity to be observed at infinity is shown as a function of age t in logarithmic scales.

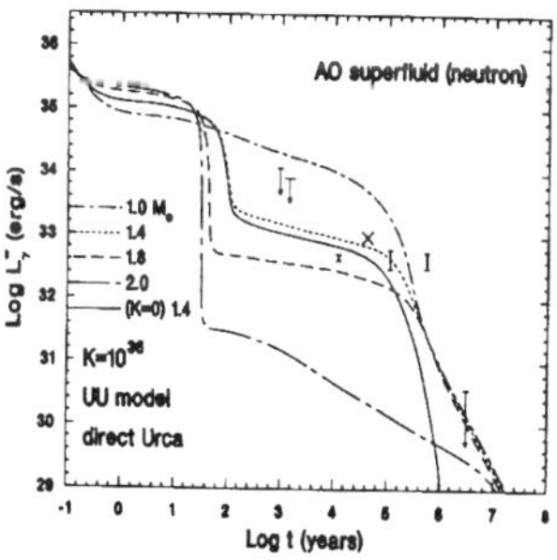

Figure 2 Observational data (*4) are compared with the direct URCA non-standard curves for a superfluid (AO model) NS with the UU (medium) EOS, and different masses as indicated. All curves except the solid curve show moderate heating effect. Other notations are the same as in Fig. 1.

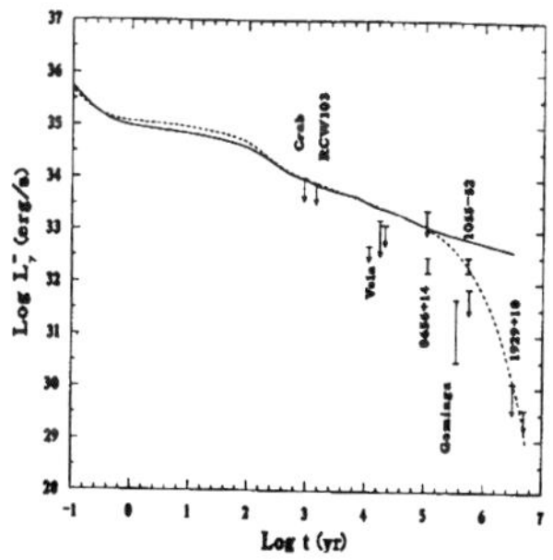

Figure 3 Various Observational data(*4) are compared with cooling curves for a magnetized NS with no superfluid particles, H = 10^{12} Gauss, the medium FP EOS and $M_A = 1.4M_\odot$, obtained with the 1D (dashed) and 2D(solid) methods. Other notations are the same as in Fig. 1.

PROBING THE NEUTRON STAR INTERIOR WITH GLITCHES

Bennett Link[1], Richard I. Epstein[2] and James M. Lattimer[3]
[1]*Montana State University Los Alamos National Laboratory*
[2]*Los Alamos National Laboratory*
[3]*State University of New York, Stony Brook*
[1]blink@dante.physics.montana.edu, [2]epstein@lanl.gov and [3]lattimer@astro.sunysb.edu

Abstract With the aim of constraining the structural properties of neutron stars and the equation of state of dense matter, we study sudden spin-ups, *glitches*, occurring in the Vela pulsar and in six other pulsars. We present evidence that glitches represent a self-regulating instability for which the star prepares over a waiting time. The angular momentum requirements of glitches in Vela indicate that $\geq 1.4\%$ of the star's moment of inertia drives these events. If glitches originate in the liquid of the inner crust, Vela's 'radiation radius' R_∞ must exceed $\simeq 12$ km for a mass of $1.4M_\odot$. The isolated neutron star RX J18563-3754 is a promising candidate for a definitive radius measurement, and offers to further our understanding of dense matter and the origin of glitches.

1. INTRODUCTION

Many isolated pulsars suffer spin jumps, *glitches*, superimposed upon otherwise gradual spin down under magnetic torque. For example, in the glitch of the Crab pulsar shown in Fig. 1, the star spun up by nearly a part in 10^7 over several days (Lyne, Smith & Pritchard 1992), corresponding to a change in rotational energy of the crust of $\sim 10^{42}$ ergs. A particularly active glitching pulsar is the Vela pulsar, which has produced more than a dozen glitches since its discovery over 30 years ago. The fractional changes in rotation rate are typically $\sim 10^{-6}$, occurring every three years on average (Downs& Krause-Polstorff 1988); McCulloch *et al.*1987); Flanagan 1989); Flanagan 1991).

Because glitching pulsars are isolated systems, glitches are thought to arise from internal torques exerted by the rotating liquid interior on the crust, whose spin rate we observe (Anderson & Itoh 1975); Ruderman

K.S. Cheng (ed.), Stellar Astrophysics, 117-126.

1976); Pines & Alpar 1985). As the star's crust is spun down by the magnetic field frozen to it, the interior liquid, which responds to the external torque indirectly through friction with the solid crust, rotates more rapidly. For example, a portion of the liquid could coast between glitches while the solid crust spins down (Fig. ??). Glitches might arise as the consequence of an instability that increases the frictional coupling between the liquid and the solid, causing angular momentum flow to the crust.

The long history of glitches in Vela makes it possible to deduce some of the properties of the interior angular momentum reservoir independent of the details of the instability that triggers these events. Here we discuss the time distribution and average angular momentum transfer rate of Vela's glitches and present evidence that glitches in Vela represent a self-regulating instability for which the star prepares over a waiting interval. We obtain a lower limit on the fraction of the star's liquid interior responsible for glitches and discuss how this result can be used to constrain the dense matter equation of state and the structural properties of neutron stars. We conclude with discussion of the nearby isolated neutron star RX J185635-3754, a promising candidate for a robust radius measurement that offers to constrain our understanding of dense matter and the origin of glitches.

2. REGULARITY OF ANGULAR MOMENTUM TRANSFER

A glitch of magnitude $\Delta\Omega_i$ requires angular momentum

$$\Delta J_i = I_c \Delta\Omega_i, \tag{1}$$

where I_c is the moment of inertia of the solid crust plus any portions of the star tightly coupled to it. Most of the core liquid is expected to couple tightly to the star's solid component, so that I_c makes up at least 90% of the star's total moment of inertia (Alpar, Langer & Sauls 1984); Abney, Epstein & Olinto 1996). [Glitches are driven by the portion of the liquid interior that is differentially rotating with respect to the crust]. The cumulative angular momentum imparted to the crust over time is

$$J(t) = I_c \bar{\Omega} \sum_i \frac{\Delta\Omega_i}{\bar{\Omega}}, \tag{2}$$

where $\bar{\Omega} = 70.4$ rad s^{-1} is the average spin rate of the crust over the period of observations. Fig. 2 shows the cumulative dimensionless angular momentum, $J(t)/I_c\bar{\Omega}$, over ~ 30 years of glitch observations of the Vela

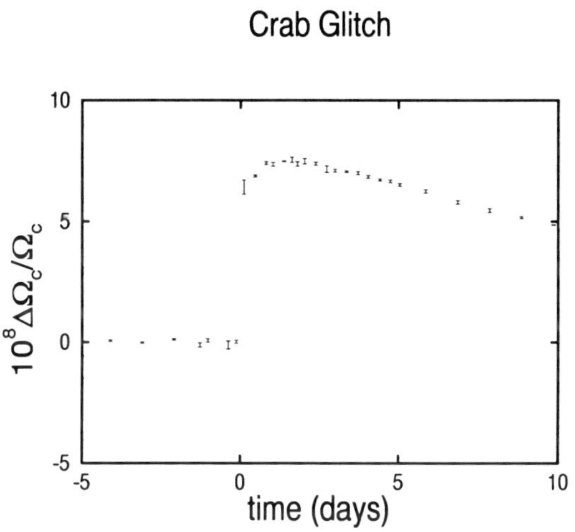

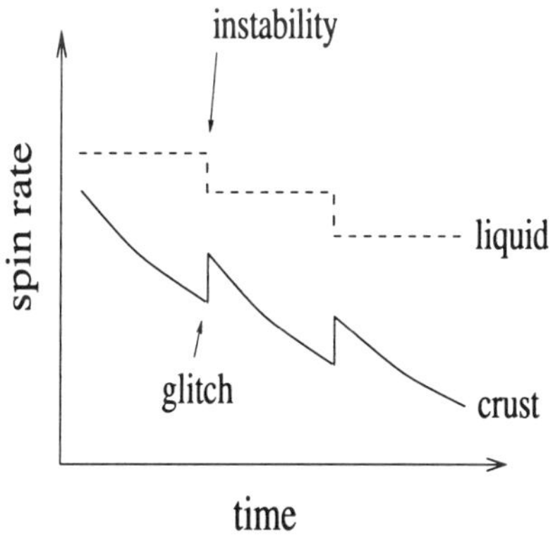

Figure 1 A glitch in the Crab pulsar. The spin rate with respect to a pre-glitch spin-down model is shown. Data are from Ref. 1.
A glitch model. Differential rotation between the interior liquid (dashed line) and the crust (solid line) develops as the crust is spun down by its external torque. At the time of a glitch, the liquid delivers angular momentum to the crust.

pulsar, with a linear least-squares fit. The average rate of angular momentum transfer associated with glitches is $I_c\bar{\Omega}A$, where A is the slope of the straight line in Fig. 2:

$$A = (6.44 \pm 0.19) \times 10^{-7}\ \mathrm{yr}^{-1}. \tag{3}$$

This rate A is often referred to as the *pulsar activity parameter* (see, *e.g.*, (Lyne 1995)).

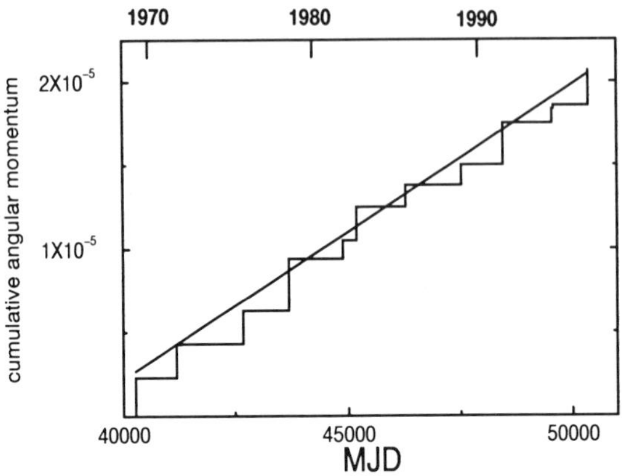

Figure 2 Cumulative dimensionless angular momentum, $J/I_c\bar{\Omega}$, imparted to the Vela pulsar's crust as a function of time. The straight line is a least-squares fit.

The angular momentum flow is remarkably regular; none of Vela's 13 glitches caused the cumulative angular momentum curve to deviate from the linear fit by more than 12%. Moreover, glitches occur at fairly regular time intervals; the standard deviation in observed glitch intervals is $0.53\langle\Delta t\rangle$, where $\langle\Delta t\rangle = 840$ d is the average glitch time interval. The probability of 13 randomly-spaced (Poisson) events having less than the observed standard deviation is only $\sim 1\%$. These data indicate that Vela's glitches are not random, but represent a self-regulating process which gives a relatively constant flow of angular momentum to the crust with glitches occurring at fairly regular time intervals.

3. THE GLITCH RESERVOIR'S MOMENT OF INERTIA

The frequent occurrence of large glitches requires that some fraction of the interior liquid (the reservoir) spins at a higher rate than the crust of the star. The average rate of angular momentum transfer in Vela's glitches (eq. 3) can be used to constrain the relative moment of inertia of the reservoir. Between glitches, the reservoir acquires excess angular momentum as the rest of the star slows under the magnetic braking torque acting on the crust. Excess angular momentum accumulates at the maximum rate if the reservoir *coasts* between glitches, without spinning down (Fig. ??). Hence, the rate at which the reservoir accumulates angular momentum capable of driving glitches is limited by

$$\dot{J}_{\rm res} \leq I_{\rm res}|\dot{\Omega}|, \tag{4}$$

where $\dot{\Omega}$ is the average spin-down rate of the crust, and $I_{\rm res}$ is the moment of inertia of the angular momentum reservoir (not necessarily one region of the star). Equating $\dot{J}_{\rm res}$ to the average rate of angular momentum transfer to the crust, $I_c\bar{\Omega}A$, gives the constraint,

$$\frac{I_{\rm res}}{I_c} \geq \frac{\bar{\Omega}}{|\dot{\Omega}|}A \equiv G, \tag{5}$$

where the *coupling parameter* G is the minimum fraction of the star's moment of inertia that stores angular momentum and imparts it to the crust in glitches. Using the observed value of Vela's activity parameter A and $\bar{\Omega}/|\dot{\Omega}| = 22.6$ Kyr, we obtain the constraint

$$\frac{I_{\rm res}}{I_c} \geq G_{\rm Vela} = 1.4\%. \tag{6}$$

A similar analysis for six other pulsars yields the results shown in Fig. 3. After Vela, the most significant limit is obtained from PSR 1737-30 which gives $I_{\rm res}/I_c \geq G_{1737} = 1\%$.

The similarity of G for the five objects of intermediate age suggests a common physical origin for glitches in these stars. The Crab pulsar and PSR 0525+21, however, appear to be unusual. It may be that the Crab's interior cannot accumulate significant excess angular momentum between glitches, perhaps as a consequence of rapid thermal creep of superfluid vortices (see, e.g., (Alpar, Anderson & Pines 1984); Link, Epstein & Baym 1993); Chau & Cheng 1993)). The value of G for PSR 0525+21 is not well determined, since only two glitches from this object have been measured.

4. IMPLICATIONS FOR THE DENSE MATTER EQUATION OF STATE

The constraint of $I_{\rm res}/I_c \geq 1.4\%$ for Vela applies regardless of where in the star glitches originate. A natural candidate for the angular momentum reservoir is the superfluid that coexists with the inner crust lattice (Anderson & Itoh 1975); Ruderman 1976); Pines & Alpar 1985), where superfluid vortex lines could pin to the nuclei and sustain a velocity difference between the superfluid and the crust. Within this interpretation, the constraint on $I_{\rm res}/I_c$ can be used to constrain the properties of matter at supranuclear density as we now demonstrate.

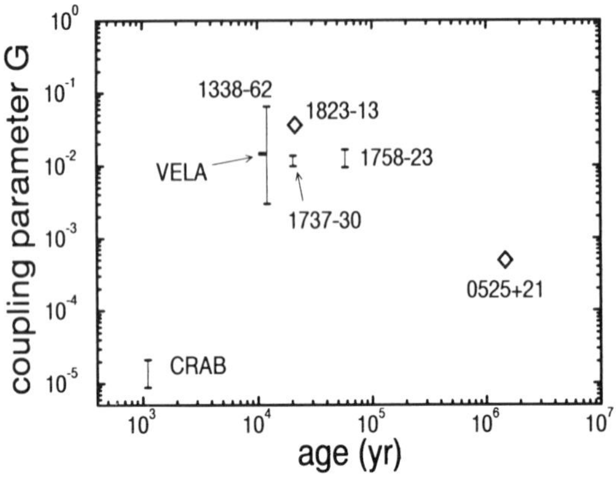

Figure 3 The coupling parameter G. The strongest constraints are obtained for Vela and PSR 1737-30, for which 13 and 9 glitches have been observed, respectively. Diamonds indicate objects with only two observed glitches, for which error bars could not be obtained. References: 0525+21 (Downs 1982), Crab (Boynton et al.1972); Lohsen 1975); Lyne & Pritchard 1987); Lyne, Pritchard & Smith 1992), Vela (Downs& Krause-Polstorff 1988); McCulloch et al.1987); Flanagan 1989); Flanagan 1991), 1338-62 (Daspi, Manchester, Johnston, Lyne & D'Amico 1992), 1737-30 (McKenna & Lyne 1990; Shemar & Lyne 1996), 1823-13 (Shemar & Lyne 1996).

The fraction of the star's moment of inertia contained in the solid crust (and the neutron liquid that coexists with it) is given approximately by (Link, Epstein & Lattimer 1999):

$$\frac{\Delta I}{I} \simeq \frac{28\pi}{3}\frac{P_t R^4}{GM^2}\left[1 + \frac{8P_t}{n_t m_n c^2}\frac{4.5 + (\Lambda - 1)^{-1}}{\Lambda - 1}\right]^{-1} . \qquad (7)$$

Here n_t is the density at the core-crust boundary, P_t is the pressure there, M and R are the stellar mass and radius, $\Lambda \equiv (1 - 2GM/Rc^2)^{-1}$ is the gravitational redshift and m_n is the neutron mass. $\Delta I/I$ is a function of M and R with an additional dependence upon the equation of state (EOS) arising through the values of P_t and n_t. P_t is the main EOS parameter as n_t enters chiefly via a correction term. In general, P_t varies over the range $0.25 < P_t < 0.65$ MeV fm^{-3} for realistic equations of state (Lattimer & Prakash). Larger values of P_t give larger values for $\Delta I/I$.

Combining eq. [7] with a lower limit on ΔI and an upper limit on P_t gives a lower limit on the neutron star radius for a given mass. In order

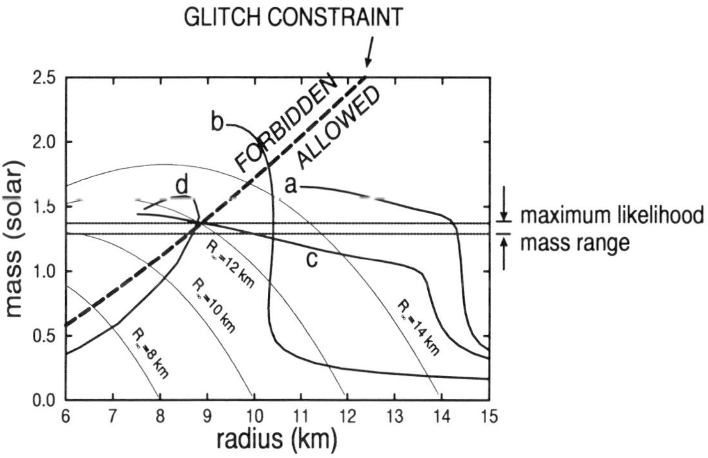

Figure 4 Limits on the Vela pulsar's radius. The heavy dashed curve delimits allowed masses and radii that are compatible with the glitch constraint $\Delta I/(I-\Delta I) \geq 1.4\%$ for $P_t = 0.65$ MeV fm^{-3}. The horizontal dashed lines indicate the mass limits for the survey of 26 radio pulsars of Ref. 16. For comparison, four mass-radius relations are shown for the equations of state of: Glendenning & Schaffner-Bielich (kaons, curve a) (Glendenning & Schaffiner-Bielich 1999), Wiringa, Fiks & Fabrocini (curve b) (Wiringa, Fiks & Fabrocini 1988), Glendenning & Schaffner-Bielich (kaons, curve c) (Glendenning & Schaffiner-Bielich 1999) and a representative strange quark star (curve d). Thin curves are contours of constant radiation radius R_∞.

to relate our observational bound on $I_{\rm res}/I_c$ to ΔI, we assume that the angular momentum reservoir is confined to the neutron superfluid that coexists with the nuclei of the inner crust. In this case, $I_{\rm res} \leq \Delta I$ and $I_c \geq I - \Delta I$. Our observational limit on $I_{\rm res}$ then gives $\Delta I/(I-\Delta I) \geq \Delta I/I_c \geq I_{\rm res}/I_c \geq 0.014$. To obtain a conservative lower limit on the neutron star radius, we take $P_t = 0.65$ MeV fm^{-3} and $n_t = 0.075$ fm^{-3}. Eq. [7], gives the heavy dashed curve in Fig. 4. This curve is given approximately by

$$R = 3.6 + 3.9M/M_\odot \,. \tag{8}$$

Stellar models that are compatible with the lower bound on $I_{\rm res}$ must fall below this line. For models above this line, the inner crust liquid constitutes less than 1.4% of the star's moment of inertia, incompatible with the observed angular momentum requirments of the Vela pulsar.

5. DISCUSSION

Vela's mass and radius are unknown, so we cannot say where it falls in Fig. 4. However, mass measurements of radio pulsars in binary systems and of neutron star companions of radio pulsars give neutron star masses that are consistent with a remarkably narrow distribution, $M = 1.35 \pm 0.04 M_\odot$ (Thorsett & Chakrabarty), indicated by the pair of horizontal dotted lines in Fig. 4. If Vela's mass is not unusual, eq. [8] constrains $R \gtrsim 8.9$ km, under the assumption that glitches arise in the inner crust superfluid. However, the quantity constrained by observations of the stellar luminosity and spectrum is not R but the larger 'radiation radius' $R_\infty \equiv \Lambda^{1/2} R = (1 - 2GM/Rc^2)^{-1/2} R$. If $M = 1.35 M_\odot$ for Vela, the above constraint gives $R_\infty \gtrsim 12$ km if glitches arise in the inner crust. For comparison, we show in Fig. 4 the mass-radius curves for several representative equations of state (heavy solid lines). Measurement of $R_\infty \lesssim 13$ km would be inconsistent with most equations of state if $M \simeq 1.35 M_\odot$. Stronger constraints could be obtained if improved calculations of nuclear matter properties give P_t significantly less than 0.65 MeV fm^{-3}.[1] For example, for $M \simeq 1.35 M_\odot$, $R_\infty \gtrsim 13$ km would be required if $P_t = 0.25$ MeV fm^{-3}. A measurement of $R_\infty \lesssim 11$ km would rule out most equations of state regardless of mass or the angular momentum requirements of glitches, and could indicate that neutron stars are not made of neutrons at all, but of strange quark matter. Explaining glitches in this case would be problematic, as strange stars have very thin crusts (Alcock, Farhi & Olinto 1986).

A black body fit to the unpulsed component of Vela's thermal emission gives $R_\infty = 3 - 4$ km (Ögelman, Finley & Zimmermann 1993). This result is difficult to interpret without knowledge of the star's atmospheric composition and magnetic field strength; atmospheric effects could increase this estimate by a factor of up ~ 6, but could also decrease it by a factor ~ 2 (Rajagopal, Romani & Miller 1997). Nevertheless, it would be interesting to check the extent to which our constraints on Vela are obeyed by other neutron stars. A promising candidate for a decisive measurement of a neutron star's radiation radius is RX J185635-3754, an isolated, non-pulsing neutron star (Walter, Wolk & Neuhauser 1996). A black body fit to the X-ray spectrum gives $R_\infty = 7.3(D/120 \text{ pc})$ km where D is the distance (known to be less than 120 pc). Taken at face value, this result would not only be inconsistent with the radius require-

---

[1]If P_t is smaller than 0.65 MeV fm^{-3}, the crust's moment of inertia would be smaller and the radius constraint more restrictive. For example, $P_t = 0.25$ MeV fm^{-3} moves the constraining contour to approximately $R = 4.7 + 4.1 M/M_\odot$.

ments of glitches in Vela - it would rule out all equations of state that do not involve strange matter. [It is possible that glitching pulsars are normal neutron stars, while RX J185635-3754 is 'strange']. However, either a non-uniform surface temperature or radiative transfer effects in the stellar atmosphere could raise this estimate significantly (An, Lattimer, Prakash & Walter). Recent HST observations of this source by F. Walter should give the proper motion and parallax, and hence, the distance. Future CHANDRA observations should yield more detailed spectral data and could establish the composition of the atmosphere if absorption lines are identified. If lines are present, atmospheric fits will give R_∞, R and M, thus restricting RX J185635-3754 to a region of Fig. 4. These data will undoubtedly further our understanding of matter at supranuclear density, and could establish whether neutron stars have properties consistent with an inner crust explanation of glitches.

Acknowledgements

We thank A. G. Lyne for providing us with glitch data for the Crab pulsar and P. M. McCulloch for Vela data. This work was performed under the auspices of the U.S. Department of Energy, and was supported in part by NASA EPSCoR Grant #291748, NASA ATP Grants # NAG 53688 and # NAG 52863, by the USDOE grant DOE/DE-FG02-87ER-40317, and by IGPP at LANL.

References

A. G. Lyne, F. G. Smith & R. S. Pritchard, Nature, **359**, 706 (1992).

J. M. Cordes, G. S. Downs & J. Krause-Polstorff, ApJ, **330**, 847 (1988); P. M. McCulloch, *et al.* Aust. J. Phys., **40**, 725 (1987); C. Flanagan, IAU Circ. No. 4695 (1989); C. Flanagan, IAU Circ. No. 5311 (1991).

P. W. Anderson & N. Itoh, Nature, **256**, 25 (1975); M. Ruderman, **203**, 213 (1976); D. Pines & M. A. Alpar, Nature, **316**, 27 (1985).

M. A. Alpar, S. A. Langer & J. A. Sauls, ApJ, **282**, 533 (1984); M. Abney, R. I. Epstein & A. Olinto, ApJ, **466**, L91 (1996).

A. G. Lyne, *Lives of the Neutron Stars*, p. 167. Ed: M. A. Alpar (Kluwer, 1995).

M. A. Alpar, P. W. Anderson & D. Pines, ApJ, **276**, 325 (1984); B. Link, R. I. Epstein, G. Baym, ApJ, **403**, 285 (1993); H. F. Chau & K. S. Cheng, K. S., Phys. Rev. B, **47**, 2707 (1993).

G. S. Downs, ApJ, **257**, L67 (1982).

P. E. Boynton *et al.* ApJ, **175**, 217 (1972); E. Lohsen, Nature, **258**, 688 (1975); A. G. Lyne & R. S. Pritchard, MNRAS, **229**, 223 (1987); A. G. Lyne, R. S. Pritchard & F. G. Smith, Nature, **359**, 706 (1992).

V. M. Kaspi, R. N. Manchester, S. Johnston, A. G. Lyne & N. D'Amico, ApJ, **399**, L155 (1992).
A. G. McKenna & A. G. Lyne, Nature, **343**, 349 (1990).
S. L. Shemar & A. G. Lyne, MNRAS, **282**, 677 (1996).
B. Link, R. I. Epstein & J. M. Lattimer, Phys. Rev. Lett., **83**, 3345 (1999).
J. M. Lattimer & M. Prakash, *in preparation.*
N. K. Glendenning & J. Schaffner-Bielich, Phys. Rev. C, **60**, 25803 (1999).
R. B. Wiringa, V. Fiks & A. Fabrocini, Phys. Rev., C, **38**, 1010 (1988).
S. E. Thorsett & D. Chakrabarty, ApJ, **512**, 288.
C. Alcock, E. Farhi & A. Olinto, ApJ, **310**, 261 (1986).
H. Ögelman, J. P. Finley & H. U. Zimmermann, Nature, **361**, 136 (1993).
M. Rajagopal, R. W. Romani & M. C. Miller, ApJ, **479**, 347 (1997).
F. Walter, S. Wolk & R. Neuhauser, Nature, **379**, 233 (1996).
P. An, J. M. Lattimer, M. Prakash, & Walter, F. *in preparation.*

PHYSICS OF NEUTRON STAR KICKS

Dong Lai
Center for Radiophysics and Space Research, Cornell University, Ithaca, NY 14853, USA
dong@spacenet.tn.cornell.edu

Abstract It is no longer necessary to "sell" the idea of pulsar kicks, the notion that neutron stars receive a large velocity (a few hundred to a thousand km s^{-1}) at birth. However, the origin of the kicks remains mysterious. We review the physics of different kick mechanisms, including hydrodynamically driven, neutrino and magnetically driven kicks.

1. INTRODUCTION

It has long been recognized that neutron stars have space velocities much greater than their progenitors'. Recent studies of pulsar proper motion give $200 - 500$ km s^{-1} as the mean 3D velocity of NSs at birth (e.g., Lyne and Lorimer 1994; Lorimer et al. 1997; Hansen & Phinney 1997; Cordes & Chernoff 1998), with possibly a significant population having velocities greater than 700 km s^{-1}. Direct evidence for pulsar velocities $\gtrsim$1000 km s^{-1} comes from observations of the bow shock produced by the Guitar Nebula pulsar (B2224+65) in the interstellar medium (Cordes et al. 1993) and studies of pulsar-supernova remnant associations (e.g., Frail et al. 1994; Kaspi 1999). A natural explanation for such high velocities is that supernova explosions are asymmetric, and provide kicks to nascent neutron stars. Support for supernova kicks has come from the detection of geodetic precession in binary pulsar PSR 1913+16 (Cordes et al. 1990; Kramer 1998; Wex et al. 1999) and the orbital plane precession in PSR J0045-7319/B-star binary (Kaspi et al. 1996) and its fast orbital decay (which indicates retrograde rotation of the B star with respect to the orbit; Lai 1996); These results demonstrate that binary break-up (as originally suggested by Gott et al. 1970; see Iben & Tutukov 1996) can not be solely responsible for the observed pulsar velocities, and that *natal kicks* are required. Evolutionary studies of neutron star binary population also imply the existence of pulsar

K.S. Cheng et al. (eds.), Stellar Astrophysics, 127–135.

kicks (e.g., Fryer et al. 1998). Finally, there are many direct observations of nearby supernovae (e.g., Wang et al. 1999) and supernova remnants which support the notion that supernova explosions are not spherically symmetric.

Despite decades of theoretical investigations, our understanding of the physical mechanisms of core-collapse supernovae remains significantly incomplete. While there is a consensus that neutrino heating of the stalled shock ($0.1 - 1$ s after bounce) plays an important role in driving the explosion, it is unclear whether the heating is sufficient to produce the observed supernova energetics; It is also unclear whether any convective motion or hydrodynamical instability is central to the explosion mechanism (e.g., Herant et al. 1994; Burrows et al. 1995; Janka & Müller 1996; Mezzacappa et al. 1998). The prevalence of neutron star kicks poses a significant mystery, and indicates that large-scale, global deviation from spherical symmetry is am important ingredient in any successful theory of core-collapse supernovae.

In this paper, we concentrate on different classes of physical mechanisms for generating neutron star kicks (§§2-4), and then briefly discuss the astrophysical/observational implications (§5).

2. HYDRODYNAMICALLY DRIVEN KICKS

The collapsed stellar core and its surrounding mantle are susceptible to a variety of hydrodynamical (convective) instabilities (e.g., Herant et al. 1994; Burrows et al. 1995; Janka & Müller 1996; Keil et al. 1996). It is natural to expect that the asymmetries in the density, temperature and velocity distributions associated with the instabilities can lead to asymmetric matter ejection and/or asymmetric neutrino emission. Numerical simulations, however, indicate that the local, post-collapse instabilities are not adequate to account for kick velocities $\gtrsim$100 km s^{-1} (Janka & Müller 1994; Burrows & Hayes 1996; Janka 1998) — These simulations were done in 2D, and it is expected that the flow will be smoother on large scale in 3D simulations, and the resulting kick velocity will be even smaller.

Global asymmetric perturbations in presupernova cores are required to produce the observed kicks hydrodynamically (Goldreich et al. 1996). Numerical simulations by Burrows & Hayes (1996) demonstrate that if the precollapse core is mildly asymmetric, the newly formed neutron star can receive a kick velocity comparable to the observed values. (In one simulation, the density of the collapsing core exterior to $0.9M_{\odot}$ and within 20° of the pole is artificially reduced by 20%, and the resulting kick is about 500 km s^{-1}.) Asymmetric motion of the exploding mate-

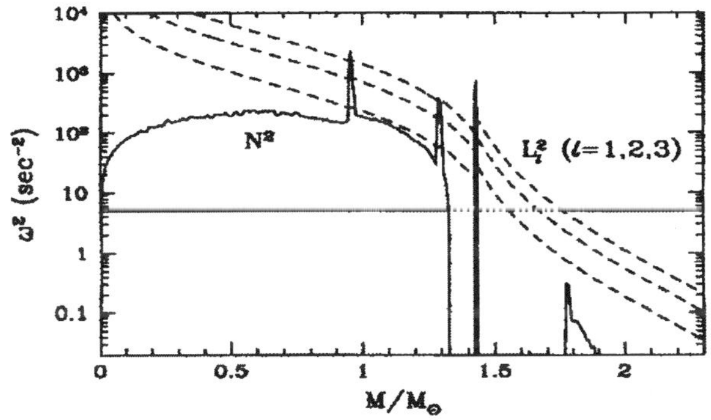

Figure 1 Propagation diagram computed for a $15M_\odot$ presupernova model of Weaver and Woosley (1993). The solid curve shows N^2, where N is the Brunt-Väisälä frequency; the dashed curves show L_l^2, where L_l is the acoustic cutoff frequency, with $l = 1,\ 2,\ 3$. The spikes in N^2 result from discontinuities in entropy and composition. The iron core boundary is located at $1.3M_\odot$, the mass-cut at $1.42M_\odot$. Convective regions correspond to $N = 0$. Gravity modes (with mode frequency ω) propagate in regions where $\omega < N$ and $\omega < L_l$, while pressure modes propagate in regions where $\omega > N$ and $\omega > L_l$. Note that a g-mode trapped in the core can lose energy by penetrating the evanescent zones and turning into an outgoing acoustic wave (see the horizontal line). Also note that g-modes with higher n (the radial order) and l (the angular degree) are better trapped in the core than those with lower n and l.

rial (since the shock tends to propagate more "easily" through the low-density region) dominates the kick, although there is also contribution (about 10 – 20%) from asymmetric neutrino emission. The magnitude of kick velocity is proportional to the degree of initial asymmetry in the imploding core. Thus the important question is: What is the origin of the initial asymmetry?

(i) Presupernova Perturbations

Goldreich et al. (1996) suggested that overstable g-mode oscillations in the presupernova core may provide a natural seed for the initial asymmetry. These overstable g-modes arise as follows. A few hours prior to core collapse, a massive star ($M \gtrsim 8M_\odot$) has gone through a successive stages of nuclear burning, and attained a configuration with a degenerate iron core overlaid by an "onion skin" mantle of lighter elements. The rapidly growing iron core is encased in and fed by shells of burning silicon and oxygen, and the entire assemblage is surrounded by a thick convection zone. The nearly isothermal core is stably stratified and supports internal gravity waves. These waves cannot propagate in the unstably stratified convection zone, hence they are trapped and give rise to core g-modes in which the core oscillates with respect to the outer parts of the star. The overstability of the g-mode is due to the "ε-mechanism" with driving provided by temperature sensitive nuclear burning in Si and O shells surrounding the core before it implodes. It is simplest to see this by considering a $l = 1$ mode: If we perturb the core to the right, the right-hand-side of the shell will be compressed, resulting in an increase in temperature; Since the shell nuclear burning rate depends sensitively on temperature (power-law index ~ 47 for Si burning and ~ 33 for O burning), the nuclear burning is greatly enhanced; This generates a large local pressure, pushing the core back to the left. The result is an oscillating g-mode with increasing amplitude.

The main damping mechanism comes from the leakage of mode energy. The local (WKB) dispersion relation for nonradial waves is

$$k_r^2 = (\omega^2 c_s^2)^{-1}(\omega^2 - L_l^2)(\omega^2 - N^2), \tag{1}$$

where k_r is the radial wavenumber, $L_l = \sqrt{l(l+1)}c_s/r$ (c_s is the sound speed) and N are the acoustic cut-off (Lamb) frequency and the Brunt-Väisälä frequency, respectively. Since acoustic waves whose frequencies lie above the acoustic cutoff can propagate through convective regions, each core g-mode will couple to an outgoing acoustic wave, which drains energy from the core g-modes (see Fig. 1). This leakage of mode energy can be handled with an outgoing propagation boundary condition in the mode calculation. Also, neutrino cooling tends to damp the mode. Since the nuclear energy generation rate depends more sensitively on temperature than pair neutrino emission (power law index ~ 9), cooling is never comparable to nuclear heating locally. Instead, thermal balance is mediated by the convective transport of energy from the shells, where the rate of nuclear energy generation exceeds that of neutrino energy emission, to the cooler surroundings where the bulk of the neutrino emission takes place. Calculations (based on the $15M_\odot$ and $25M_\odot$ presupernova models of Weaver & Woosley 1993) indicate that a large number of g-modes are overstable, although for low-order modes (small l and n) the results depend sensitively on the detailed structure and burning rates of the presupernova models (Lai & Goldreich 2000b, in preparation).

Our tentative conclusion is that overstable g-modes can potentially grow to large amplitudes prior to core implosion, although a complete understanding of the global pre-collapse asymmetries is probably out of reach at present, given the various uncertainties in the presupernova models (see Bazan & Arnett 1998 for complications due to convective shell burning in presupernova stars).

(ii) Amplification of Perturbation During Core Collapse

Core collapse proceeds in a self-similar fashion, with the inner core shrinking subsonically and the outer core falling supersonically at about half free-fall speed (Goldreich & Weber 1980; Yahil 1983). The inner core is stable to non-radial perturbations because of the significant role played by pressure in its subsonic collapse. Pressure is less important in the outer region, making it more susceptible to large scale instability. A recent stability analysis of Yahil's self-similar collapse solution (which is based on Newtonian theory and a polytropic equation of state $P \propto \rho^{\Gamma}$, with $\Gamma \sim 1.3$) does not reveal any unstable global mode before the proto-neutron star forms (Lai 2000). However, during the subsequent accretion of the outer core (involving 15% of the core mass) and envelope onto the proto-neutron star, nonspherical perturbations can grow according to $\delta\rho/\rho \propto r^{-1/2}$ or even $\delta\rho/\rho \propto r^{-1}$ (Lai & Goldreich 2000). The asymmetric density perturbations seeded in the presupernova star, especially those in the outer region of the iron core, are therefore amplified (by a factor of 5-10) during collapse. The enhanced asymmetric density perturbation may lead to asymmetric shock propagation and breakout, which then give rise to asymmetry in the explosion and a kick velocity to the neutron star (Goldreich et al. 1996; Burrows & Hayes 1996).

3. NEUTRINO DRIVEN KICKS

The second class of kick mechanisms rely on asymmetric neutrino emission induced by strong magnetic fields. The fractional asymmetry α in the radiated neutrino energy required to generate a kick velocity $v_{\rm kick}$ is $\alpha = Mv_{\rm kick}c/E_{\rm tot}$ ($= 0.028$ for $v_{\rm kick} = 1000$ km s^{-1}, neutron star mass $M = 1.4\,M_\odot$ and total neutrino energy radiated $E_{\rm tot} = 3 \times 10^{53}$ erg).

(i) Effect of Parity Violation

Because weak interaction is parity violating, the neutrino opacities and emissivities in a magnetized nuclear medium depend asymmetrically on the directions of neutrino momenta with respect to the magnetic field, and this can give rise to asymmetric neutrino emission from the proto-neutron star. Chugai (1984) (who gave an incorrect expression for the electron polarization in the relativistic, degenerate regime) and Vilenkin (1995) considered neutrino-electron scattering, but this is less important than neutrino-nucleon scattering in determining neutrino transport in proto-neutron stars. Dorofeev et al. (1985) considered neutrino emission by Urca processes, but failed to recognize that in the bulk interior of the star the asymmetry in neutrino emission is cancelled by that associated with neutrino absorption (Lai & Qian 1998a).

Horowitz & Li (1998) suggested that large asymmetries in the neutrino flux could result from the cumulative effect of multiple scatterings of neutrinos by slightly polarized nucleons (see also Janka 1998; Lai & Qian 1998a). However, it can be shown that, although the scattering cross-section is asymmetric with respect to the magnetic field for individual neutrinos, detailed balance requires that there be no cumulative effect associated with multiple scatterings in the bulk interior of the star where thermal equilibrium is maintained to a good approximation (Arras & Lai 1999a; see also Kusenko et al. 1998). For a given neutrino species, there is a drift flux of neutrinos along the magnetic field in addition to the usual diffusive flux. This drift flux depends on the deviation of the neutrino distribution function from thermal equilibrium. Thus asymmetric neutrino flux can be generated in the outer region of the proto-neutron star (i.e., above the neutrino-matter decoupling layer, but below the neutrinosphere) where the neutrino distribution deviates significantly from thermal equilibrium. While the drift flux associated with ν_μ's and ν_τ's is exactly canceled by that associated with $\bar{\nu}_\mu$'s and $\bar{\nu}_\tau$'s, there is a net drift flux due to ν_e's and $\bar{\nu}_e$'s. Arras & Lai (1999b) found that the asymmetry parameter for the ν_e-$\bar{\nu}_e$ flux is dominated for low energy neutrinos ($\lesssim$15 MeV) by the effect of ground (Landau) state electrons in the absorption opacity, $\epsilon_{\rm abs} \simeq 0.6 B_{15}(E_\nu/1\ {\rm MeV})^{-2}$, where $B_{15} = B/(10^{15}\ {\rm G})$, and for high energy neutrinos by nucleon polarization ($\sim \mu_m B/T$). Averaging over all neutrino species, the total asymmetry in neutrino flux is of order $\alpha \sim 0.2\epsilon_{\rm abs}$, and the resulting kick velocity $v_{\rm kick} \sim 50\, B_{15}$ km s^{-1}. There is probably a factor of 3 uncertainty in this estimate. To firm up this estimate requires solving the neutrino transport equations in the presence of parity violation for realistic proto-neutron stars.

(ii) Effect of Asymmetric Field Topology

A different kick mechanism relies on the asymmetric magnetic field distribution in proto-neutron stars (see Bisnovatyi-Kogan 1993; However, he considered neutron decay, which is not directly relevant for neutrino emission from proto-neutron stars). Since the cross section for ν_e ($\bar{\nu}_e$) absorption on neutrons (protons) depends on the local magnetic field strength due to the quantization of energy levels for the e^- (e^+) produced in the final state, the local neutrino fluxes emerged from different regions of the stellar surface are different. Calculations indicate that to generate a kick velocity of $\sim$ 300 km s^{-1} using this mechanism alone would require that the difference in the field strengths at the two opposite poles of the star be at least 10^{16} G (Lai & Qian 1998b). Note that unlike the kick due to parity violation [see (i)], this mechanism does not require the magnetic field to be ordered, i.e., only the magnitude of the field matters.

(iii) Exotic Neutrino Physics

There have also been several interesting ideas on pulsar kicks which rely on nonstandard neutrino physics. It was suggested (Kusenko &

Segre 1996) that asymmetric ν_τ emission could result from the Mikheyev-Smirnov-Wolfenstein flavor transformation between ν_τ and ν_e inside a magnetized proto-neutron star because a magnetic field changes the resonance condition for the flavor transformation. This mechanism requires neutrino mass of order 100 eV. Another similar idea (Akhmedov et al. 1997) relies on both the neutrino mass and the neutrino magnetic moment to facilitate the flavor transformation. More detailed analysis of neutrino transport (Janka & Raffelt 1998), however, indicates that even with favorable neutrino parameters (such as mass and magnetic moment) for neutrino oscillation, the induced pulsar kick is much smaller than previously estimated (i.e., $B \gg 10^{15}$ G is required to obtain 100 km s^{-1} kick).

4. EM RADIATION DRIVEN KICKS

For completeness, we mention the post-explosion "rocket" effect due to electromagnetic (EM) radiation from off-centered magnetic dipole in the pulsar (Harrison & Tademaru 1975). In this mechanism, the neutron star velocity comes at the expense of its spin kinetic energy, which is radiated away asymmetrically via EM braking; The kick is attained on the timescale of the initial spindown time of the pulsar (i.e., this is not a "natal" kick). The neutron star velocity changes according to $M\dot{v} = \epsilon L/c$, where L is the EM braking power, and ϵ is the asymmetry parameter. Typically, $\epsilon \sim 0.1(\Omega s/c)(\mu_\phi/\mu_z)$, where Ω is the spin and s is the "off-center" displacement of the dipole; In fact, there is theoretical maximum, $\epsilon = 0.16$, achieved for $\mu_R = 0$, $\mu_\phi/\mu_z = 0.63(\Omega s/c)$ (where μ_R, μ_ϕ, μ_z are the three cylindrical components of the dipole). The kick velocity is along the spin axis, and $v_{\rm kick} \simeq 600\,(\bar{\epsilon}/0.1)(\nu_0/1\ {\rm kHz})^2$ km s^{-1} (where ν_0 is the initial spin). Clearly, Even if the neutron star were born with maximum rotation rate *and* ϵ were maintained at near the maximum value, the kick velocity would still be at most a few hundred km s^{-1}. Given that most pulsars were born rotating slowly (see Spruit & Phinney 1998), we conclude that "EM rocket" cannot be the main mechanism for pulsar kicks.

5. DISCUSSION

Statistical studies of pulsar population have revealed no correlation between $v_{\rm kick}$ and magnetic field strength, or correlation between the kick direction and the spin axis (e.g., Lorimer et al. 1995; Cordes & Chernoff 1998; Deshpande et al. 1999). Given the large systematic uncertainties, the statistical results, by themselves, cannot reliably constrain any kick mechanism (see Cordes & Chernoff 1998). For example, the magnetic field strengths required for the neutrino-driven mechanisms are $\gtrsim 10^{15}$ G, much larger than the currently inferred dipolar surface fields of typical radio pulsars; the internal magnetic fields of neutron stars and their evolution remain clouded in mystery; and several different mechanisms may contribute to the observed kick velocities.

It is of interest to note that soft gamma repeaters ("magnetars": neutron stars with observed magnetic fields $\gtrsim 10^{14}$ G; see Thompson & Duncan 1996) have very high velocities, $\gtrsim 1000 - 2000$ km s^{-1} (e.g., Kaspi 1999; Marsden et al. 1999). Such a high velocity may well require superstrong magnetic fields ($\gtrsim 10^{16}$ G) to be present in the proto-neutron stars, although hydrodynamical effects remain a viable kick mechanism if enough presupernova asymmetry can be generated. It has recently been suggested that MHD jets can play an important role in supernovae (Khokhlov et al. 1999), but the origin of the jets is unknown, nor it is clear why the two opposite jets are so different (a necessary condition to produce a kick).

I thank my collaborators Phil Arras, Peter Goldreich and Yong-Zhong Qian for their important contribution and insight. This work is supported by NASA grant NAG 5-8356 and by a fellowship from the Alfred P. Sloan foundation.

References

Akhmedov, E. K., Lanza, A., & Sciama, D. W. 1997, Phys. Rev. D, 56, 6117

Arras, P., & Lai, D. 1999a, ApJ, 519, 745 (astro-ph/9806285)

Arras, P., & Lai, D. 1999b, Phys. Rev. D60, 043001 (astro-ph/9811371)

Bisnovatyi-Kogan, G. S. 1993, Astron. Astrophys. Trans., 3, 287

Burrows, A., Hayes, J., & Fryxell, B.A. 1995, ApJ, 450, 830

Burrows, A., & Hayes, J. 1996, Phys. Rev. Lett., 76, 352

Chugai, N.N. 1984, Sov. Astron. Lett., 10, 87

Cordes, J.M., Romani, R.W., & Lundgren, S.C. 1993, Nature, 362, 133

Cordes, J.M., Wasserman, I., & Blaskiewicz, M. 1990, ApJ, 349, 546

Cordes, J.M., & Chernoff, D.F. 1998, ApJ, 505, 315

Dorofeev, O.F., et al. 1985, Sov. Astron. Lett., 11, 123

Frail, D.A., Goss, W.M., & Whiteoak, J.B.Z. 1994, ApJ, 437, 781

Fryer, C., Burrows, A., & Benz, W. 1998, ApJ, 498, 333

Goldreich, P., Lai, D., & Sahrling, M. 1996, in "Unsolved Problems in Astrophysics", ed. J.N. Bahcall and J.P. Ostriker (Princeton Univ. Press)

Hansen, B.M.S., & Phinney, E.S. 1997, MNRAS, 291, 569

Herant, M., et al. 1994, ApJ, 435, 339

Horowitz, C.J., & Li, G. 1998, Phys. Rev. Lett., 80, 3694
Janka, H.-T., & Müller, E. 1996, A&A, 306, 167
Janka, H.-T. 1998, in "Neutrino Astrophysics", ed. M. Altmann et al.
Janka, H.-T., & Raffelt, G.G. 1998, Phys. Rev. D59, 023005
Kaspi, V.M., et. al. 1996, Nature, 381, 583
Kaspi, V.M. 1999, astro-ph/9912284
Keil, W., Janka, H.-Th., Müller, E. 1996, ApJ, 473, L111
Khokhlov, A.M., et al. 1999, astro-ph/9904419
Kramer, M. 1998, ApJ, 509, 856
Kusenko, A., & Segré, G. 1996, Phys. Rev. Lett., 77, 4872 (also astro-ph/9811144)
Kusenko, A., Segré, G., & Vilenkin, A. astro-ph/9806205
Lai, D. 1996, ApJ, 466, L35
Lai, D. 2000, ApJ, in press.
Lai, D., & Goldreich, P. 2000, ApJ, in press (astro-ph/9906400)
Lai, D., & Qian, Y.-Z. 1998a, ApJ, 495, L103 (erratum: 501, L155)
Lai, D., & Qian, Y.-Z. 1998b, ApJ, 505, 844
Lorimer, D. R., Bailes, M., & Harrison, P. A. 1997, MNRAS, 289, 592
Lyne, A.G., & Lorimer, D.R. 1994, Nature, 369, 127
Mazzacappa, A., et al. 1998, ApJ, 495, 911.
Morse, J.A., Winkler, P.F., & Kirshner, R.P. 1995, AJ, 109, 2104
Spruit, H., & Phinney, E.S. 1998, Nature, 393, 139
Thompson, C., & Duncan, R. C. 1996, ApJ, 473, 322
Vilenkin, A. 1995, ApJ, 451, 700
Wang, L., Howell, D.A., Höflich, P., & Wheeler, J.C. 1999, astro-ph/9912033
Weaver, T.A., & Woosley, S.E. 1993, Phys. Rep., 227, 65
Wex, N., Kalogera, V., & Kramer, M. 1999, astro-ph/9905331.

COLLAPSING STRANGE QUARK MATTER IN VAIDYA GEOMETRY-A POSSIBLE SOURCE OF GAMMA RAY BURSTS

T. Harko[1] and K.S. Cheng[1]
[1] *Department of Physics, The University of Hong Kong, Pokfulam, Hong Kong, P. R. China.*

Abstract We investigate the possibility that gamma-ray bursts could be naked singularities formed as a result of the gravitational collapse of strange stars. During the collapse to a naked singularity strange stars could release an amount of energy as high as $4 \times 10^{54} erg$ visible to an external observer as a powerful burst of radiation . The possible physical mechanisms of the energy emission are also considered.

Gamma-ray bursts (GRBs) are cosmic gamma ray emissions with typical flux of 10^{-5} to $5 \times 10^{-4} ergcm^{-2}$ with the rise time as low as $10^{-4}s$ and the duration of bursts from 10^{-2} to $10^3 s$ [1]. The distribution of the bursts is isotropic and they are believed to have a cosmological origin, recent observations suggesting that GRBs might originate at extra-galactic distances [1]. The large inferred distances imply isotropic energy losses as large as 3×10^{53} for GRB 971214 and $3.4 \times 10^{54} erg$ for GRB 990123 [2]. The origin of the bursts is not yet known, the proposed models involving merger of binary neutron stars [3], [4], capture of neutron stars by black holes [5], differentially rotating neutron stars [6],[7] or neutron star-quark star conversion [8] etc. There are still many open problems concerning GRBs that have been summarized by Tavani [9] in the four crisis contemporary astrophysics have to face: 1) the energy crisis, which need an explanation of the huge amount of energy emitted during bursts 2) the spectral crisis, related to the fact that there is no evidence for beaming or spectral distortions caused by the inverse Compton scattering in large radiation energy density media 3) the afterglow crisis, the longevity of optical transients detectable up to six months after the burst and the erratic behavior of the radio emission and ,finally, 4) the population crisis from data indicating that only hard and long GRBs show a strong deviation from an Euclidian brightness distribution, just the opposite

K.S. Cheng et al. (eds.), Stellar Astrophysics, 137–142.

of what expected from extragalactic models without cosmological evolution. From these questions the most important is the problem of the source of the large energy emission during the bursts.

A promising recent model describing gamma-ray bursts is the one based on neutron star-quark star conversion [8]. It is largely believed today that strange quark matter, consisting of u-, d-, and s quarks is the most energetically favorable state of baryon matter. Witten [10] specified the two ways of formation of the strange matter: the quark-hadron phase transition in the early universe and conversion of neutron stars into strange ones at ultra high densities. Quark bag models in the theories of strong interactions suppose that the breaking of physical vacuum takes place inside hadron. As a result the vacuum energy densities inside and outside a hadron become essentially different and the vacuum pressure on a bag wall B (the bag constant) equilibrate the pressure of quarks thus stabilizing the system (see [11] for a recent review of quark matter properties).

Strange quark matter is expected to form in an astrophysical setting as a result of the conversion of neutron matter to strange matter, the conversion time being less than 1s [8]. Neutron stars with $1.4M_{\odot}$ must accrete matter of the order of $0.4M_{\odot}$ -$0.6M_{\odot}$ in order that their central density reaches the deconfinement density. Once this condition is satisfied strange-matter seeds are formed in the interiors of neutron stars. After this seed is formed the whole neutron star is converted to a strange star. The resulting total thermal energy of the strange star is given by $E_{th}\tilde{\ }5 \times 10^{51} \left(\frac{\rho}{\rho_0}\right)^{\frac{2}{3}} R_6^3 T_{11}^2 erg$ [8], where ρ is the average mass density, R_6 the stellar radius in units of 10^6 and T_{11} the temperature in units of $10^{11}K$. The thermal energy of the strange star can be as large as $E_{th}\tilde{\ }5 \times 10^{52} erg$ and it can be finally radiated through non-thermal processes in shocks [8].

The neutron-matter strange matter conversion can be realized as a result of accretion in binaries with low-mass companions. In some binary systems the amount of matter accreted could be as high as $0.5M_{\odot}$ [12]. As a result some of the millisecond pulsar may be strange stars having masses around $2M_{\odot}$. But there are no physical reasons to believe that once a strange star is formed the accretion mechanisms are stopped. Once the mass of the strange star exceeds $2M_{\odot}$ no physical process is able to produce enough pressure to avoid the collapse of the star.

In order to study the collapse of a mixture of radiation and quark matter we choose the Vaidya geometry permitting the incorporation of the effects of radiation and which offers a more realistic background than static geometries, where all back reaction is ignored.

In ingoing Bondi coordinates (u, r, θ, ϕ) and with advanced Eddington time coordinate $u = ct + r$ (r decreasing towards the future) the line element describing the radial collapse of a coherent stream of radiation and strange matter is

$$ds^2 = -\left[1 - \frac{2Gm(u,r)}{c^2 r}\right] du^2 + 2dudr + r^2\left(d\theta^2 + \sin^2\theta d\phi^2\right) \quad (0.1)$$

with $r \geq 0$ an affine parameter along the null generators of the $u = const.$ null hypersurface. $m(u, r)$ is called the mass function and gives the gravitational mass within a given radius r.

For a collapsing mixture of radiation with energy density $\mu(u, r)$ and strange quark matter obeying the MIT bag equation of state $p = \frac{1}{3}(\rho c^2 - 4B)$ (with ρ the energy density of strange matter and p its thermodynamic pressure) the Einstein gravitational field equations have the following solution [13]:

$$m(u,r) = \frac{c^2}{G}\left[F(u) + C(u)r^{\frac{1}{3}} + \frac{4\pi GB}{3c^2}r^3\right] \quad (0.2)$$

$$\mu(u,r) = \frac{c^2}{4\pi G r^2}\left[\frac{dF(u)}{du} + \frac{dC(u)}{du}r^{\frac{1}{3}}\right] \quad (0.3)$$

$$\rho(u,r) = \frac{c^2}{4\pi G r^2}\left[\frac{1}{3}C(u)r^{-\frac{2}{3}} + \frac{4\pi GB}{3c^2}r^3\right] \quad (0.4)$$

$$p(u,r) = \frac{c^2}{12\pi G r^2}\left[\frac{1}{3}C(u)r^{-\frac{2}{3}} + \frac{4\pi GB}{3c^2}r^3\right] \quad (0.5)$$

The arbitrary integration functions $F(u)$ and $C(u)$ describe the injection and initial distribution of the mass respectively.

By assuming $F(u) = \frac{\alpha u}{2}$, $C(u) = \frac{\beta u^{\frac{2}{3}}}{2}$, with $\alpha > 0$, $\beta > 0$ constants, the equation of the radially outgoing, future-directed null geodesic originating at the singularity $r = 0$ will be uniquely defined and exist at the singular point $r = 0$ if and only if the algebraic equation $\alpha X_0^2 + \beta X_0^{\frac{5}{3}} - X_0 + 2 = 0$ has at least one positive root $X_0 > 0$. For $\beta = 0$ we obtain the condition of formation of a naked singularity in the Vaidya metric, $\alpha < \frac{1}{8}$ [14].A positive root of the geodesic equation leads to the possibility of formation of a naked singularity as a result of the collapse. Naked singularities are gravitational singularities that are not covered by a horizon. Near the singularity the space-time curvature and the gravitational tidal forces grow very strongly. During the collapse naked singularities could emit powerful bursts of radiation visible to a distant external observer situated far away from the sight of the collapse [14].

The constant α is defining the accretion rate of the collapse at $r=0$ and is given (in CGS units) by $\frac{dm}{dt}\mid_{r=0}=\frac{\alpha}{2}\frac{c^3}{G}\approx 2\times\alpha\times 10^{38}gs^{-1}$, β, describing the initial mass distribution of the collapsing star, can be obtained from $\frac{dm}{dr}\mid_{t=0}=\frac{c^3\beta}{6G}+4\pi Br^2$.

As an astrophysical application of the strange matter collapse we consider the possibility that gamma - ray bursts could in fact be energy emission during the collapse of strange quark matter ending with the formation of a naked singularity. An estimation of the energy emitted during the collapse shows that it is of the same order of magnitude as the one measured during $\gamma-$ ray bursts, this mechanism providing a valuable explanation for this phenomenon.

In order to calculate the energy emitted during the strange matter collapse we shall admit that the luminosity of the collapsing object should not exceed the rate of collapsing matter energy [14]. The variation of the mass of the neutral quark fluid during collapse is given (in CGS units) by

$$\Delta\frac{Gm(u,r)}{c^2}=\left[\frac{\alpha}{2}\Delta u+\frac{\beta}{3}\left(\frac{r}{u}\right)^{\frac{1}{3}}\Delta u+\frac{\beta}{6}\left(\frac{u}{r}\right)^{\frac{2}{3}}+4\pi Br^2\Delta r\right] \quad (0.1)$$

The variation of the advanced time coordinate u is given by the approximate expression $\Delta u=c\Delta t+\Delta r=c\Delta t\left(1+\frac{1}{c}\frac{\Delta r}{\Delta t}\right)\approx c\Delta t\left(1+\frac{v_f}{c}\right)\approx c\Delta t$, where we defined $v_f=\frac{\Delta r}{\Delta t}\approx\frac{R}{t_{ff}}=\frac{2\times\alpha\times R\times 10^{38}}{M}$ as being the speed of the collapsing fluid as measured by a local observer. Then we obtain $\Delta m=\left[\frac{\alpha}{2}+\frac{\beta}{2}\left(\frac{v_f}{c}\right)^{\frac{1}{3}}+4\pi Br^2\frac{v_f}{c}\right]\frac{c^3}{G}\Delta t$.

The energy radiated in unit time is given by

$$\frac{\Delta E_r}{\Delta t}\approx\frac{1}{2}\left[\alpha+\beta\left(\frac{v_f}{c}\right)^{\frac{1}{3}}\right]\frac{c^5}{G}\approx 2\times\left[\alpha+\beta\left(\frac{v_f}{c}\right)^{\frac{1}{3}}\right]\times 10^{59}ergs^{-1} \quad (0.2)$$

where we have neglected the rate of variation of the energy due to the change in the radial coordinate r inside the collapsing object.

The values of the constants α and β are fixed by the condition that the geodesic equation has at least one positive root. This restriction leads to a maximum value of α around 0.12 and to a corresponding $\beta=0.02$. With this choice we have two positive roots $X_0^{(1)}=3.56$ and $X_0^{(2)}=4.74$.

We shall now apply equation (ref1), giving the rate of energy emitted during the gravitational collapse of a mixture of strange fluid and radiation, to the case of the gamma ray bursts (GRBs). By considering that the energy is emitted during a time interval that can be obtained as $t_{ff}=\frac{M\times 10^{-38}}{2\times\alpha}s$ (this time interval is of the same order of magnitude with the Newtonian free falling time $t_{ff} = \frac{R^{3/2}}{\sqrt{2GM}}$), we obtain for the total emitted energy:

$$\Delta E_\gamma \approx 2 \times \left[\alpha + \beta \left(\frac{v_f}{c}\right)^{1/3}\right] \times \frac{M}{2 \times \alpha} \times 10^{21} erg \qquad (0.3)$$

For $\alpha = 0.12$,$\beta = 0.02$,$R = 10^6 cm$,$M = 2M_\odot = 4 \times 10^{33} g$ and $v_f = 6 \times 10^9 cm/s$ we obtain a value of the emitted energy $\Delta E_\gamma \approx 4 \times 10^{54} erg$, in a very close agreement with the recent observational determinations obtained for GRB990123 [2].

Under the effect of collapse the strange star will heat up to a temperature of the order of $T\tilde{\ }10^{13} K$, higher than that occurring in super nova explosions. The main energy loss mechanism of the super-heated naked strange singularity would be neutrino radiation, originating from the neutrinosphere formed around the naked singularity. The neutrinos and anti neutrinos interact with protons and neutrons via the URCA processes $n + \nu_e \rightarrow p + e^-$, $p + \bar{\nu}_e \rightarrow n + e^+$ [15]. At temperatures higher than the nuclear Fermi temperature $kT^{F,N} = \left(\frac{6\pi^2}{g}\right)^{2/3} \left(\frac{\hbar}{2m}\right) \left(\frac{N}{V}\right)^{2/3}$ that can also be expressed in the form $T_{11}^{F,N} = 1.47 \times 10^{-3} \rho_9^{2/3}$ [16] the integrated optical neutrino depth is unity. Hence the deposition energy can be estimated as $E \approx (1 - e^{-\tau}) \Delta E_\gamma \approx 2.8 \times 10^{54} ergs$. The process $\gamma + \gamma \rightarrow e^+ + e^-$ will generate a fireball that will expand outward. The expanding shell interacts with the inter stellar medium surrounding the strange naked singularity and the kinetic energy is finally radiated through non-thermal processes in shocks [8],[17].

Hence we obtain the following simple physical picture of the gamma ray burst emission. After the inner core of the accreting neutron star has been compressed to super nuclear densities the neutron matter -quark matter phase transition takes place. The accreted material together with the in falling stellar material leads to an increase in the mass of the quark star and when this becomes greater than $2M_\odot$ the collapse starts and the accreting quark star ends in a naked singularity.

The energy released during the collapse of the strange matter and radiation into a naked singularity has the same order of magnitude as that observed in the case of gamma ray bursts. This strongly suggests the possibility that gamma ray bursts could be massive compact objects collapsing in a naked singularity in a cosmological environment. The model can also explain some other observed features of the bursts like multiple emission.

References

T.Piran astro-ph/9801001
S.R.Kulkarni et al, Nature 398 (1999) 389
T.Piran, Astrophys.J.389 (1992) L45
E. S. Phinney,Astrophys.J.380 (1991) L17
B. Carter, Astrophys.J.391 (1992) L67
W.Kluzniak and M Ruderman, Astrophys.J.505 (1998) L113
Z.G. Dai and T. Lu ,Phys.Rev.Lett. 81 (1998) 430
K.S.Cheng and Z.G.Dai, Phys.Rev.Lett. 77 (1996) 1210
M.Tavani, astro-ph/9812422
E. Witten , Phys. Rev D 30 (1984) 272
K.S.Cheng, Z.G. Dai and T. Lu , Int. J. Mod. Phys. D 7 (1998) 139
R. Sari and T. Piran , Astrophys.J. 485 (1997) 270
T.Harko and K.S.Cheng, Phys. Lett.A , in press
S.K.Chakrabarti and P.S.Joshi,Int.J.Mod.Phys.D 3 (1994) 647
H.A. Bethe and J.A.Wilson, Astrophys.J. 295 (1985) 14
P.Meszaros and M.J.Rees, Astrophys.J.397 (1992) 570
P.Meszaros and M.J.Rees, Astrophys.J.405 (1993) 278

CONVERSION OF NEUTRON STARS INTO STRANGE STARS

KIN-WANG NG
Institute of Physics, Academia Sinica
Nankang, Taipei, Taiwan

1. INTRODUCTION

There have been a great deal of interest in the conversion of neutron stars into "strange" stars (Baym *et al.*1985; Olinto1987; Olinto1991; Olesen & Madsen1991; Heiselberg *et al.*1991; Horvath & Benvenuto1988; Benvenuto *et al.*1991). As Witten (Witten1984) first proposed, nuclear matter may not be the most stable form of matter. Rather, "strange" matter, matter consisting of equal numbers of up, down and strange quarks, is. The average energy per nucleon was shown by Fahri and Jaffe (Fahri & Jaffe1984) to be lower for this "strange" matter than for nuclear matter. It is believed that if a lump of strange matter comes in contact with a neutron star by whatever means possible, the neutron star will quickly be converted into a strange star. The question is how fast and under precisely what conditions this conversion will take place.

There are currently two methods being used to analyze this conversion process. The first is due to Olinto (Olinto1987) who uses a non-relativistic diffusion model. As such, theirs is a slow combustion model, with the burn front propagating at a speed of approximately 10 m/sec. This is determined primarily by the rate at which one of the down quarks inside the neutron is converted through a weak decay to a strange quark: $d + u \rightarrow s + u$. The second method was introduced by Horvath and Benvenuto (Horvath & Benvenuto1988) who model the conversion as a detonation. Their conversion rate is several orders of magnitude faster than that predicted by Olinto. In this Proceedings, we report our study on the neutron star burning problem.

We shall not *a priori* assume that the conversion process is due to either a slow combustion or a detonation. In fact, one of the purposes of our analysis is to establish under what conditions one will have a slow burning of the star or a detonation. Rather, we shall use the standard

K.S. Cheng et al. (eds.), Stellar Astrophysics, 143–148.

theory of relativistic combustion to analyze the conversion of the star which depends solely on the equations of state in the two media and the conservation of energy-momentum and baryon number across the combustion front. It is a kinematic analysis and has the advantage of not only being independent of the details of how the neutrons are absorbed into the strange matter, which at this point is not well understood, but also independent of any assumptions as to the rate at which the conversion will take place. Although the specific method by which strange matter is formed during the burning is not addressed by our analysis, any such strangeness formation processes should conserve energy density and baryon number, however, and would be included within our analysis. Since the only equation of state for the strange matter is the MIT bag model, any analysis of the conversion of the neutron star into a strange star will be dependent, to a certain extent, on the value of the bag constant, which is not known to any great degree of accuracy. We shall assume the density of neutron matter near the burning front to be a constant, and this density varies in the neutron star. We therefore calculated the velocity of the conversion front for a wide range of values of the bag constant and the density of neutron matter at various temperatures.

2. METHODOLOGY

We shall effectively be working in one spatial dimension, with the strange matter and the nuclear matter being seperated by a well defined planar combustion wavefront (Landau & Lifshitz1959). It has been shown that this wavefront is quite abrupt (Horvath & Benvenuto1988; Olinto1991), meaning that the thickness δ of the wavefront is much smaller than the size of the neutron star. We shall work in the frame which is at rest with the combustion wavefront. The energy-momentum tensor for the whole system is

$$T_{\mu\nu} = (\epsilon + p)u_\mu u_\nu - p g_{\mu\nu}, \tag{1}$$

where u_μ is the covariant fluid velocity, ϵ is the energy density and p is the pressure. The baryonic number density is $N_\mu = n u_\mu$. We will be using the ideal gas approximation for the quarks in the strange matter with all the quark masses $m_q = 0$ and the strong coupling constant $\alpha_s = 0$. In this zeroth order approximation the fraction of electrons in the strange quark matter $Y_e \equiv n_e/n_q = 0$ and the energy loss by neutrino emission from the quark matter vanishes (Iwamoto1982). Even though the process of absorbing neutrons is very exothermic, with each absorbed neutron releasing $\sim$10 MeV of energy (Fahri & Jaffe1984), we shall consider that the reaction, as a whole, conserves energy with the excess energy released in the absorption process going into heating up the star as a whole. In more realistic situations where m_s, α_s and Y_e are not zero, thermal energy may be lost by emission of neutrinos produced in URCA processes and the assumption of energy conservation may no longer hold. However, we have showed that our method of analyzing the burning process will be applicable because neutrino emission is relatively small at low temperatures and at high temperatures, neutrino trapping becomes efficient (Cho *et al.*1994). Then $T_{\mu\nu}$ is conserved across the combustion front, and it can be shown that (Landau & Lifshitz1959),

$$v_q^2 = \frac{(p_q - p_n)(\epsilon_n + p_q)}{(\epsilon_q - \epsilon_n)(\epsilon_q + p_n)}, \qquad v_n^2 = \frac{(p_n - p_q)(\epsilon_q + p_n)}{(\epsilon_n - \epsilon_q)(\epsilon_n + p_q)}, \tag{2}$$

where v_q and v_n are the velocities relative to the burn front of the quarks and the neutrons in the strange matter and in the neutron star respectively. v_n is also the velocity of the front relative to the star. (Variables with subscript q (n) will denote quantities measured in the strange (nuclear) matter.) Next, baryon number must be conserved across the wavefront, from which we obtain

$$n_q^2 = n_n^2 \frac{(\epsilon_q + p_q)(\epsilon_q + p_n)}{(\epsilon_n + p_n)(\epsilon_n + p_q)}. \tag{3}$$

For the strange matter, we shall assume a zero strange quark mass and take the MIT bag model equation of state:

$$\begin{aligned} e_q &= \frac{19}{12}\pi^2 T^4 + \frac{9}{2}T^2\mu^2 + \frac{9}{4\pi^2}\mu^4 + B, \\ p_q &= \frac{19}{36}\pi^2 T^4 + \frac{3}{2}T^2\mu^2 + \frac{3}{4\pi^2}\mu^4 - B, \\ n_q &= T^2\mu + \frac{1}{\pi^2}\mu^3, \end{aligned} \tag{4}$$

where T, μ, and B are the temperature, chemical potential of each quark, and bag constant respectively. For the nuclear matter there are various equations of state that we can choose. The two that we shall work with are the zero-temperature Bethe-Johnson (BJ) equation of state (Shapiro & Teukolsky1983):

$$\begin{aligned} \epsilon_n &= (236 n_n^{1.54} + m_n) n_n \text{ MeV fm}^{-3}, \\ p_n &= 364 n_n^{2.54} \text{ MeV fm}^{-3}, \end{aligned} \tag{5}$$

where m_n is the mass of the neutron in MeV, and the zero-temperature ideal Fermi-Dirac (FD) neutron gas (Landau & Lifshitz1980). The velocities v_q and v_n may now be determined for any given values of B,

n_n and T. Here we require $0.3 \text{ fm}^{-4} < B$ so that nuclei with high atomic numbers would be stable against decay into non-strange quark matter (Fahri & Jaffe1984). In addition, requiring the star to be hydrodynamically stable against small perturbations limits $n_n < 1.5 \text{ fm}^{-3}$ for the hard BJ equation of state, and $n_n < 5 \text{ fm}^{-3}$ for the soft FD equation of state (Shapiro & Teukolsky1983).

At this point we need to introduce some terminology (Landau & Lifshitz1959). Let v_q^s (v_n^s) be the speed of sound in the strange (nuclear) matter. $v_q^s = \frac{1}{\sqrt{3}}$ while

$$v_n^s = [\frac{n_n^{1.54}}{1.01 + 0.648 n_n^{1.54}}]^{\frac{1}{2}}, \tag{6}$$

is the speed of sound for the BJ equation of state. When $v_q < v_q^s$ and $v_n < v_n^s$, the burning is called a *deflagration* or a slow combustion. When $v_q \leq v_q^s$ and $v_n > v_n^s$, it is called a *detonation.* When $v_q > v_q^s$ and $v_n < v_n^s$, the burning is *absolutely unstable*, meaning that in the presence of any small perturbation the wavefront will no longer remain as a well defined plane and the model itself fails. When $v_q > v_q^s$ and $v_n > v_n^s$, the burning is a fast combustion without detonation which may involve either very efficient heat transfer in the unburnt gas, or reactions which are initially exothermic but are endothermic in their final stages.

3. RESULTS

We find that the conversion of a neutron star into a strange star is *never* due to a detonation, although, except for some judicious choices of parameters, it will occur extremely rapidly. If the fast combustion is stable, then some presently unknown exotic mechanism must be present which either allows for a conversion process which must be first exothermic and then endothermic or else the heat conversion must allow for very efficient heat transfers (Cho *et al.*1994). Although, it has been suggested that the conversion of neutron matter into strange matter is either direct, through two-flavor quark matter (Olinto1991), or through a quantum tunnelling process (Benvenuto *et al.*1991). However, it is unlikely that the conversion has involved reactions which are initially exothermic but endothermic in their final stages. Next, it has been theorized that the core of a neutron state may contain a superfluid state, which permits very efficient heat transfers. It is not clear whether or not the superfluid state may exist near the burn front. Because v_n is typically close to the speed of light, it may be faster than the critical velocity of the superfluid.

There is one other mechanism which may allow for very effcient heat transfers which was first suggested by Horvath and Benvenuto (Horvath & Benvenuto1988). Because the Reynolds numbers for both fluids are extremely large, it may be that the heat transfer mechanism is not due to thermal conduction but rather by turbulent convection of the two materials. This will allow for much more efficient heat transfers. Unlike Horvath and Benevcnuto who argued that the burning will quickly reach detonation, we find that because v_q is always greater than the speed of sound in the strange matter, detonation will never occur.

Most of the parameter space which would allow for the burning of the star is taken up by the absolutely unstable burning region. In this region of the parameter space where (condition for absolute unstable burning) the burn front is absolutely unstable, meaning that any perturbation of the planar burn front will very quickly grow and there will be a turbulent mixing of the nuclear and strange matter. Modeling the burn front as an infinite flat plane is no longer realistic and other methods need to be used to analyze the burning.

We have also analysed a case in which the equation of state of the neutron gas is of BJ type and $T = 10$ MeV. The results are similar. Moreover, we could not find any choices of B and n_n for $T = 10$ MeV and $m_s = 0$ which would reproduce Olinto's result. As was pointed out in Olinto (Olinto1991), however, because the burning of the star is inherently a dynamical system, a metastable two-flavour quark state may exist in the burn front. In this case, extremely relativistic velocities may result. The use of the non-relativistic equation would no longer be valid. This is actually borne-out in our calculations which show that the burn front should propagate at some fraction of the speed of light, if it occurs at all. Therefore, to reconcile our results with Olinto's, a metastable two-flavour quark state in the burn front should be present.

Preliminary results for the case of finite strange quark mass have shown no qualitative changes to the parameter space. Non-zero strong coupling corrections should have similar results.

4. CONCLUSIONS

We find that the conversion of the neutron star into a strange star will never be due to slow combustion or detonation and in fact may not even be describable using a simple infinite planar model of the burn front. If it happens at all, then the conversion could be a fast combustion without detonation. The basic culprit is v_q which we have found to be greater than the speed of sound in the strange matter for most of the parameter space. As such, detonation will never take place; but then the conditions for stable burning will also be difficult to satisfy. We note, however, that all models of the burning, including our own, used the infinite plane description of the burn front. The neutron star is a sphere, however, and it may be more realistic to use a spherical burn front to model the burning. As the boundary conditions for a spherical geometry is quite different from the infinite plane geometry, namely that the strange matter must be at rest after the burning, a parameter space which will allow for stable burning may be found. If not, however, then the conversion of the neutron star will most probably be absolutely unstable and some way would need to be found to model the turbulent mixing of the strange and nuclear matter. We are in the process of analyzing the burning using a spherical geometry.

References

Baym, G., Kolb, E. W., McLerran, L. and Walker, T. P. (1985) Phys. Lett. B **160**, 181.

Benvenuto, G., Horvath, J. E. and Vucetich, H. (1991) Nucl. Phys. B (Proc. Suppl.) **24**, 125.

Cho, H. T., Ng, K.-W. and Speliotopoulos, A. D. (1994) Phys. Lett. B **326**, 111.

Fahri E. and Jaffe, R. L. (1984) Phys. Rev. D **30**, 2379.

Heiselberg, H., Baym, G. and Pethick, C. J. (1991) Nucl. Phys. B (Proc. Suppl.) **24**, 144.

Horvath, E. and Benvenuto, O. G. (1988) Phys. Lett. B **213**, 516.

Iwamoto, N. (1982) Ann. Phys. **141** 1.

Landau, L. D. and Lifshitz, E. M. (1959) *Fluid Mechanics*, Chapters 14 & 15 (Pergamon Press, New York).

Landau, L. D. and Lifshitz, E. M. (1980) *Statistical Physics, Part 1, 3rd ed.*, Chapter 5 (Pergamon Press, New York).

Olesen, L. and Madsen, J. (1991) Nucl. Phys. B (Proc. Suppl.) **24**, 170.

Olinto, A. V. (1987) Phys. Lett. B **192**, 71.

Olinto, A. V. (1991) Nucl. Phys. B (Proc. Suppl.) **24**, 103.

Shapiro, L. and Teukolsky, S. A. (1983) *Black Holes, White Dwarfs, and Neutron Stars*, Chapter 8 (John Wiley & Sons, New York).

Witten, E. (1984) Phys. Rev. D **30**, 272.

DO STRANGE STARS EXIST IN THE UNIVERSE?

IGNAZIO BOMBACI

Dipartimento di Fisica, Università di Pisa, and INFN Sez. di Pisa, via Buonarroti, 2, I-56127 Pisa, Italy

Abstract Definitely, an affirmative answer to this question would have implications of fundamental importance for astrophysics (a new class of compact stars), and for the physics of strong interactions (deconfined phase of quark matter, and strange matter hypothesis). In the present work, we use observational data for the newly discovered millisecond X-ray pulsar SAX J1808.4-3658 and for the atoll source 4U 1728-34 to constrain the radius of the underlying compact stars. Comparing the mass–radius relation of these two compact stars with theoretical models for both neutron stars and strange stars, we argue that a strange star model is more consistent with SAX J1808.4-3658 and 4U 1728-34, and suggest that they are likely strange star candidates.

1. INTRODUCTION

The possible existence of a new class of compact stars, which are made entirely of deconfined u,d,s quark matter (*strange quark matter* (SQM)), is one of the most intriguing aspects of modern astrophysics. These compact objects are called strange stars. They differ from neutron stars, where quarks are confined within neutrons, protons, and eventually within other hadrons (*e.g. hyperons*). The investigation of such a possibility is relevant not only for astrophysics, but for high energy physics too. In fact, the search for a deconfined phase of quark matter is one of the main goals in heavy ion physics. Experiments at Brookhaven National Lab's Relativistic Heavy Ion Collider (RHIC) and at CERN's Large Hadron Collider (LHC), will hopefully clarify this issue in the near future.

The possibility that strange stars do exist is based on the so called *strange matter hypothesis*, formulated by Witten (1984) (see also Bodmer, 1971). According to this hypothesis, strange quark matter, in equilibrium with respect to the weak interactions, could be the true ground

K.S. Cheng et al. (eds.), Stellar Astrophysics, 149–160.

state of strongly interacting matter rather than ^{56}Fe, *i.e.* the energy per baryon of SQM must fulfil the inequality

$$\left(\frac{E}{A}\right)_{SQM} \leq \frac{E(^{56}Fe)}{56} \simeq 930\ MeV, \tag{1}$$

at the baryon density where the pressure is equal to zero.

If the strange matter hypothesis is true, then a nucleus with A nucleons, could in principle lower its energy by converting to a strangelet (a drop of SQM). However, this process requires a very high-order simultaneous weak interactions to convert about a number A of u and d quarks of the nucleus into strange quarks. The probability for such a process is extremely low [1], and the mean life time for an atomic nucleus to decay to a strangelet is much higher than the age of the Universe. On the other hand, a step by step production of s quarks, at different times, will produce hyperons in the nucleus, *i.e.* a system (hypernucleus) with a higher energy per baryon with respect to the original nucleus. In addition, finite size effects (surface and shell effects) place a lower limit (A $\sim$ 10–100) on the baryon number of a stable strangelet even if bulk SQM is stable (Farhi & Jaffe, 1984). Thus, according to the strange matter hypothesis, the ordinary state of matter, in which quarks are confined within hadrons, is a metastable state.

The success of traditional nuclear physics, in explaining an astonishing amount of experimental data, provides a clear indication that quarks in a nucleus are confined within protons and neutrons. Thus, the energy per baryon $(E/A)_{ud}$ of u,d quark matter (nonstrange quark matter) must be higher than the energy per baryon of nuclei

$$\left(\frac{E}{A}\right)_{ud} \geq 930\ MeV + \Delta, \tag{2}$$

being $\Delta \sim 4$ MeV a quantity which accounts for the lower energy per baryon of a finite chunk ($A \sim 250$) of nonstrange quark matter with respect to the bulk ($A \to \infty$) case (Farhi & Jaffe, 1984). These stability conditions (eq.s (1) and (2)) in turn may be used to constrain the parameters entering in models for the equation of state (EOS) of SQM. As we show below, the existence of strange stars is allowable within the uncertainties inherent in perturbative Quantum Chromo-Dynamics (QCD). Thus *strange stars may exist in the Universe.*

---

[1] It is proportional to G_F^{2A}, being G_F the Fermi constant, and assuming a number A of simultaneous weak processes.

2. THE EQUATION OF STATE FOR STRANGE QUARK MATTER

From a basic point of view the equation of state for SQM should be calculated solving the equations of QCD. As we know, such a fundamental approach is presently not doable. Therefore one has to rely on phenomenological models. In this work, we discuss two phenomenological models for the EOS of strange quark matter. The first one is a well known model related to the MIT bag model (Chodos *et al.* 1974) for hadrons. The second one is a new model developed by Dey *et al.* (1998).

At very high density SQM behaves as a relativistic gas of weakly interacting fermions. This is a consequence of one of the basic features of QCD, namely asymptotic freedom. To begin with consider the case of massless quarks, and consider gluon exchange interactions to the first order in the QCD structure constant α_c. Under these circumstances the EOS of β–stable SQM can be written in the parametrical form:

$$\varepsilon = K n_B^{4/3} + B, \qquad P = \frac{1}{3} K n_B^{4/3} - B, \qquad K \equiv \frac{9}{4}\pi^{2/3}\left(1 + \frac{2\alpha_c}{3\pi}\right)\hbar c \quad (3)$$

ε being the energy density, and P the pressure. Eliminating the baryon number density n_B one gets:

$$P = \frac{1}{3}(\varepsilon - 4B) \quad (4)$$

Here B is a phenomenological parameter which represents the difference between the energy density of "perturbative vacuum" and true QCD vacuum. B is related to the "bag constant" which in the MIT bag model for hadrons (Chodos *et al.* 1974) gives the confinement of quarks within the hadronic bag. The density of zero pressure SQM is just $\rho_s = 4B/c^2$. This is the value of the surface density of a bare strange star. Taking a non-vanishing value for the mass m_s of the strange quark, the EOS becomes more involved (see *e.g.* Farhi & Jaffe, 1984) with respect to the simple expression (4). However, for m_s = 100–300 MeV, equation (4) is less than 5% different from the "exact" case for $m_s \neq 0$. In summary, in this model for the equation of state for SQM there are three phenomenological parameters, namely: B, m_s, and α_c. It is possible to determine ranges in the values of these parameters in which SQM is stable, and nonstrange quark matter is not (Farhi & Jaffe, 1984). For example, in the case of non–interacting quarks ($\alpha_c = 0$) one has $B \simeq$ 57–91 MeV/fm^3 for $m_s = 0$, and $B \simeq$ 57–75 MeV/fm^3 for m_s = 150 MeV.

The schematic model outlined above becomes less and less trustworthy going from very high density region (asymptotic freedom regime) to lower densities, where confinement (hadrons formation) takes place. Recently, Dey *et al.* (1998) derived a new EOS for SQM using a "dynamical" density-dependent approach to confinement. The EOS by Dey *et al.* has asymptotic freedom built in, shows confinement at zero baryon density, deconfinement at high density. In this model, the quark interaction is described by a colour-Debye-screened inter-quark vector potential originating from gluon exchange, and by a density-dependent scalar potential which restores chiral symmetry at high density (in the limit of massless quarks). The density-dependent scalar potential arises from the density dependence of the in-medium effective quark masses M_q, which, in the model by Dey *et al.*(1998), are taken to depend upon the baryon number density according to

$$M_q = m_q + 310 \cdot sech\left(\nu \frac{n_B}{n_0}\right) \qquad \text{(MeV)}, \qquad (5)$$

where $n_0 = 0.16 \text{ fm}^{-3}$ is the normal nuclear matter density, $q(= u, d, s)$ is the flavor index, and ν is a parameter. The effective quark mass $M_q(n_B)$ goes from its constituent masses at zero density, to its current mass m_q, as n_B goes to infinity. Here we consider two different parameterizations of the EOS by Dey *et al.*, which correspond to a different choice for the parameter ν. The equation of state SS1 (SS2) corresponds to $\nu = 0.333$ ($\nu = 0.286$). These two models for the EOS give absolutely stable SQM according to the strange matter hypothesis.

3. STRANGE STAR CANDIDATES

To distinguish whether a compact star is a neutron star or a strange star, one has to find a clear observational signature. There is a striking qualitative difference in the mass–radius (MR) relation of strange stars with respect to that of neutron stars (see Fig. 1). For strange stars with "small" ($M << M_{max}$) gravitational mass, M is proportional to R^3. In contrast, neutron stars have radii that decrease with increasing mass. This is a consequence of the underlying interaction between the stellar constituents which makes "low" mass strange stars self-bound objects (see *e.g.* Bombaci 1999) contrary to the case of neutron stars which are bound by gravity [2]. As we know, there is a minimum mass for a neutron star ($M_{min} \sim 0.1\ M_\odot$). In the case of a strange star, there is essentially no minimum mass. As the central density $\rho_c \to \rho_s$ (surface density), a strange star (or better a strangelet for very low baryon number) is a

[2] As an idealized example, remember that pure neutron matter is not bound by nuclear forces.

self–bound system, until the baryon number becomes so low that finite size effects destabilize it.

3.1 SAX J1808.4-3658

The transient X-ray burst source SAX J1808.4-3658 was discovered in September 1996 by the BeppoSAX satellite. Two bright type-I X-ray bursts were detected, each lasting less than 30 seconds. Analysis of the bursts in SAX J1808.4-3658 indicates that it is 4 kpc distant and has a peak X-ray luminosity of 6×10^{36} erg/s in its bright state, and a X-ray luminosity lower than 10^{35} erg/s in quiescence (in't Zand 1998). The object is nearly certainly the same as the transient X-ray source detected with the Proportional Counter Array (PCA) on board the Rossi X-ray Timing Explorer (RXTE) (Marshall, 1998). Coherent pulsations at a period of 2.49 milliseconds were discovered (Wijnands & van der Klis 1998). The star's surface dipolar magnetic moment was derived to be less than 10^{26} G cm^3 from detection of X-ray pulsations at a luminosity of 10^{36} erg/s (Wijnands & van der Klis 1998), consistent with the weak fields expected for type-I X-ray bursters and millisecond radio pulsars (MS PSRs) (Bhattacharya & van den Heuvel 1991). The binary nature of SAX J1808.4-3658 was firmly established with the detection of a 2 hour orbital period (Chakrabarty & Morgan 1998) as well as with the optical identification of the companion star (Roche *et al.* 1998). SAX J1808.4-3658 is the first pulsar to show both coherent pulsations in its persistent emission and X-ray bursts, and by far the fastest-rotating, lowest-field accretion-driven pulsar known. It presents direct evidence for the evolutionary link between low-mass X-ray binaries (LMXBs) and MS PSRs. SAX J1808.4-3658 is the only known LMXB with an MS PSR.

A mass–radius (MR) relation for the compact star in SAX J1808.4-3658 has been recently obtained by Li *et al.* (1999a) [3] using the following two requirements. (*i*) Detection of X-ray pulsations requires that the inner radius R_0 of the accretion flow should be larger than the stellar radius R. In other words, the stellar magnetic field must be strong enough to disrupt the disk flow above the stellar surface. (*ii*) The radius R_0 must be less than the so-called co-rotation radius R_c, *i.e.* the stellar magnetic field must be weak enough that accretion is not centrifugally inhibited:

$$R_0 \lesssim R_c = [GMP^2/(4\pi^2)]^{1/3}. \tag{6}$$

[3] see also Burderi & King (1998), Psaltis & Chakrabarty (1999).

Here G is the gravitation constant, M is the mass of the star, and P is the pulse period. The inner disk radius R_0 is generally evaluated in terms of the Alfvén radius R_A, at which the magnetic and material stresses balance (Bhattacharya & van den Heuvel 1991): $R_0 = \xi R_A = \xi[B^2R^6/\dot{M}(2GM)^{1/2}]^{2/7}$, where B and $\dot{M}$ are respectively the surface magnetic field and the mass accretion rate of the pulsar, and ξ is a parameter of order of unity almost independent of $\dot{M}$ (Li 1997, Burderi & King 1998). Since X-ray pulsations in SAX J1808.4-3658 were detected over a wide range of mass accretion rate (say, from $\dot{M}_{\rm min}$ to $\dot{M}_{\rm max}$), the two conditions (*i*) and (*ii*) give $R \lesssim R_0(\dot{M}_{\rm max}) < R_0(\dot{M}_{\rm min}) \lesssim R_{\rm c}$. Next, we assume that the mass accretion rate $\dot{M}$ is proportional to the X-ray flux F observed with RXTE. This is guaranteed by the fact that the X-ray spectrum of SAX J1808.4-3658 was remarkably stable and there was only slight increase in the pulse amplitude when the X-ray luminosity varied by a factor of ~ 100 during the 1998 April/May outburst (Gilfanov *et al.* 1998, Cui *et al.* 1998, Psaltis & Chakrabarty 1999). Therefore, Li *et al.* (1999a) get the following upper limit of the stellar radius: $R < (F_{min}/F_{max})^{2/7}R_{\rm c}$, or

$$R < 27.5\left(\frac{F_{min}}{F_{max}}\right)^{2/7}\left(\frac{P}{2.49\ ms}\right)^{2/3}\left(\frac{M}{M_\odot}\right)^{1/3}\ \text{km}, \tag{7}$$

where $F_{\rm max}$ and $F_{\rm min}$ denote the X-ray fluxes measured during X-ray high- and low-state, respectively, $M_\odot$ is the solar mass. Note that in writing inequality (7) it is assumed that the pulsar's magnetic field is basically dipolar (see Li *et al.* 1999a for arguments to support this hypothesis) [4].

Given the range of X-ray flux at which coherent pulsations were detected, inequality (7) defines a limiting curve in the mass–radius plane for SAX J1808.4-3658, as plotted in the dashed curve in Fig. 1. The authors of ref. (Li *et al.* 1999a) adopted the flux ratio $F_{\rm max}/F_{\rm min} \simeq 100$ from the observations that during the 1998 April/May outburst, the maximum $2-30$ keV flux of SAX J1808.4-3658 at the peak of the outburst was $F_{\rm max} \simeq 3 \times 10^{-9}\,{\rm erg\,cm^{-2}\,s^{-1}}$, while the pulse signal became barely detectable when the flux dropped below $F_{\rm min} \simeq 2 \times 10^{-11}\,{\rm erg\,cm^{-2}\,s^{-1}}$ (Cui *et al.* 1998, Psaltis & Chakrabarty 1999). The dashed line $R = R_{\rm s} \equiv 2GM/c^2$ represents the Schwartzschild radius - the lower limit of the stellar radius to prevent the star collapsing into a black hole. Thus

[4] see also Psaltis & Chakrabarty (1999) for a study of the influence on the MR relation for SAX J1808.4-3658 of a quadrupole magnetic moment, and of a *non-standard* disk–magnetosphere interaction model.

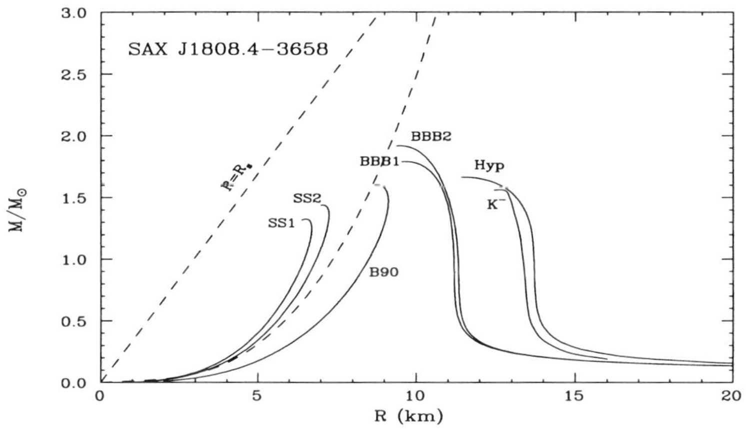

Figure 1 Comparison of the mass–radius relation of SAX J1808.4 -3658 determined from RXTE observations with theoretical models of neutron stars and of strange stars. See text for more details.

the allowed range of the mass and radius of SAX J1808.4-3658 is the region confined by these two dashed curves in Fig. 1.

In the same figure, we report the theoretical MR relations (solid curves) for neutron stars given by some recent realistic models for the EOS of dense matter (see Li *et al.* 1999a for references to the EOS models). Models BBB1 and BBB2 are relative to "conventional" neutron stars (*i.e* the core of the star is assumed to be composed by an uncharged mixture of neutrons, protons, electrons and muons in equilibrium with respect to the weak interaction). The curve labeled Hyp depicts the MR relation for a neutron star in which hyperons are considered in addition to nucleons as hadronic constituents. The MR curve labeled K^- is relative to neutron stars with a Bose-Einstein condensate of negative kaons in their cores. It is clearly seen in Fig. 1 that none of the neutron star MR curves is consistent with SAX J1808.4-3658. Including rotational effects will shift the MR curves to up-right in Fig. 1 (Datta *et al.* 1998), and does not help improve the consistency between the theoretical neutron star models and observations of SAX J1808.4-3658. Therefore SAX J1808.4-3658 is not well described by a neutron star model. The curve B90 in Fig. 1 gives the MR relation for strange stars described by the schematic EOS (4) with B = 90 MeV/fm^3. The two curves SS1 and SS2 give the MR relation for strange stars calculated with the EOS by Dey et al. (1998). Figure 1 clearly demonstrates that a strange star model is more compatible with SAX J1808.4-3658 than a neutron star one.

3.2 4U 1728-34

Recently, Li et al. (1999b) investigated possible signatures for the existence of strange stars in connection with the newly discovered phenomenon of kilohertz quasi–periodic oscillations (kHz QPOs) in the X-ray flux from LMXB (for a review see van der Klis 2000). Initially, kHz QPO data from various sources were interpreted assuming a simple *beat–frequency model* (see *e.g.* Kaaret & Ford 1997). In many cases, two simultaneous kHz QPO peaks ("twin peaks") are observed. The QPO frequencies vary and are strongly correlated with source flux. In the beat–frequency model the highest observed QPO frequency ν_u is interpreted as the Keplerian orbital frequency ν_K at the inner edge of the accretion disk. The frequency ν_l of the lower QPO peak is instead interpreted as the beat frequency between ν_K and the neutron star spin frequency ν_0, which within this model is equal to the separation frequency $\Delta\nu \equiv \nu_u - \nu_l$ of the two peaks. Thus $\Delta\nu$ is predicted to be constant. Nevertheless, novel observations for different kHz QPO sources have challenged this simple beat–frequency model. The most striking case is the source 4U 1728-34, where it was found that $\Delta\nu$ decreases significantly, from 349.3 ± 1.7 Hz to 278.7 ± 11.6 Hz, as the frequency of the lower kHz QPO increases (Méndez & van der Klis 1999). Furthermore, in the spectra observed by the RXTE for 4U 1728-34, Ford & van der Klis (1998) found low-frequency Lorentian oscillations with frequencies between 10 and 50 Hz. These frequencies as well as the break frequency (ν_{break}) of the power spectrum density for the same source were shown to be correlated with ν_u and ν_l.

A different model was recently developed by Osherovich & Titarchuk (1999) (see also Titarchuk & Osherovich 1999), who proposed a unified classification of kHz QPOs and the related observed low frequency phenomena. In this model, kHz QPOs are modeled as Keplerian oscillations under the influence of the Coriolis force in a rotating frame of reference (magnetosphere). The frequency ν_l of the lower kHz QPO peak is the Keplerian frequency at the outer edge of a viscous transition layer between the Keplerian disk and the surface of the compact star. The frequency ν_u is a hybrid frequency related to the rotational frequency ν_m of the star's magnetosphere by: $\nu_u^2 = \nu_K^2 + (2\nu_m)^2$. The observed low Lorentzian frequency in 4U 1728-34 is suggested to be associated with radial oscillations in the viscous transition layer of the disk, whereas the observed break frequency is determined by the character-

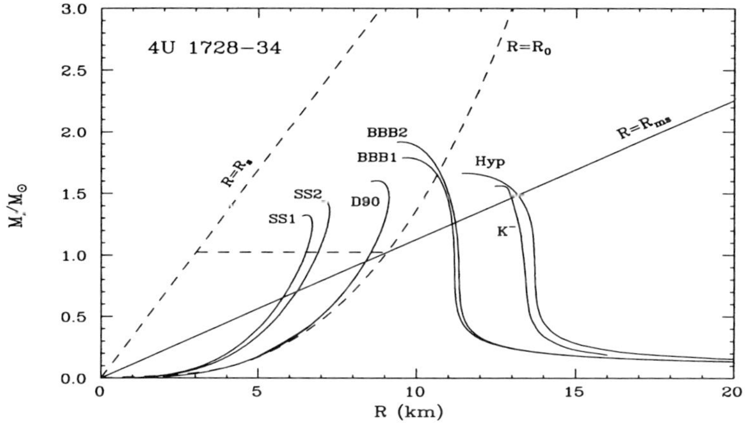

Figure 2 Comparison of the MR relation of 4U 1728 34 determined from RXTE observations with theoretical models of neutron stars and of strange stars. The range of mass and radius of 4U 1728-34 is allowed in the region outlined by the dashed curve $R = R_0$, the horizontal dashed line, and the dashed line $R = R_s$. The solid curves represents theoretical MR relations for neutron stars and strange stars.

istic diffusion time of the inward motion of the matter in the accretion flow (Titarchuk & Osherovich 1999). Predictions of this model regarding relations between the QPO frequencies mentioned above compare favorably with recent observations for 4U 1728-34, Sco X-1, 4U 1608-52, and 4U 1702-429.

The presence of the break frequency and the correlated Lorentzian frequency suggests the introduction of a new scale in the phenomenon. One attractive feature of the model by Titarchuk & Osherovich (1999) is the introduction of such a scale in the model through the Reynolds number for the accretion flow. The best fit for the observed data was obtained by Titarchuk & Osherovich (1999) when

$$a_k = (M/M_\odot)(R_0/3R_\mathrm{s})^{3/2}(\nu_0/364\,\mathrm{Hz}) = 1.03, \tag{8}$$

where M is the stellar mass, R_0 is the inner edge of the accretion disk [5], R_s is the Schwarzschild radius, and ν_0 is the spin frequency of the star. Given the 364 Hz spin frequency of 4U 1728-34 (Strohmayer et al.

[5] In the expression for a_k reported in Titarchuk & Osherovich (1999), one has $x_0 = R_0/R_\mathrm{s}$, where R_0 is erroneously indicated as the neutron star radius (Titarchuk, private communication).

1996), the inner disk radius can be derived from the previous equation. Since the innermost radius of the disk must be larger than the radius R of the star itself, this leads to a mass-dependent upper bound on the stellar radius,

$$R \leq R_0 \simeq 8.86\, a_k^{2/3} (M/M_\odot)^{1/3}\, \text{km}, \tag{9}$$

which is plotted by dashed curve in Fig. 2.

A second constraint on the mass and radius of 4U 1728-34 results from the requirement that the inner radius R_0 of the disk must be larger than the radius of the last stable circular orbit $R_{\rm ms}$ around the star:

$$R_0 \geq R_{\rm ms}. \tag{10}$$

To make our discussion more transparent, neglect for a moment the rotation of the compact star. For a non-rotating star $R_{\rm ms} = 3R_{\rm s}$, then the second condition gives:

$$R_0 \geq 3R_{\rm s} = 8.86\ (M/M_\odot)\ \text{km}. \tag{11}$$

Therefore, the allowed range of the mass and radius for 4U1728-34 is the region in the lower left corner of the MR plane confined by the dashed curve ($R = R_0$), by the horizontal dashed line, and by the Schwartzschild radius (dashed line $R = R_s$). In the same figure, we compare with the theoretical MR relations for non-rotating neutron stars and strange stars, for the same models for the EOS considered in Fig. 1. It is clear that a strange star model is more compatible with 4U 1728-34 than a neutron star one. Including the effects of rotation (ν_0 =364 Hz) in the calculation of the theoretical MR relations and $R_{\rm ms}$, does not change the previous conclusion (Li *et al.* 1999b).

4. FINAL REMARKS

The main result of the present work (*i.e.* the likely existence of strange stars) is based on the analysis of observational data for the X-ray sources SAX J1808.4-3658 and 4U 1728-34. The interpretation of these data is done using *standard* models for the accretion mechanism, which is responsible for the observed phenomena. The present uncertainties in our knowledge of the accretion mechanism, and the disk–magnetosphere interaction, do not allow us to definitely rule out the possibility of a neutron star for the two X-ray sources we discussed. For example, making *a priori* the *conservative* assumption that the compact object in SAX J1808.4-3658 is a neutron star, and using a MR relation similar to our eq. (7) Psaltis & Chakrabarty (1999) try to constrain disk–magnetosphere interaction models or to infer the presence of a quadrupole magnetic moment in the compact star.

SAX J1808.4-3658 and 4U 1728-34 are not the only LMXBs which could harbour a strange star. Recent studies have shown that the compact objects associated with the X-ray burster 4U 1820-30 (Bombaci 1997), the bursting X-ray pulsar GRO J1744-28 (Cheng *et al.* 1998b) and the X-ray pulsar Her X-1 (Dey *et al.* 1998) are likely strange star candidates. For each of these X-ray sources (strange star candidates) the conservative assumption of a neutron star as the central accretor would require some particular (possibly *ad hoc*) assumption about the nature of the plasma accretion flow and/or the structure of the stellar magnetic field. On the other hand, the possibility of a strange star gives a simple and unifying picture for all the systems mentioned above. Finally, strange stars have also been speculated to model γ-ray bursters (Haensel *et al.* 1991, Bombaci & Datta 2000) and soft γ-ray repeaters (Cheng & Dai 1998a).

Acknowledgements

I thank my colleagues J. Dey, M. Dey, E.P.J. van den Heuvel, X.D. Li, and S. Ray with whom the ideas presented in this talk were developed. I am grateful to the Organizing Committee of the Pacific Rim Conference on Stellar Astrophysics for inviting me and for financial support. Particularly, I thank Prof. K.S. Cheng for the warm hospitality, and for many stimulating discussions during the conference. It is a pleasure to acknowledge fruitful and stimulating discussions with Prof. G. Ripka during the workshop Quark Condensates in Nuclear Matter, held at the ECT* in Trento.

In memory of Bhaskar Datta

I dedicate this paper to my great friend and colleague Bhaskar Datta, who passed away on december 3rd 1999 in Bangalore.

References

Bhattacharya, D., & van den Heuvel, E. P. J. 1991, Phys. Rep., 203, 1

Bodmer, A. R. 1971, Phys. Rev. D, 4, 1601

Bombaci, I., 1997, Phys. Rev. C, 55, 1587

Bombaci, I. 1999, Neutron stars' structure and nuclear equation of state, in M. Baldo (ed.), *Nuclear methods and the nuclear equation of state*, World Scientific, Singapore, pp. 381-457

Bombaci, I., & Datta, B, 2000, Astrophys. J. 530, L69

Burderi, L., & and King, A.R. 1998, Astrophys. J. 505, L135

Chakrabarty, D., & Morgan, E. H. 1998, Nature 394, 346

Cheng, K.S, & Dai, Z.G., 1998a, Phys. Rev. Lett. 80, 1998
Cheng, K.S, Dai, Z.G., Wai, D.M. & Lu, T. 1998b, Science 280, 407
Chodos, A. *et al.*, 1974, Phys. Rev. D, 9, 3471
Cui, W., Morgan, E.H., & Titarchuk, L. 1998, Astrophys. J. 504, L27
Datta, B., Thampan, A.V., & Bombaci, I. 1998, Astron. Astrophys. 334, 943
Dey, M., Bombaci, I., Dey, J., Ray, S., & Samanta, B. C. 1998, Phys. Lett. B, 438, 123; erratum, 1999 Phys. Lett. B 467, 303
Ford, E. & van der Klis, M. 1998, Astrophys. J. , 506, L39
Gilfanov, M., Revnivtsev, M., Sunyaev, R., & Churazov, E. 1998, Astron. Astrophys. 338, L83
Haensel, P., Paczynski, B., & Amsterdamski, P. 1991, Astrophys. J. 375, 209
in't Zand, J. J. M. *et al.*, Astron. Astrophys. 331, L25 (1998)
Kaaret, P., & Ford, E.C., 1997, Science, 276, 1386
Li, X.-D. 1997, Astrophys. J. 476, 278
Li, X.-D., Bombaci. I., Dey, M., Dey, J., & van den Heuvel, E. P. J., 1999a, Phys. Rev. Lett., 83, 3776
Li, X.-D., Ray S., Dey, J., Dey, M., & Bombaci. I. 1999b, Astrophys. J. , 527, L51
Marshall, F.E., 1998, IAU Circ. No. 6876
Méndez, M. & van der Klis, M. 1999, Astrophys. J. , 517, L51
Osherovich, V. & Titarchuk, L. 1999, Astrophys. J. , 522, L113; 523, L73
Psaltis, D., & Chakrabarty, D. 1999, Astrophys. J. , 521, 332
Roche, R. *et al.*, 1998, IAU Circ. No. 6885
Strohmayer, T. E., Zhang, W., Swank, J. H., Smale, A., Titarchuk, L., Day, C., & Lee, U. 1996, Astrophys. J. , 469, L9
Titarchuk, L. & Osherovich, V. 1999, Astrophys. J. , 518, L95
van der Klis, M., 2000, Ann. Rev. Astr. Astrophys. (to appear Sept. 2000). arXiv:astro-ph/0001167
Wijnands, R., & M. van der Klis, M., 1998, Nature 394, 344
Witten, E. 1984, Phys. Rev. D, 30, 272

OSCILLATIONS OF QUARK STARS

C. W. YIP, M. C. CHU and P. T. LEUNG
Department of Physics, The Chinese University of Hong Kong, Shatin, N.T., Hong Kong.

Abstract We study the relativistic quadrupole oscillations ($l = 2$) of strange quark stars by calculating their quasi-normal modes, and we thus explore the possibility of using these oscillation frequencies and damping times to differentiate among relativistic polytropic stars, neutron stars, pure light-quark stars, and hybrid neutron stars with quark cores. We found that quark and strange-quark stars are more efficient gravitational-waves radiators than pure neutron stars. Also, the damping times of various modes are rather sensitive to the details of the quark model such as the Bag constant, quark-gluon interactions, and quark mass corrections.

The state of matter at high temperature and/or density is not yet clearly known from QCD, though there are good reasons to suspect that some phase transitions occur at a critical temperature or density and nucleonic matter "melts" into a quark-gluon plasma (1). While such a state has not been identified in laboratory conditions, the possibility that quark matter exists in the cores of neutron stars (NS) is exciting and has profound implications to the astrophysics of compact stars. The search for astrophysical signatures of quark stars - either the light-quark stars made up of u and d quarks or the so-called strange-quark stars consisting of a mixture of u, d, s quarks - is difficult, because there are very few probes that can reveal the state of matter in the stellar core. In this talk we present results on the quadrupole oscillation modes of quark stars, which, because of their coupling to gravitational waves, may provide an observational tool to study the stellar interior (2) and a way to differentiate between quark stars and neutron stars. This differentiation is an important issue in astrophysics. If at high enough density, strange quark matters are more stable than ^{56}Fe (3; 4), then many of the pulsars may actually be strange-quark stars.

K.S. Cheng et al. (eds.), Stellar Astrophysics, 161–168.

Previous works on the oscillations of strange quark stars in the literature were concerned with the radial case only but not the nonradial one. For example, Cutler *et al.* (5) calculated the relativistic radial oscillation frequencies of a quark star using the MIT Bag Model equation of state (EOS), and Datta *et al.* ((6)) redid the calculation using a more realistic EOS taking into account quark-gluon interactions and quark mass corrections. Such calculations give the range of the eigenfrequencies of radial pulsations of quark stars, which can be compared with those of neutron stars. However, the radial oscillations of stars will not generate gravitational waves. Thus, nonradial oscillation modes should contain information more relevant to future gravitational-waves observations (7).

Einstein's theory of relativity predicts that nonradial stellar oscillations are damped due to the emission of gravitational waves, and thus their frequencies are complex; the oscillation modes are therefore quasinormal modes (QNM's).

We have studied three kinds of quark star models, characterized by three different EOS's: a pure light-quark star (LQS), a hybrid neutron star with a light-quark core and a neutron crust (HLQS), and a strange star with a core made up of u, d, s quarks and a neutron crust (HSS). In the latter case, we further studied two EOS's with different Bag constant for long-range quark-gluon interaction, and we name such models HSSI and HSSII. Whereas in these simple quark EOS's, the value of the bag constant determines whether strange quark matter is more stable than light-quark matter, and it probably is (4), we nevertheless calculate the oscillation modes for both light-quark stars and strange stars, since a comparison between these two classes of stars is still of theoretical interest given the current uncertainties in the state of QCD matter at high density.

We study the oscillation modes of several classes of model stars: light-quark stars entirely made up of quark matters (LQS1 – LQS6) described by the MIT Bag Model EOS (8), light quark stars with crusts made up of normal nuclear matter (HLQS1 – HLQS7), and strange quark stars made up of symmetric uds quark matter (6) with two different bag constants (HSSI1-7, HSSII1-7). As a comparison to the above quark star models, we also calculate the oscillation modes of relativistic polytropic stars as well as those of "traditional" neutron stars (AU1-6) constructed with the AV14 + UVII EOS tabulated in Ref. (9).

To describe small oscillations of a star, one considers small perturbations in the metric. It is natural and popular to decompose a general metric perturbation into terms with definite (odd or even) parities:

$$ds^2 = (ds^2)_o + h_{\mathrm{a}} + h_{\mathrm{p}} \quad , \tag{1}$$

where $(ds^2)_o$ is the background unperturbed metric, and $h_{\rm a}$ and $h_{\rm p}$ are the parity-odd and even metric perturbations respectively. The linearized general relativistic equation governing the axial stellar oscillations is (10)

$$\left[\frac{d^2}{dr_*^2} + \omega^2 - V_{rw}(r_*)\right] \psi_{rw}(r_*) = 0 \ , \tag{2}$$

where the tortoise coordinate r_* is dcfined as

$$r_* \ \equiv \ \int_0^r e^{(-\phi+\lambda)/2} dr \ , \tag{3}$$

and

$$V_{rw}(r_*) \ \equiv \ \frac{e^\phi}{r^3}\left[l(l+1)r + 4\pi r^3(\rho - P) - 6m(r)\right] \tag{4}$$

is an effective potential, which diverges as $l(l+1)/r^2$ as $r \to 0$. Outside the star, Eq. (2) reduces to the Regge-Wheeler wave equation with the effective potential

$$V_{rw}(r_*) = \left(1 - \frac{2M}{r}\right)\left[\frac{l(l+1)}{r^2} - \frac{6M}{r^3}\right] \ , \tag{5}$$

where M is the total mass of the star, and the tortoise coordinate in this region ($r \geq R$) can be simplified to

$$r_* = r + 2\,M\,\ln\left(\frac{r}{2M} - 1\right) \ , \tag{6}$$

This potential reaches a maximum value at an intermediate value of r and vanishes as $1/r^2$ as $r \to \infty$. The effective potentials both inside and outside the star are positive. As $r_* \to +\infty$, the wavefunction must have the familiar outgoing-wave form

$$\psi_{rw} \to e^{-i\omega r_*} \ . \tag{7}$$

Stellar oscillations observing polar parity in linearized general relativity are governed by four coupled first-order ordinary differential equations (ODE's) (10). Outside the star, the system is described by the Zerilli equation (10)

$$\left[\frac{d^2}{dr_*^2} + \omega^2 - V_z(r_*)\right] \psi_z(r_*) = 0 \ , \tag{8}$$

and V_z is the Zerilli potential, defined as

$$V_z(r_*) \equiv \frac{1 - 2M/r}{r^3(nr + 3M)^2}\left[2n^2(n+1)r^3 + 6n^2Mr^2 + 18nM^2r + 18M^3\right] \ . \tag{9}$$

Here $n = (l-1)(l+2)/2$, and M is the total mass of the star.

The quasi-normal modes of a system are defined as the oscillations of the system that satisfy the regularity boundary conditions at $r = 0$ and also the outgoing-wave boundary conditions at spatial infinity. We calculate these QNM's using the method of Leaver's series (11) because it is analytic, simple to implement and valid in the complex frequency domain we considered. Details of our calculations can be found in Ref. (12).

Since axial oscillations are not coupled with the motions of stellar fluid, only one kind of oscillation mode exists, which corresponds to the oscillations of the spacetime, called the gravitational-wave mode, or w-mode. We have calculated two families of w-modes of quark stars and the results are shown in Figure 4 for a central density of 1.6×10^{15} gcm^{-3}. For the first family of w-modes, called simply the w-modes or "curvature modes" (13), the oscillation frequencies and the damping rates *decrease* along the sequence of HSSII4, HSSI4, HLQS4 and LQS4, which is due to a corresponding *increase* of the masses of the stars (from $1.63 - 2.02$ solar mass), resulting from an increase of the stiffness of the EOS. Thus, we conclude that inclusion of the quark-gluon interactions and quark masses in a pure light-quark star would cause an increase the oscillation frequencies as well as the damping rates of a star. Also, we notice that the oscillation frequencies and damping rates of HSSI4 are similar to those of the neutron star AU4, because their masses are about the same.

It is also interesting to compare the oscillation modes of a hybrid neutron star to those of a quark star. As expected, the few lowest w-modes which have long wavelengths are not sensitive to the properties of the thin crust. The existence of the crust only affects higher modes that have wavelengths short enough to resolve the crust of the order of one kilometer. We see from the results of w-modes (see Figure 4 again) of LQS and HLQS that only the *seventh* (or other higher order) modes are sensitive to the crust.

The fundamental modes of oscillation, the so-called f-modes, of quark stars are shown in Figure 4. We notice that both the oscillation frequencies increase as the central densities of the quark and strange quark stars increase, following the same behaviour for realistic neutron stars in some regime of central densities (AU1 – AU4). Morever, for less compact quark, strange-quark and neutron stars, the damping rate increases as the density of the star increases. This means that a star radiates gravitational waves more efficiently as it becomes more compact. However, above some critical value of M/R, the behaviour changes. The damping rates of f-modes decrease again. This is because as a star becomes more compact than some critical value, the exterior potential can "trap" the gravitational waves efficiently.

This change in the behaviour of the damping rates is found to occur around $M/R \sim 0.3$ in homogeneous stars (14) and polytropic stars (15). For the realistic neutron stars we construct, it happens around $M/R = 0.2657$ (AU4). However, this behaviour does not occur for all quark and strange quark stars we consider, even though some of them are more compact. Thus, we conclude that, at least for the case of LQS and HLQS, the quark stars and hybrid stars are more efficient in radiating gravitational-waves than neutron stars. Furthermore, we notice that the damping rates of all quark and strange quark stars are larger compared with neutron stars, for the case of small compactness, or of central density $< 1.6 \times 10^{15}$ gcm^{-3} (i.e. the 4th models of each case.). If quark and strange quark stars described by the EOS's we use indeed exist in nature, their central densities would not be larger than that of the 5th model in each case, i.e. $\rho_c \sim 2 - 3 \times 10^{15}$ gcm^{-3}, because otherwise they would become unstable against radial oscillations.

For a fixed central density $\rho_c = 6 \times 10^{14}$ gcm^{-3}, we found the periods $T = 2\pi/\mathrm{Re}(\omega)$ of f-mode oscillations of quark stars, polytropic stars and realistic neutron stars to be very similar (around 0.6 ms), whereas their damping times $\tau = 1/\mathrm{Im}(\omega)$ are 0.3, 1.2, and 1.9s respectively, differing by a factor of at least 4. The effects of the quark core in a neutron star are clear. The damping time is however rather sensitive to the value of the Bag constant. For example, changing B from 56 MeVfm^{-3} to 67 MeVfm^{-3} induces an increase of the damping time by a factor of three (comparing results for HSSI and HSSII). Given the large uncertainties in the quark matter EOS, it seems difficult to look for signals of quark stars through the f-mode damping times. Fortunately, the profile of f-modes of realistic neutron stars are different from those of quark and strange quark stars, especially for a central density that lies between $1.6 - 3.0 \times 10^{15}$ gcm^{-3}.

We studied the relativistic nonradial oscillations of quark stars and hybrid neutron stars with quark and strange quark cores by calculating their quasi-normal modes. We found significant differences in the oscillation frequencies of quark stars and hybrid neutron stars with quark cores for the higher axial and polar w-modes. Also, the damping times for all modes are rather sensitive to the Bag constant, quark-gluon interactions, and quark masses corrections. From the results in f-modes we conclude that quark and strange-quark stars are more efficient gravitational-waves radiators than realistic neutron stars.

This work is partially supported by a Hong Kong Research Grants Council grant CUHK 4189/97P and a Chinese University Direct Grant 702/1. Also, we thank W. M. Suen and Y. T. Liu for various discussions.

References

McLerran L. D. and Svetitsky B. (1981), Phys. Lett. **98B**, 195; Kuti J., Polónyi J., and Szlachányi K. (1981), Phys. Lett. **98B**, 199; Engels J., Karsch F., Satz H. and Montvay I. (1981), Phys. lett. **101B**, 89.

Haensel P., Zdunick J. L. and Schaeffer R. (1989), Astron. Astrophys., **217**, 137.

Bodmer A. R. (1971), Phys. Rev. D **4**, 1601.

Witten E. (1984), Phys. Rev. D **30**, 272.

Cutler C., Lindblom L. and Splinter R. J. (1990), Ap. J., **363**, 603.

Datta B., Sahu P. K., Anand J. D. and Goyal A. (1992), Phys. Lett., **B283**, 313.

Abramovici A. A. *et al.* (1992), Science **256**, 325.

Chodes A., Jaffe R. L., Johnson K., Thorn C. B. and Weisskopf V. F. (1974), Phys. Rev. D **9**, 3471.

Wiringa R. B. and Fiks V. (1988), Phys. Rev. C **38**, 1010.

Lindblom L. and Detweiler S. L. (1983), Ap. J. Suppl., **53**, 73; Detweiler S. L. and Lindblom L. (1985), Ap. J., **292**, 12.

Leaver E. W. (1985), Proc. Roy. Soc. London A, **402**, 285; Leaver E. W. (198 J. Math. Phys., **27**, 1238.

Yip C. W., Chu M.-C., and Leung P.T. (1999), Ap. J., **513**, 849.

Andersson N., Kojima Y. and Kokkotas K. D. (1996), Ap. J., **462**, 855.

Detweiler S. L. (1975), Ap. J., *197*, 203.

Andersson N. and Kokkotas K., "Pulsation modes for increasingly relativistic polytropes", report gr-qc/9706094.

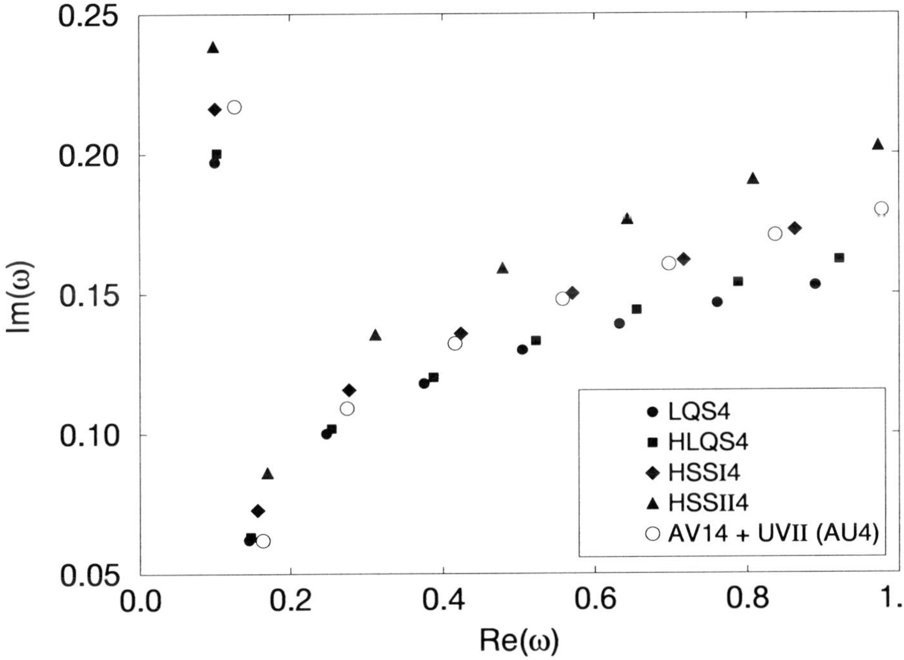

Figure 1 The axial w-mode (both w- and w_{II}-modes) QNM spectra for quark stars and hybrid neutron stars. In all of them $\rho_c = 1.6 \times 10^{15}$ gcm^{-3}. The frequencies are shown in km^{-1} $(G = c = 1)$.

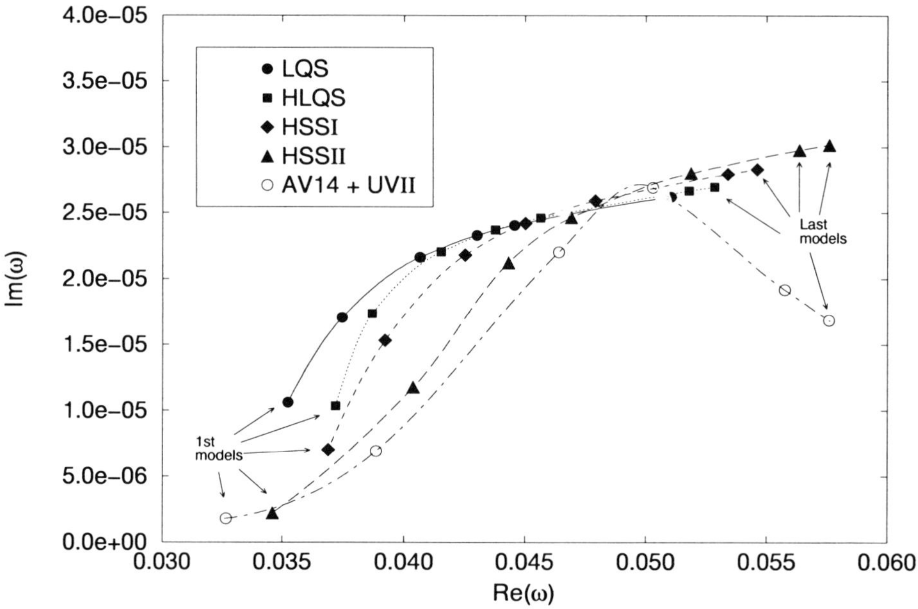

Figure 2 The fundamental mode (f-mode) of oscillations of quark stars and hybrid neutron stars with different central densities ρ_c.

A NEW INDICATOR FOR THE SPECTRAL STATES AND STATE TRANSITIONS IN CYGNUS X-1

Linqing Wen, Wei Cui and Hale V. Bradt
Center for Space Research, MIT, Cambridge, MA 02139 USA
lqw@space.mit.edu, cui@space.mit.edu, hale@space.mit.edu

Abstract Based on data from the *RXTE* All-Sky Monitor, we found that, on time scales of days to weeks, the X-ray intensity of Cyg X-1 is correlated with the spectral hardness of the source in the soft state, but is anti-correlated with the latter in the hard state. The correlation shows interesting evolution during the state transition of 1996 May–August. The entire episode consists of three distinct phases: (1) a $\sim$ 20-d transition phase from the hard state to the soft state, where the correlation changes from negative to positive, (2) a 50-d soft state with a steady positive correlation, and (3) a 20-d transition back to the hard state. The existence and time scales of these three phases are consistent with the timing results from the RXTE/PCA observations. This result confirms the conclusion that the soft X-ray flux alone is not a reliable indicator for a true soft state. We argue that the flux-hardness correlation likely provides a better indicator for defining the spectral states of Cyg X-1.

Keywords: binaries: general — stars: individual (Cygnus X-1) — X-rays: stars

1. INTRODUCTION

Cyg X-1 is identified as a binary system of 5.6-day orbital period which contains an O9.7 Iab supergiant and a compact object that is believed to be a black hole ((Bolton 1972); (Webster & Murdin 1972)). The observed intense X-ray flux from this system is thought to be produced close to the black hole in an accretion disk which emits soft X-ray photons and in a hot corona ($T \sim 10^8$–10^9 K) that inverse-Compton scatters low energy photons to higher energies (e.g., (Liang, & Nolan 1984); (Tanaka & Lewin 1995) and references therein). The accretion flow from the supergiant is probably intermediate between Roche-lobe overflow and stellar wind accretion (e.g., (Gies & Bolton 1986)).

K.S. Cheng et al. (eds.), Stellar Astrophysics, 169–174.

The observations of Cyg X-1 indicate that the system usually assumes one of the two states, the hard state and the soft state. Most of the time, Cyg X-1 stays in the hard state where its 2–10 keV luminosity is relatively low and the energy spectrum is hard. Every few years, the system undergoes a transition to the soft state and stays there for weeks to months before returning to the hard state. During the transition to the soft state, the 2–10 keV luminosity increases, often by a factor of more than 4, and the energy spectrum becomes softer. There is a strong anti-correlation relation between the low energy ($\lesssim$ 10 keV) and high-energy X-ray fluxes with the total 1.3–200 keV luminosity remaining unchanged to within $\sim$ 15% ((Zhang et al. 1997)). The transition between the states, as indicated by dramatic change of soft X-ray flux, requires from less than a day to more than a week (see reviews by (Oda 1977); (Liang, & Nolan 1984) and references therein).

A number of models have been proposed to explain the spectral evolution during the state transition. For instance, Ichimaru (1977) suggested that the physical conditions of the accreted gas near the disk outer-boundary could drive the disk into either an optically thick state or an optically thin state, which corresponds to the soft and hard state respectively. Based on the effects of the blackhole rotation on the strength of the ultrasoft disk component, Zhang, Cui, & Chen (1997) argued that the state transitions of Cyg X-1 may be simply due to a temporary disk reversal between a retrograde disk (hard state) and a prograde disk (soft state), which can occur in wind accretion systems.

In the magnetic flare model ((Di Matteo, Celotti, & Fabian 1999)), the soft (hard) state corresponds to a lower (higher) scale height of magnetic flares above the accretion disk. The flares are energized by the reconnection of magnetic flux tubes rising from the accretion disk due to magnetic buoyancy instability. In the soft state, intense flares close to the disk greatly enhance the soft photon field which results in the soft spectrum. In the hard state, the flare is triggered high above the disk, the system is photon-starved and results in hard Comptonized energy spectrum. In the framework of advection-dominated accretion flows (ADAFs), Esin et al. (1998) argued that the states are uniquely determined by the mass accretion rate $\dot{m}$. In the hard state, $\dot{m}$ is relatively low, the optically thin ADAF region is large and thus dominates the X-ray emission. In the soft state, the $\dot{m}$ is so high that the inner ADAF zone becomes very small and the X-ray emission is dominated by the optically thick geometrically thin disk which extends to the marginally stable orbit. More data are needed to distinguish these models.

Here, we report the results from a quantitative study of the correlation between the soft X-ray intensity and the spectral properties of the source on a time scale of days to weeks. We show, for the first time, that the correlation is positive for the soft state but is negative for the hard state. We argue that this correlation relation provides a more refined indicator for defining the spectral states than the soft X-ray intensity.

2. OBSERVATION AND DATA

The All Sky Monitor on board the Rossi X-ray Timing Explorer (*RXTE*) ((Bradt, Rothschild, & Swank 1993)) has been monitoring the sky routinely since 1996 February. The ASM consists of three Scanning Shadow Cameras, each consisting of a coded mask and a position-sensitive proportional counter. A linear least squares fit to the shadow patterns from a 90-s observation by one of the three cameras of the ASM yields the source intensity in three energy bands (1.5–3, 3–5, and 5–12 keV). The intensity is usually given in units of the count rate expected if the source were at the center of the field of view in one of the cameras; in these units, the 1.5–12 keV Crab nebula flux is about 75 ASM ct/s. The estimated errors of the source intensities include the uncertainties due to counting statistics and a systematic error of 1.9% determined empirically from the light curves of the Crab Nebula. A given source is typically observed $\sim$ 15 times a day. A detailed description of the ASM and the light curves can be found in Levine et al. (1996) and Levine (1998).

We analyzed the ASM intensity and two hardness ratios HR1 and HR2 of Cygnus X-1 over a four-year period (1996 March–1999 August). We define HR1 as the ratio of the ASM count rates in the 3–5 keV band to that in the 1.5–3 keV band, and HR2 is the ratio of the count rate in the 5–12 keV band to that in the 3–5 keV band. The hardness ratios provide a rough measure of the X-ray spectrum of the source.

3. ANALYSIS AND RESULTS

The correlation between the ASM intensity of Cyg X-1 and the hardness ratios was quantified by means of a non-parametric method first proposed by Spearman (the Spearman ranking method; (Press et al. 1992)). For each data set, the value of each data point was replaced by the value of its ranks among all other data points, that is, for N data points, the smallest value would be replaced with value 1 and the largest with N. The tied values were assigned the mean of the ranks they would have if they were to be slightly different. The correlation coefficient r_s is defined as,

$$r_s = \frac{\sum_i (R_i - \overline{R})(S_i - \overline{S})}{\sqrt{\sum_i (R_i - \overline{R})^2}\sqrt{\sum_i (S_i - \overline{S})^2}}, \tag{1}$$

where R_i and S_i are the assigned ranks for the data points in each of the two data sets. The significance of a non-zero value was tested by computing:

$$t = r_s \sqrt{\frac{N-2}{1-r_s^2}} \tag{2}$$

which is distributed approximately as Student's distribution with $N-2$ degrees of freedom. The main advantage of the Spearman-ranking method is that the significance level of the correlation does not depend on the the original probability distribution of the data. Because of this, the significance level of the correlation can be reliably computed even if the number of the data points used is small. Despite of some loss of information in replacing the original values by ranks, the method is reliable in the sense that when a correlation is demonstrated to be present non-parametrically, it is really there. In our calculation, the coefficients were calculated from data sets of every 5-day duration. We excluded a few data segments with less than 30 data points to avoid spurious r_s values.

The results are shown in Figure 1. There is a distinct positive correlation on a time scale of 5 days during the soft state with an average $r_s \sim 0.7$, which corresponds to a false alarm possibility $< 10^{-30}$. Note this is in contrast to the global anti-correlation of the ASM intensity and hardness ratios between the two states. For the entire $\sim$ 1000 days of the hard state, the coefficients are negative (partly shown in Figure 1), even though the correlation strength ($rs \sim -0.2$) is not as large or significant as that of the soft state. The entire episode of the 1996 state transition can be divided into three distinct phases: (1) a $\sim$ 20-d transition phase from the hard state to the soft state, where the correlation coefficients went from negative to positive, (2) a 50-d soft state with a steady positive correlation, and (3) a 20-d transition phase back to the hard state. We chose the start and end time of the transitional period somewhat arbitrarily, i.e., when the hardness ratio was roughly at the mid-point between the mean levels of the hard and soft states.

4. DISCUSSION

Our results show that the transition between the hard and soft states lasted for about 20 days during the 1996 state transition of Cygnus X-1 as opposed to $\lesssim$ 7 days indicated by the change of the soft X-ray flux.

This agrees remarkably well with the results of detailed PCA timing analyses ((Cui et al. 1997a); (Cui et al. 1997b)). It is well-known that for Cyg X-1, the power density spectra (PDS) is characterized by a white noise component at low frequencies and a high-frequency cut-off for the hard state, and by a power-law component in the soft state. The results from the PCA timing analyses showed that, during the 1996 state transition of Cyg X-1, the characteristic of the PDS evolved from the hard-state, through an intermediate stage, to the soft state as the system finally settled into the soft state. The inferred transition period agrees with our phase 1. The reverse was observed during the transition back to the hard state, which agrees with our definition of phase 3. We also note that Li, Feng, & Chen (1999) have analyzed the PCA data for the intensity-hardness correlation relations in the two states. Their results are similar to ours, but on shorter time scales of 0.01s–100s and for different energy bands (they used intensity in the 2-60 keV band and hardness ratio of the 13–60 keV to the 2–6 keV band).

In conclusion, we suggest that the spectral states of Cyg X-1 are better defined by the correlation between the soft X-ray flux and the spectral hardness of the source (than by the soft flux alone). Our results clearly indicate that a $\sim$ 20 days transition phase is required for the system to settle into and to leave the soft state.

Acknowledgements

We are very grateful to the entire *RXTE* team at MIT for their support. Support for this work was provided in part by NASA Contract NAS5-30612.

References

Bolton, C. T. 1972, Nature, 235, 271

Bradt, H. V., Rothschild, R. E., & Swank, J. H. 1993, Astronomy and Astrophysics, 97, 355

Cui, W., Heindl, W. A., Rothschild, R. E., Zhang, S. N., Jahoda, K., & Focke, W. 1997a, ApJ, 474, L57

Cui, W., Zhang, S. N., Focke, W., & Swank, J. H. 1997b, ApJ, 484, 383

Di Matteo, T., Celotti, A. & Fabian, A. C. 1999, MNRAS, 304, 809

Esin, Ann A., Narayan, Ramesh, Cui, Wei, Grove, J. Eric & Zhang, Shuang-Nan 1998, ApJ, 505, 854

Gies, D. R. & Bolton, C. D. 1986b, ApJ, 304, 389

Ichimaru, S. 1977, ApJ, 214, 840

Li, T. P., Feng, Y. X. & Chen, L. 1999, ApJ. 521, 789

Liang, E. P. & Nolan, P. L., 1984, Space Science Reviews, 38, 353

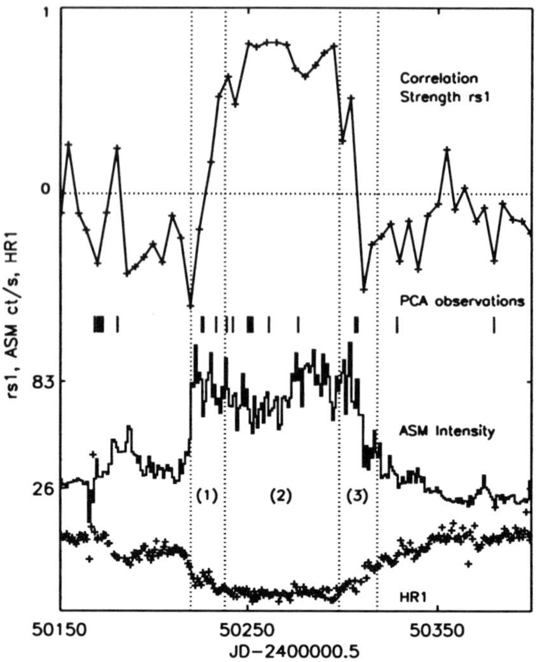

Figure 1 In the soft state, the 1-12 keV ASM intensity is strongly correlated with the hardness ratio ($rs_1 \sim 0.7$). The entire transition can be divided into three distinct phases, indicated as (1), (3) for the transition phases and (2) for the steady soft state. In the hard state, the correlation is negative. Also shown are the ASM intensity and hardness ratio HR1 and the PCA observations indicated in vertical lines.

Levine, A. M., Bradt, H., Cui, W., Jernigan, J. G., Morgan, E. H., Remillard, R., Shirey, R. E., & Smith, D. A. 1996, ApJL, 469, L33

Levine, A. M. 1998, in Nuclear Physics B (Proc. Suppl.), 69/1-3, 196

Oda, M. 1977, Space Science Reviews, 20, 757

Press, W. H., Teukolsky, S. A., Vetterling, W. T., & Flannery, B. P. 1992, Numerical Recipes, 2nd ed. (Cambridge: Cambridge University Press), 569

Tanaka, Y. & Lewin,W. H. G. 1995, in X-ray Binaries, ed. Lewin, W. H. G.,Van Paradijs, J., & Van den Heuvel, E. P. J. (Cambridge: Cambridge University Press), 126

Webster, B. L., & Murdin, P. 1972, Nature, 235, 37

Zhang, S. N., Cui, W., Harmon, B. A., Paciesas, W. S., Remillard, R. E., & Van Paradijs, J. 1997, ApJL, 477, L95

Zhang, S. N., Cui, Wei & Chen, Wan 1997, ApJL, 482, L155.

DOES ROTATION ALONE DETERMINE WHETHER AN A-TYPE STAR'S SPECTRUM IS ABNORMAL OR NORMAL?

HELMUT A. ABT & DARYL W. WILLMARTH
Kitt Peak National Observatory
Box 26732, Tucson, AZ 85726-6732, USA

Abstract The principal abnormalities among A-type stars are the metallic-line (Am) and peculiar A (Ap) stars. In these, diffusion (Michaud 1980) occurring in stars rotating at less than about 120 $km\ s^{-1}$ allows certain elements to sink and others to rise into the atmosphere. Observationally, all Am and Ap stars are observed to have rotational velocities less than that limit and most of the normal stars have higher rotational velocities. But a large study (Abt & Morrell 1995) showed that 15 – 20% of the normal stars had lower rotational velocities. Could it be that rotation alone does not determine whether a slow rotator has an abnormal or normal spectrum?

We determined from published K-line measures that about 17% of the "normal" A2-F0 stars are really undetected abnormal stars. This conclusion is supported by a current search for binaries among 119 stars. Among the A0-A1 stars our inability to always distinguish luminosity class IV from V stars caused some (slowly-rotating) IVs to be included among the Vs. We conclude that rotation alone is sufficient to determine whether a star will have an abnormal or normal spectrum.

1. INTRODUCTION

Let us first describe the kinds of abnormal stars that occur among Population I A-type stars. Those are summarized in Table 1.

The Am stars (Titus & Morgan 1940) are usually in binaries (Abt 1961) so the rotational velocities are reduced by tidal interaction below 120 $km\ s^{-1}$, which then allows diffusion (Michaud 1980) to act and produce abnormal atmospheric abundances. In the Ap stars their strong magnetic fields, rather than duplicity, evidently reduces the rotational velocities so that extreme diffusion effects can occur. In the λ

K.S. Cheng et al. (eds.), Stellar Astrophysics, 175–180.

Table 1 Kinds of Abnormal A-type Stars

Kind	Abnormality	Rotation	Binary?	Mechanism
Metallic-line (Am)	Mild over- and under-abundances	$V < 120\ k/s$	Usually	Diffusion
Peculiar A (Ap)	Extreme over- and under-abundances	$V < 120\ k/s$	Rarely	Diffusion
λ Boo and 4481 weak	Mild under-abundances	All $O < V < 300$	Rarely	Accretion

Bootis stars (and milder cases where only Mg II 4481 is weak), the accretion/diffusion model by Turcotte & Charbonneau (1993) may work. Because the observations indicate that that process is independent of rotational speed, we will not discuss it further.

Abt & Morrell (1995) (hereafter AM95) listed MK classifications, projected rotational velocities ($V \sin\ i$), and Mg II 4481 line strengths for about 2000 stars. Stars can have small values of $V \sin\ i$ either because of small rotational velocities, V, or small angles between the rotational axes and lines of sight, i. To convert from a distribution in $V \sin\ i$ to one in V, we will assume random orientations of axes.

2. THE A5-F0 STARS

The resulting distribution of rotational velocities for the A5-F0 stars is given in Fig. 1. We see that all the 17 Ap and 116 Am stars have rotational velocities less than about 120 $km\ s^{-1}$ and most of the normal class V stars have larger rotational velocities but about 17% of the normal stars have $V < 120\ km\ s^{-1}$.

Visual MK classification does not involve quantitative measurements, so subtle abnormalities can be missed. In fact, three standards (Sirius, Vega, and α Dra) initially used by AM95 were subsequently found to be abnormal in abundance.

Most Ap and Am stars have weak Mg Il 4481 but that is not a unique characteristic of abnormal stars (AM95). However Henry & Hesser (1971) built a special spectrograph to measure Ca II K-line strengths and found that all Am (and probably Ap) stars have weak K lines. Among the 15 Henry & Hesser stars thought by AM95 to be normal, three (20$\pm$14%) have weak K lines. All three have $V \sin i < 120\ km\ s^{-1}$, which is a necessary but not sufficient condition for them to be abnormal. We do not have radial velocity measures for those three stars to

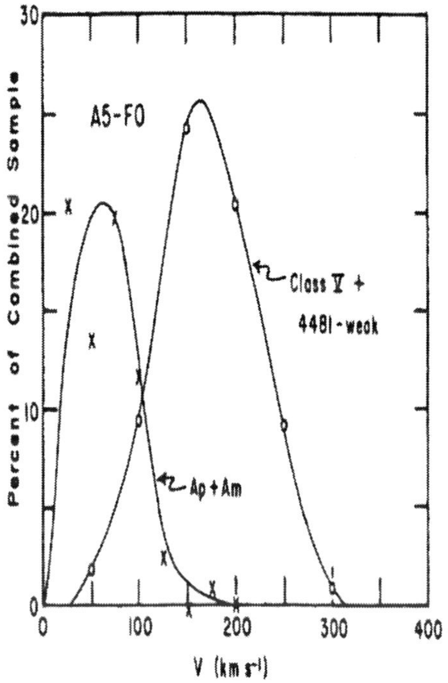

Figure 1 The rotational velocity distributions, corrected for random orientations of rotational axes, for 17 Ap and 116 Am stars on the left and 234 A5-F0 V stars on the right.

determine whether they are binaries. However new radial velocities in this program for 25 normal A5-F0 V stars from the AM95 list and with $V \sin i < 100\ km\ s^{-1}$ show that at least three ($12 \pm 8\%$) are binaries with periods less than 100 days. Among the late A stars, Abt (1965) (see Fig. 2) found that whereas Am stars have orbital periods from less than 1 day to many years, no normal stars occur in binaries with periods less

than 100 days. [If a star were in a short-period binary, tidal interactions would slow its rotation and it would become an Am star.] Therefore again we have evidence that roughly 17% of the "normal" A5-F0 V stars in AM95 are really Am stars. Those would be found among the slow rotators.

We conclude so far that the reason for the overlap in rotational velocity distributions among the A5-F0 V stars is that about 17% are really undetected Am stars.

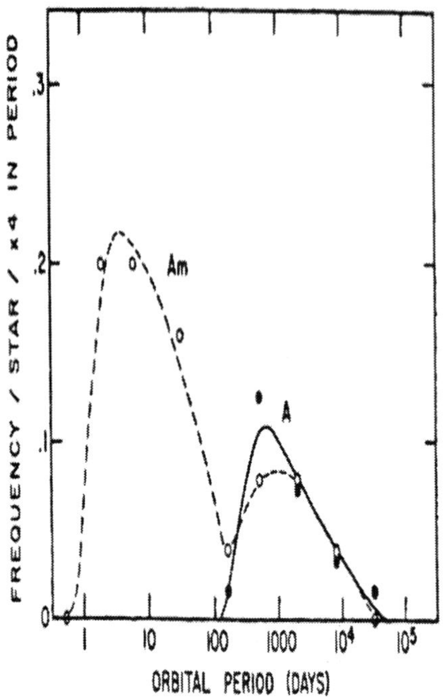

Figure 2 The orbital periods for 20 Am binaries (circles and dashed line) and for 15 normal A4-F2 IV or V binaries (filled circles and solid line).

3. THE A2-A4 STARS

The A2-A4 stars (see Fig. 3) show the same kind of distribution in rotational velocity with essentially none of the 29 Ap and nine Am stars having $V > 120\ km\ s^{-1}$ and most of the normal class V stars having greater rotational velocities, but about 19% have low rotational velocities like the Ap and Am stars.

Using the same methods, we find that four of the 26 stars ($15 \pm 7\%$) common to AM95 and Henry & Hesser have weak K lines. All four have $V \sin i < 120\ km\ s^{-1}$, a necessary but not a sufficient condition that they are Ap or Am stars. The current study of 38 A2-A4 V stars with $V \sin i < 100\ km\ s^{-1}$ shows that at least four of them have orbital periods less than 100 days. Therefore $11 \pm 6\%$ of them are probably not normal stars.

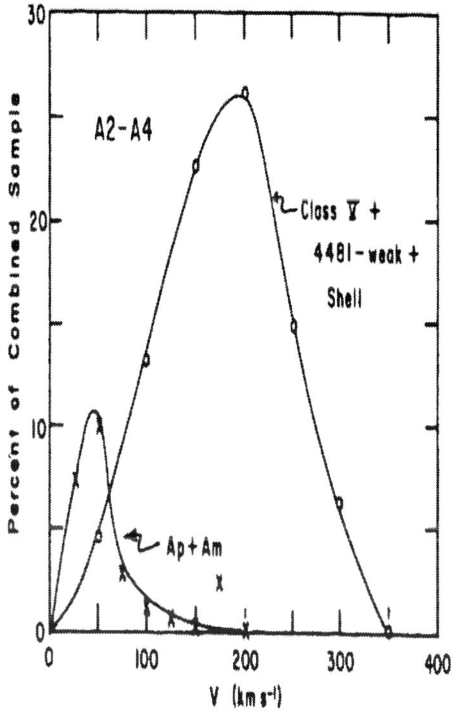

Figure 3 The rotational velocity distributions, corrected for random orientation of rotational axes, for 29 Ap and 9 Am stars on the left and 242 A2-A4 V stars plus 31 4481-weak stars on the right.

These results are consistent with the overlap among the A2-A4 stars also being due to our not recognizing all of the abnormal stars in this region.

4. THE A0-A1 STARS

The rotational velocity distributions for the A0-A1 stars (shown in AM95) also has an overlap between the normal and abnormal stars. However, the explanation used for the A2-F0 stars does not work for the A0-A1 stars because none of the normal class V stars have weak K lines.

We note that:

1. The class IV stars have much lower rotational velocities (mean of 62 $km\ s^{-1}$) than class V stars (mean of 142 $km\ s^{-1}$) and

2. It is somewhat difficult to distinguish class IV from V stars.

Could it be that some of the normal "class V" stars are really of class IV and that is why they have smaller rotational velocities?

We used Hipparchos parallaxes as an independent way to tell class IV from V stars. Because the parallaxes of these stars average 0.010", not much larger than the measuring errors, parallaxes for individual stars are not meaningful. But if we determine means for many stars, the means are significant. For instance, for A0 IV and A0 V stars the mean absolute visual magnitudes are $M_v = 0.20 \pm 0.22$ (s.e.) and 0.49 ± 0.07 mag, respectively. For Al IV and Al V stars the means are 0.46 ± 0.09 and 0.79 ± 0.08 mag., respectively. Consider the rotational velocities of stars whose parallaxes indicate $M_v < 0.63$ and $M_v > 0.63$ mag., where $M_v = 0.63$ mag. is an average value for A0-Al IV and V stars. The mean rotational velocities are 132 and 152 $km\ s^{-1}$, respectively. When we derive the rotational velocity distribution for the normal A0-A1 stars with the fainter absolute magnitudes ($M_v > 0.63$), the overlap nearly disappears. Therefore we suspect that in our sample of A0-A1 V stars, we accidentally included some IVs that have low rotational velocities.

Our overall conclusion is that the overlaps of 15 – 20% in rotational velocities in A-type stars is that we failed to fully separate the abnormal and normal stars, or the class IV and V stars, and that rotation alone does determine whether a star will be a normal or abnormal one.

References

Abt, H. A. 1961, ApJS, 6, 37
Abt, H. A. 1965, ApJS, 11, 429
Abt, H. A. & Morrell, N. 1. 1995, ApJS, 99, 135 (AM95)
Henry, R. C., & Hesser, J. E. 1971, ApJS, 23, 421
Michaud, G. 1980, AJ, 85, 589
Titus, J., & Morgan, W. W. 1940, ApJ, 92, 256
Turcotte, S., & Charbonneau, P. 1993, ApJ, 413, 376

THE DISTRIBUTION OF SEPARATIONS OF BINARIES.

ARCADIO POVEDA, CHRISTINE ALLEN and M.A. HERRERA
Instituto de Astronomía, Universidad Nacional Autónoma de México, Circuito Exterior, Ciudad Universitaria, 04510 México, D. F. México

Abstract We examine the distribution of separations (major semiaxes) between components in wide binaries for various samples of double stars: IDS, binaries in the solar vicinity, binaries in young clusters (Hyades and Orion Nebula Cluster) and binaries among high-velocity, low-metallicity stars. For all the groups studied, we find that binaries follow Öpik's distribution of separations in the interval $40 \lesssim a\ (AU) \lesssim s_c(t)$, where $s_c(t)$ is an age- dependent critical separation.

We conclude that the primeval distribution of separations among wide binaries follows Öpik's function $f(s) \propto 1/s$ up to separations $s_c(t)$ which, for the youngest groups, extend to more than 20 000 AU; the critical separation $s_c(t)$ for the oldest binaries in the solar vicinity is 2400 AU. Kuiper's Gaussian representation of the distribution in log a seems to be the result of the superposition of two different distributions: an increasing one for the close binaries ($a \lesssim 40$ AU) and a decreasing one (Öpik's) for the wide binaries each one with its respective noise due to small number statistics. We argue that one should not attempt to force a single distribution on two groups of binaries that have different formation mechanisms: close binaries that originated from disk fragmentation and wide binaries that were formed by independent condensations.

1. INTRODUCTION

The present day distribution of the elements of wide binaries gives important information relative to the initial conditions of star formation and their subsequent dynamical evolution. In fact, the frequency distribution of the masses of the components, the multiplicity of the field stars, the relation of close binaries to wide binaries, the distribution of major semiaxes for each type, and the kinematics of various groups of binaries, are all characteristics that keep memory of the process of star formation and of the subsequent dynamical interactions, both internal and external.

K.S. Cheng et al. (eds.), Stellar Astrophysics, 181–190.

In the present paper we are concerned with the distribution of major semiaxes of wide binaries; In fact, several functions have been proposed in the past to describe the distribution of the periods, major semiaxes, or separations. We identify two main distributions: Öpik's (1924) frequency distribution of separations $f(s) \propto 1/s$, and Kuiper's (1942) distribution, which is expressed as a Gaussian in log a (or log P).

In previous work, the present authors (Poveda 1988, Poveda, Allen, and Herrera, 1997, Allen, Poveda and Herrera, 2000) have presented various samples of stars that follow Öpik's frequency distribution of separations. On the other hand, Duquennoy and Mayor (1991), based on a sample of high precision radial velocities for stars in Gliese's et al. catalogue (1991), found the frequency distribution of log P to be a Gaussian with the maximum at log P (days) ≈ 4.5.

The purpose of the present paper is to clarify this apparent contradiction between the two representations of $f(a)$. We will argue that the distribution of separations need not be unique for all types of binaries (close and wide) and for all ranges of separations and that for wide binaries the primeval distribution follows Öpik's relation up to a critical separation $s_c(t)$ which, due to dynamical interactions with massive objects, evolves toward smaller values, depending on the nature of the perturbing objects. However, the ascending branch of the Gaussian in log P is an appropriate description of the distribution for close binaries.

2. TWO DIFFERENT GROUPS OF BINARIES

In a classical paper, Abt and Levy (1976) showed that the mass frequency distribution of the secondaries of a sample of G-type binaries followed van Rhijn luminosity function (Salpeter's mass function) for those binaries with periods larger than about 100 years ($a \gtrsim 25$ AU), while for those binaries with smaller periods the distribution of masses of the secondaries was more like $f(m) \propto m^{1/2}$. That is, while for wide binaries the less massive the secondary the more frequent it is, for the short period ones the smaller the mass of the secondary the less frequent it is. This important result has been later confirmed by other authors (see for instance Mazeh and Goldberg 1992, and Mazeh et al. 1992).

The fact that the secondaries have two different mass spectra provides a strong indication that we are facing two different processes of star formation, one for the short-period binaries ($a \lesssim 40$ AU) and another for the wide binaries. Since wide binaries follow Salpeter's mass distribution, we take this as evidence that these binaries are the result of the random pairing of two independent condensations (stars) each one extracted from the Salpeter distribution. On the other hand, the close binaries, because of the small scale of the major semiaxes, are entirely compatible with the scale of preplanetary disks, and thus they are likely to be the result of fragmentation of circumstellar disks. Numerical simulations of disk fragmentation (Boss 1995, and references therein) support this explanation for the formation of close binaries. Having a different origin, each one of these groups of binaries may follow a different distribution of separations. For example, binaries that stem from disk fragmentation cannot have major semiaxes larger than the radius of preplanetary disks (at most a few hundred AU), while binaries that result from independent condensations in a dense cloud may certainly have much larger major semiaxes, of the order of Jeans' scale in the cores of molecular clouds, i.e., tens of thousands of AU; binaries stemming from independent condensations may not be able to form (or survive) at separations smaller than a few tens of AU, because they would be destroyed by the process of disk formation and fragmentation into stars.

The previous considerations supply good reasons to abandon the concept of having a single, time independent, frequency distribution of separations to describe both groups of binaries. It seems more fruitful to study, instead, the distribution of separations of each group separately (for details see Allen et al. 1997, and Poveda et al. 1997); moreover, for the wide binaries we recognize that there ought to be a subgroup of very wide and weakly bound binaries that, because of external gravitational perturbations by massive objects that dissociate them, will deviate from Öpik's distribution at separations larger that $s_c(t)$.

Our present work is based on the premise of a single primeval frequency distribution of separations for the wide binaries (stemming from independent condensations), which is valid in the interval $40 \lesssim a\ (AU) \lesssim s_c(t)$, where $s_c(t)$ is a function of the age of the binary and of the ambient conditions (number density of perturbers, relative velocity between the binary and the perturbers, etc.)

In the following section we will review the frequency distribution of separations of wide binaries in various groups of stars of different ages.

3. THE DISTRIBUTION OF SEPARATIONS OF WIDE BINARIES

To establish the distribution of major semiaxes for wide binaries requires data for a large number of binaries for which orbital elements are known. Unfortunately, the periods of wide binaries (visual and common-proper-motion pairs) are so long that no orbit determination is possible and therefore no precise major semiaxes are available for them.

In order to circumvert the above limitations, and in view of the objective of the present discussion, we will take advantage of the following theorems, valid in a statistical sense:

1. For wide binaries of known distances, Couteau (1960) derived the following relation:

$$E(\log\ a) - E(\log\ s) = 0.146, \tag{1}$$

   where a is the major semiaxis, s the projected separation on the sky of the components of a binary, and E the expectation value of the variable in parentheses.

   The above relation was derived by Couteau on theoretical grounds. He tested its observational validity against a catalogue of 410 orbits and found the constant to be 0.150. It is not too different from the empirical relation derived earlier by Kuiper (1942), where the constant had a value of 0.11.

   Couteau's relation allows us to obtain a statistical estimation of the major semiaxes of a binary from the observed angular separation s of the components. When the distance to a given pair is known, the angular separation can be converted to a linear separation (in AU), and hence the estimate of a is also known in AU. We have determined the expected major semiaxes in AU for a large number of wide binaries both in the solar vicinity, and in galactic clusters.

   There are, however, quite a number of binaries for which no distances are known, for instance, binaries from the IDS, from the WDS, or from the LDS. For these objects we can still investigate the form of the frequency distribution of separations by taking advantage of a theorem first suggested by Ambartsumian:

2. If the distribution $f(s)$ of the projected angular separations of a group of binaries located at very different distances from the observer is $f(s) \propto s^{-n}$, then the distribution of their "true" projected separations r has also the form $f(r) \propto r^{-n}$.

   With the help of the above theorems we have been able to determine the form of the distribution of separations of binaries from catalogues like the IDS and the LDS (Poveda 1988).

   (a) The IDS Catalogue.
   A study of the binaries in the IDS catalogue, after elimination of a large number of optical pairs, has shown that the distribution of separations has the form $f(s) \propto s^{-1}$ (Poveda 1988). The close companions of the triples in the same sample also

follow Öpik's distribution, while the distant companions of the same triples show a much slower decay with separation.

(b) The Solar Vicinity.

A search for wide binaries in Gliese's catalogue (Poveda et al. 1994) produced a list of 334 wide binaries and multiples for which projected separations in AU are available; moreover, for these systems, an age classification[1] has been given, thus making it possible to compare the distribution of separations for the youngest and for the oldest systems. From the analysis of the solar vicinity binaries we found not only that the distribution of separations follows Öpik's function, but that the maximum separation $s_c(t)$ for which Öpik's distribution holds is $s_c(t) = 2400$ AU for the oldest systems, and $s_c(t) = 8000$ AU for the youngest. An analysis of the velocity dispersions for these groups, yields average ages of 7 and 2.5 Gy respectively. In Figure 1, we display a comparison of the cumulative distribution of the logarithms of the major semiaxes of the oldest binaries in the solar vicinity against Öpik's relation, which on the plane $N - \log\ a$ is a straight line. Note that the data begin to depart from Öpik's relation at $a \gtrsim 2400$ AU. For details, see Poveda et al. (1997).

Weinberg et al. (1987) have modeled the evolution of binaries subject to the perturbations by encounters with molecular clouds. From Figure 6 in their paper it follows that systems with separations smaller that 2000 AU are practically unperturbed after 10^{10} *years*, while systems with original separations of 13 000 AU have a probability of survival of only 0.1.

Thus, we can conclude that for the mixture of stars in the solar vicinity the primeval distribution of separations follows Öpik's, and has remained intact up to 2400 AU throughout the age of the galactic disk; for younger binaries in the solar vicinity, Öpik's distribution holds up to much larger separations.

(c) The Youngest Systems

Larson (1995) has compiled a number of binaries found by various authors in the Taurus-Auriga region; his analysis shows that the surface density of companions follows the relation

[1]The age assignment was based on photometric, spectroscopic, and kinematic characteristics; for details, see Poveda et al. 1994.

$\Sigma(\theta) \propto \theta - 2.15$ where θ is the angular separation of components. This relation is essentially Öpik's distribution, which in Larson's paper extends up to 8000 AU; this distribution may extend to even larger separations, but in Larson's Figure 1 the distribution of separations of stars in clusters appears superposed to that of binaries and may mask the widest binaries.

An analysis of the surface distribution of Luyten's Hyades stars (1981) has yielded some 53 wide binaries, whose distribution of separations also follows Öpik's relation, up to separations of 20 000 AU (Nigoche 1999). A similar analysis of the Orion Nebula Cluster (Orion variables plus $H\alpha$ emission stars) has produced some 80 wide binaries which again follow Öpik's distribution to separations even larger than 20 000 AU (ibid.). The youth of the latter cluster, where stars have hardly moved from their birth place in the time elapsed (10^6 years), confirms the idea that Öpik's distribution of separations is the primeval one, and that it extends to well beyond 20 000 AU. On the other hand, pre-main sequence binaries in southern star forming regions also follow closely the

(d) High-Velocity, Metal-Poor Binaries

The present authors (Allen et al. 2000) have searched for common-proper-motion companions to the more than 1200 high-velocity, metal-poor stars observed by Schuster and Nissen (1988, 1989). The comparison of the proper motions of the above high-velocity stars with Luyten's NLTT stars has yielded 122 wide binaries belonging mostly to the galactic thick disk or halo. The distribution of separations for this group of binaries follows Öpik's relation, but now up to separation $s_c(t) \lesssim 10000$ AU. In Figure 2 we display the comparison of the cumulative distribution of the logarithms of the major semiaxes of the high-velocity, metal-poor stars against Öpik's distribution. Note that the data begin to depart from Öpik's relation at $a \gtrsim 10000$ AU.

4. DISCUSSION AND CONCLUSIONS

In the present communication, we have examined a variety of wide binaries in different groups and environments, with ages ranging from extremely young (10^6 y) to very old (14×10^9 y). We have consistently found that they follow Öpik's relation. Moreover, the critical separations $s_c(t)$ change with age in a manner concordant with the expectations, in the sense that the older the binaries, the smaller $s_c(t)$. Even the high-velocity, metal-poor binaries confirm this behavior, because the larger value found for the critical separation, $s_c(t)$ = 10000 AU, can be understood as a result of the fact that most of these objects have orbits that take them away from the galactic disk, where the number density of perturbers (molecular clouds) is expected to be largest.

Duquennoy and Mayor have proposed that the distribution of the logarithms of the periods of solar vicinity binaries, both close and wide, is Gaussian. However, in the plane $N - \log\ P$ Öpik's distribution is a horizontal straight line; if we plot Öpik's distribution in Duquennoy and Mayor's Figure 7 for major semiaxes in the interval $40 \lesssim a(AU) \lesssim 4000$ we see that Öpik's distribution is entirely consistent with Duquennoy and Mayor histogram for the few wide binaries they plot in this interval.

From the above discussion it follows that Kuiper's Gaussian function is the result of combining two different distributions of major semiaxes, an increasing one for close binaries and a decreasing one for wide binaries. The intersection of both distributions at about a = 25 -40 AU produces the observed maximum of the Gaussian distribution.

The distribution of major semiaxes for close binaries seems well described by the ascending branch of the Gaussian in Duquennoy and Mayor's Figure 7; we can speculate that in the absence of the disrupting effects of the inner part of the Öpik distribution, the ascending branch of Duquennoy and Mayor function may well extend to larger values compatible with the largest observed scales of the preplanetary disks.

The models of binary formation by fragmentation of circumstellar disks ought to take in consideration the distribution of separations as given by the ascending branch of Duquennoy and Mayor's Gaussian function. In the case of independent condensations the statistical three-body treatment by Valtonen (1997) seems to give a reasonable dynamical explanation for Öpik's distribution.

Summing up, the answer to the question: Öpik or Kuiper? appears to be:

- For close binaries (a $\lesssim$ 40 AU) : Kuiper.

- For wide binaries (a $\gtrsim$ 40 AU) : Öpik.

Acknowledgements

We thank A. Nigoche and A. Hernández for valuable assistance, and Conacyt for parcial financial support for this research throught Grant Nr. 25502-E.

References

Abt, H.A. and Levy, S. (1976), ApJS, 30, 273.

Allen, C., Poveda, A. and Herrera, M. A. (1997), in Docobo et al. (eds.): Visual Double Stars: Formation, Dynamics and Evolutionary Tracks, Dordrecht: Kluwer, p. 133.

Allen, C., Poveda, A. and Herrera, M. A. (2000), A&A, in press.

Boss, A.P. (1995), RMexAA, Conference Series, 1, 165.

Couteau, P. (1960), J. des Observateurs, 43, No. 3.

Duquennoy, A. and Mayor, M. (1991), A&A, 248, 485.

Gliese, W. and Jahreiss, H. (1991), in Sel. Astr. Cat., Vol. I, NSSC, NASA, GSFC (GJ91). Kuiper, G.P. (1942), ApJ, 95, 201.

Larson, R.V. (1995), MNRAS, 272, 213.

Luyten, W.J. (1981), University of Minnesota, Minneapolis.

Öpik, E. (1924), Tartu Obs. Publ, 25, No. 6.

Mazeh, T. and Goldberg, D. (1992), ApJ, 394, 592.

Mazeh, T., Goldberg, D., Duquennoy, A. and Mayor, M. (1992), ApJ, 401, 265.

Nigoche, A. (1999), Master's Thesis, UNAM.

Poveda, A. (1988), Ap. & Sp. Sci., 142, 67.

Poveda, A., Herrera, M.A., Allen, C., Cordero, G., and Lavalley, C. (1994), RMexAA, 28, 43.

Poveda, A., Allen, C. and Herrera, M. A. (1997), in Docobo et al. (eds.): Visual Double Stars: Formation, Dynamics and Evolutionary Tracks, Dordrecht: Kluwer, p. 191.

Schuster, W. and Nissen, P. (1988), A&AS, 73, 225.

Schuster, W. and Nissen, P. (1989b), A&A, 222, 69.

Valtonen, M. (1997) in Docobo et al. (eds.): Visual Double Stars: Formation, Dynamics and Evolutionary Tracks, Dordrecht: Kluwer, p. 241.

Weinberg, M.D., Shapiro, S.L. and Wasserman, I. (1987), ApJ, 312, 367.

Zinnecker, H. and Brandner, W. (1997) in Docobo et al. (eds.): Visual Double Stars: Formation, Dynamics and Evolutionary Tracks, Dordrecht: Kluwer, p. 115.

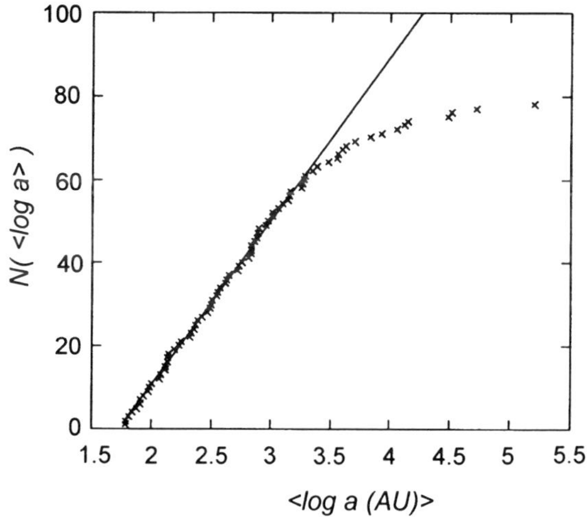

Figure 1 The cumulative distribution $N(<\log\ a>)$ of the logarithms of the major semiaxes for the oldest binaries in the solar vicinity; on the plane $N(<\log\ a>)$ vs. $<\log\ a>$ Öpik's relation is a straight line. The cumulative function $N(<\log\ a>)$ is shown for 78 binaries, where every cross is a binary. The best linear fit to the plotted data corresponds to straight line in the interval $60 \lesssim a(AU) \lesssim 2409$. The corresponding Kolmogorov-Smirnov goodness-of-fit parameter is Q = 0.999 (See Poveda 1997 for details). In this Figure we clearly see that the data begin to depart from Öpik's relation (the straight line) at $a \gtrsim 2400$ AU.

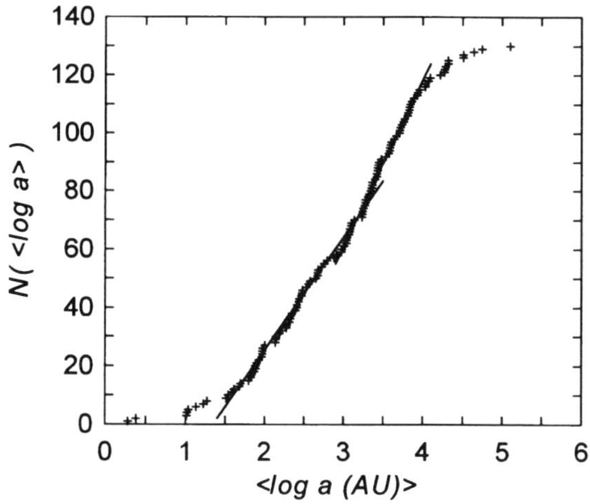

Figure 2 As in Figure 1, we show here the cumulative distribution, $N(< \log\ a >)$ of the logarithms of the major semiaxes for 130 high-velocity, metal-poor binaries. (For details see Allen et al., 2000). In this Figure we clearly see that the data begin to depart from Öpik's relation at $a \gtrsim 10000$ AU.

CORONAL ACTIVITY OF NEAR-CONTACT BINARIES

J. Scott Shaw[1] and Erin Smith[2]
[1]*Department of Physics and Astronomy University of Georgia Athens, GA. 30602, USA,*
[2]*Department of Astronomy University of Texas Austin, TX 78712, USA*
[1]jss@hal.physast.uga.edu

1. INTRODUCTION

The subclass of Near-Contact Binary stars (NCB) was defined by Shaw (1990). They are eclipsing binaries which 1) have periods of less than one day, 2) exhibit strong tidal actions, 3) are less than 0.1 orbital radius apart at their facing surfaces, but they are not in contact. They also have an A- or F- type primary component and a much cooler (by one or two spectral types) secondary, and most have at least one component filling its Roche lobe. There is evidence that they are evolutionary precursors to A-type UMa sources and that they are in the early stages of mass transfer.

Since NCB contain a late type star as their cool component, they are ideal candidates for activity due to strong magnetic fields. The cooler component is forced to rotate synchronously and thus has an unnaturally rapid rotation rate that enhances its magnetic activity. This, in turn, produces strong chromospheric emission, spots, coronal activity, x-rays, flares, etc. on the cooler components. In the optical most of this activity is swamped by the light of the primary. In order to detect the activity one needs to observe in the x-ray, ultraviolet or radio. Radio observations have yet to be done. However, Shaw (1992) found enhanced chromospheric emission from the cooler component using the IUE satellite and this study extends previous work on NCB in the x-ray regime.

2. X-RAY OBSERVATIONS

This paper is an extenuation of a similar survey conducted by Shaw, Caillault and Schmitt (1996) which investigated NCB in the ROSAT All Sky Survey. Several other studies of the x-ray properties of close

K.S. Cheng et al. (eds.), Stellar Astrophysics, 191–195.

binary systems have been done as well. W-type UMa systems have been studied by Cruddace and Dupree (1984) and by McGale, et. al (1996) who used the Einstein and ROSAT observatories respectively; RS CVn and BY Dra systems have been studied by Dempsey, et. al (1993a, 1993b, 1994, 1997) with ROSAT; and Algol systems have been studied by Singh et. al. (1995, 1996) with ROSAT and ASCA. In the RS CVn, BY Dra and W-type W UMa systems, both components may emit x-rays, and it is believed that A- type W UMa systems emit x-rays via a common envelope. In NCB, only the cool component is believed to be the source of the x-rays, since earlier surveys of A stars have shown them to be relatively inactive in the x-ray spectrum. If this is the case, the x-ray flux of such a NCB system might exhibit periodic behavior as the secondary rotates and is eclipsed by the primary. Unfortunately, because of the relative weakness of NCB as x-ray sources, few studies of the periodicity of the x-ray emission have been done. Shaw et al. (1994) showed that FO Virginis did indeed exhibit variable x-ray emission, but they lacked sufficient data to show the effect for other systems. By using the pointed ROSAT observations, we were able to obtain five light curves, including a revised version of the light curve for FO Virginis. In this paper we present the results of our search in the archive of ROSAT pointed observation for NCB.

Of the more than 178 NCB's searched for, 36 were in the telescope's field of view. Two NCB, FO Virginis and V1010 Ophiuchi were targets. Once the NCB's position coincided with a ROSAT field, photometry was performed using a 2.5' radius aperture. If the number of source counts exceeded 3 times the background error, it was considered a detection, and count rates and luminosities were found. If we did not detect the NCB, we calculated the 3 sigma level upper limit on the count rate and luminosity. Table 1 contains the 10 NCB found in the ROSAT pointed observations.

3. RESULTS

The detected luminosities all ranged between the values of log $L_x =$ 29.08 to 31.39 $ergs\ s^{-1}$. The average luminosity of the detected NCB was log $L_x = 30.69 \pm .3\ ergs\ s^{-1}$. All these values are slightly higher than the average luminosity $L_x = 30.09 \pm .1\ erg\ s^{-1}$, found by the search of NCB in the ROSAT All Sky Survey, which detected a total of 14 Near Contact Binary stars (Shaw et. al, 1996). One explanation for this that the NCB are variable stars and they may have been observed at minimum by the RASS while the longer observation times of the pointed

Table 1 NCB Detected in the ROSAT Pointed Observations

NCB	Distance parsecs	Period days	Counts	Rate cts/sec	Error ct rate	log L_x $ergs\ s^{-1}$
V1010 Oph(HRI)	74	0.6614	25754	0.1068	0.0436	30.50
V1010 Oph (PSPC)	"	"	8993	0.4747	0.0079	30.50
FO Vir	84	0.776	8301	0.0636	0.0041	29.64
BV Eri	111	0.5977	9415	0.0074	0.0027	29.08
DO Cas	147	0.6847	33890	0.0079	0.0023	29.51
HS Aqr	180	0.7101	18154	0.0934	0.0021	30.68
GQ Dra	219	0.7659	18563	0.0069	0.0019	29.81
V Crt	360	0.702	4779	0.0112	0.0028	30.67
RZ Dra	400	0.5509	2554	0.0111	0.0054	30.78
V401 Cyg	410	0.5827	2300	0.0106	0.0044	30.77
V Tri	760	0.5852	44954	0.0036	0.0015	31.39

Table 2 Mean X-ray Luminosities for Coronae in Binary Stars

Binary	log $< L_x >$ $erg\ s^{-1}$	VEM (hot)	VEM (cool)	Primary Sp Type	Secondary Sp Type
NCB (RASS)	$30.09 \pm .10$			A-F IV-V	G-K V*
NCB(pointed)	$30.66 \pm .20$	10.7	11.6		
A-type W UMA	$29.17 \pm .21$			A-FV	A-F V*
W-type W UMA	$29.85 \pm .06$	2.8	3.4	G-K V*	G-K V*
RS CVn	$30.35 \pm .09$	11.9	56.9	G-K IV-V	G-K III-IV*
Algol	$30.9 \pm .4$			B-A V	G-K IV*

VEM =volume emission measure of the hot and cool star
* source of x-rays

observations may have 'averaged out' this effect. The x-ray fluxes of the NCB are compared in Table 2 to other x-ray active close binaries.

4. LIGHT CURVES

For five of the 10 detected NCB there was enough data to search for periodic variations in the x-ray flux. FO Vir and V1010 Oph were both directly observed by Shaw et al. (1994). The other three systems were

DO Cas, V Tri, and HS Aqr. For each NCB, x-ray intensity was plotted versus the orbital phase of the binary system. For three systems, V1010 Oph, FO Vir and DO Cas, the x-ray intensity was lowest near 0.8 phase! If the cool component itself were the major source of the x-rays, then the count rate should reach a minimum at a phase of 0.5 when the cool component is eclipsed by the x-ray quiet hot star. Instead, the minimum is at 0.8 phase. The ROSAT HRI light curve for V1010 Oph seen in Figure 1 illustrates the phenomenon. The other light curves behave in much the same manner, all reaching minimum at 0.75 to 0.85 phase, and have a maximum near phase 0.25. The most likely explanation is that a majority of the x-ray flux is coming from a coronal prominence on the trailing side of the secondary. This prominence would not be entirely eclipsed by the primary at a phase of 0.5, but would be eclipsed at a phase of 0.75

Two stars did not exhibit such periodic behavior, HS Aqr and V Tri. HS Aqr has been shown to be near magnetic saturation and thus its x-ray activity extends over the entire surface of the cooler component. V Tri was barely detected except for a large flare event in one image. The remaining x-ray light curve was not bright enough to show any periodic variation.

5. CONCLUSIONS

More study of the x-ray properties of Near-Contact Binary stars is needed to determine both their long-term luminosities and their periodic behavior. Because these stars are considered possible evolutionary precursors to the W and A-type W UMa stars, their x-ray behavior is important to explore.

The periodic effects seen in the light curves are not conclusive enough to yield anything but the most general of models. More complete x-ray light curves of several stars are needed to give more conclusive results.

References

Cruddace, R. G., Dupree, A. K. 1984, ApJ, 277, 263

Dempsey, R. C., Linsky, J. L., Fleming, T. A., Schmitt, J. H. M. M. 1993a, ApJS, 86,599

Dempsey, R.C., Linsky, J. L., Schmitt, J. H. M. M., Fleming, T. A. 1993b, ApJ, 413, 333

Dempsey, R. C., Linsky, J. L., Fleming, T. A., Schmitt, J. H. M. M. 1994, in ASP Conf. Ser. 64, Cool Stars, Stellar Systems and the Sun, Eighth Cambridge Workshop, ed. J.-P. Caillault (San Franciso;ASP),74

Dempsey, R. C., Linsky, J. L. Fleming,T. A., Schmitt, J. H. M. M. 1997, ApJ,478,358
McGale,P. A., Pye, J. P. Hodgkin, S. T. 1996, MNRAS, 280, 627
Shaw, J. S. 1990, in Active Close Binaries, ed. C. Ibanoglu (Kluwer, Dordrect),241
Shaw, J. S. 1992, in Evolutionary Processes in Interacting Binary Stars, ed. Y. Kondo et al.,(Kluwer, Dordrect),373
Shaw, J. S. 1994, in Evolutionary Links in the Zoo of Interacting Binaries, eds. F. d'Antona et. al., Mem. Astr. Soc. Ital. 65, 95
Shaw, J. S., Caillault, J.-P., Peltier, S. J. 1994,, in ASP Conf. Ser. 64, Cool Stars, Stellar Systems and the Sun, Eighth Cambridge Workshop, ed. J.-P. Caillault (San Franciso;ASP), 143
Shaw, J. S., Caillault, J.-P., and Schmitt, J. H. M. M 1996, ApJ, 461, 961
Singh, K. P., Drake, S. A. 1995, ApJ, 445, 840
Singh, K. P., Drake, S. A., White, N. E. 1996, AJ, 111

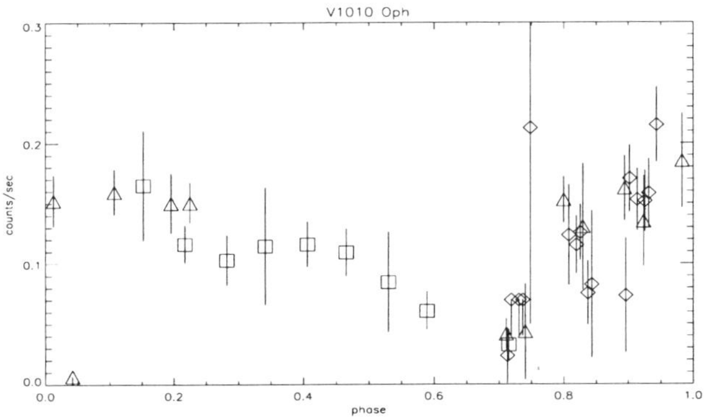

Figure 1 X-Ray Light Curve of the NCB V1010 Ophiuchi. Different symbols indicate observational epochs separated by one orbital period or more. Vertical lines represent $\pm1\sigma$ error.

ISO-SWS STUDIES OF WOLF-RAYET STARS: NEON ABUNDANCES

Karel A. van der Hucht[1], Patrick W. Morris[2], Luc Dessart[3], Paul A. Crowther[4], Allan J. Willis[4], D. John Hillier[5] and Peredur M. Williams[6]

[1] *Space Research Organization Netherlands, Sorbonnelaan 2, NL-3584 CA Utrecht, the Netherlands*

[2] *SIRTF Science Center, IPAC, California Institute of Technology, M/S 100-22, 1200 E. California Blvd, Pasadena, CA 91125, USA*

[3] *Département de physique, Université Laval and Observatoire du Mont Mégantic, Québec QC G1k 7P4, Canada*

[4] *Dept. of Physics and Astronomy, University College London, Gower St., London WC1E 6BT, UK*

[5] *Dept. of Physics and Astronomy, University of Pittsburgh, Pittsburgh, PA 15260, USA*

[6] *Institute for Astronomy, University of Edinburgh, Royal Observatory, Edinburgh EH9 3HJ, UK*

Abstract *ISO*-SWS data of one WN star and four WC stars are presented. The *ISO*-SWS data of WR 147 (WN8(h)+B0.5 V) show the [CaIV] 3.21 μm, [SIV] 10.5 μm and [NeIII] 15.5 μm fine-structure lines. We obtain a terminal wind velocity of $\sim$950 km s^{-1} from the helium recombination lines and the observed infrared fine-structure lines. Previously derived interstellar extinction values of $E_{\rm B-V} \simeq 4.1$ mag at $R_{\rm V} \simeq 2.7$ are confirmed. The distance to WR 147 is revised slightly upwards to 650 pc, based on galactic and LMC WN8-9 calibrators in the K-band. The observed spectrum exhibits clear evidence for clumping in the wind of the WN8 star. Using a non-LTE, line-blanketed model that allows for clumping, we find that a volume filling factor of 10% nicely matches the observed helium line-profiles, though we cannot exclude filling factors in the range of 4% to 25%. This leads to a mass-loss rate range of (1.5–3.7)$\times 10^{-5}$M$_\odot$yr^{-1}, substantially lower than derived for WR 147 from optical or radio observations under assumptions of homogeneity. Additionally including the effects of line-blanketing, the wind performance factor, $\dot{M}v_\infty/(L_*/c)$, is reduced to $\sim$2.5. Hydrogen is severely depleted, but present at 9% by mass. Observations of the [Ca IV] 3.21 μm, [S IV] 10.5 μm, and [Ne III] 15.5 μm lines allow us to place constraints on the surface abundances of these elements. In all cases we find good agreement with cosmic abundances (adjusted to a hydrogen-poor envi-

K.S. Cheng et al. (eds.), Stellar Astrophysics, 197–210.

ronment) and thus with evolutionary predictions for WN stars, which should not yet exhibit the products of core-burning stages beyond the CNO cycle before reaching the WC phase.

ISO-SWS data reveal the [NeIII] line in the WC stars WR 11, WR 90, WR 135, and WR 146. In addition, WR 11 shows also the [Ne II] 12.8 μm, [SIV] 10.5 μm and [SIII] 18.7 μm lines. Using a method up-dated from Barlow *et al.* (1988) to account for clumped winds, we derive for WR 11 Ne/He = 3-4 $\times 10^{-3}$ and S/He = 6 $\times 10^{-5}$ by number. Ne/S in WR 11 is a factor 8 larger than the cosmic value. However, the observed Ne/He ratio is a factor 2 lower than predictions of current evolutionary models for massive stars. WC star carbon abundances span a range $0.08 \leq C/He \leq 0.25$ by number, while hydrogen is absent.

1. INTRODUCTION

1.1 THE WN STAR WR 147

WN8 stars are reputed to be the most active subclass of Wolf-Rayet (WR) stars because of their high levels of spectroscopic and photometric variability (Lamontagne & Moffat 1987; Robert *et al.* 1989; Marchenko *et al.* 1994, 1998; Antokhin *et al.* 1995; Crowther *et al.* 1995). Visual light-curve amplitudes of 0.05–0.1 mag are intermediate between those of OIII-I stars and the microvariational amplitudes of LBVs, according to Marchenko *et al.* (1998), and may be accompanied by variations in the absorption components of optical P-Cygni profiles (Marchenko *et al.* 1998), or by narrow emission sub-peaks propagating outwards from emission-line centers (Lépine & Moffat 1999). These typically occur on time-scales of days. Besides the high variability, other phenomenological similarities exist between some WN8 stars and LBVs, such as ejecta-type ring nebulae, low binary frequency, and spatial distribution, suggesting that WN8 stars may be descendants from LBVs. This is plausible according to models of massive star evolution (Langer *et al.* 1994; Maeder 1996).

The stochastic line-profile variations have been interpreted to be associated with inhomogeneities, propagating outward in the rapidly expanding envelopes surrounding the stellar cores (*e.g.*, Moffat 1996, 1999). It is believed that *all* hot star winds show stochastic line-profile variations, indicating some hierarchy of wind clumping. If clumping affects the structure of a WR wind, and observations indicate that it does (*e.g.*, Marchenko *et al.* 1998), then it should lead to a revision of the stellar and wind parameters spectroscopically derived using "standard" model assumptions, namely, that the outflows are spherical and homogeneous (Hillier 1991, 1995; Moffat & Robert 1994; Hamann & Koesterke 1998a; Nugis *et al.* 1998). The derived value of the mass-loss rate $\dot{M}$ is especially sensitive to these assumptions. According to the way in which clumping is parameterized in the latest non-LTE wind models, the filling factor f in a clumped wind depends on the shape of the wind velocity curve $v(r)$, and $\dot{M}/\sqrt{f}$ becomes a fitting parameter (Hillier & Miller 1998). Hamann & Koesterke (1998a) find reasonably good fits to the emission-line profiles, particularly in the electron scattering wings, in spectra of three WN stars using a filling factor of 1/4. This decreases the mass-loss rates derived from homogeneous wind models by a factor of two overall.

The impact of clumping in the spectroscopic analyses of WR winds should be clearly evident with the WN8 stars, if their spectroscopic activity is indeed a manifestation of wind inhomogeneities. The second nearest WR star, WR 147 (AS 431), is a very long-period ($\sim 10^3$ yr) binary, with WN8 and OB-type components. Williams *et al.* (1997) have resolved clumps in the wind of the WN8 star on a scale of ~ 20 mas using *MERLIN*, so the spectrum of this star seems ideal for study using the latest non-LTE, line-blanketed wind models that incorporate clumping into the description of the wind structure. Despite its close proximity of $d = 0.63$ kpc, however, WR 147 is difficult to observe because of unusually high IS extinction, with $A_V \simeq 12$ mag based on the near-infrared colors of the WN8 star (Churchwell *et al.* 1992). With high angular-resolution *VLA* observations they confirmed the presence of the companion, $\sim 0\farcs6$ to the north (WR 147N) of the optical position of the WN8 star (WR 147S), previously resolved by *MERLIN* (Moran *et al.* 1989). IR images by Williams *et al.* (1997) set the projected separation at $0\farcs64$, *i.e.*, at least 400 AU at a distance of 0.63 kpc. WR 147N is a source of non-thermal radio emission, interpreted as a colliding wind effect (Churchwell *et al.* 1992; Williams *et al.* 1997; Contreras & Rodríguez 1999), and is classified as B0.5V by Williams *et al.* (1997), consistent with the ratio of visual brightnesses of the two sources measured with *HST*-WFPC2 (Niemela *et al.* 1998). WR 147S is the thermal main component and is reported to have varied in 3.6 cm flux by ~25% between 1995 and 1996 (Contreras & Rodríguez 1999). Variability of WR 147S at 6 cm has been reported by Watson *et al.* (1999). The 3.6 cm *VLA* observations by Contreras & Rodríguez (1999) also confirm the presence of inhomogeneities in the outflow of WR 147S. Combined variations of WR 147 (N+S) have been reported by van der Hucht *et al.* (1995), Skinner *et al.* (1999), and Setia Gunawan *et al.* (2000).

Prior to this study, the stellar and wind parameters of WR 147S were estimated by Churchwell *et al.* (1992), using *VLA* measurements of the free-free radio continuum (see also Nugis *et al.* 1998), and by Hamann *et al.* (1995) and Hamann & Koesterke (1998b) using a standard model analysis of the optical spectrum.

A preliminary description of *ISO*-SWS observations of WR stars was given by van der Hucht *et al.* (1996), who estimated the terminal velocity of the WN8 wind to be $965 \pm 20\,\mathrm{km\,s^{-1}}$ from the outer wind fine-structure lines. Morris *et al.* (1998) later estimated the Ne/He abundance ratio from the observed [Ne III] 15.5 μm line flux and the assumption of H/He = 0 based on a recombination analysis over the *ISO*-SWS wavelength range, and found evidence for a Ne/He enhancement by a factor of 2–3. This is in apparent conflict with evolutionary predictions for abundances at the surfaces of WR stars (*cf.* Maeder 1991; Meynet 1999). To make this determination, however, Morris *et al.* relied on standard model results by Hamann & Koesterke (1998b) for the value of $\dot{M}$, rescaled to the reddening and distance values given by Churchwell *et al.* (1992). It was further realized by Morris *et al.* (1998) that the ionic abundance estimates are very sensitive to the value of $\dot{M}$, which is in general an uncertain quantity for WR stars due to distance and reddening uncertainties. In view of the unexpected Ne/He value, its sensitivity to the value of $\dot{M}$, and the evidence for clumping in the wind of WR 147S, we were highly motivated to seek new solutions for the star's physical properties using the widest available range of spectroscopy and the latest non-LTE, line-blanketed wind models in which the assumption of homogeneity can be relaxed. The results will be demonstrative of the impact of wind clumping in spectroscopic analyses for the purpose of determining chemical abundances and the mass-loss rates on which they rely.

1.2 WC STARS

The properties of a large sample of carbon sequence (WC-type) WR stars have now been quantitatively derived using detailed models, accounting for non-LTE effects, spherical geometry and an expanding atmosphere (Howarth & Schmutz 1992; Koesterke & Hamann 1995). However, recent studies have demonstrated that clumping (Moffat *et al.* 1988; Hillier 1991, 1996; Schmutz 1997) and line-blanketing (Schmutz 1997; Hillier & Miller 1998) may have a significant effect on the derived physical properties of WR stars.

Overall, evolutionary predictions for massive stars (*e.g.*, Meynet *et al.* 1994) are in good agreement with the observed properties of WR stars. Van der Hucht & Olnon (1985) derived a Ne/He ratio for γ^2 Vel (WR 11, HD 68273, WC8+O7.5III-V) from *IRAS* observations analyzed with a formalism of Aitken *et al.* (1982). The result was found to be in good agreement with expectations. However, Barlow *et al.* (1988) identified a numerical flaw in the formalism used and added new ground-based observations to reveal $Ne/He = 1.0 \pm 0.35 \times 10^{-3}$, a factor of six lower than predicted by evolutionary models. Is this discrepancy due to failure of evolutionary models, peculiarities of the γ^2 Vel binary system, or incorrect assumptions for the stellar properties of the WC8 star? The combination of *ISO*-SWS observations of a larger sample of WC stars, plus recent progress in quantitative modelling of WR stars, provides an answer to this question.

ISO-SWS spectroscopy of WR 11 was presented by van der Hucht *et al.* (1996), who quoted excellent agreement of the fine-structure Ne line fluxes with *IRAS* and ground-based observations. Morris *et al.* (1998) subsequently re-estimated the Ne/He ratio using contemporary information on the stellar distance ($d = 0.26$ kpc, van der Hucht *et al.* 1997; Schaerer *et al.* 1997) and mass-loss rate (Stevens *et al.* 1996) of WR 11, which revealed a considerable Ne enrichment. Willis *et al.* (1997) used *ISO*-SWS data of WR 146 (WC6+O8), again revealing significantly enriched Ne. We combine the above results with *ISO*-SWS observations of the WC stars WR 90 (HD 156385, WC7) and WR 135 (HD 192103, WC8).

For details of this study, see Morris *et al.* (2000) and Dessart *et al.* (2000).

2. RESULTS FOR THE WN STAR WR 147

ISO-SWS infrared spectrophotometry of the WN8+BO.5V binary WR 147 has been analyzed using non-LTE methods to deduce the physical properties of the WN8 star. The assumption of wind homogeneity has been relaxed, warranted by strong observational evidence for clumping in the winds of WN8 stars as the most spectroscopically and photometrically variable among WR subtypes, and by the evidence for clumping from radio observations of WR 147 in particular. Our analyses represent the first detailed spectroscopic study of a WN8 star using a wind model that incorporates clumping and line-blanketing, and the first to include mid-infrared spectroscopy. The results can be summarized as follows:
(i) Extinction. The best fit between the observed optical to near-IR flux distribution (de-reddened using curves of Cardelli *et al.* 1989 at different values of R_V) and our theoretical energy distribution leads to a colour excess $E_{B-V} = 4.1$ mag at $R_V = 2.7$ ($A_V = 11.2$ mag). These are in good agreement with estimates by Churchwell *et al.* (1992). The value of R_V indicates extinction properties that are atypical of dust in the ISM. This is clearly underscored by the mid-IR extinction curve we derived from

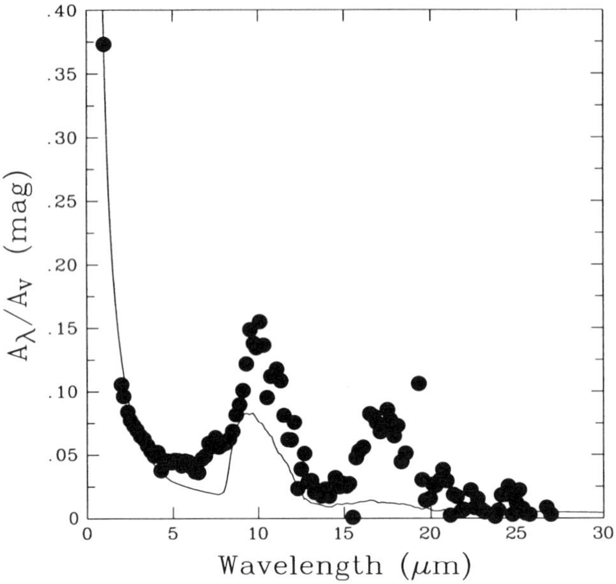

Figure 1 WR 147: IR extinction curve (filled dots), assuming that both the theoretical WR flux distribution and *ISO*-SWS flux calibration are appropriate, compared with the curve of Cohen (solid line, priv. comm), normalized to A_λ/A_V.

the *ISO*-SWS spectrum, which deviates from an average ISM extinction curve at $\lambda > 5\,\mu$m, especially in the silicate bands at 10 and 18 μm. The nature of the extinction is very puzzling due to the atypical shape of the 10 μm absorption profile, and the close proximity of the system.

(ii) Distance. Using WR 105 (WN9h) and HD 33133 (WN8h) as calibrators in the K-band, we obtain a distance of 650 pc to WR 147, in good agreement with Churchwell *et al.* (1992). The absolute magnitude of the system is $M_V = -6.4$ mag, and individual components have $M_V(\text{WN8}) = -6.3$ mag and $M_V(\text{B0.5V}) = -4.1$ mag, and are presently separated by 420 AU. We derived the bolometric luminosity and stellar mass of the B0.5V star, confirming estimates by Williams *et al.* (1997).
(iii) Clumping. Optical and mid-IR line profiles exhibit excesses of flux in the red electron scattering wings. A filling factor $f = 0.1$ achieves good agreement between observed and synthetic profiles (emphasizing the scattering wings), but values between 0.04 and 0.25 are very acceptable. The value of f or β (in the velocity law) could not be tuned with the overall optical-IR distribution due to reddening uncertainties.

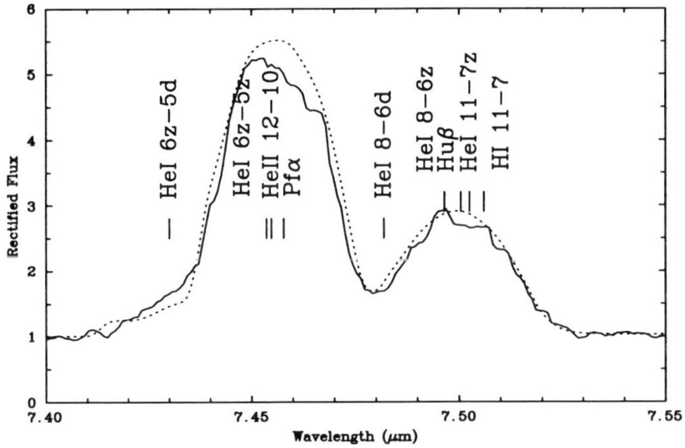

Figure 2 WR 147: comparison between the synthetic profile around Pf α (dotted line) and the *ISO*-SWS observations (solid line). Emission lines contributing to the synthetic profile are indicated.

Nonetheless, the preference on f is towards the lower end on the basis of comparison between observed radio fluxes and those predicted over the range of $\dot{M}/\sqrt{f}$, leading to $f(\text{opt–IR}) \simeq 0.04$, $f(\text{radio}) \simeq 0.1$, and $\dot{M} \simeq 1.5 \times 10^{-5}\,\mathrm{M_{\odot} yr^{-1}}$, taking the lower wind density of the radio-emitting regime into account. This mass-loss rate is approximately 2.5 to 5 times lower than estimated from radio observations and optical spectroscopy on the assumption that $f = 1.0$.

(iv) Line-blanketing. The inclusion of line-blanketing by metal ions may have contributed to lowering the derived mass-loss rate by altering the ionization structure and individual line strengths. The bolometric luminosity is not substantially altered, in agreement with tests of other WNL stars (Herald *et al.* 1999), so that the performance number $\dot{M}v_{\infty}/(L_{*}/c)$ is reduced from about 10 (Hamann *et al.* 1995) to ~2.5 mainly as a consequence of allowing for clumping.

(v) Elemental abundances. Hydrogen is found to be present at 9% by mass from the best-fit wind model. Fine-structure lines of [Ca IV], [S IV], and [Ne III] in the *ISO*-SWS spectrum were used to estimate surface abundance ratios of these elements, using an integration technique developed by Dessart *et al.* (2000), that follows the basic approach by Barlow *et al.* (1988) for normalization by the mass-loss rate. In the ratio with He, the abundances are in good agreement with cosmic values for the H-depleted environment of WR 147S. Small differences are probably due to uncertainties on the value of $\dot{M}$ rather than our neglect of unseen ionization stages, since ionization potentials of the observed ionization

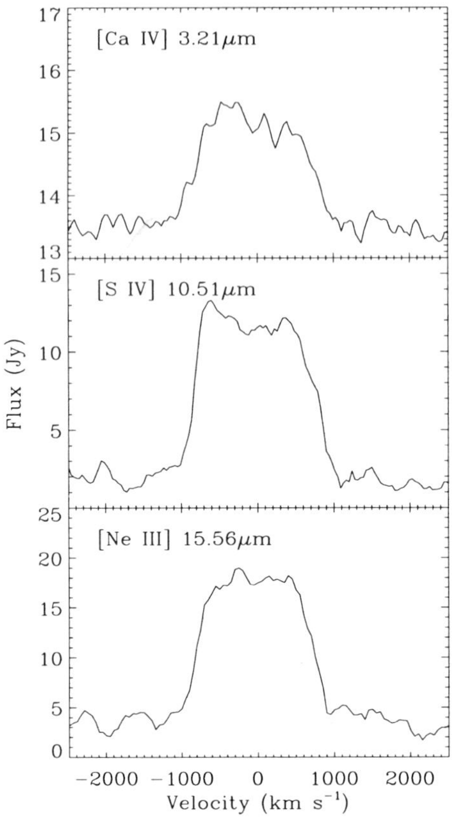

Figure 3 WR 147: mid-IR fine-structure transitions of Ca IV, S IV, and Ne III observed with *ISO*-SWS . Half-widths at zero intensity indicate a terminal wind velocity of $v_\infty \simeq$ 950 km s^{-1}.

stages make others somewhat unlikely, whilst Ne/Ca and Ne/S (which divides out the $\dot{M}$ uncertainty) are in excellent agreement with cosmic values.

In conclusion, the sensitivity of the derived value of the mass-loss rate to clumping can have a significant impact on the testing of evolutionary predictions. Previous to this study, abundances of ^{22}Ne are predicted (Maeder 1991) to reach two orders of magnitude over the solar value at the surfaces of galactic WC stars through a chain of core nucleosynthesis reactions involving α-capture by ^{14}N, β-decay of ^{18}F, and α-capture by ^{18}O. No Ne production is involved in the CN or ON branches of the CNO cycle, and thus Ne enhancements are not expected at the surfaces of WN stars, which tend to show CNO-equilibrum surface abundances (*e.g.*, Willis 1991). Most recently, however, Meynet (1999) and Maeder (1999) have presented new models of massive rotating stars that show how rotational mixing resulting from shear instabilities explains that the observed surface abundances of WN stars are not a result of mass-loss that uncovers the CNO-processed layers of the core, but instead as a result of diffusive mixing. The results also show that enhancements of N (a producte of H-burning) may be observed simultaneously with C and Ne enhancements (He-burning products), suggestive of the WN/WC transition stars (Meynet 1999). No WN/WC stars with flux levels within the range of *ISO*-SWS sensitivities could be observed to investigate the Ne abundances in this transition phase, but Morris *et al.* (1998) had reported indications of Ne enhancement (roughly factor of two above expectations) at the surface of WR 147S, a possible result of rotational mixing. The estimate relied on literature values of the mass-loss rate for the WN8 star, which to this point have been determined under the assumption of homogeneity. While it was emphasized that the overabundance is within the range of corresponding mass-loss rate uncertainties (see especially Morris *et al.* 1999), the results of this study resolve the overabundance as an overestimate arising from the neglect of clumping.

The consequences of this study on testing the latest evolutionary abundance predictions, then, are two-fold: *(a)* the normal Ne abundance at the surface of the WN8 star provides an endorsement that advanced burning products are not exposed via rotational mixing prior to the WN/WC or WC stages; and *(b)* allowances must be made for inhomogeneities in the line-forming regions of WR winds.

3. RESULTS FOR THE WC STARS

We have performed quantitative analyses of four WC5–8 stars, using models that account for line-blanketing and clumping. Comparisons between synthetic spectra and de-reddened UV to mid-IR observations are excellent, with few modelling deficiencies identified. Stellar parameters support previous determinations (*e.g.*, Koesterke & Hamann 1995), except that the incorporation of line-blanketing yields higher stellar luminosities, while clumping indicates lower wind performance numbers, supporting the conclusions of Hillier & Miller (1999) for WR111 (WC5). Future studies will derive properties of WC-type stars, at both earlier (WO) and later (WC9) spectral type, and investigate whether predicted ionizing properties are consistent with nebular observations (*e.g.*, Crowther *et al.* 1999).

Table 1 Derived Ne and S abundances. The ionization balance is assumed to consist of He^{+}, C^{2+} and O^{2+}. WR numbers are from the WR catalogues of van der Hucht *et al.* (1981, 1988). γ_i is derived from the integration method of Dessart *et al.* (2000).

WR	sp. type	γ_e	Z	μ 10^{-5}	$\gamma_{S^{2+}}$ 10^{-5}	$\gamma_{S^{3+}}$ 10^{-5}	S/He 10^{-4}	$\gamma_{Ne^{+}}$ 10^{-4}	$\gamma_{Ne^{2+}}$ 10^{-4}	Ne/He
11	WC8+O7III-V	1.137	1.189	5.14	1.9	3.1	5.9	5.4	21.6	31.9
90	WC7	1.219	1.287	5.84	–	–	–	$\leq$4.0	22.6	$\leq$34.0
135	WC8	1.138	1.189	5.21	–	–	–	$\leq$5.9	32.0	$\leq$44.0
146	WC6+O8	1.083	1.117	4.70	–	–	–	$\leq$2.8	22.8	$\leq$28.2

Using a revised formulation of the Barlow *et al.* (1988) technique to account for wind clumping, we derive Ne abundances of Ne/He =3–4×10^{-3} by number, seven times higher than the cosmic value adjusted

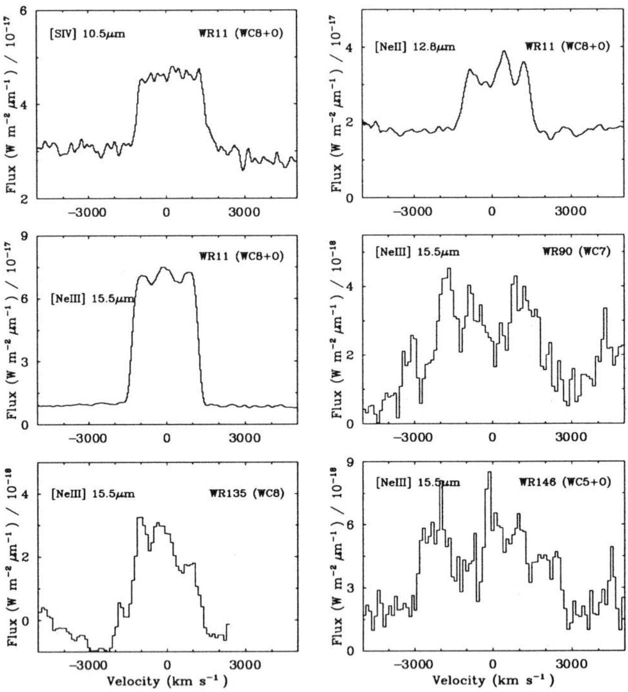

Figure 4 *ISO*-SWS observations of the fine-structure emission lines [S IV] 10.5 μm, [Ne II] 12.8 μm and [Ne III] 15.5 μm in WR 11, WR 90, WR 135 and WR 146.

for the H-depleted WC environment, supported by Ne/S = 50 for WR 11 from S fine-structure lines. The Ne enrichment is a factor of ~ 2 times lower than predictions of current theoretical models. However, differences in volume filling factors between the (high density) UV/ optical line formation regions and (low density) mid-IR fine-structure forming regions represent the greatest source of uncertainty in current Ne/He abundance determinations. Nevertheless, Ne/S provides an independent confirmation of the Ne enrichment since it is *independent* of outer wind filling factors. Future large ground and space-based telescopes that are optimised for the IR, will allow Ne and S line-flux measurements and abundance determinations for more distant Wolf-Rayet stars. Of particularly interest are abundances in carbon and oxygen-rich WO stars.

Acknowledgements

This work is based on observations with *ISO*, an ESA project with the participation of ISAS and NASA, and with instruments funded by ESA Member States, especially the PI countries: France, Germany, the Netherlands and the United Kingdom. KAvdH thanks the LKBF for a travel grant.

References

Aitken, D.K., Roche, P.F., Allen, D.A. 1982, *MNRAS* 200, 69P

Antokhin, I., Betrand, J.-F., Lamontagne, R., Moffat, A.F.J., Matthews, J. 1995, *AJ* 109, 817.

Barlow, M.J., Roche, P., Aitken, D.A. 1988, *MNRAS* 232, 821

Cardelli, J.A., Clayton, G.C., Mathis, J.S. 1989, *ApJ* 345, 245

Churchwell, E., Bieging, J.H., van der Hucht, K.A., Williams, P.M., Spoelstra, T.A.Th., Abbott, D.C. 1992, *ApJ* 393, 329

Contreras, M.E., Rodríguez, L.F. 1999, *ApJ* 515, 762

Crowther, P.A., Smith, L.J., Hillier, D.J., Schmutz, W. 1995, *A&A* 293, 427

Crowther, P.A., Pasquali, A., De Marco, O., Schmutz, W., Hillier, D.J., de Koter, A. 1999, *A&A* 350, 1007

Dessart, L., Crowther, P.A., Hillier, D.J., Willis, A.J., Morris, P.W., van der Hucht, K.A. 2000, *MNRAS* in press

Hamann, W.-R., Koesterke, L., Wessolowski, U. 1995, *A&A* 299, 151

Hamann, W.-R., Koesterke, L. 1998a, *A&A* 335, 1003

Hamann, W.-R., Koesterke, L. 1998b, *A&A* 333, 251

Herald, J.E., Schulte-Ladbeck, R.E., Hillier, J.D. 1999, in: K.A. van der Hucht, G. Koenigsberger & P.R.J. Eenens (eds.), Wolf-Rayet Phenomena in Massive Stars and Starburst Galaxies, Proc. IAU Symp. No. 193 (San Francisco: ASP), p. 242

Hillier, D.J. 1991, *A&A* 247, 455

Hillier, D.J. 1995, in: K.A. van der Hucht, G. Koenigsberger & P.R.J. Eenens (eds.), Wolf-Rayet Phenomena in Massive Stars and Starburst Galaxies, Proc. IAU Symp. No. 193 (San Francisco: ASP), p. 163

Hillier, D.J., 1996, in: J.-M. Vreux *et al.* (eds.), Wolf-Rayet stars in the Framework of Stellar Evolution, Proc. 33rd Liège Int. Ast. Coll., Université de Liège, p. 509

Hillier, D.J., Miller, D.L. 1998, *ApJ* 496, 407

Hillier, D.J., Miller, D.L. 1999, *ApJ* 519, 354

Howarth, I.D., Schmutz, W. 1992, *A&A* 261, 503

van der Hucht, K.A., Conti, P.S., Lundström, I., Stenholm, B. 1981, *SSR* 28, 227

van der Hucht, K.A., Olnon, F.M. 1985, *A&A* (Letters) 149, L17

van der Hucht, K.A., Hidayat, B., Admiranto, A.G., Supelli, K.R., & Doom, C. 1988, *A&A* 199, 217

van der Hucht, K.A., Williams, P.M., Spoelstra, T.A.Th., Swaanenvelt, J.P. 1995, in: K.A. van der Hucht & P.M. Williams (eds.), in: Wolf-Rayet Stars: Binaries, Colliding Winds, Evolution, Proc. IAU Symp. No. 163 (Dordrecht: Kluwer), p. 559

van der Hucht, K.A., Morris, P.W., Williams, P.M. *et al.* 1996, *A&A* (Letters) 315, L193

van der Hucht, K.A., Schrijver H., Stenholm, B. *et al.* 1997, *New Astronomy* 2, 245

Koesterke, L., Hamann, W.-R. 1995, *A&A* 299, 503

Lamontagne, R., Moffat, A.F.J. 1987, *AJ* 94, 1008

Langer, N., Hamann, W.-R., Lennon, M., Najarro, F., Pauldrach, A.W.A., Puls, J. 1994, *A&A* 290, 819

Lépine, S., Moffat, A.F.J. 1999, *ApJ* 514, 909

Maeder, A. 1991, *A&A* 242, 93

Maeder, A. 1996, in: J.-M. Vreux *et al.* (eds.) Wolf-Rayet Stars in the Framework of Stellar Evolution, Proc. 33rd Liège Int. Astroph. Coll. (Université de Liège), p. 39

Maeder, A. 1999, in: K.A. van der Hucht, G. Koenigsberger & P.R.J. Eenens (eds.), Wolf-Rayet Phenomena in Massive Stars and Starburst Galaxies, Proc. IAU Symp. No. 193 (San Francisco: ASP), p. 177

Marchenko, S.V., Antokhin, I.I., Bertrand, J.-F. *et al.* 1994, *AJ* 108, 678

Marchenko, S.V., Moffat, A.F.J., Eversberg, T. *et al.* 1998, *MNRAS* 294, 642

Meynet, G., Maeder, A., Schaller, G., Schaerer, D., Charbonnel, C. 1994, *A&AS* 103, 97

Meynet, G. 1999, in: K.A. van der Hucht, G. Koenigsberger & P.R.J. Eenens (eds.), Wolf-Rayet Phenomena in Massive Stars and Starburst Galaxies, Proc. IAU Symp. No. 193 (San Francisco: ASP), p. 218

Moffat, A.F.J., Drissen, L., Lamontagne, R., Robert, C., 1988, *ApJ* 334, 1038

Moffat, A.F.J., Robert, C. 1994, *ApJ* 421, 310

Moffat, A.F.J. 1996, in: J.-M. Vreux *et al.* (eds.) Wolf-Rayet Stars in the Framework of Stellar Evolution, Proc. 33rd Liège Int. Astroph. Coll. (Université de Liège), p. 199

Moffat, A.F.J. 1999, in: K.A. van der Hucht, G. Koenigsberger & P.R.J. Eenens (eds.), Wolf-Rayet Phenomena in Massive Stars and Starburst Galaxies, Proc. IAU Symp. No. 193 (San Francisco: ASP), p. 278

Moran, J.P., Davis, R.J., Spencer, R.E., Bode, M.F., Taylor, A.R. 1989, *Nature* 340, 449

Morris, P.W., van der Hucht, K.A., Willis, A.J., Williams, P.M. 1998, in: L.B.F.M. Waters, C. Waelkens, K.A. van der Hucht & P.A. Zaal (eds.), *ISO*'s View on Stellar Evolution, Proc. International Conf. 1-4 July Noordwijkerhout, The Netherlands, *ApSS* **255**, 157

Morris, P.W., van der Hucht, K.A., Willis, A.J., Dessart, L., Crowther, P.A., Williams, P.M. 1999, in: K.A. van der Hucht, G. Koenigsberger & P.R.J. Eenens (eds.), Wolf-Rayet Phenomena in Massive Stars and Starburst Galaxies, Proc. IAU Symp. No. 193 (San Francisco: ASP), p. 77

Morris, P.M., van der Hucht, K.A., Crowther, P.A., Hillier, D.J., Dessart, L., Williams, P.M., Willis, A.J. 2000, *A&A* 353, 624

Niemela, V.S., Shara, M.M., Wallace, D.J., Zurek, D.R., Moffat, A.F.J. 1998, *AJ* 115, 2047

Nugis, T., Crowther, P.A., Willis, A.J. 1998, *A&A* 333, 956

Robert, C., Moffat, A.F.J., Bastein, P., Drissen, L., St-Louis, N. 1989, *ApJ* 347, 1034

Schaerer, D., Schmutz, W., Grenon, M. 1997, *ApJ* (Letters) 484, L153

Schmutz, W. 1997, *A&A* 321, 268

Setia Gunawan, D.Y.A., de Bruyn, A.G., van der Hucht, K.A., Williams, P.M. 2000, in preparation

Skinner, S.L., Itah, M., Nagase, F., Zhekov, S.A. 1999, *ApJ* 524, 394

Stevens, I.R. Corcoran, M.F., Willis, A.J. *et al.* 1996, *MNRAS* 283, 589

Watson, S.K., Davis, R.J., Williams, P.M. 1999, in: K.A. van der Hucht, G. Koenigsberger & P.R.J. Eenens (eds.), Wolf-Rayet Phenomena in Massive Stars and Starburst Galaxies, Proc. IAU Symp. No. 193 (San Francisco: ASP), p. 396

Williams, P.M., Dougherty, S.M., Davis, R.J., van der Hucht, K.A., Bode, M.F., Setia Gunawan, D.Y.A. 1997, *MNRAS* 289, 10

Willis, A.J. 1991, in: K.A. van der Hucht & B. Hidayat (eds.) , Wolf-Rayet Stars and Interrelations with Other Massive Stars in Galaxies, Proc. IAU Symp. No. 143 (Dordrecht: Kluwer), p. 265

Willis, A.J., Dessart, L., Crowther, P.A., Morris, P.W., Maeder, A., Conti, P.S., van der Hucht, K.A. 1997, *MNRAS* 290, 371

SHORT TIME SCALE VARIATION OF A NEAR-CONTACT BINARY SYSTEM, GR TAURI

Soonthornthum, B.[1], Yang, Y.[2], Qingyao, L.[2], Tanvanichkul, B.[1]
[1] *Astronomy Research Unit, Sirindhorn Observatory and Department of Physics, Faculty of Science, Chiang Mai University, Chiang Mai 50200, THAILAND.*
[2] *Yunnan Observatory, Chinese Academy of Science, Kunming, P.R. of CHINA.*

Abstract A near-contact binary system, GR Tauri, was observed photometrically in V wavelength band in November 1995 at the Yunnan Observatory . The asymmetric light curve was found in this observation, where the maximum at phase 0.25 is brighter than the maximum at phase 0.75 by 0.02 magnitude .

Analysis for its physical properties and orbital elements was done by using Wilson-Devinney code. The model found is a semi-detached binary system with the secondary component almost filling its critical Roche lobe while the primary component fills it completely. The results were compared with the previous investigations (Yamasaki,1984, Hanzl,1990, La'zaro,1995 and Fang,1994). The evidence of a short time scale variation of this near-contact binary system was found.

Keywords: near-contact binary system, O'Connell effect, Roche lobe.

1. INTRODUCTION

GR Tauri (BD +20° 0685) was discovered, through photographic survey, to be an eclipsing binary system by Strohmeier et al. (1957). Its orbital period was determined to be $0^d.474021$(Stromeier et al(1963)) and the spectral type of A9 was assigned (Gotz and Wenzel(1962)). Yamazaki et al(1984) found, by B and V photoelectric and spectroscopic observations, that GR Tauri is a totally eclipsing system with a photometric asymmetry where the secondary component almost filling its Roche lobe and the primary component was completely filled. The system was found to be a near-contact binary system which spectral type was classified to be A5V. La'zaro et al (1995) made photometric observations in B,V,J,K bands and new elements of GR Tauri were calculated

K.S. Cheng et al. (eds.), Stellar Astrophysics, 211–219.

using Wilson-Devinney code. The results yielded symmetric light curves where, this time, the primary component almost filling the Roche lobe with secondary component fills it completely.

Near-contact binary systems are divided into 2 subclasses (Shaw,1990) as follows :

1. V1010 Ophiuchi binary system which both components are slightly evolved. The primary component is completely filled Roche Lobe where the secondary component is inside. The light curve is always asymmetric where Max I (phase 0.25) is always brighter than Max II (phase 0.75).

2. FO Virginis binary system which both components may be very evolved. The primary component is inside Roche Lobe where the secondary component is completely filled. The light curve shows no photometric asymmetry.

From the previous studies, it was found noticeably that GR Tauri shows properties of these two subclasses of near-contact binary system. More investigation was done in this research work for the explanation of the changes within the short period time scale on the properties of this near-contact binary system.

The asymmetry of a light curve such that one maximum is higher than the other is called the O'Connell effect . Therefore, the O'Connell effect may always be found on the V1010 Ophiuchi binary system, but never be found on the FO Virginis binary system. Photometric asymmetries in the light curves of near-contact binary systems may be explained by the mass transfer from the hot component onto the hemisphere of the cool component.

2. OBSERVATIONS:

In this study, GR Tauri was observed using photoelectric photometer in V wavelength band with the 60-cm Cassegrain reflector at the Yunnan observatory. An EMI 6256B photomultiplier with a Corning 3389 of stock thickness for the yellow region was used. A circular diaphragm of 32 arcsec and an integration time of 20 seconds were also used. Observations were done in 3 consecutive nights during November 27-29, 1995 and 347 data points were obtained for a complete light curve. The comparison and check stars used were BD $+20^o$ 684 and BD $+29^o$ 642 respectively. Data of GR Tau, comparison and check stars are shown in Table 1.

Figure 1 and 2 show star chart of GR Tauri and its light curve observed in this study.

Table 1 Data of GR Tauri, Comparison and Check Stars

Star	BD	R.A.(1950)	Dec(1950)	V	B-V	U-B
GR Tauri	$+20^o685$	3h 58m 08s	$+20^o17'.1$	10.26	+0.32	+0.24
Comparison	$+20^o684$	3h 57m 45s	$+20^o16'.4$	8.05	+0.21	+0.09
Check	$+20^o642$	3h 57m 45s	$+20^o09'.8$	9.39	+0.53	+0.05

Table 2 The observed time of minima of GR Tauri

HJD	E	(O-C)
2450050.4149	12742	0.0216
2450051.2665	12744	-0.0297

The epoch used in calculating times of minima is
JD(Hel) 2444573.1071±0.0002
(p.e.) and the period P = 0.42984±0.00001(p.e.) day (Yamasaki et.al., 1984). From the analysis of all published time of minima,including also times of minima obtained in this study, using the linear least squares fitting yields the improved ephremeris of binary's minima as :

$$MinI = 2444573.09668 + 0^d.429865E \quad (1)$$

The measurement of the observed period of GR Tauri in this study is $0^d.4295841$ and the observed time of minima (Min I) are shown in Table 2.

3. LIGHT CURVE ANALYSIS :

Photometric observational data of GR Tauri were analyzed using Wilson-Devinney synthetic light curve program (Wilson, 1992). The total number of 119 normal points were obtained by weighting the raw data.

In calculating the photometric solution by DC program, the following initial values were adopted : average surface temperature of the primary component $T_1 = 7500\ K$ (Lazaro,1995) with the average surface temperature of the secondary component was estimated by comparing depths of primary and secondary minima, gravity darkening coefficients $g_1 = 1.0$, $g_2 = 0.32$ and bolometric albedos $A_1 = 1.0$, $A_2 = 0.5$ were according to radiative and convective envelopes of primary and secondary components respectively, limb darkening coefficients $x_1 = 0.51$, $x_2 = 0.83$ were obtained from tables of Al-Naimiy (1978) in V.

The adjustable parameters allowed were orbital inclination (i), average surface temperature (T_2), surface potentials of primary (Ω_1) and secondary (Ω_2) components, relative monochromatic luminosity of primary star (L_1) where relative monochromatic luminosity of secondary star (L_2) was calculated automatically after L_1. The mass ratio q = 0.2, 0.3, 0.4, 0.5 were used in calculations.

For each mass ratio, mode 2 (detached system) was used initially in the calculation. For q=0.2, the final solution was converged to mode 3 (contact system). However, for q = 0.3, 0.4, 0.5, the final solutions were all converged to mode 4 (semi-detached system). More precise values of q = 0.22, 0.24, 0.26, 0.28 were used in finding the best solution. We finally found that q = 0.26 gave the minimum value of sum of residues ,Σ.

In final calculations, the mass ratio (q) was also allowed as an adjustable parameter with q = 0.26 was used as the initial input value. Best solutions of GR Tauri were shown in Table 3 and LC program was used to calculate its synthetic light curve. Figure 3 shows the comparison between observational and synthetic light curves.

4. RESULTS AND DISCUSSION

From the analysis of the light curve of GR Tauri using DC program, it was found that the system is a near-contact binary system with the secondary component almost filling its critical Roche lobe while the primary component fills it completely. The asymmetric light curve was also found as shown in Figure 3. Hence, GR Tauri, in this study, shows properties of V1010 Ophiuchi subclass.

Observations of V1010 Ophiuchi with IUE satellite(Shaw and Guinan, 1990), Lyman α emission was seen at 0.25 and 0.75 phase with the total flux at 0.25 phase was over twice that at 0.75 phase. It was concluded that there may be mass transfer from hot primary component onto the hemisphere of the cool secondary component causing extra light emitted from the secondary component. However, for GR Tauri, the period change was not observed in this study.

Comparison with previous investigations, GR tauri was found to have properties of both subclasses of near-contact binary system.

Table 3 Solutions of GR Tauri.

Parameters	Computed Values
i (degrees)	83.770
g_1	1
g_2	0.32
T_1	7500
T_2	4675
A_1	1
A_2	0.50
Ω_1	2.3740
Ω_2	2.4339
q	0.25919

Table 4 Eclipses depths and symmetric/asymmetric properties of GR Tauri

Depth of Primary Minima	Depth of Secondary Minima	$V(0.75) - V(0.25)$	Year of Observations	References
0.46	0.17	0.03	1980	Yamasaki(1984)
0.37	0.13	~0.03	1989	Hanzl(1990)
0.39	0.13	0.00	1992	La'zaro(1995)
0.48	0.17	0.05	1993	Fang(1994)
0.47	0.17	0.03	1995	This study

From Table 4, the system shows the variation in a short period time-scale between a state of asymmetric light curves (Yamasaki,1984, Fang, 1994 and this study) and a state of symmetric light curves (Hanzl, 1990, La'zaro,1995). It was also noticed that the deeper minima were likely to be found at the asymmetric state than at the symmetric state which may be caused by the extra light on the surface of the star.

This work was supported by National Research Council of Thailand (NRCT) and National Natural Science Foundation of China (NSFC) under cooperative research project between Faculty of Science, Chiang Mai University and Yunnan Observatory, Academia Sinica.

References

Al-Naimiy, H.M.,1978, ApJS,53, 181.

Fang,M., Zhai, D., Gao,Z.,Zhang,J. and Zhang,X., 1994, IBVS, No. 4082.

Gotz, W. and Wenzel, W., 1962, Mitt. Veranderl Sterne, No.628 Bamberg, 5, No.17.

Hanzl, D., 1990, IBVS, No. 3423.

La'zaro, C., Niarchos, P., Rovithis, P., Rovithis-Livaniou, E., Arevalo, M.J. and Antonopoulou, E., 1995, Astron. J., 110(4), 1796-1808.

Shaw, J.S., 1990, in Active Close Binaries, edited by C. Ibanoglu(Kluwer, Dordrecht), 241-251.

Strohmeier, W., Kippenhahn, R. and Geyer, E.,1957, Kleine Veroff. Bamberg, No.18.

Stromier, W., Knigge, R. and Ott, H., 1963, Veroff. Remeis-Sternwarte. Bamberg, 5, No.17.

Wilson, R.E., 1992, in Documentation of Eclipsing Binary Computer Model, University of Florida.

Yamasaki, A., Okazaki, A. and Kitamura, M., 1984, PASJ, 36, 175.

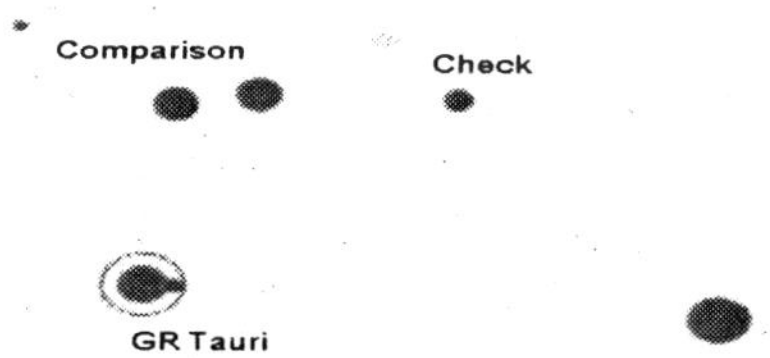

Figure 1 Star Chart of GR Tauri

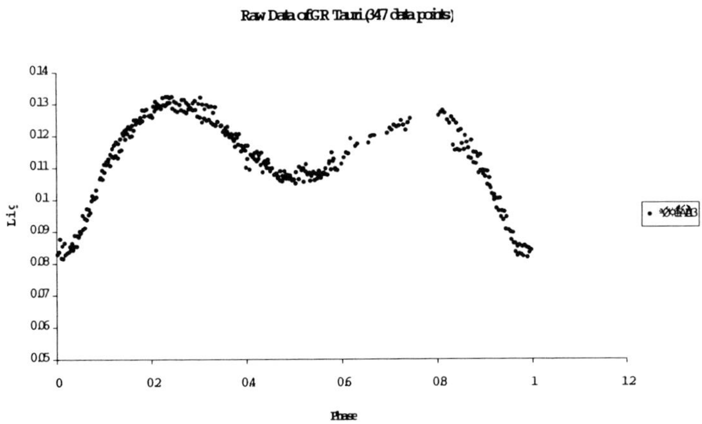

Figure 2 Light curve of GR Tauri

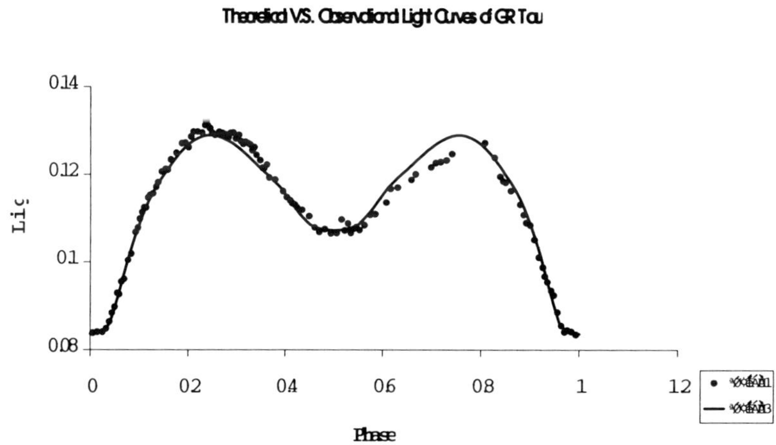

Figure 3 Comparison between comparison and synthetic light curves.

Progress

GENERAL RELATIVISTIC HYDRODYNAMIC EQUATION IN MULTIPLE SYSTEM

Chongming Xu *, Xuejun Wu* and Erwin Brüning †

Abstract In this paper, general relativistic hydrodynamic equation in multiple system is discussed by means of DSX scheme. It might be useful to calculate multipoles moments in every local coordinates of N-body systems. Especially for binary system, it is practical important for us to discuss the coalescing compact binary systems. Mostly we discuss in the perfect fluid. For nonperfect fluid we will point out the way to be extended. At the end, we also point out the problem to be improved.

1. INTRODUCTION

General relativistic hydrodynamics is always an important topics in classical general relativity (Lichnarowicz, 1967; Taub, 1978; Weinberg, 1980). The importance is not only from theoretical pointview, but also in the practical application, e.g., compact star and precise measurement in geophysics. For first order post-Newtonian (1PN) hydrodynamics it has been investigated by many authors (e.g. Chandrasekhar, 1965; Blanchet, Damour & Schäfer, 1990; Will, 1993). But we should point out all of the calculation are in one global coordinate, or we should call them as one system hydrodynamics. In some case, e.g. in binary or N-body system, it is convenient to introduce a local coordinate system for each object to calculate local multipole moments.

Before 1991, there existed no suitable reference system to convert the coordinate systems between global one to local one or vice versa. Only after 1991, Damour, Soffel and Xu (DSX) suggested a complete scheme (Damour, Soffel & Xu, 1991; 1992; 1993; 1994, separately defined as DSX paper I-IV) to solve 1PN problems in celestial mechanics. In the

*Nanjing Normal University, Nanjing, 210097, China; Present address (up to Sept. 2001) is: Lohrmann Observatory, TU Dresden, Mommsenstrasse 13, D-01062 Dresden, Germany
†University of Durban Westville, Durban South Africa

K.S. Cheng et al. (eds.), Stellar Astrophysics, 221–227.

DSX paper I it was the first time to obtain a complete reference systems by which we could establish the hydrodynamics in multiple systems. In DSX scheme, in each body the infinite series is a bilinear form exhibiting the coupling of all B-D multipole moments of the body to the 1PN tidal moments felt by the body. But for the time evolution of multipole moments we still do not know in DSX scheme. Because it is dependent on the equation of state for every local body and the equation of hydrodynamics in each body. This situation is not very harmful to the completeness of the general relativistic celestial mechanics, since in celestial mechanics the local multipole moments can be measured by various of satellites, also the most of planets are not very far to a rigid body. But in some problem, e.g. in the coalescing compact binary systems (Xu, Wu, Schäfer, 1997). The time evolution of multipole moments have to be considered.

In DSX scheme, they define sources Σ and Σ^i only, where

$$\Sigma = \frac{T^{00}+T^{ss}}{c^2} + O(4) \tag{1}$$

$$\Sigma^i = \frac{T^{0i}}{c} + O(3)\,. \tag{2}$$

Besides the reference systems has been solved already in DSX scheme, the equations for every local coordinate and global coordinate have the similar form also. In DSX scheme, they have shown the 1PN hydrodynamic equation (Euler equation), but in their formula T^{ab} is not expressed by an explicite form. Therefore the equation can not be used as the time evoluting equation.

In this paper, we will follow the DSX scheme. All of the notation and convention are the same with them. To perfect fluid we can express energy-momentum tensor T^{ab} in terms of Σ and Σ^a (4 function) and write down the hydrodynamics equation in multiple coordinate. Certainly in general case $T^{\mu\nu}$ (10 components) can not be expressed by means of Σ and Σ^a. We also point out main topics to extend to non-perfect fluid.. With these equations and the equation of state we can obtain the time evolution of the multipole moments in principle. In some meaning, this work is a supplement to DSX scheme.

2. PN HYDRODYNAMIC EQUATION OF PERFECT FLUID

First, we consider PN hydrodynamic equation of perfect fluid in multiple coordinate systems. The so-called hydrodynamic equations are just PN energy equations and Euler equations. In DSX scheme , they are

(see DSX paper I)

$$\frac{\partial}{\partial T}\Sigma + \frac{\partial}{\partial X^a}\Sigma^a = \frac{1}{c^2}\frac{\partial}{\partial T}T^{bb} - \frac{1}{c^2}\Sigma\frac{\partial}{\partial T}W + O(4) \tag{3}$$

$$\frac{\partial}{\partial T}\left[\left(1+\frac{4W}{c^2}\right)\Sigma^a\right] + \frac{\partial}{\partial X^b}\left[\left(1+\frac{4W}{c^2}\right)T^{ab}\right] = F^a(T, X^a) + O(4) \tag{4}$$

where

$$F^a = \Sigma E_a + \frac{1}{c^2}B_{ab}\Sigma^b \tag{5}$$

$$E_a = \partial_a W + \frac{4}{c^2}\partial_T W_a \tag{6}$$

$$B_{ab} = -4(\partial_a W_b - \partial_b W_a) \tag{7}$$

Σ and Σ^a are defined in equation (1) and (2), W and W_a are potential and vector potential separately. These equations are valid for local coordinate A as well as for any local coordinate in DSX scheme, since they have the same form. Even in global coordinate the form of equation is similar except substituting Σ, Σ^a, W and W_a into σ, σ^i, w and w_i. Because in global coordinate the form of perfect fluid is well known already. Therefore, at first we obtain the relation between σ and σ^i and energy-momentum tensor in global coordinate $\bar{T}^{\mu\nu}$, then we just convert these relation into local coordinate. Perfect fluid in global coordinate has the form of energy-momentum tensor (Will 1993):

$$\bar{T}^{\mu\nu} = \left(\rho(1+\frac{\Pi}{c^2}) + \frac{p}{c^2}\right)u^\mu u^\nu + g^{\mu\nu}p + O(4) \tag{8}$$

where Π, p and u^μ are the specific density of internal kinetic and themal energy, pressure and four velocity respectively. Its components are

$$\bar{T}^{00} = c^2\rho\left(1+\frac{\Pi}{c^2}+\frac{v^2}{c^2}+\frac{2w}{c^2}\right) + O(4) \tag{9}$$

$$\bar{T}^{0i} = c\rho v^i\left(1+\frac{\Pi}{c^2}+\frac{v^2}{c^2}+\frac{2w}{c^2}+\frac{p}{\rho c^2}\right) + O(3) \tag{10}$$

$$\bar{T}^{ij} = \rho v^i v^j\left(1+\frac{\Pi}{c^2}+\frac{v^2}{c^2}+\frac{2w}{c^2}+\frac{p}{\rho c^2}\right) + p\delta_{ij}\left(1-\frac{2w}{c^2}\right) + O(4) \tag{11}$$

where v^i and v are the components of 3-dimension velocity and its value. Considering the definition of σ and σ_i in DSX scheme, it is

$$\sigma = \frac{\bar{T}^{00}+\bar{T}^{ss}}{c^2} = \rho\left(1+\frac{\Pi}{c^2}+\frac{2v^2}{c^2}+\frac{2w}{c^2}+\frac{3p}{\rho c^2}\right) + O(4) \tag{12}$$

$$\sigma^i = \frac{\bar{T}^{0i}}{c} = \rho v^i \left(1 + \frac{\Pi}{c^2} + \frac{v^2}{c^2} + \frac{2w}{c^2} + \frac{p}{\rho c^2}\right) + O(4) \tag{13}$$

From eqs.(12) and (13), it is easy to get the expression of 3-velocity

$$v^i = \frac{\sigma^i}{\sigma}\left(1 + \frac{v^2}{c^2} + \frac{2p}{\sigma c^2}\right) + O(4) \tag{14}$$

here v^2 can be expressed by σ_i and σ in Newtonian level also

$$\frac{v^2}{c^2} = \frac{\sigma_i \sigma_i}{\sigma^2 c^2} + O(4) \tag{15}$$

and p is a scalar function to be determined from the equation of state in which the energy density $\rho(1 + \Pi/c^2)$ is related with σ by eq.(12). Then we can express the eq.(9)-(11) again by means of σ and σ_i:

$$\bar{T}^{00} = c^2 \sigma \left(1 - \frac{v^2}{c^2} - \frac{3p}{\sigma c^2}\right) + O(2) \tag{16}$$

$$\bar{T}^{0i} = c\sigma^i + O(1) \tag{17}$$

$$\bar{T}^{ij} = \frac{\sigma_i \sigma_j}{\sigma}\left(1 + \frac{v^2}{c^2} + \frac{2p}{\sigma c^2}\right) + p\delta^{ij}\left(1 - \frac{2w}{c^2}\right) + O(4) \tag{18}$$

Combining with eq.(15), the equation of state $p = p(\rho)$ and Einstein field equation to solve w, w^i, σ and σ^i can fully express $\bar{T}^{\mu\nu}$ in perfect fluid. Since in local coordinate the form of formulae has the same as the one in global coordinate except substituting the global quantities $\sigma\sigma^i$, w, w_i and $\bar{T}^{\mu\nu}$ into local one $\overset{(A)}{\Sigma}\overset{(A)}{\Sigma^a}$, $\overset{(A)}{W}$, $\overset{(A)}{W_a}$ and $\overset{(A)}{T^{\alpha\beta}}$. Normally we neglect subscript (A). Therefore in every local coordinate (A) we have

$$T^{00} = c^2 \Sigma \left(1 - \frac{\Sigma^a \Sigma^a}{\Sigma c^2} - \frac{3p}{\Sigma c^2}\right) + O(2) \tag{19}$$

$$T^{0a} = c\Sigma^a + O(1) \tag{20}$$

$$T^{ab} = \frac{\Sigma^a \Sigma^b}{\Sigma}\left(1 + \frac{\Sigma^c \Sigma^c}{\Sigma^2 c^2} + \frac{2p}{\Sigma c^2}\right) + p\delta^{ab}\left(1 - \frac{2W}{c^2}\right) . \tag{21}$$

¿From the last equation, it is also easy to obtain

$$\frac{T^{bb}}{c^2} = \frac{\Sigma^b \Sigma^b}{\Sigma c^2} + \frac{3p}{c^2} . \tag{22}$$

Since p is a scalar, they are not changing with converting from global to local one. Substituting eq.(22) into (3), we get energy equation

$$\dot{\Sigma} + \Sigma^a_{,a} = \frac{1}{c^2}\left(\frac{2\dot{\Sigma}_b \Sigma_b}{\Sigma} - \frac{\dot{\Sigma}\Sigma_b \Sigma_b}{\Sigma^2} + 3\dot{p} - \Sigma \dot{W}\right) + O(4) . \tag{23}$$

Substituting eq.(19)-(21) into (4), we have

$$\frac{\partial}{\partial T}\left[\left(1+\frac{4W}{c^2}\right)\Sigma^a\right]+\frac{\partial}{\partial X^b}\left[\left(1+\frac{4W}{c^2}\right)\left(\frac{\Sigma^a\Sigma^b}{\Sigma}\left(1+\frac{\Sigma^c\Sigma^c}{\Sigma^2c^2}+\frac{2p}{\Sigma c^2}\right)\right.\right.$$
$$\left.\left.+p\delta^{ab}\left(1-\frac{2W}{c^2}\right)\right)\right]=\Sigma W_{,a}+\frac{1}{c^2}\left(4\Sigma\dot{W}_a-4\partial_{[a}W_{b]}\Sigma^b\right)\ . \tag{24}$$

This is the PN hydrodynamic equation in local coordinate (A). In principle we can write down hydrodynamic equation in any local coordinate and global coordinate within DSX scheme. Now we have nine unknown function W, W^a, Σ, Σ^a and p and nine equations: energy equation, PN Euler equation, the equation of state and four field equation (DSX paper I eq.(3.18)). Therefore we can get the time evolution of $\Sigma(t)$ and $\Sigma^a(t)$, then the time evolution of B-D moment can be also obtained. In perfect fluid DSX scheme is completeness.

3. THE TOPICS ON PN HYDRODYNAMIC EQUATION IN NONPERFECT FLUID

Many years ago, we already know the general expression of $T_{\alpha\beta}$ for a thermally conducting, viscous and compressible fluid (Greenberg, 1971; Ehlers, 1993)

$$\begin{aligned}\bar{T}^{\alpha\beta} &= \rho(1+\frac{\Pi}{c^2})u^\alpha u^\beta+p\gamma^{\alpha\beta}+\lambda(u^{(\alpha;\beta)}-\frac{1}{c^2}\dot{u}^{(\alpha}u^{\beta)})\\ &- \frac{1}{3}\beta\theta\gamma^{\alpha\beta}+2u^{(\alpha}q^{\beta)}\end{aligned} \tag{25}$$

here $\gamma_{\alpha\beta}$ is projection operator ($\gamma_{\alpha\beta}=g_{\alpha\beta}+\frac{u_\alpha u_\beta}{c^2}$); λ is coefficient of bulk viscosity; $\theta=u^\mu_{;\mu}$ expansion scalar; q_β is heat flux ($q_\beta=-kh^\gamma_\beta(T_{0,\gamma}+T_0\dot{u}_\gamma)$, T_0 is the temperature, k is the thermal conductivity). All of p, π, λ, β and k are scalar, especially, λ, β, k are considered as constants, which are dependent on the character of fluid. In the following parts we only consider the quasi-adiabatic situation, then

$$q_\beta=0\ . \tag{26}$$

The quasi-adiabatic situation means: (a) the time scale of variation of the entropy t_0 is larger than the length scale of object l divided by the speed of sound v_s, i.e. $t_0\gg l/v_s$; (b) there is no shock wave. For most case it is satisfied, but the explosion (e.g. supernova) is an exception. Therefore we discuss the viscous and compressible fluid only. In eq.(25), we can calculate $\gamma^{\mu\nu}$, θ, u^α, $\dot{u}^\alpha$ and $u^{(\alpha;\beta)}-g^{\alpha\beta}u^{(\alpha}{}_{;\beta)}$ (here round parenthesis means symmetric indeces e.g. $u^{(\alpha}{}_{;\beta)}=\frac{1}{2}(u^\alpha{}_{;\beta}+u^\beta{}_{;\alpha})$) by meansof 3-velocity v^i, w, w^i and their derivative. Then we could express $\bar{T}^{\alpha\beta}$ interms of ρ, p, λ, β, v^a, w and w_i. ¿From $\bar{T}^{\alpha\beta}$ we might obtain σ and σ^i (similar to eq.(12) and (13)). ¿From σ and σ^i we got v^i in first order approximation (similar to (14)). Substituting v^i and ρ by means of σ, σ^i, p, λ, β, w and w^i and their derivative, we get a new express of $\bar{T}^{ij}$ in which there are no v^i and ρ, only σ, σ^i, p, λ, β, w and w^i. Directly convert $\bar{T}^{\mu\nu}$, σ, σ^i, w and w^i into $T^{\alpha\beta}$, Σ, Σ^a, W and W^a in local coordinate. Then we obtain T^{ab} and substitute it into eq.(3) and (4). We can get hydrodynamic equations of nonperfect fluid in every local coordinate. Since the formulae are quite long and the space limitation of this paper, here we present concept only. The detail results will be published elsewhere.

4. CONCLUSION

In perfect fluid we got the hydrodynamic equations (eq.(23) an (24)) in multiple systems. For nonperfect fluid (eq.(25)), we pointed the way to get the hydrodynamic equations in multiple systems. These work might be useful in relativistic binary systems.

Here, we should point out the inconvenience in our treatment. In higher order part (c^{-2}) of eq.(23) and (24) are nonlinear to Σ and Σ^i, the mathematical treatment may be complex in numerical calculation.

Acknowledgements

This work is supported by Chinese National Science Foundation (19873008 and 19835040)

References

Blanchet, L., Damour, T & Schäfer, G. (1989), MNRAS **242**, 289
Chandrasekhar, S. (1965), Astrophys. J., **142**, 1488
Damour, T., Soffel, M. and Xu, C. (1991), Phys. Rev. D**43**, 3273
Damour, T., Soffel, M. and Xu, C. (1992), Phys. Rev. D**45**, 1017
Damour, T., Soffel, M. and Xu, C. (1993), Phys. Rev. D**47**, 3124
Damour, T., Soffel, M. and Xu, C. (1994), Phys. Rev. D**49**, 618
Ehlers, J. (1993), Gen. Rel. & Grav. **25**, 1225
Greenberg, P.J. (1971), Astrophys. J. **164**, 569
Lichnerowicz, A. (1967), Relativistic Hydrodynamics and Magnetohydrodynamics, *Lectures on the Existence of Solutions*, W.A. Benjamin, New York
Taub, A.H. (1978), Ann. Rev. Fluid Mech. **10**, 301

Weinberg, S., Gravitation and Cosmology, John Wiley, Inc. New York, 1980

Will, C.M. (1993), Theory and experiment in gravitational Physics, Cambridge Univ. Press, Cambridge, G.B.

Xu, C., Wu, X. and Schäfer, G. (1997), Phys. Rev. D**55**, 528

EXPLANATION OF THE O'CONNELL EFFECT OF A CLOSE BINARY

QINGYAO LIU
and YULAN YANG
Yunnan Observatory, Chinese Academy of Sciences, Kunming, China
United Laboratory of Optical Astronomy, Chinese Academy of Sciences, China
National Astronomical Observatories, Chinese Academy of Sciences, China
bily@public.km.yn.cn

Abstract Based on the hypothesis that the circumstellar matter of a binary system is captured by its components a theoretical model for explaining the O'Connell effect of the close binary is given in this article. The results inferred from the model show that the late-type and/or short-period binaries can easily produce the obvious O'Connell effect and the occurrence of the O'Connell effect has no relation to the types of binaries. These conclusions are in accordance with the observed results. The observed O'Connell effects of six binary systems are explained by the model. For the three W-subtype W UMa binaries (i.e. YY Eri, BX Per and SW Lac), assuming that the density of the circumstellar matter captured by the primary is equal to that captured by the secondary, the densities of the circumstellar matter captured by the components are calculated and through further calculation it can be obtained that the amount of the O'Connell effect calculated is close to the one observed. As for the three A-subtype W UMa systems (i.e. CN And, FG Hya and AU Ser), assuming that the density of the circumstellar matter captured by one of the components is different from that captured by the other, the densities of the circumstellar matter captured by them are respectively calculated and the O'Connell effect theoretically calculated is compared with that observed, with the results showing that the consistency of the theoretically calculated O'Connell effect with the observed one is better than that obtained in the former case.

Keywords: close binary-W UMa binary-circumstellar matter-O'Connell effect

1. INTRODUCTION

The photometrically observed results show that there is a difference between the brightness of the two light maximum phases of the light

K.S. Cheng et al. (eds.), Stellar Astrophysics, 229–241.

curves of some contact or near-contact binary systems. O'Connell found in 1951, that the height of one of the two maxima of the light curve of the eclipsing binary RX Cas did not equal that of the other and in Abhyankar (1959) found the similar phenomenon for AO Cas. Later on, Milone (1969) called this phenomenon as the O'Connell effect. The fine observations which have been carried out since 1960s show that the O'Connell effect generally exist in the short-period and/or late-type contact or near-contact binaries. If the first maximum (following the primary eclipse) of the light curve is brighter than the second maximum (following the secondary eclipse), the O'Connell effect is called as the positive one, whereas if the first is fainter than the second, the O'Connell effect is regarded as the negative one. The amounts of the O'Connell effects of many systems are changeable, and even for some systems the O'Connell effect sometimes changes from the positive to the negative and vice versa.

The explanation of the O'Connell effect has been one of the celebrated difficult problems in research on the close binary systems. Various explanations of the O'Connell effect have so far been proposed and among them there mainly are three models, i.e. the Coriolis force effect, spot activities and mass exchange between the two components. Zhou and Leung (1993) have pointed out that the O'Connell effect of a contact binary system is caused by the Coriolis force produced from the interaction between the rotation and circulation of the system. Acted by the Coriolis force, the direction of the circulation within the common envelope of the contact binary results in the asymmetry with the respect to the equatorial plane, thereby causing the asymmetry (or O'Connell effect) of the light curve of the contact binary. However, the O'Connell effect of a non-contact binary without circulation can hardly be explained by this model. Binnendijk (1960) was the first to explain the asymmetry of the light curve of the binary AH Vir by means of spot activities, and later on, many researchers all explained the O'Connell effect of late-type binaries using the spot activities (Bell, et al., 1990, Linnell, et al., 1989). But, as pointed out by Maceroni & Van 't Veer (1993), there is much uncertainty in the explanation by means of the spot activity model since the perfect fits of the some observed light curve can be obtained by means of various spot models. In addition, the scale of the O'Connell effect often is much larger than that of the spot activity, and therefore, if the spot model is used to fit, the observed spot area should be 30% to 50% of the component surface, with such a large spot being hard to understand. Shaw (1994) has thought that the O'Connell effect of a near-contact binary is caused by the hot spots which are formed when the mass flows from one component to the other. But, the O'Connell effect of a contact binary with the common envelope is hardly explained by means of the hot spot model of mass exchange. In a word, at present, there is still no satisfactory model to be used to explain the O'Connell effect.

The authors attempt to explain the O'Connell effect of a close binary by using the interaction of the components with the circumstellar matter. The model for explaining the O'Connell effect is given in Section 2, the comparison of the model with the observations is described in Section 3 and some discussions are presented in Section 4.

2. THEORETICAL MODEL

As there are various complicated interactions between the two components of a close binary, violent activities at the surfaces of the components and small disturbances of the sizes and shapes of the Roche lobes, so it is reasonably considered that the surface matter of the two components is continuously ejected leaving the components and forms the so-called circumstellar matter with various states of motion surrounding the binary system. Assume that the average velocity of part of the circumstellar matter with respect to the mass center of the binary system is zero and the average density of the matter is ρ. In the course of the orbital movement of the two components, on the average, the circumstellar matter is respectively captured by the components at the orbital speeds of them, and only the hemisphere of each component towards the direction of the orbital motion can capture the circumstellar matter. Assume that all the kinetic energy of the matter captured by the components is turned into the thermal energy, thereby heating the atmosphere of the hemisphere of the matter captured by one of the components so that there appears the difference between the surface temperatures of the two hemispheres of each component. As for an observer, the heated hemispheres of the primary and secondary components are alternately observed. Since the primary and secondary components are differently heated, there appear different observed brightnesses at the phase 0.25 and the phase 0.75, i.e. the so-called O'Connell effect.

In order to describe quantitatively this model, assume that one of the two components approximates to a sphere and its rotation and revolution synchronize. A moving rectangular coordinate system with rotation velocity of a binary system is set up, with the coordinate plane being parallel to the axis of rotation of the binary, and the mass center is taken to be the origin of the coordinate system, the rotation direction of the system to be the y axis and the direction from the primary to the secondary to be the x axis (see Fig. 1).

Assuming that a and b are respectively the nodes of the surface of the primary and the secondary with the side of the x axis close to the mass center, then the coordinates of the center of the primary are [-(a+R_1), 0] and the coordinates of the center of the secondary are (b+R_2,0), were R_1 and R_2 are the radii of the primary and secondary, respectively, with the solar radius being used as the unit.

According to the definition of the mass center of a binary system, the following equations can be inferred,

$$a = \frac{-Aq - R_1(1+q)}{1+q} \tag{1}$$

$$b = \frac{A - R_2(1+q)}{1+q} \tag{2}$$

were A is the distance between the two components in the solar radius and q is the mass ratio of the two components and q= $\frac{m_2}{m_1}$. The distribution of the orbital velocity of the surface of a component is written as follows:

$$V(x) = \frac{2\pi x}{p} \tag{3}$$

where p is the orbital period of the binary in a day. assume that ds=2ydx, within the unit time the mass of the matter captured by the surface unit ds is

$$\partial m = \frac{4\pi \rho y x dx}{p} \tag{4}$$

when it sweeps over the space, and the kinetic energy is

$$dE = 0.5\partial m V(x)^2 = \frac{8\pi^3 \rho y x^3 dx}{p^3} \tag{5}$$

Assume that each of the two components is in the shape of a sphere, and therefore the projections of the surfaces of the two components on the XY plane are two circles. Thus, as far as the primary is concerned, there should be

$$y = \sqrt{R_1^2 - (x + \frac{Aq}{1+q})^2} \tag{6}$$

and as for the secondary

$$y = \sqrt{R_2^2 - (x - \frac{A}{1+q})^2}. \tag{7}$$

Assume that within the unit time the kinetic energy of the matter captured by the components is completely turned into the thermal energy and therefore the increased luminosities of the primary and secondary are respectively written as follows,

$$\Delta L_1 = \frac{8\pi^3\rho}{p^3}\int_a^{a-2R_1}\sqrt{R_1^2-(x+\frac{Aq}{1+q})^2x^3dx} \tag{8}$$

$$\Delta L_2 = \frac{8\pi^3\rho}{p^3}\int_b^{b+2R_2}\sqrt{R_2^2-(x-\frac{A}{1+q})^2x^3dx} \tag{9}$$

After the integration we have

$$\Delta L_1 = \frac{\pi^4 AqR_1^2\rho}{p^3(1+q)}(3R_1^2+\frac{4A^2q^2}{(1+q)^2}) \tag{10}$$

$$\Delta L_1 = \frac{\pi^4 AR_2^2\rho}{p^3(1+q)}(3R_2^2+\frac{4A^2}{(1+q)^2}) \tag{11}$$

The radiation luminosities of the two components are respectively shown as follows:

$$L_1 = R_1^2T_1^4 \tag{12}$$

$$L_2 = R_2^2T_2^4 \tag{13}$$

Assuming that the binary system rotates counterclockwise, the bolometric radiation magnitude difference observed between at the first maximum phase (the maximum after the primary eclipse) and at the second maximum phase (the maximum after the secondary eclipse) should be

$$\Delta m = -2.5Log\frac{0.5(L_1+L_2)+\Delta L_1}{0.5(L_1+L_2)+\Delta L_2} \tag{14}$$

As ΔL_1 is not equal to ΔL_2, so Δm does not zero, or there should appear the so-called O'Connell effect.

3. TEST BY MEANS OF OBSERVATIONS

The above-mentioned model can be tested by means of the comparison with observations.

(1) Easily appearing the O'Connell effect in late-type and/or short-period binaries

One of the important observation fact is that the O'Connell effect can be observed in W UMa binaries and near-contact binaries. This observational fact can be just right explained by means of the above-mentioned model of ours. The Kepler Law may be written as follows,

$$A^3 = 74.5p^2m_1(1+q) \tag{15}$$

where A is the distance between the two components in solar radius, m_1 the mass of the primary in the solar mass and p the orbital period in the day. According to the definition of the relative radius of a component, we have

$$r_{1,2} = \frac{R_{1,2}}{A} \tag{16}$$

substituting Eqs. (10), (11), (12), (13), (15) and (16) into Eq. (14), one may obtain

$$\Delta m = -2.5Log \frac{\frac{T_1^4}{m_1}p(1+q)^2(1+\frac{L_2}{L_1}) + 149\pi^4 q\rho\left[3r_1^2\left(1+q^2\right)+4q^2\right]}{\frac{T_1^4}{m_1}p(1+q)^2(1+\frac{L_2}{L_1}) + 149\pi^4(\frac{R_2}{R_1})^2\rho\left[3r_2^2\left(1+q\right)^2+4\right]} \tag{17}$$

Eq. (17) shows that the late-type and/or short-period binaries easily exert the O'Connell effect, by reason that $\frac{T_1^4}{m_1}$ and p of a late-type and/or short-period binary are numerically less (see Table 1, where the data are cited from Alen (1973)), thereby showing much more the observed effects produced by the interaction of the circumstellar matter with the two components on the whole.

To explain this conclusion, an example is cited as follows: Assume that the primary is one of the main sequence stars, the orbital period of the binary is p=1 day, the mass ratio q=0.4, the luminosity ratio $\frac{L_2}{L_1} = 0.03$, the radius ratio $\frac{R_2}{R_1} = 0.5$, $r_1 = 0.4$ and $r_2 = 0.2$, the theoretical values of the effective surface temperatures and masses of the main sequence stars with different spectral type are used and the corresponding Δm is calculated by means of Eq. (17), and then draw Figure 2. If the spectral type of the binary is G5V, the corresponding $T_1^4/m_1 = 0.888$, and the above-mentioned assumed values of the mass ratio, radius ratio and relative radius of the components are still adopted, draw Figure 3, which shows Δm with respect to different orbital periods.

The observational fact that the O'Connel effect is produced in the late-type and/or short-period binaries is naturally explained by Figs. 2 and 3 obtained from our model. Fig. 2 and 3 show that the O'Connell effect is easily observed in the binaries with an orbital period less than a day and a spectral type later than F0, and the W UMa binary is the very one.

(2) The theory of the O'Connell effect of three W subtype W UMa binaries and the comparison with the observations.

Given the average density ρ of the circumstellar matter captured by the two components of the binary, the measure, Δm, of the theoretical O'Connell effect can be calculated by means of Eq. (17), thereby comparing the calculated Δm with the observed one. Unfortunately, we

Table 1 $\frac{T^4}{m}$ of the main sequence stars with different spectral types

Sp.	O5	B0	B5	A0	A5	F0	F5	G0	G5	K0	K5	M0	M5
$\frac{T^4}{m}$	56.81	30.56	7.89	2.62	2.21	1.56	1.28	1.06	0.88	0.65	0.37	0.28	0.26

have not know how to determine the density ρ. In the researches on the late-type contact binaries and near-contact binaries there is anther problem difficult to understand, i.e. the so-called over-luminosity problem, or the observed luminosity of a binary greater than the theoretical luminosity of a main-sequence component with the same mass. Assuming that the over-luminosity is from the $\Delta L_1+ \Delta L_2$ in our model, the over-luminosity $\Delta L_1+ \Delta L_2$ can be expressed by the difference between the observed luminosities of the components and the theoretical luminosity of the main sequence components with the same masses. From Eqs. (10) and (11) the density ρ can be obtained, and the theoretical O'Connell effect Δm can be further calculated by means of Eq. (17) and then compared with the observed.

YY Eri is a W-subtype W UMa binary with the spectral type being G5V and the orbital period being 0.3215 day. Batten et al. (1989) have published the spectroscopic orbital solutions for this system, and Maceroni et al. (1994) have carried out the photoelectric photometry and analysis. At the V band, $\Delta m(Max.I - Max.II) = -0.^m04$. According to the relevant parameters of the binary given by Maceroni et al.(1996), $m_1 = 1.02m_\odot$, $m_2 = 0.44m_\odot$ and the sum of the luminosities of the two components $L_1 + L_2 = 1.12L_\odot$ based on the mass-luminosity relation of a main sequence star while the observed sum of the luminosities is $1.20L_\odot$, i.e. the observed sum is $(\Delta L_1 + \Delta L_2) = 0.08L_\odot$ greater than the theoretical one. According to our model, this over-luminosity can be comprehended as the result of the interaction of the circumstellar matter with the components. From Eqs. (10) and (11), one may obtain $\rho = 3.6 \times 10^{-12} g/cm^3$, and from Eq.(17) one may also obtain $\Delta m(Max.I - Max.II) = -0.07$ bolometric magnitude, close to the the practically measured -0.04 V magnitude.

BX Peg is a W-subtype W UMa binary with the spectral type being G4V and the orbital period being 0.2804 day. The photoelectric photometry and analysis of this binary were published by Kaluzny(1984). According to the relevant parameters published by Maceroni et al.(1996) and our model, one may obtain $\rho = 2.89 \times 10^{-12} g/cm^3$ and $\Delta m(Max.I-$

$MaxII) = 0 - 0.04$ bolometric magnitude, close to the observed -0.02 V magnitude.

SW Lac also is a W-subtype W UMa binary with the spectral being G8V and the orbital period being 0.3207 day. The researched results of the photoelectric photometry and spectroscopic study of this binary were given by Zhai and Lu (1989). According the relevant parameters published by Maceroni et al. (1996) and through the model calculation, one may obtain $\rho = 2.52 \times 10^{-12} g/cm^3$ and $\Delta m(Max.I - Max.II) = 0.02$ bolometric magnitude close to the practically measured 0.04 V magnitude.

(3) The comparison of the theory with the observations of the O'Connell effect for three A-subtype W UMa binaries.

As far as the following A-subtype W UMa binaries (i.e. CN And, FG Hya and AU Ser), assume that the density of the circumstellar matter captured by the primary is different from that by the secondary, and the densities of the circumstellar matter captured by the components can be calculated from Eqs. (10) and (11), respectively, based on the over-luminosity (the difference between the observed luminosity and the theoretical luminosity of a main sequence star with the same mass) of each individual of the two components, thereby comparing the theoretical O'Connell effect calculated from Eq.(17) with the observed one.

CN And is an A-subtype W UMa binary, with the spectral type of F8 and the orbital period of 0.4628 day. The researched results of the photometry of the system were published by Rafert et al.(1985). According to the relevant parameters of the binary published by Maceroni et al. (1996), the over-luminosity of the primary is $\Delta L_1 = 0.409 L_\odot$ and the one of the secondary is $\Delta L_2 = 0.372 L_\odot$. Assuming that the density of the circumstellar matter captured by the primary is different from that by the secondary and ρ_1 stands for the density in Eq.(10), corresponding to $\Delta L_1 = 0.409 L_\odot$, one may obtain $\rho_1 = 3.79 \times 10^{-11} g/cm^3$. If ρ_2 stands for the density in Eq.(11), corresponding to $\Delta L_2 = 0.372 L_\odot$, one may obtain $\rho_2 = 1.40 \times 10^{-11} g/cm^3$. The O'Connell effect of the binary calculated from Eq.(17) is $\Delta m(Max.I - Max.II) = -0.032$ bolometric magnitude, almost in accordance with the observed value -0.04 magnitude at B band.

FG Hya is an A-subtype W UMa binary, with the spectral type being G0 and the orbital period being 0.3278 day. The photoelectric photometry and analysis for this system were published by Twigg et al.(1979) and Yang et al.(1990), respectively. According to the relevant parameters of the binary published by Maceroni et al.(1996), the over-luminosity of the primary reaches $\Delta L_1 = 0.256 L_\odot$ and that of the secondary is $\Delta L_2 = 0.289 L_\odot$. Assuming that the density of the circumstellar matter captured by the primary is different from that by the secondary, one may obtain $\rho_1 = 6.86 \times 10^{-11} g/cm^3$ and $\rho_2 = 2.16 \times 10^{-11} g/cm^3$. From Eq.(17), $\Delta m(Max.I - Max.II) = 0.026$ bolometric magnitude can be obtained, which is consistent with the observed value of 0.02 magnitude in B band.

AU Ser also is an A-subtype W UMa binary, with the spectral type being G5 and the orbital period being 0.3865 day. The photoelectric photometry and the solution of the binary were published by Kaluzny et al.(1986), and the spectroscopic observations and analysis of this system were published by Hrivnak (1993). The mass ratios obtained from the analyses of the photometry and spectroscopic observations are 0.800 and 0.710, respectively, and the mass ratio of q=0.75 is adopted in the present analyses. According to the relevant parameters of the binary published by Maceroni et al.(1996), the over-luminosity of the primary is $\Delta L_1 = 0.120 L_\odot$ and the one of the secondary is $\Delta L_2 = 0.090 L_\odot$. Assuming that the densities of the circumstellar matter captured by the primary and the secondary are different from each other, one may obtain $\rho_1 = 8.90 \times 10^{-12} g/cm^3$ and $\rho_2 = 4.32 \times 10^{-12} g/cm^3$. The O'Connell effect calculated from Eq.(17) is $\Delta m(Max.I - Max.II) = -0.042$ bolometric magnitude, in accordance with the observed value of -0.05 magnitude in B band.

4. DISCUSSION

The satisfactory explanation of the O'Connell effect of light curves of close binaries has not been obtained ever since half of a century. The theoretical model for explaining the O'Connell effect obtained in the present article by proceeding from the assumption that the components of a binary capture the circumstellar matter is satisfactory. (1). The observed fact that only the late-type and/or short-period binaries can have the O'Connell effect is quite naturally explained by means of the model. Since no restriction is imposed on the spectral type of the components and orbital period of the systems when the model is make, our model is applicable to all the binaries. However, the results inferred from the model show that the late-type and/or short-period binaries can easily produce the obvious O'Connell effects, which is just consistent with the observations. (2). The observed O'Connell effects of the six binaries are successfully explained. As far as three W-subtype W UMa binaries (YY Eri, BX Peg and SW Lac) are concerned, assume that the densities of the circumstellar matter captured by the primary and secondary are equal to each other and the densities of the circumstellar matter captured by the components are calculated from the difference between the observed luminosities of the two components and the theoretical luminosities of the main sequence components with the same masses, thereby comparing the calculated O'Connell effect with the observed ones, with the results showing that the theoretical O'Connell effect are in accordance with the observed ones very well. As far as three A-subtype W UMa binaries (CN And, FG Hya and AU Ser), assume that the densities of the circumstellar matter captured by the two components are different from each other and the densities of the circumstellar matter captured by the two components are calculated by each individual over-luminosities (i.e. the difference between the observed luminosity and the theoretical luminosity of a main sequence star with the same mass) of the two components, respectively, thereby comparing the theoretical O'Connell effect with the observed one. The results show that the consistence of the theoretical O'Connell effect with the observed one is much better than that obtained in the former case.

Two principal assumption are proposed in the present model for the convenience of calculation and for the sake of the lack of an understanding of the state of the circumstellar matter. The first assumption is that each of the two components should be a globoid, which is for the convenience of calculation. Actually, the shape of one of the two components has a certain deviation from a globoid because of the gravitational action between the two components, the fast rotation of the components and so forth. Strictly speaking, the shape of the components as defined in Roche geometry should be used. However, the error with an order of magnitude can not be produced calculatively by the spherical approximation and it is considered that the observed O'Connell effect is expressed by the magnitude difference, thus the error introduced from the spherical approximation should not have obvious display on the magnitude because of the logarithmic relation. The second assumption is the state of the circumstellar matter. For lack of the recognition of the circumstellar matter at present, perhaps the relevant assumptions proposed in the model can only be tested by the observations. Owing to the complexity and variability of the source of the circumstellar matter, a natural inference is that the O'Connell effect should also change according to the present model, which is in accordance with the observations because the observations show that O'Connell effect is certainly variable and changes at random.

As the density and dynamic state of the circumstellar matter are approximately assumed, so the concrete value of the O'Connell effect of a single binary obtained through calculation is also only an approximate one. However, the consistence of the calculated theoretical value with the observed value shows that the present model is possibly correct in principle. On the other hand, if our model is correct, the density of the circumstellar matter and the variation of the density can be inferred from the measurement of the O'Connell effect of a binary. Therefore, this may also be a way to the understanding of some characteristics of the circumstellar matter of a binary.

Acknowledgements

The authors would like to express our gratitude for the support from the Chinese National Science Foundetion Committee and the Chinese Academy of Sciences.

References

Abhyankar, K.D. 1959, ApJ Suppl., 4, 157
Alen, C.W. 1973, in Astrophysical Quantities, p.251-254
Batten, A.H., Fletcher, J.M., Mc Carthy, D.G. 1989, Publ. D.A.O., XVII
Bell, S.A., Rainger, P.P., and Hilditch, R.W. 1990, MNRAS, 247, 632
Binnendijk, L. 1960, AJ, 65, 358
Hrivnak, B.J. 1993, in New Frontiers in Binary star research, K.C.Lueng and I.S.Nha, eds, PASPC 38, p.269
Kaluzny, J. 1984, Acta Astron. 34, 217
Kaluzny, J. 1986, Acta Astron. 36, 113
Linnell, A.P. & Olson E.C. 1989, ApJ, 343, 909
Maceroni, C., & van't Veer F. 1993, A&A, 277, 515
Maceroni, C., Vilhu, O., van't Veer F., and Van Hamme W. 1994, A&A, 288, 529
Maceroni, C., & van't Veer F. 1996, A&A, 311, 523
Milone, E.F. 1969, in Non-period Phenomena in Variable Star, L. Detre ed. p.457
Rafert, J.B., Markworth, N.L., and Michaels, E.J. 1985, PASP, 97, 310
Shaw, J.S. 1994, Mem.S.A.It., 65. 1
Twigg, L.W. 1979, MNRAS, 189, 907
Yang, Y., Liu, Q., Zhang, Y., Wang, B., and Lu, L. 1991, Acta Astron. Chin.,32, 326
Zhai, D., & Lu, W. 1989, Chin. Astron Astrophys., 9. 208
Zhou, D., & Lueng, C.K., 1993, Astron. Soc. Conf. Ser., 38, 206

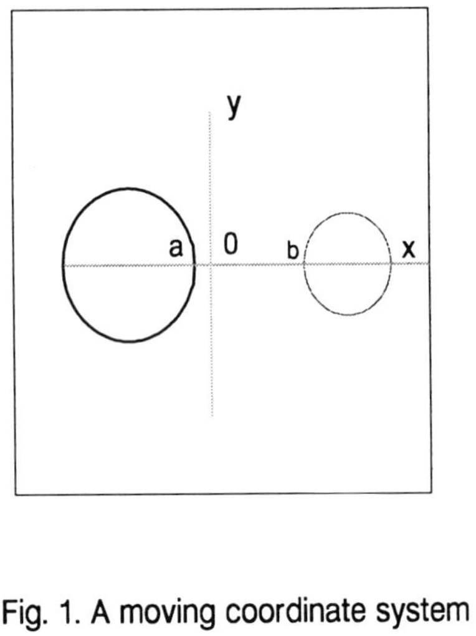

Figure 1 A moving coordinate system

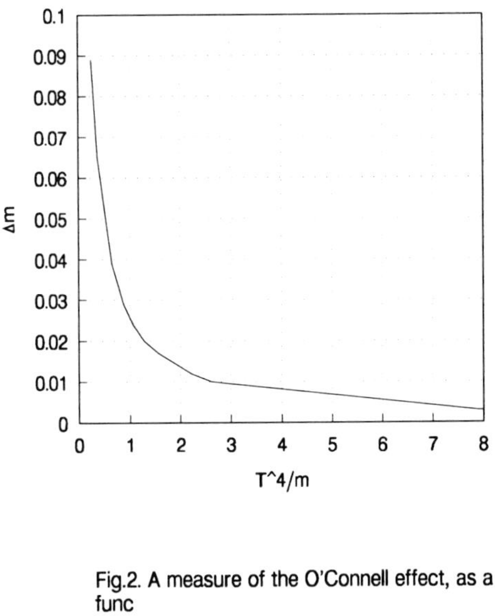

Figure 2 A measure of the O'Connell effect, as a function of mass and temperature of the primary of a binary

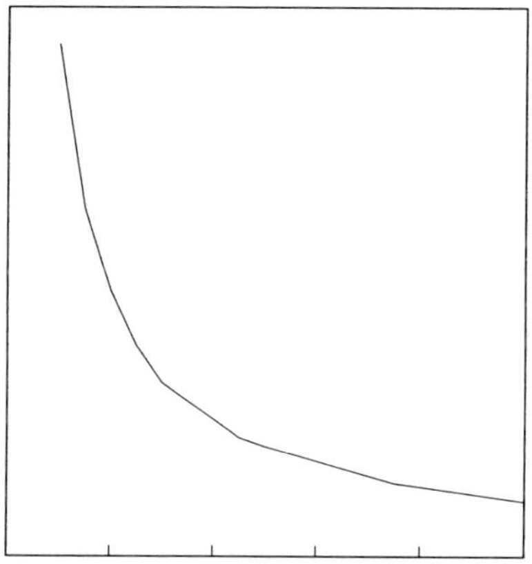

Figure 3 A meare of the O'Connell effect, as a function of the orbital period of a system

EFFECTS OF THE BLANDFORD-ZNAJEK PROCESS ON EVOLUTION OF RADIAL STRUCTURE OF BLACK HOLE ACCRETION DISKS

D. X. Wang

Department of Physics, Huazhong University of Science and Technology, Wuhan, 430074, China

1. INTRODUCTION

It is well known that the Blandford-Znajek (BZ) process is an effective mechanism for extracting rotating energy and angular momentum of a central black hole (BH) surrounded by accretion disk (Blandford & Znajek 1977; Macdonald & Thorne 1982 henceforth MT). The BZ extracting mechanism always works together with accretion of matter onto BHs: Leaving the inner edge of the accretion disk, the disk plasma gradually spirals and carries its frozen-in magnetic field onto the BH horizon. The strength of the BZ process depends crucially on the strength of the magnetic field, $B_\perp$, normal to the BH horizon. Recently, it was pointed out that the values of $B_\perp$ were overestimated substantially in previous works, and the BZ extracting power (hereafter the BZ power) is not as strong as imagined previously (Ghosh & Abramowicz 1997; Livio, Ogilvie & Pringle 1999). The overestimate of $B_\perp$ arises from the incomplete argument on the BZ process, in which the Maxwell pressure, $B_\perp^2/8\pi$, was assumed to correspond to an equilibrium with the maximum total pressure in the inner parts of the accretion disk rather than the Maxwell pressure near the inner edge. The modified BZ power P in radiation-pressure dominated case can be expressed as (Wang 1999a):

$$P = (4/65)k(1-k)a_\star^2 E_{in} M_0 \tag{1}$$

Eq.(1) will be used to investigate the effects of the BZ process on the evolution of the radial structure of BH accretion disks in this paper.

K.S. Cheng et al. (eds.), Stellar Astrophysics, 243–248.

2. A CHARACTERISTIC FUNCTION DESCRIBING THE BZ EFFECTS ON THE PARAMETERS OF BH ACCRETION DISKS

Kerr BH accretion disk can be regarded as a system with two parameters, BH mass M and its spin parameter, $a_\star \equiv J/M^2$. The basic equations governing the evolution of BH accretion disks in the BZ process are expressed as follows (Moderski & Sikora 1996; Wang, Lu & Yang 1998 henceforth WLY; Wang 1998):

$$(dM/dt)_{BZ} = E_{in}\dot{M}_0 - P \tag{2}$$

$$(dJ/dt)_{BZ} = L_{in}\dot{M}_0 - P/\Omega_F \tag{3}$$

Taking $k \equiv \Omega_F/\Omega_H = 0.5$, which corresponds to the maximum BZ power (MT) in eq.(1) and incorporating eqs.(2) and (3), we derived the rate of change of $a_\star$ (WLY):

$$(da_\star/dt)_{BZ} = da_\star/dt - (2P/Ma_\star)(1+q)^2 \tag{4}$$

where $q = \sqrt{1-a_\star^2}$, $da_\star/dt$ and $(da_\star/dt)_{BZ}$ are the rates of change of $a_\star$ in the pure accretion and the BZ process, respectively.

Generally speaking, a parameter Q of Kerr BH accretion disk can be expressed as the following form:

$$Q = M^\alpha \xi(a_\star) \tag{5}$$

The rates of change of Q in a pure accretion and the BZ process are expressed by eqs. (6) and (7), respectively.

$$dQ/dt = \alpha\xi(a_\star)M^{\alpha-1}(dM/dt) + M^\alpha(d\xi/da_\star)(da_\star/dt) \tag{6}$$

$$(dQ/dt)_{BZ} = \alpha\xi(a_\star)M^{\alpha-1}(dM/dt)_{BZ} + M^\alpha(d\xi/da_\star)(da_\star/dt)_{BZ} \tag{7}$$

Substituting eqs.(2) and (4) into eq.(7), we have

$$(dQ/dt)_{BZ} = (dQ/dt) - PM^{\alpha-1}[\alpha\xi + (2/a_\star)(d\xi/da_\star)(1+q)^2] \tag{8}$$

Defining $\triangle(dQ/dt) \equiv (dQ/dt)_{BZ} - dQ/dt$, we can rewrite eq.(8) as

$$\triangle(dQ/dt) = PM^{\alpha-1}\eta \tag{9}$$

where in eq.(9)

$$\eta \equiv -\xi[\alpha + \frac{2(1+q)^2}{a_\star}\frac{dln\xi}{da_\star} \tag{10}$$

Table 1 Some Parameters of BH Accretion Disks and the Corresponding Expressions of η

Q	α	ξ	η	$\Delta(dQ/dt)$
M	1	1	-1	$P\eta < 0$
$a_\star$	0	$a_\star$	$-2(1+q)^2/a_\star$	$PM^{-1}\eta < 0$
J	2	$a_\star$	$-4(1+q)/a_\star$	$PM\eta < 0$
Ω_H	-1	$\frac{a_\star}{2(1+q)}$	$-\frac{q^2+q+2}{2a_\star q}$	$PM^{-2}\eta < 0$
r_H	1	1+q	$(q^{-1}+1)(q+2)$	$P\eta > 0$
r_{ms}	1	χ_{ms}^2	$\chi_{ms}^2[\frac{2(4\chi_{ms}-3a_\star)}{3a_\star(\chi_{ms}-a_\star)^2}(1+q)^2-1]$	$P\eta > 0$
r_{mb}	1	χ_{mb}^2	$\chi_{mb}^2[\frac{2(1+q)^2}{a_\star\chi_{mb}(\chi_{mb}-1)}-1]$	$P\eta > 0$
T_{bh}	-1	$\frac{q}{4\pi(1+q)}$	$\frac{(q^2+2q+2)}{4\pi q(1+q)}$	$PM^{-2}\eta > 0$
S_{bh}	2	$2\pi(1+q)$	$4\pi(1+q^{-1})$	$PM\eta > 0$

$\Delta(dQ/dt)$ represents the effects of the BZ process on the rates of change of the parameter Q, which has the same sign as η. Therefore η can be regarded as a characteristic function describing the effects of the BZ process on the evolution of the parameters of BH accretion disks. It is easy to find from eq.(10) that both value and sign of η are decided uniquely by the expression of Q. Some parameters of BH accretion disks and the corresponding expressions of η are calculated and listed in Table 1.

It is easy to check that the results listed in Table 1 are all consistent with those in our previous works (WLY; Wang 1998; Wang 1999a, 1999b). Inspecting Table 1 and eq.(9), we find the following effects of the BZ process on the evolution of the parameters of BH accretion disks:

- All values of $\Delta(dQ/dt)$ listed in Table 1 are proportional to the BZ power P;
- All values of $\Delta(dQ/dt)$ involving the BH spin parameters are negative, which arises from extracting of BH angular momentum in the BZ process;

- All values of $\triangle(dQ/dt)$ involving the radial and thermodynamic parameters are all positive. Inspecting the expressions of these parameters, we find a common feature: Reducing BH angular momentum $a_\star$ makes these parameters increase, while reducing BH mass M makes them decrease. The positive values of $\triangle(dQ/dt)$ implies that the effects of extracting BH angular momentum are stronger than those due to reducing BH mass.

3. EVOLUTION CHARACTERISTICS OF RADIUS OF INNER EDGE OF BH ACCRETION DISKS

As an example, we are going to investigate the BZ effects on the evolution of the radius of the inner edge of BH accretion disks. Radius of inner edge of a thin disk is $r_{ms} = M\chi_{ms}^2$, where χ_{ms} is a dimensionless radial parameter depending on $a_\star$ (Novikov & Thorne 1973). The rates of change of r_{ms} in pure accretion and the BZ process can be expressed by equations (11) and (12), respectively.

$$\begin{cases} dr_{ms}/dt = h(a_\star)\dot{M}_0 \\ h(a_\star) = -\frac{(4\chi_{ms}-3a_\star)}{\sqrt{3}} \end{cases} \tag{11}$$

$$\begin{cases} (dr_{ms}/dt)_{BZ} = H(a_\star)\dot{M}_0 \\ H(a_\star) \equiv h(a_\star)[1 + \frac{a_\star^2}{65} - \frac{2a_\star(4\chi_{ms}-3a_\star)(1+q)^2}{195(\chi_{ms}-a_\star)^2}] \end{cases} \tag{12}$$

$H(a_\star)$ and $h(a_\star)$ versus $a_\star \equiv a/M$ are shown by the upper and lower curves, respectively, in Fig. 1(a) and (b):

Figure 1 The curves of $H(a_\star)$ (the upper) and $h(a_\star)$ (the lower) versus $a_\star$, (a) $0 < a_\star < 0.998$; (b) $0.950 < a_\star < 0.998$

Inspecting eqs.(11),(12) and Fig. 1(a) and (b), we obtain the following evolution characteristics of r_{ms}:

- For the same accretion rate $\dot{M}_0$ we have $dr_{ms}/dt < (dr_{ms}/dt)_{BZ}$, and the difference between the two corresponding rates increases as $a_\star$.
- $h(a_\star)$ is always negative in the whole evolution process, which means that r_{ms} decreases monotonously in the pure accretion.
- The values of $H(a_\star)$ changes from negative to positive at the turning points, $a_\star^{c1} \approx 0.997$, which means that r_{ms} decreases (increases) before (after) this turning points, and attains a minimum at this point.
- Substituting the expression for the radius of the inner edge of a thick disk, $r_{mb} = M\chi_{mb}^2$, we can obtain the similar evolution characteristics: r_{mb} decreases (increases) before (after) a turning points, and attains a minimum at $a_\star^{c2} \approx 0.999$.

4. EVOLUTION OF RADIUS OF INNER EDGE OF BH ACCRETION DISKS AND RUNAWAY INSTABILITY

In 1983, Abramowicz et al. pointed out that sufficiently massive tori with constant angular momentum l are all runaway unstable (ACN). It was argued, if the relativistic Roche lobe shrinks sufficiently enough, the disk matter which was bounded before the mass transfer will become unbounded, falling catastrophically into the BH on a dynamical time scale. This is the runaway instability. It seems that the two effects are involved contrarily in the runaway instability: the self-gravity makes tori runaway unstable, while the BH spin stabilizes them (Wilson 1984; Khanna & Chakrabarti 1992).

According to the results of this paper, radius of inner edge r_{in} always decreases monotonously in pure accretion onto a Kerr BH. It means that the bulk of disk matter will be bounded by the critical surface with the cusp located at r_{mb}, the monotonous decrease of r_{in} avoids the runaway instability. It is consistent with Wilson's result for the Kerr BH surrounded by a thick disk. However, this result will be changed significantly by the BZ mechanism. Considering that the BH spin is depressed by the BZ mechanism, we think that the stabilizing effects of BH spin will be reduced. As argued above, the BZ mechanism makes r_{ms} and r_{mb} increase after $a_\star$ attains the corresponding turning points, $a_\star^{c1} \approx 0.997$ and $a_\star^{c2} \approx 0.999$, respectively. In the BZ process, the bulk of disk matter which was bounded by the critical surface would be unbounded due to the increasing of r_{ms} or r_{mb} , and this in turn makes disk matter fall catastrophically into the BH on a dynamical time scale. Therefore we conclude the runaway instability will happen in the BZ process, provided that the central BH evolves very close to the extreme Kerr BH.

The runaway instability of BH accretion disks would be used to explain the strongest γ-ray bursts occuring at the cosmological distance.

References

Abramowicz M. A., Calvani M., Nobili L., 1983, Nature, 302, 597 (ACN)
Blandford R. D., Znajek R. L., 1977, MNRAS, 179, 433
Ghosh P., Abramowicz M. A., 1997, MNRAS, 292, 887
Khanna R., Chakrabarti S. K., 1992, MNRAS, 259, 1
Livio M., Ogilvie G. I., Pringle J. E., 1999, ApJ, 512, L100
Macdonald D., Thorne K. S., 1982, MNRAS, 198, 345 (MT)
Moderski R., Sikora M ., 1996, MNRAS, 283, 854
Novikov, I. D., Thorne, K. S., In : C. Dewitt, eds., Black Holes, New York: Gordon and Breach, 1973, 345
Wang D. X., Lu Y., Yang L. T., 1998, MNRAS, 294, 667 (WLY)
Wang D. X., 1998, Gen. Rel. Grav. 30, 1025
Wang D. X., 1999a, A&A, 347, 1069; 1999b, Acta Phys. Sinica. 48, 1552 (in Chinese)
Wilson D. B., 1984, Nature, 312, 620

THEORY OF ACCRETION DISKS AND APPLICATIONS IN ASTROPHYSICS

L.T. Yang[1], Y.Lu [1,2], and S.X.Ding[1]
[1]*Department of Physics, Huazhong Normal University, Wuhan, 430079, China*
[2]*Beijing Astronomical Observatory, Chinese Academy of Sciences, Beijing, 100012, China*

Abstract Accretion disks are important for many astrophysics phenomena and objects such as: quasars, active galatic nuclei, X-ray binary systems, young stellar objects and gamma ray bursts etc. Here, we review the main contents of disk's theory and some recent progress associated with the applications, including the thin and thick/tori disks, emitted spectrum and radiation mechanism, self-gravitating and advection disks, instabilities of accretion disks. Finally, we give a summary discussion.

Keywords: Accretion discs – Theory – Applications

1. INTRODUCTION

The theory of accretion disks is important for astrophysical phenomena , including quasars, active galactic nuclei, X-ray binary system, young stellar objects, and γ-ray bursts etc. (Pring 1981; Papaloizou & Lin 1995; Lin & Papaloizou 1996)

There are two main processes during the materials accreted into the central stars (or black holes). One is that the most materials are accreted into the central objects, another is that few materials bring angular momentum of the fluid to go out. Since the only energy source is the gravitational potential, this means that the viscosity converts the gravitational potential energy into radiation in an efficient manner.

2. BASIC EQUATIONS OF THIN DISKS

For an axisymmetric thin disk, in cylindrical coordinates, the surface density, angular velocity Ω , radial velocity v_r and temperature T can be derived from the standard continuity, momentum and energy equations

K.S. Cheng et al. (eds.), Stellar Astrophysics, 249–256.

after integration over z (Blumenthal et al. 1984):

$$\frac{\partial \Sigma}{\partial t} + \frac{1}{r}\frac{\partial}{\partial r}(r\Sigma v_r) = 0 \tag{1}$$

$$\Sigma r^3 \frac{\partial \Omega}{\partial t} + \Sigma r v_r \frac{\partial}{\partial r}(r^3\Omega) - \frac{\partial}{\partial r}(\Sigma \nu r^3 \frac{\partial \Omega}{\partial r}) = 0 \tag{2}$$

$$\Sigma \frac{\partial v_r}{\partial t} + \Sigma v_r \frac{\partial v_r}{\partial r} - \Sigma \Omega^2 r + \frac{GM\Sigma}{r^2} - 2\frac{\partial}{\partial r}(HP)$$
$$-2\frac{\partial}{\partial r}(\Sigma\nu \frac{\partial v_r}{\partial r}) - 2\nu\Sigma \frac{\partial}{\partial r}(\frac{v_r}{r}) + \frac{2}{3}\frac{\partial}{\partial r}[\frac{\Sigma \nu}{r}\frac{\partial}{\partial r}(r v_r)] = 0 \tag{3}$$

$$C_v[\Sigma \frac{\partial T}{\partial t} + \Sigma v_r \frac{\partial T}{\partial r} - (\Gamma_3 - 1)T(\frac{\partial \Sigma}{\partial t} + v_r \frac{\partial \Sigma}{\partial r})]$$
$$-\Sigma\nu(r\frac{\partial \Omega}{\partial r})^2 + 2F_s + \frac{2}{r}\frac{\partial}{\partial r}(rHF_r) = 0 \tag{4}$$

here,$\Sigma = \int_{-\infty}^{\infty} \rho dz \approx 2\rho H$, $F_s = \frac{2acT^4}{3\kappa\Sigma}$, $F_r = \frac{4acT^3}{3\kappa\rho}\frac{\partial T}{\partial r}$, $\beta = \frac{P_{gas}}{P_{gas}+P_{rad}}$

3. α -MODEL

The most difficulty is the uncertainly as to the nature and magnitude of the viscosity, Shakura & Sunyaev (1973) introduced a dimensionless parameter α as:

$$\nu = \alpha c_s H \tag{5}$$

Then, the have gotten the solution from those basic equations and the left panel of Fig 1 shows the outline of thin disk. The luminosity of accretion disks is roughly:

$$L = \int 2\pi F_z r dr = \frac{GM}{r}\dot{M} \tag{6}$$

For a black hole $L \approx \dot{M}c^2$.

4. EMITTED SPECTRUM AND RADIATION MECHANISM

If the disk is optically thick, in the sense that each element of the disk radiates as a black body with T(R), the problem simplifies enormously. We then have :

$$Q^- = \sigma T^4 \tag{7}$$

and get spectrum S_v as:

$$S_v \propto \nu^{1/3} \tag{8}$$

This is sometimes regarded as a characteristic accretion disk spectrum. If the predominant disk viscosity is magnetic then this could lead to the formation of a strongly radiating hot corona above and below the disk. A variety of cooling mechanisms is avilable to potential disk model, including bremsstrahlung, Compton scattering of cool disk photons, and cyclotron or synchrotron radiation from the hot electrons in the reconnecting magnetic field. Coronal disk models have been applied to X-ray bursts. Cygnus X-1, and to the production of γ -rays in Cygnus X-3 (Liang 1997a; Thorne & Price 1975; Meszaros et al. 1977).

Tidal interaction of a central black hole could play a key role in the evolution of the stellar cluster and the fuelling of central engine of QSOs and AGNs (Woodward et al. 1992; Diener et al. 1997). Recently Cheng and Lu (1999) have investigated the possibility of inactive qussars which have harboured rapidly spinning black holes embedded in a dense star cluster as the hosts of γ-ray bursts.

5. GEOMETRICALLY THICK DISK/TORUS

If the radiative process is inefficient relative to the local dissipation of energy then the disk may be thick (See the right panel of Fig. 1). Thick disks can have luminosities 100 times above the Eddington limit (Abramowicz et al.1980; Paczynsky & Wiita 1980). For a $10^8 M_\odot$ black hole, this limit is about 10^{48}erg s^{-1}. The result is useful to explain the luminosity of QSOs and AGNs.

The thick torus around stellar mass black holes may be possible sources of energy for the γ-ray bursts at comsmological distances (Narayan et al. 1992; Mochkovitch 1993; Wiita et al. 1994; Jaroszynski 1996). For thick disks, one can compute $\frac{dh}{dr}$ as (Abromowicz et al. 1980):

$$-\frac{dh}{dr} = \left(\frac{\partial p/\partial r}{\partial p/\partial z}\right)_{z=h} = \left[\frac{\partial\phi/\partial r - L^2(r)/r^3}{\partial\phi/\partial z}\right]_{z=h} \tag{9}$$

The result is:

$$h(r) = \left\{\left[\frac{1}{GM}\int_{r_{in}}^{r} L^2(r)r^{-3}dr - (r_{in}^2 + h_{in}^2)^{1/2}\right]^{-2}\right\}^{1/2} \tag{10}$$

Equations (9), (10) are important to resolve the thick accretion disk.

6. SELF-GRAVITATING ACCRETION DISKS

When the density of disks is greater than the density of the central objects, the self-gravity of the disk may play an important role. Paczynski (1978) proposed a self-gravitating disk around a supermassive black hole model. They assume that:

$$g_s = 2\pi G\Sigma_z \tag{11}$$

then

$$\frac{dp}{dz} = -(\frac{GM}{r^3}z + 2\pi G\Sigma_z)\rho \tag{12}$$

They defined a parameter A as:

$$A = \frac{g_s}{g_z} = \frac{2\pi\Sigma_z r^3}{M_z} \tag{13}$$

Finally, they got the radiate luminosity as 1.5×10^{47}erg s^{-1} for $10^{10}M_{\odot}$ of a black hole and $10^9 M_{\odot}$ of a disk mass.

7. ADVECTION DISKS

As we know, in the standard α-model, the viscosity heating balances by radiative cooling. However, if the radiative cooling is not efficient, some energy will be advected inward, the advection will be nonnegligible. Recently years, the advection disks have been studied when disk is either optically thick or optically thin by a number of authors (Narayan & Yi 1995; Igumenshcher et al. 1996; Wu & Li 1996). The key assuming for advection disk is:

$$C_v[\Sigma v_r \frac{\partial T}{\partial r} - (\Gamma_3 - 1)T v_r \frac{\partial \Sigma}{\partial r}] - q\Sigma\nu(r\frac{\partial \Omega}{\partial r})^2 = 0 \tag{14}$$

If the disk is radiative cooling dominated. Q is nearly zero, and if it is advection dominated, q is nearly 1.

8. STABILITY OF ACCRETION DISKS

The stability properties of disks have become an important area in accretion disk theory. The instabilities of accretion disk have been successfully explained the observed phenomena of the variabilities and luminosities for various astronomical objects, such as proto-planetary nebula, dwarf novae, X-ray binaries, AGNs and quasars.

There are a lot of kinds of instabilities in accretion disks such as: convective instability, viscous instability, thermal instability, gravitational instability, hydro-magnetic instability, runaway instability, pulsational overstability and so on.

8.1 SOME CRITERIA

The criterion for convection instability is:

$$\frac{dlgT}{dlgp} > \frac{r-1}{r} \tag{15}$$

For viscosity is

$$\frac{\partial \nu \Sigma}{\partial \Sigma} < 0 \tag{16}$$

For thermal instability is :

$$\frac{\partial lg\nu}{\partial lgT_c} > \frac{\partial lgF_z}{\partial lgT_c} \tag{17}$$

For gravitational instability, when

$$\frac{\Omega c_s}{2\pi^2 G\Sigma} < 1 \tag{18}$$

the disk may be unstable.

8.2 PULSATIONAL OVERSTABILITY

There is also a possible mode of pulsational overstability. In this case, one looks for instabilities in which oscillations on the orbit time scale grow in amplitude because of the effects of viscosity. (Lin & Papaloizou 1996; Kato 1978; Blumenthal et al. 1984 Okuda & Mineshige 1992; Chen & Taam 1995). Recently years, our institute members have studied systematicly the pulsational oscillation instabilities of an isothermal magnetized accretion disk (including non-isothermal, $B(B_r, B_\phi, B_z)$, radial-azimuthal(Yang et al. 1997, 1995; Wu et al. 1994; Lu et al. 1996; Ding et al. 1998; Yu et al. 1994).

8.3 A LOCAL STABILITY ANALYSIS OF AN ISOTHERMAL DISK WITH THREE-DIMENSIONAL MAGNETIC FIELDS

Here we shall present a broad outline of a pulsational oscillation instabilities of an isothermal disk with three-dimensional magnetic field. (Yang et al. 1997) Basic equations:

$$\frac{ln\rho}{dt} + \triangle \cdot \rho\nu = 0 \tag{19}$$

$$\frac{\partial \Sigma}{\partial t}+\frac{1}{r}\frac{\partial(rv_r)}{\partial r}+\frac{1}{r}\frac{\partial(\Sigma v_\phi)}{\partial \phi}+\frac{\partial(|sigmav_z)}{\partial z}=0 \tag{20}$$

$$\frac{\partial v_r}{\partial t}+v_r\frac{\partial v_r}{\partial r}-\frac{v_\phi^2}{r}+r\Omega_k^2+\frac{1}{\Sigma}\frac{\partial W}{\partial r}+\frac{W}{\Sigma}\frac{\partial ln\Omega_k}{\partial lnr}$$
$$-\frac{B_z}{4\pi\rho}\frac{\partial B_r}{\partial z}+\frac{B_\phi^2}{4\phi\rho r}+\frac{B_z}{4\pi\rho}\frac{\partial B_z}{\partial r}+\frac{B_\phi}{4\pi\rho}\frac{\partial B_\phi}{\partial r}=0 \tag{21}$$

$$\frac{\partial v_\phi}{\partial t}+v_r\frac{\partial v_\phi}{\partial r}+\frac{v_\phi v_r}{r}-\frac{\alpha c_s^2}{\Sigma r^2}\frac{\partial}{\partial r}(\Sigma\Omega_k^{-1}r^3\frac{\partial\Omega}{\partial r})$$
$$-\frac{1}{4\pi\rho}B_z\frac{\partial B_\phi}{\partial z}-\frac{1}{4\pi\rho}B_r\frac{\partial B_\phi}{\partial r}-\frac{1}{4\pi\rho}B_z\frac{B_\phi B_r}{r}=0 \tag{22}$$

$$\frac{\partial B_z}{\partial z}+\frac{1}{r}\frac{\partial}{\partial r}(B_r r)=0 \tag{23}$$

$$\frac{\partial B_r}{\partial t}+\frac{\partial(v_z B_r-v_r B_z)}{\partial z}=0 \tag{24}$$

$$\frac{\partial B_\phi}{\partial t}+\frac{\partial(v_r B_\phi-v_\phi B_r)}{\partial r}+\frac{\partial(v_z B_\phi-v_\phi B_z)}{\partial z}=0 \tag{25}$$

$$\frac{\partial B_z}{\partial t}-\frac{1}{r}\frac{\partial}{\partial r}[r(v_z B_r-v_r B_z)]=0 \tag{26}$$

$$W=c_s^2\Sigma \tag{27}$$

We assume perturbations in the form

$$\delta Q=Q_0 exp[i(\omega t-k_r r-k_z z)] \tag{28}$$

for the quantities $\rho, P, v_r, v_\phi, B_r, B_\phi, B_z$ in the perturbations. Through laborious deducing, we can get the first-order perturbation equations and the dispersion relation.

Finally, our isothermal weak magnetized accretion disks with perturbation oscillation instability might be applied to explain the periodic variability of FU Orionis and T Tauri stars.

8.4 RUNAWAY INSTABILITY

Toroidal fluid configuration (or thick disks) around holes have been suggested as modes of quassars, other active galactic nuclei, some X-ray binaries, and γ-ray bursts. The question of stability of such configurations attracted considerable because of the important in Astrophysics. Papalaizou & Pringle (1984, 1985); Blaes (1985), Goodman et al. (1988) studied the instability of torus in seccession.

Abramowicz et al.(1983) presented the concept of runaway instability. The runaway instability is very fast catastrophic mass exchange. This instability operates in close binaries when the more massive star overflows its Roche lobe. The Roche lobe radius shrinks due to the mass

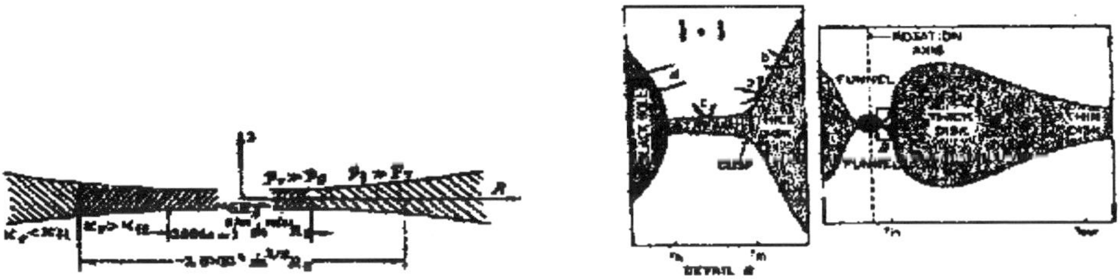

Figure 1

exchange more rapidly than the radius of the star. The star keeps overflowing its Roche lobe and continuously loss mass. This instability is important for many astronomical objects mentioned above, and many authors studied it up to now. (Abramowicz et al. 1998; Daigene et al. 1997; Nishida et al. 1996 ; Lu et al. 1999)

9. SUMMARY

In section 2-8, we have presented the main contents of accretion disks, including the basic equations of thin disks, α-model, thick disks/torus, self-gravitating disks, advection disks and instability of accretion disks. We are sure the theory of accretion disks would be continuously one of the most interesting and fruitful theory in Astrophysics in future.

Acknowledgements

This project was supported by the National Natural Science Foundation of China

References

Abramowicz M. A., Calvani, M., Nobili L. (1980). APJ, 242: 772
Abramowicz M. A., Karas V., Lanza A. (1998). A&A, 331: 1143
Blaes O. M. (1985). MNRAS, 216: 553
Blumenthal G. R, Yang L.T , Lin D. N. C.(1984). APJ, 287: 774
Cheng K. S, Lu Y.(1999). APJ, Submitted(astro/ph-9906265)
Chen X. M, Tamm R. E.(1995). APJ, 441: 354
Diener P. et al. (1997). APJ, 479: 164
Daigne F, Mochkovitch R.(1997). MNRAS, 285: L15

Ding S. X., Yang L. T., Wu S. P. (1998). Acta.Astro.Sinica, 18:352
Goodman J., Narayan R. (1988). MNRAS, 231: 97
Igumenshosev I. V., Chen X. M., Abramowicz M. A. (1996). MNRAS, 278: 236
Jaroszynski M. (1996). A&A, 305: 839
Kato S. (1978). MNRAS, 185: 629
Lin D. N. C., Papaloizou J. C. B. (1986). Ann.Rev.A&A, 34: 703
Liang E. P. T. (1977a). APJ, 211: L67
Lu Y., Cheng K. S., Yang L. T. , Zhang L. (1999). MNRAS, in press
Lu Y., Yang L. T., Wu S. P., Yang P. B. (1996). APSS, 240: 305
Meszaros P., Meyer P., Pringle J. E.(1997). Nature, 268: 420
Mochkovitch R et al. (1993). Nature, 361: 236
Narayan R., Paczynski B., Prian T. (1992). APJ, 395: L83
Narayan R.,Yi I. (1995). APJ, 452: 710
Nishida S et al. (1996). MNRAS, 278: L41
Okuda T., Mineshige S. (1991). MNRAS, 249: 684
Pring J. E. (1981). Ann.Rev.A&A, 19: 137
Papaloizou J.C.B., Lin D.N.C. (1985). Ann.Rev. A&A, 33: 504
Paczynaski B., Wiita P.J. (1980). A&A, 88: 23
Paczynski B. (1978). Acta Astro., 28.2: 91
Shakura N.I., Sunyaev R.A. (1973). A&A, 24: 337
Thorne K.S., Price R.H. (1975). APJ, 195: L101
Woodward J.W., Sankaran S., Tohline J.E. (1992). APJ, 394: 248
Witt H.J et al. (1994). APJ, 422: 291
Wu X.B., Li Q.B. (1996). APJ, 469: 776
Wu X.B., Yang L.T. (1994). APJ, 432: 672
Yang L.T., Henning Th., Lu Y., Wu S.P. (1997). MNRAS, 288: 965
Yang L.T., Wang D.X., Wu S.P., Liu Z.D.(1995). A&A, 303: 635
Yu W. F., Yang L.T., Wu X.B. (1994). MNRAS, 270: 131

A CELLULAR AUTOMATON MODEL FOR THE POWER LAW NOISE OF AN ACCRETION DISC

H. F. Chau and W. S. Tang
Department of Physics, University of Hong Kong, Pokfulam Road, Hong Kong

Abstract Power law fluctuations in the power spectra are commonly observed in accreting systems such as low mass x-ray binaries and accreting black-holes. We therefore believe that the physical origin of this fluctuation is quite universal and robust. Here, we report a new model for the $1/f^\beta$ noise in the X-ray power spectrum of an accretion disc using ideas from cellular automaton and self-organized criticality.

Keywords: Accretion Disc, Cellular Automaton, Power Law Noise, X-ray Fluctuation

1. INTRODUCTION

The $1/f^\beta$ power law fluctuation in the power spectra are commonly observed in astrophysical accretion discs. The typical value of the spectral index β ranges from 1 to about 2. (See, for example, McHardy & Czerny 1987.) The common occurrence of this power law fluctuation leads us to believe that its physical origin is universal and robust. In 1988, Bak et al. introduced the notion of self-organized criticality (SOC). They claimed that many physical systems in nature will organize under their own dynamics to states that exhibit power law spectra. They demonstrated their idea using a toy model known as the sandpile model. It is now believed that one of the necessary conditions for a system to exhibit SOC is the separation of time scales in all the underlying physical processes that drive the system (Turcotte 1999).

Soon after the discovery of SOC, Lu and Hamilton (1991) proposed a SOC model for solar flares. Later on, Mineshige et al. (1994a) proposed a cellular automaton SOC model to explain the $1/f^\beta$ fluctuations in X-rays from accretion discs of black holes. And in their subsequent works, Mineshige and his collaborators further refined their model and per-

K.S. Cheng et al. (eds.), Stellar Astrophysics, 257–265.

formed extensive numerical studies (Mineshige et al. 1994b; Mineshige et al. 1995; Yonehara and Mineshige 1996). The power spectrum exponent β obtained in their original model is about 1.8. Taking into account of the viewing angle, Abramowicz and Bao (1994) pointed out that β can be further lowered to about 1.4 for a large inclination angle. Recently, Yonehara et al. (1997) further extended their model to explain the flickering of cataclysmic variables.

While the model of Mineshige and his co-workers successfully reproduces the $1/f^{\beta}$ X-ray power spectra fluctuations with $\beta \geq 1.4$, it is apparently not possible to reproduce power spectrum fluctuations with β close to 1. Here we report a new cellular automaton model that is able to produce values of β that are close to 1 in the power spectra of accreting systems. But before we go on, let us briefly review the model of Mineshige and his co-workers.

2. MINESHIGE ET AL.'S MODEL

In their model, the disc is first divided into N concentric rings each with the same width Δr. They then further divide each ring into N equal arcs with constant angular extend $\Delta\theta \equiv 2\pi/N$ from the center of the disc. They call each such arc a grid point. Clearly, a grid point can be labelled by the coordinates (i, j) where i and j are the radial and angular positions respectively (Mineshige et al. 1994a). To model the accretion and fluctuation processes in the disc, Mineshige et al. (1994a) applied the following cellular automaton rules to the grid points:

1. Each grid point is assigned with a certain (non-negative) value denoting the mass of material of that grid point. Since we are interested in the asymptotic behavior of the disc, the initial values of grid points is not important.

2. To capture the continuous stochastic inflow of mass from the outermost region of the disc, a fixed amount of mass is added to a randomly chosen grid point in the outermost ring. That is, mass is added to a grid point (N, j).

3. If the mass of a grid point is larger than a fixed critical value, $M_{crit}(i, j)$, mass flow will occur. M_{crit} corresponds to the maximum amount of mass that can be stored in the grid point. As the area of the grid point depends on the radius of the ring, $M_{crit}(i, j)$ is directly proportional to the distance $r(i)$ of the grid point from the center of the disc. When a mass flow occurs, a fixed amount of mass flows to the three nearest grid points in the adjacent inner ring. This process is carried out instantaneously in every grid

point. This procedure is repeated until the mass of each grid point is less than or equal to the critical mass of that grid.

4. The half of the gravitational potential energy released by the mass flow E_g is recorded. We assume that the total X-rays luminosity in each turn is of the order E_g.

5. The whole process of stochastic mass injection, critical mass flow and X-ray release is repeated over and over again.

Mineshige et al. (1994a) found that the exponent of power spectrum β is equal to 1.8 and the power law extends over about 1.5 decades. Later on, they also included the effect of Keplerian rotation of the disc (Mineshige et al. 1994b) and discovered that such rotation has little effect on the value of β. Then, Mineshige and his co-workers introduced mass diffusion in the disc and is able to bring the value of β down to about 1.6 (Mineshige et al. 1995; Yonehara and Mineshige 1996). Furthermore, they modified their griding method. Instead of dividing the disc into rings of equal width, the width of the ith ring is chosen in such a way that the gravitational potential energy release by moving a small mass blob inward by one ring is a constant. By doing so, the energy released by large events close to the inner edge of the disc is reduced. Consequently, they are able to reduce the value of β to about 1.1. However, the $1/f^{1.1}$ power law only extend over about one decade. The low frequency region of the power spectrum is still flat.

3. OUR NEW MODEL

We have performed an extensive study on the effects of griding method, mass diffusion, various kinds of stochastic injection, differential disc rotation and critical mass flow triggering mechanisms. Our results strongly suggest that the power law index β is sensitive to only the griding method and mass diffusion (Tang 1999). And in order to reproduce the $\beta \approx 1$ power law index observed in some sources, we propose the following cellular automaton model (Tang 1999):

1. The disc is divided into grid points by $3N$ equal angular divisions together with N radial divisions as shown in Fig. 1. (In our simulation, we choose $N = 100$. Dividing the angular grid three times finer than the radial grid is essential. Otherwise, repeated application of step 4 below may lead to (the unphysical) mixing of local density triggered mass flow in the inner part of the disc.) The radial divisions are chosen so that the difference in gravitation potential energy between two successive rings is constant. Thus, the

radial thickness of the ring Δr is smaller in the inner part of the disc.

2. The critical density to trigger inward mass flow is constant over the disc. Thus, the corresponding critical mass $M_{crit}(i, j)$ of a grid point (i, j) is proportional to r^3 where r is the distance of the grid point from the center of the accreting star.

3. To model stochastic mass injection, a small blob of material with mass M_{add} is added to a randomly chosen grid point in the outermost ring at the beginning of each time step. We choose M_{add} to be a thousandth the critical mass of the outermost ring.

4. To model local density triggered inward mass flow, we demand that if the mass of the grid (i, j) is larger than its critical value $M_{crit}(i, j)$, mass flow will occur at that grid point. And in this case, mass will redistribute amongst the nearby grid points according to the following rule:

$$\left\{ \begin{array}{rcl} M_{i,j} & \longrightarrow & M_{i,j} - \Delta m(i,j)_1 \\ M_{i-1,[j+t\Delta\omega_i]} & \longrightarrow & M_{i-1,[j+t\Delta\omega_i]} + 0.5\Delta m(i,j)_1 \\ M_{i-1,[j+t\Delta\omega_i]\pm 1} & \longrightarrow & M_{i-1,[j+j\Delta\omega_i]\pm 1} + 0.25\Delta m(i,j)_1 \end{array} \right. \tag{1}$$

where $\Delta m(i, j)_1 = 0.01 M_{i,j}$, t is the time elapsed since the beginning of the simulation, and $\Delta\omega_i$ is the differential Keplerian angular rotation between the ith and $(i-1)$th ring. Note that parallel update is used should more than one grid point satisfies the above local density triggered mass flow condition. (Note that unlike the models of Mineshige et al., the amount of inward mass flow here is proportional to the mass contained in the triggering grid point.)

5. To model randomized mass flow, a grid point is randomly chosen in each ring. Such a chosen grid point will experience mass flow even if its local density is less than the corresponding local triggering density. In this case, mass is redistributed according to Eq. (1) except that $\Delta m(i, j)_1$ is replaced by $\Delta m(i, j)_2 = 0.05 M_{i,j}$. Again, parallel update is used.

6. As material falls into the compact object, we assume that half of the gravitational potential energy gained is released almost instanteously. We record the energy released in this way in each time step.

7. Each ring then rotates by an angle according to its Keplerian velocity. Thus, the ith ring rotates by an angle $\Delta\omega_i$ with respected to the $(i-1)$th ring.

8. Then, the whole process starting from stochastic mass injection is repeated over and over again. The obtain the power spectrum, we ignore the initial transient luminosity and concentrate only on the long term behavior of the system.

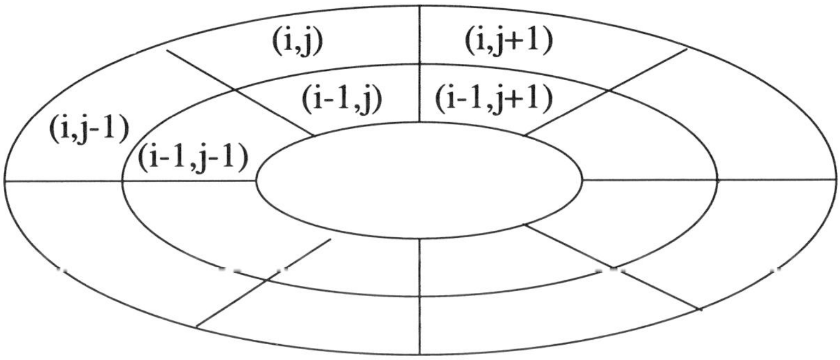

Figure 1 The griding method used in our model as well as that in Mineshige et al. (1994ab).

In our simulation, we choose the distances of the innermost and outermost ring to be 10^7 m and 10^8 m from the center of the accreting object respectively. Moreover, the time elapsed in each time step Δt is set to be the free-fall time across a outermost ring. That is, $\Delta t \approx 5$ s. We can vary the following two parameters in the system, namely, the values of $\Delta m(i,j)_1/M_{i,j}$ and $\Delta m(i,j)_2/M_{i,j}$. The results are tabulated in Table 1.

The typical light curve and the power spectrum of model A are shown in Figs. 2 and 3 respectively. As we can see from Fig. 3, model A gives an overall power spectral index $\beta \approx 1.1$. Nevertheless, when we examine the power spectrum more closely, it is clear that the power spectrum consists of two parts. The slope of the low frequency part of the spectrum is about -1.3 when that of the high frequency part is about -0.8. The transition occurs at about 3×10^{-3} Hz.

To see how the power spectrum varies as a function of the relative importance between stochastic mass inflow and local density triggered mass inflow, we changes the values of $\Delta m(i,j)_1/M_{i,j}$ and $\Delta m(i,j)_2/M_{i,j}$. And as shown in Fig. 4, as we reduce the importance of the stochastic

Model	$\Delta m(i,j)_1/M_{i,j}$	$\Delta m(i,j)_2/M_{i,j}$	β
A	0.01	0.05	1.1
B	0.01	0.01	1.2
C	0.01	0.0025	1.4

Table 1 The scaling exponent β for various parameters in our model.

mass inflow by reducing the value $\Delta m(i,j)_2$, the resultant power spectrum fits better to a power law with β increases slightly to about 1.2.

One would then tempt to think that an even better power law is obtained in the absence of stochastic mass inflow. However, this is not the case. Fig. 5 is the power spectrum obtained when we almost turn off the stochastic mass inflow. In doing so, the slope in the low-frequency region almost flatten out and at the same time, the slope in the high-frequency region decreases to about -1.4. Thus, we conclude that stochastic mass

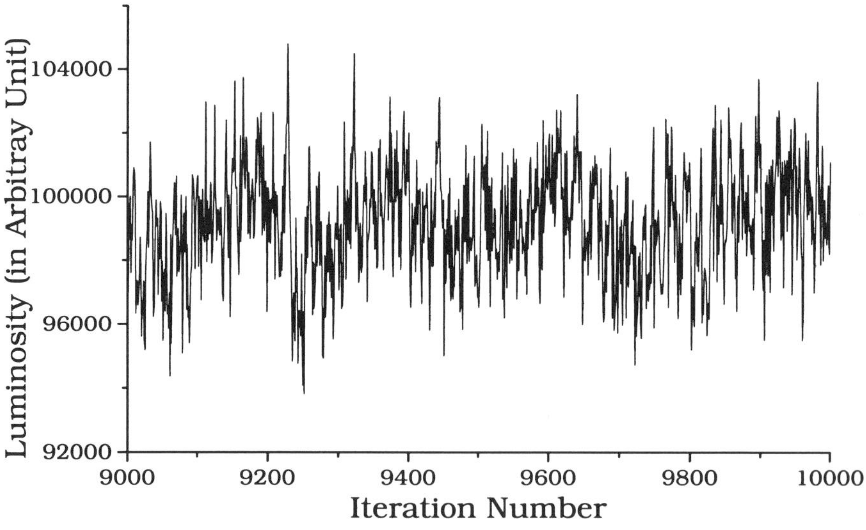

Figure 2 Typical light curve of model A.

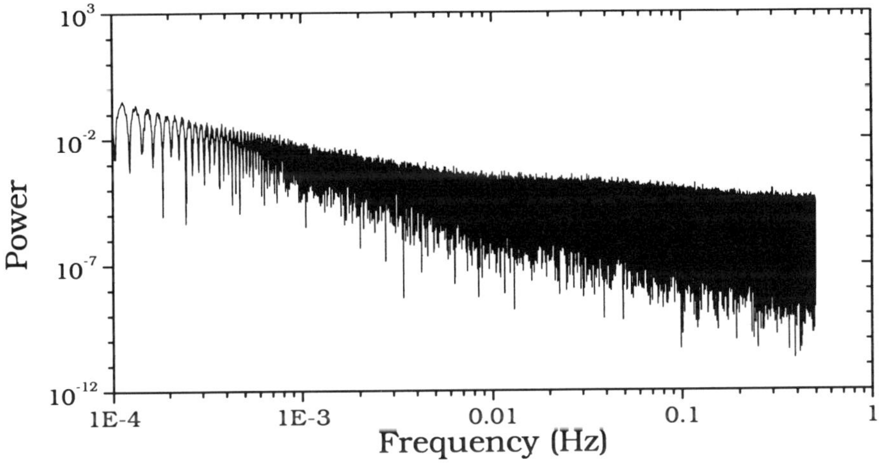

Figure 3 The power spectrum of model A.

inflow and local density triggered mass inflow are both important in obtaining the $1/f^{1.1}$ X-ray power spectrum.

4. DISCUSSIONS

In summary, we have introduced a new cellular automaton model for the X-ray power law fluctuation of accretion disc based on the idea of self-organized criticality. We found that in order to obtain an X-ray power spectrum with the power law index β close to 1, we must divide our radial grids in such a way that the gravitational potential energy gain per radial grid is a constant. Besides, both the stochastic mass inflow as well as the local density triggered mass inflow are essential in maintaining the power law spectrum.

Acknowledgements

The work by W. S. Tang is supported by a postgraduate studentship of the University of Hong Kong.

References

Abramowicz, M., & Bao, G., PASJ 46, 523 (1994).
Bak, P., Tang, C., & Wiesenfeld, K., PRA 38, 364 (1988).

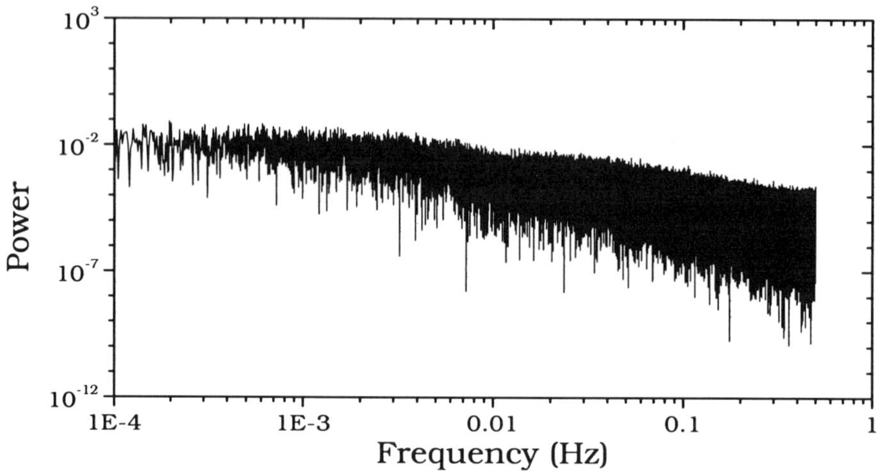

Figure 4 The power spectrum of model B.

Li, E. T., & Hamilton, R. J., ApJLett 380, L89 (1991).
McHardy, I., & Czerny, B., Nature 325, 696 (1987).
Mineshige, S., Ouchi, N. B., & Nishimori, H., PASJ 46, 97 (1994).
Mineshige, S., Takeuchi, M., & Nishimori, H., ApJLett 435, L125 (1994).
Takeuchi, M., Mineshige, S., & Negoro, H., PASJ 47, 617 (1995).
Tang, W.S., "Study of Power Spectrum Fluctuation in Accretion Disc by Cellular Automaton", M. Phil. thesis, University of Hong Kong (1999).
Turcotte, D. L., RepProgPhys 62, 1377 (1999).
Yonehara, A., & Mineshige, S., ApJ 486, 388 (1996).
Yonehara, A., Mineshige, S., & Welsh, W. F., ApJ 486, 388 (1997).

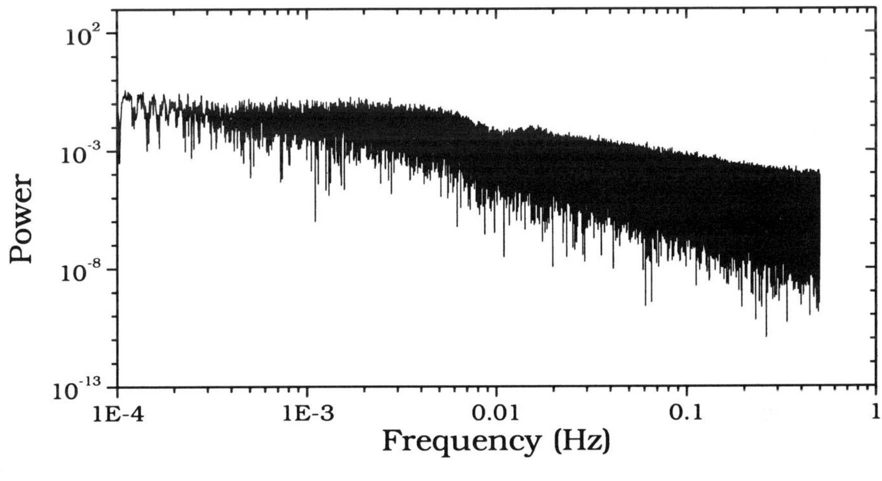

Figure 5 The power spectrum of model C.

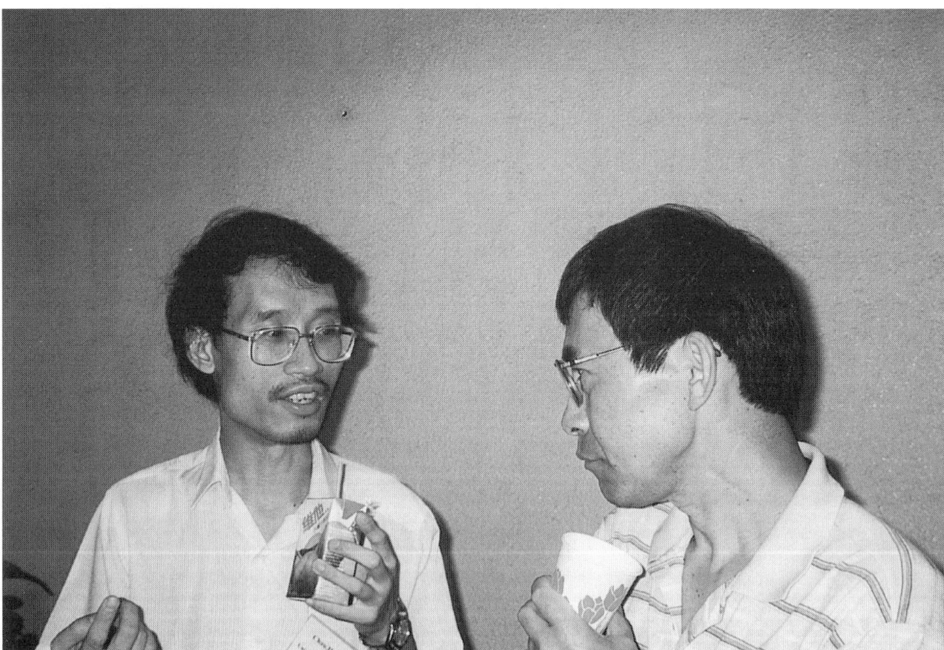

X-RAY SPECTRAL VARIATIONS OF X-RAY NOVA XTE J1550-564 DURING THE RISING PHASE OF THE 1998 OUTBURST

Xue-Bing Wu[1], S.N. Zhang[2], Yangsen Yao[2], Xuejun Su[2], Wei Cui[3] and Wan Chen[4]

[1]*Beijing Astronomical Observatory, Chinese Academy of Sciences, Beijing 100012, China,*

[2]*Physics Department, University of Alabama in Huntsville, Huntsville, AL35899, USA*

SD50, NASA/Marshall Space Flight Center, Huntsville, AL 35812, USA,

[3]*Center for Space Research, MIT, Cambridge, MA 02139, USA,*

[4]*Dept. of Astronomy, University of Maryland, College Park, MD 20742, USA*

NASA/Goddard Space Flight Center, Code 661, Greenbelt, MD 20771, USA

[1]wuxb@bao.ac.cn

Abstract We present the spectral analysis results of the RXTE/PCA observations on X-ray nova XTE J1550-564 during the 13-day rising phase of its outburst in September 1998. The spectra are dominated by a hard component which contributes more than 90% of total X-ray flux in the 2-20 keV range. The soft component is very weak and only becomes significant when the source approaches the giant flare. Considering that a large part of soft photons are probably scattered by the hot electrons in the corona, we adopted a recent established theoretical model to derive the corrected normalization of the disk blackbody (soft) component; the corrected normalization parameter is found to be almost constant in the late stage of rising phase and well match the value in the soft state, provided that the inclination angle of the system is about 75°. The combination of X-ray spectral transition and the slow decline of the optical brightness supports that XTE J1550-564 may experience an outside-in burst caused by accretion disk instability. The 13-day rising time can be explained by the viscous time-scale provided the disk instability was triggered at a distance of more than 100 times of gravitational radius away from the central object.

Keywords: Accretion, accretion disks, black hole physics, X-rays: stars

K.S. Cheng et al. (eds.), Stellar Astrophysics, 267–273.

1. INTRODUCTION

The X-ray nova XTE J1550–564 was discovered with the All Sky Monitor (ASM) aboard the *Rossi X-ray Timing Explorer* (RXTE) on 1998 September 7 (Smith *et al.* 1998). The position quickly led to the identifications of the possible optical and radio counterparts (Orosz, Bailyn, & Jain 1998, Campbell-Wilson *et al.* 1998), which were later confirmed by a much improved error box on the position of XTE J1550-564, derived from an ASCA observation (Mashall *et al.* 1998).

During the intensity rise after the outburst, BATSE observations (Wilson *et al.* 1998) revealed the significant X-ray spectrum softening, with photon index changing from 2.1 to 2.7. The broad emission lines in its optical spectrum also showed typical features of a soft X-ray transient source in outburst (Castro-Tirado *et al.* 1998; Sánchez-Fernández *et al.* 1999). Low frequency QPOs were detected with BASTE (20-50 keV) in the range 55-480 mHz before September 10 (Finger *et al.* 1998) and with RXTE later in the range 0.08-8 Hz (Cui *et al.* 1999). On September 19-20, XTE J1550-564 experienced a giant flare and reached 6.8 Crab at 2-10 keV, making it to be the brightest transient source observed yet with RXTE (Remillard *et al.* 1998). The significant radio and hard X-ray flares were also detected with the MOST at 843 MHz and with BATSE at 20-100 keV, while only a small optical response was reported (Jain *et al.* 1999). After the giant flare it went through a 4-day rapid decay, a 30-day intensity plateau and another 10-day rapid decay. A high frequency QPO near 185 Hz was also discovered during the giant flare and the plateau states (McClintock *et al.* 1998; Remillard *et al.* 1999). After December 3, 1998, XTE J1550-564 brightened again and experienced another long and soft outburst. It returned to the quiescent state after May 1999. Both the characteristic X-ray spectral properties and high frequency QPOs strongly support that XTE J1550-564 is very likely a black hole candidate.

In this paper we report the spectral analysis results of RXTE/PCA observations during the rising phase of the major X-ray outburst in September 1998. The related physics about the radiation transfer and the outburst mechanism will be also discussed.

2. SPECTRAL ANALYSIS AND RADIATION TRANSFER CORRECTION

Our spectral observations of XTE J1550-564 using the PCA instrument aboard RXTE cover the entire rising phase of the outburst at a typical rate of once or twice per day for one RXTE orbit. The PCA data were taken in the standard 2 format and the response matrix for each PCU(0-4) was obtained from the response files distributed in 1998 January. The spectrum was fit over the energy range 2.5-20 keV after including a systematic error in the count-rates of 1% and background subtracted using the standard background models. In this paper, we report the results obtained by using only PCU 0 & 1 for the spectral fitting. Both PCUs were fit simultaneously using *XSPEC* package. The spectra of first two observations in a later RXTE run were also added. The detailed results of late observations were reported by Sobczak *et al.* (1999).

Very frequently, the disk blackbody plus power law model is used to fit the two components of X-ray spectra observed in X-ray binaries. However, if we believe that the hard component is produced by the inverse Compton scattering of low-energy photons emitted from accretion disk by high energy electrons in a hot corona, a sharp cut-off must be present in the hard component at the low-energy region. Neglecting this low-energy cut-off by applying a simple power-law model would underestimate the flux in the soft component. Therefore, in this paper we adopt disk blackbody (*diskbb*) plus thermal Comptonization (*compTT*) model in *XSPEC* to fit the spectra. We also consider a Fe emission line with a central energy around 6.5 keV and a fixed width of 1.2 keV. The Hydrogen column density value inferred from the recent ASCA observation on XTE J1550-564, $N_H = 0.86 \times 10^{22} cm^{-2}$, was fixed through our analyses. A sphere geometry of Comptonization corona was assumed in our spectra fitting.

The fit parameters, QPO frequencies and X-ray fluxes in 2-20 keV are plotted in Figure 1. Obviously the spectra of XTE J1550-564 are dominated by the hard component, which contributes more than 90% of the observed X-ray flux. The total X-ray flux and its disk blackbody and Comptonization components increase steadily and are well correlated with the QPO frequency. We can also observe a tendency of anti-correlation between the inner disk temperature and the disk blackbody normalization (which directly gives the inner disk radius once the distance and inclination angle are known). The low energy and high energy cutoff temperatures in Comptonization model are roughly constant during the whole rising phase. The optical depth of the Comptonization area decreases gradually from 6 to 1 in five days during the initial outburst. When approaching the giant flare, the inner disk temperature, optical depth, normalization of hard component, as well as the X-ray flux, all increase sharply, while the disk blackbody normalization and the high energy cutoff parameter KT decrease.

Our direct spectral modeling produces the disk blackbody normalization much smaller than the values in the soft state. However, we cannot

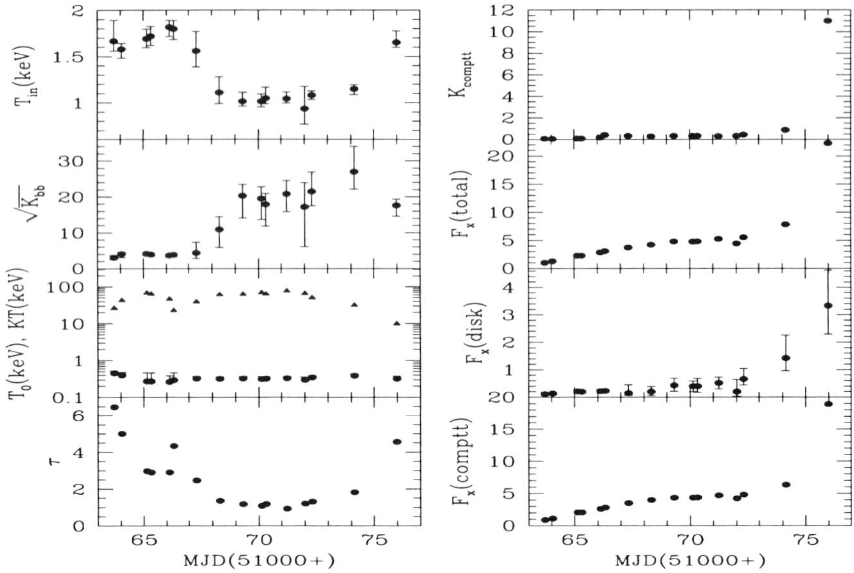

Figure 1 Spectral parameters and fluxes for PCA observations of XTE J1550–564 in rising phase fitted by the disk blackbody plus thermal Comptonization model. KT is plotted as triangle in the left panel. Only the error boxes of disk blackbody parameters and the upper limit of T_0 are shown.

conclude the dramatic variation of inner disk radius if the correction due to radiation transfer is not made. Recently we have established a simple model to make such correction by involving thermal Compotonization process in a spherical corona surrounding the cold disk (Zhang *et al.* 1999). Using this model, we can obtain the intrinsic disk blackbody normalization. In figure 2 we give our results for the cases of four different inclination angles. It is clear that except in the initial outburst the corrected disk blackbody normalization vary very little during the rising phase if the size of Comptonizition corona is about several hundreds of r_g (gravitational radius defined as $r_g = GM/c^2$) and the inclination angle is large. Especially when the inclination angle is around 75°, the corrected disk blackbody normalization match well the value in the soft state. This suggests that the inner disk radius keeps almost constant even when the spectral state changes.

3. DISCUSSION

Our results clearly show that XTE J1550-564 experienced a state transition from a low/very hard state to a very high/hard state during the rising phase. We also confirm the tight correlation between the 0.08-

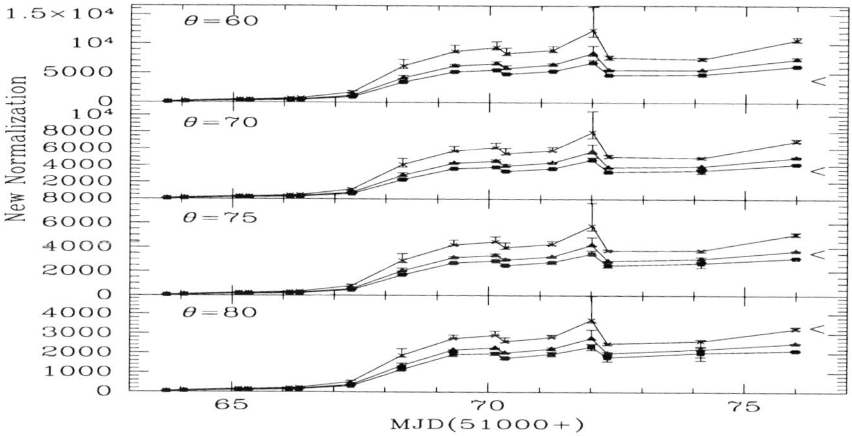

Figure 2 Comptonization-corrected disk blackbody normalizations of XTE J1550–564 in the rising phase. The results are obtained for four different inclination angles. The lines with crosses, filled triangles and filled squares correspond to R_0/r_g =50, 200 and 800 respectively (R_0 is the radius of hot corona). The label "¡" represents the corrected normalization in the soft state.

14Hz QPO and the total X-ray flux as reported by Cui *et al.* (1999). Obviously the similar correlation exists between QPO frequency and the flux of the hard component, which dominated the total flux during the rising phase.

In contrast to the steadily increase of X-ray flux, optical magnitude decline quite slowly during the rising phase of XTE J1550-564 (Jain *et al.* 1999; Sánchez-Fernández *et al.* 1999). The similar anti-correlation between optical and X-ray fluxes has been found for GRO J1655-40 after its initial outburst in April 1996 (Hynes *et al.* 1998). However, the optical brightening has been observed 6 days prior to the onset of X-ray outburst for GRO J1655-40 (Orosz *et al.* 1997), which was explained with the inward propagating of heat front caused by the "outside-in" instability in the accretion disk (Hameury *et al.* 1997). Although the optical brightening of XTE J1550-564 has not been detected prior to the X-ray outburst, the initial optical magnitudes observed on MJD 51065 (two days after the X-ray outburst) indicate that XTE J1550-564 was indeed much brighter in optical. It is quite possible that the optical outburst occurred before the X-ray brightening. Therefore the "outside-in" instability of accretion disk perhaps also happened for XTE J1550-564. If the heat front propagates inward slowly on a viscous timescale described by $t_{visc} = 30(\frac{\alpha}{10^{-2}})^{-1}(\frac{M}{M_\odot})^{-1/2}(\frac{R}{10^7 cm})^{7/2}(\frac{\dot{M}}{10^{18} g/s})^{-2}$ (Belloni *et al.* 1997), the 13-day rise-time of XTE J1550-564 can be explained provided R (the outburst starting radius) larger than 100 r_g and some typical disk parameters for black hole X-ray binaries, $\alpha = 0.1$, $M = 10M_\odot$ and $\dot{M} = 10^{18} g/s$, are adopted. It might imply that the disk instability was triggered at least 100 times of gravitational radius away from the central object.

In this paper we adopted a recently established model to correct the disk blackbody normalization considering that the hard component is produced by inverse Compton scattering of low-energy photons emitted from accretion disk with hot electrons in a corona. Such a radiative transfer correction is crucial to obtain the physical properties of accretion disks. The corrected disk blackbody normalizations for XTE J1550-564 indicate that the inner disk radius may not vary too much even when the spectral state changes.

References

Belloni, T., Mendez, M., King, A. R., van der Klis, M., & van Paradijs, J. 1997, ApJ, 488, 109

Campbell-Wilson, D., McIntyre, V., Hunstead, R., & Green, A. 1998, IAU Circ.7010

Castro-Tirado, A.J., Duerbeck, H.W., Hook, I., Yan, L., 1998, IAU Circ.7013

Cui, W., Zhang, S.N., Chen, W., Morgan, E.H., 1999a, ApJ, 512, L43

Finger, M.H., Dieters, S.W., Wilson, R.B., 1998, IAU Circ.7010

Hameury, J.-M., Lasota, J.-P., McClintock, J. E. & Narayan, R. 1997, ApJ, 489, 234

Hynes, R. I., Haswell, C. A., Shrader, C. R., Chen, W., Horne, K, Harlaftis, E. T., O'Brien, K., Hellier, C. & Fender, R. P. 1998, MNRAS, 300, 64

Jain, R., Bailyn, C. D., Orosz, J. A., Remillard R. A., & McClintock, J. E. 1999, ApJ, 517, L131

McClintock, J.E., Sobczak, G.J., Remillard, R. A., Morgan, E.H., Levine, A. M., 1998, IAU Circ.7025

Marshall, F.E., Smith, D.A., Dotani, T., Ueda, Y., 1998, IAU Circ.7013

Orosz, J., Bailyn, C., & Jain, R. 1998, IAU Circ.7009

Orosz, J., Remillard, R. A., Bailyn, C. D., McClintock, J. E. 1997, ApJ, 478, L83

Remillard, R. A., McClintock, J. E., Sobczak, G. J., Bailyn, C. D., Orosz, J. A., Morgan, E. H., & Levine, A. M. 1999, ApJ, 517, L127

Remillard, R. A., Morgan, E.H., McClintock, J.E., Sobczak, G.J., 1998, IAU Circ.7019

Sánchez-Fernández, C., *et al.* 1999, A&A, 348, L9

Smith, D. A. *et al.* 1998, IAU Circ.7008

Sobczak, G. J., McClintock, J. E., Remillard, R. A., Levine, A. M., Morgan, E. H., Bailyn, C. D., & Orosz, J. A. 1999, ApJ, 517, L121
Wilson, C. A., Harmon, B. A., Paciesas, W. S., & McCollough, M. L. 1998, IAU Circ.7010
Zhang, S. N., Wu, X.-B., Sun, X., Yao, Y. S., Cui, W., Chen, W. 1999, in preparation

JUMBO

A NEW PARAMETER IN ACCRETION DISK MODEL

Feng Yuan
Department of Astronomy, Nanjing University, Nanjing 210093, China
fyuan@nju.edu.cn

Abstract Taking optically thin accretion flows as example, we calculate the dynamics and the emergent spectra of accretion flows with different outer boundary conditions (OBCs) and find that OBC plays an important role in accretion disk model. This is because the accretion equations describing the behavior of accretion flows are a set of *differential* equations, therefore, accretion is intrinsically an initial-value problem. We argue that optically thick accretion flow should also show OBC-dependent behavior. The results means that we should seriously consider the initial physical state of the accretion flow such as its angular momentum and its temperature. An application example to Sgr A* is presented.

1. INTRODUCTION

It has long been assuming that the parameters describing the accretion flow include the accretion rate, the mass of the central black hole, the viscosity parameter, and the parameter describing the strength of the magnetic field in the accretion flow. Once these parameters are given, we can obtain almost all the information of the accretion flow including the dynamics and the emergent spectrum. However, the set of equations describing the accretion flow are nonlinear differential equations, therefore it is intrinsically an initial-value problem. Thus the outer boundary condition (OBC) possibly plays an important role.

On the other hand, the complicated astrophysical environments make the physical states of the accreting gas at the outer boundary, such as its temperature and angular momentum, various. For example, in semi-detached binary system, where the critical Roche lobe is filled up and the accretion of matter takes place through the inner Lagrangian point, the angular momentum of the accreted gas should be high; while in detached binary system the accretion matter is stellar winds therefore

K.S. Cheng et al. (eds.), Stellar Astrophysics, 275–280.

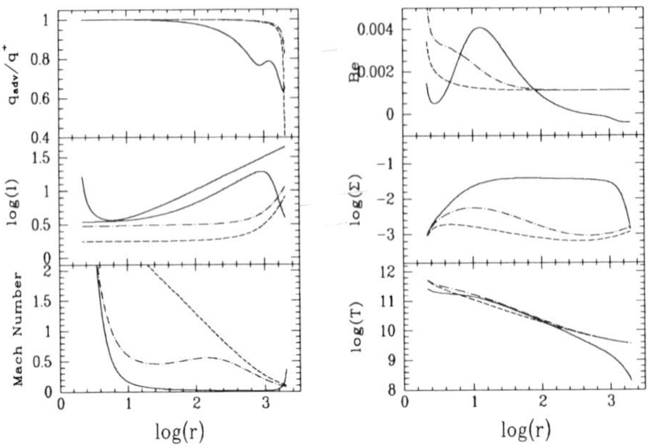

Figure 1 Solutions for one temperature global solutions with different OBCs for $M = 10M_\odot$, $\dot{M} = 10^{-3}\dot{M}_{\rm E}$ and $\alpha = 10^{-2}$. The solid, dot-dashed and dashed lines represent $(T_{\rm out}, \lambda_{\rm out}(\equiv v/c_s \equiv v/\sqrt{p/\rho}))$=$(2\times10^8{\rm K}, 0.4)$, $(3.6\times10^9{\rm K}, 0.08)$ and $(3.6\times 10^9{\rm K}, 0.107)$ respectively. The units of Σ, T are g cm^{-2} and K, r, Be and l are in $c = G = M = 1$ units. Mach number is simply defined as v/c_s. The upper-left plot represents the ratio of the advected energy to the viscous dissipated energy. Adopted from Yuan (1999).

their angular momenta are much lower (Illarionov & Sunyaev 1975). In the case of elliptical galaxy, the specific angular momentum of the matter captured by the black hole located at the center of the galaxy is widely assumed to be much lower than the Keplerian value and the temperature may be as high as local virial temperature; while in spiral galaxies the situation may be on the opposite (e.g. Sparke & Shu 1980). The complexity of astrophysical environments makes it important to investigate the role of OBC in accretion disk model.

2. THE ROLE OF OBC IN OPTICALLY THIN ACCRETION FLOWS

In previous papers (Yuan 1999; Yuan et al. 1999), taking optically thin accretion onto a black hole as an example, we calculated the dynamics and the emergent spectrum of one- and two-temperature accretion plasma by self-consistently solving the radiation hydrodynamical equations. For the one temperature case, only bremsstrahlung emission and its Comptonization are considered, while for the two temperature case, synchrotron emission and its Comptonization are also included. We concentrated on the role of OBC by setting the same "general parameters" such as accretion rate, viscosity parameter and black hole mass while

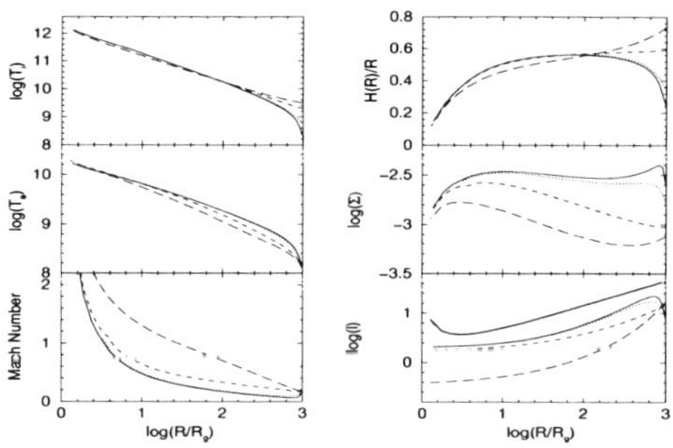

Figure 2 Solutions for two temperature global solutions with different $T_{\mathrm{out},i}$. The solid line (type I solution) is for $T_{\mathrm{out},i} = 2 \times 10^8 K$, the dotted line (type I) for $T_{\mathrm{out},i} = 6 \times 10^8 K$, the dashed line (type II) for $T_{\mathrm{out},i} = 2 \times 10^9 K$ and the long-dashed line (type III) for $T_{\mathrm{out},i} = 3.2 \times 10^9 K$. Other OBCs are $T_{\mathrm{out},e} = 1.2 \times 10^8 K$ and $\lambda_{\mathrm{out}} = 0.2$. The outer boundary is set at $r_{\mathrm{out}} = 10^3 r_{\mathrm{g}}$. Other parameters are $\alpha = 0.1, \beta = 0.9, M = 10^9 M_{\odot}$ and $\dot{M} = 10^{-4} \dot{M}_{\mathrm{Edd}}$. The units of Σ and T are g cm^{-2} and K. Adopted from Yuan et al. (1999).

adopting different OBCs. We adopted the temperature T_{out} and the ratio of the radial velocity to the local sound speed λ_{out} (or, equivalently, the angular velocity Ω_{out}) at a certain outer boundary r_{out} as the outer boundary conditions and found that in both cases, the topological structure and the profiles of angular momentum and surface density of the flow differ greatly under different OBCs, as shown by Figures 1 (for a one-temperature plasma) and 2 (for a two-temperature plasma; only the ions temperature $T_{\mathrm{out},i}$ varies, for other cases, see Yuan et al. 1999). In terms of the topological structure and the profile of the angular momentum, three types of solutions are found. When T_{out} is relatively low, the solution is of type I. When T_{out} is relatively high and the angular velocity Ω_{out} is higher than a critical value Ω_{crit}, the solution is of type II. Both types I and II possess small sonic radii, but their topological structures and angular momentum profiles are different. When T_{out} is high but the angular velocity is lower than Ω_{crit}, the solution becomes of type III, characterized by a much larger sonic radius. Similar transition has been found previously in the context of adiabatic (inviscid) accretion flow by Abramowicz & Zurek (1981). In that case, they found that when the specific angular momentum of the flow decreased across a critical value, a transition from a disk-like accretion pattern (with small sonic radii) to a Bondi-like one (with large sonic radii) would happen (Abramowicz & Zurek 1981; Lu & Abramowicz 1988). Here in this paper we find that this transition still exist when the flow becomes viscous, confirming the prediction of Abramowicz & Zurek (1981). Figure 3 shows the emergent spectrum of the solutions presented in Figure 2. Considering that

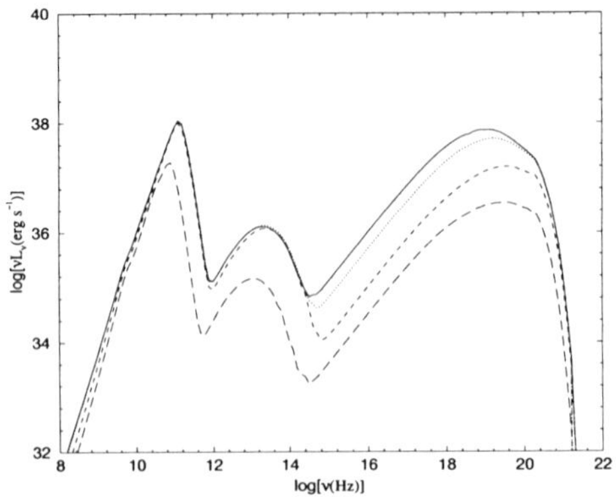

Figure 3 The corresponding spectra of the solutions shown in Figure 2. Adopted from Yuan et al. (1999).

they possess the same "general" parameters the discrepancy among the spectra completely caused by the difference of OBC is impressive.

3. AN ILLUSTRATIVE APPLICATION TO SGR A*

As an illustrative example, we apply the above results to the compact radio source Sgr A* located at the center of our Galaxy. Advection-dominated accretion flow (ADAF) model has been turned out to be a great success to explain its low luminosity and spectrum (Narayan, Yi & Mahadevan 1995; Narayan et al. 1998). However, there exists a discrepancy between the mass accretion rate favored by ADAF models in the literature and that favored by the three dimensional hydrodynamical simulation, with the former ($\sim 6.8 \times 10^{-5} \dot{M}_{\rm Edd}$; see Quataert & Narayan 1999) being 10-20 times smaller than the latter ($\sim 9 \times 10^{-4} \dot{M}_{\rm Edd}$; see Coker & Melia 1997). By seriously considering the outer boundary condition of the accretion flow, we find that due to the low specific angular momentum of the accretion gas, the accretion in Sgr A* should belong to type III which possesses a very large sonic radius. This accretion pattern can significantly reduce the discrepancy between the mass accretion rate, as Figure 4 shows (see Yuan et al. 1999 for details).

4. DISCUSSION

The present study is concentrated on the low-$\dot{M}$ case where the differential terms in the equation such as the energy advection play an important role therefore the effect of OBC are most obvious. How about the role of OBC when the flows become optically thick? In this case,

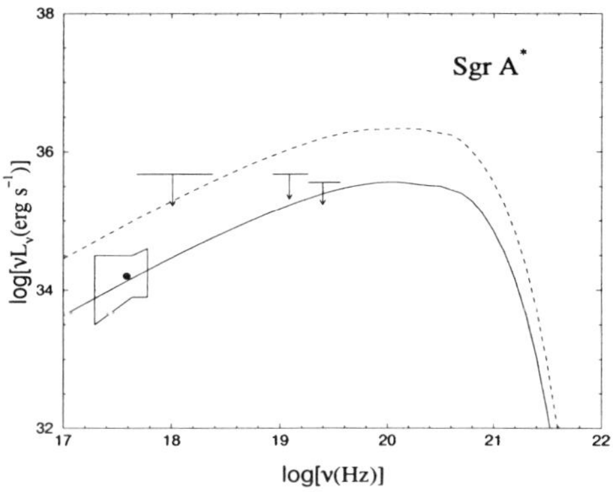

Figure 4 The X-ray spectrum of Sgr A*. The observational data are compiled by Narayan et al. (1998). The spectra represented by the solid and the dashed lines are produced by the accretion flows with the same accretion rate $\dot{M} = 4 \times 10^{-4} \dot{M}_{\rm Edd}$ but different angular momentum, $\Omega_{\rm out} = 0.15\Omega_{\rm K}$ for the solid line and $\Omega_{\rm out} = 0.46\Omega_{rmK}$ for the dashed line. Due to the difference of the angular momentum of the flow at the outer boundary, the X-ray flux differs by a factor ~ 8. Adopted from Yuan et al. (1999).

the electron and the ion possess the identical temperature due to the strong couple between them and the local viscous dissipation and radiation loss terms in the energy balance play an important role. As a result, the temperature profile is mainly determined *locally* rather than *globally* as in the case of optically thin flows. Thus, the discrepancy of the temperature caused by OBC will lessen rapidly with the decreasing radii from the outer boundary. This is also the reason why the temperature profiles of one-temperature plasma and ions in Figures 1 and 2 converge rapidly with decreasing radii. However, from our calculation to the one-temperature accretion flow whose temperature is also principally determined locally (Yuan 1999), we predict that the optically thick accretion flow should still present OBC-dependent behavior in, e.g., the angular momentum and the Mach number profiles, and, when the angular momentum of the accretion flow is less than a certain critical value, the sonic radius should become very large. Although these conjectures need the confirmation of detailed calculation, we note that the angular momentum profile of slim disk model (see Figure 3 of Abramowicz et al. 1988), and a recent numerical simulation (Igumenshchev, Illarionov & Abramowicz 1999) seems to support this point.

Why the role of OBC in accretion disk models has been long neglected? In the standard thin disk model of Sakura & Sunyaev (1973), all the differential terms in the equations are neglected and the differential equations are reduced into an algebraic one which don't entail any boundary conditions at all. In the later works on the global solutions for slim disks (Matsumoto et al. 1984; Abramowicz et al. 1988; Chen & Taam 1993) and optically thin advection-dominated accretion flows (Narayan, Kato & Honma 1997; Chen, Abramowicz & Lasota 1997), some authors did investigate the role of OBC, but failed to find its importance. The main reason is that for optically thick accretion flows (slim disk) or *one-temperature* optically thin accretion flow, the local viscous dissipation plays an important role in the energy equation, so the effect of OBC lessen rapidly away from the outer boundary. In addition, the angular momentum in their outer boundary condition was always somewhat large. This might be the reason why they didn't find the solutions with very large sonic radii.

References

Abramowicz, M.A., Czerny, B., Lasota, J.P., & Szuszkiewicz, E., 1988, ApJ, 332, 646

Abramowicz, M., A., Zurek, W.H., 1981, ApJ, 246, 31

Chen, X., Abramowicz, M. A., & Lasota, J.-P. 1997, ApJ, 476, 61

Chen, X. & Taam, R. 1993, ApJ, 412, 254

Coker, R., & Melia, F., 1997, ApJ, 488, L149

Igumenshchev, I.V., Illarionov, A.F. & Abramowicz, M.A. 1999, ApJ, 517, L55

Illarionov, A.F., & Sunyaev, R.A. 1975, A&A, 39, 185

Lu, J.F., & Abramowicz, M.A. 1988, Acta Ap. Sin., 8, 1

Matsumoto, R., Kato, S., Fukue, J., & Okazaki, A.T. 1984, PASJ, 36, 71

Narayan, R., Kato, S., & Honma, F. 1997, ApJ, 476, 49

Narayan, R., Mahadevan, R., Grindlay, J.E., Popham, R. & Gammie, C., 1998, ApJ, 492, 554

Narayan, R., Yi, I., Mahadevan, R. 1995, Nature, 374, 623

Quataert, E., & Narayan, R., 1999, ApJ, 520, 298

Shakura, N.I., & Sunyaev, R.A., 1973, A&A, 24, 337

Sparke, L.S., & Shu, F.H. 1980, ApJ, 241, L65

Yuan, F. 1999, ApJ, 521, L55

Yuan, F., Peng, Q.H., Lu, J.F., & Wang, J.M. 1999, submitted to ApJ

PHOTOMETRIC PROJECT FOR ZERO-AGE CONTACT AND NEAR CONTACT BINARY SYSTEMS

Kam-Ching Leung
Behlen Observatory Department of Physics and Astronomy University of Nebraska

1. OBSERVATIONS NEEDED

There have been major advances in the development of computing modeling codes for analyzing photometric light curves in the last couple decades. As a result, a large number of photometric solutions are available in astronomical journals. Many contact systems have been discovered. However, most observers have preference to selection relatively short period systems. Most of the contact systems found are located in the vicinity of the main sequence. It suggests that they are most likely the result of mass exchange taking place during the core hydrogen-burning phase of the massive component. Very few contact systems are found very far from the main sequence. It is believed that these systems may be the result of mass exchange taking place during the shell hydrogen-burning phase of stellar evolution. Contact systems that are located far away from the main sequence are having relatively long periods. Due to the selection effect long period systems are extremely seldom observed. Therefore, observational data on more advanced evolutionary contact systems are lacking. We would like to suggest that concentrated effort should be made to study these very long period systems.

Another group of contact binary that we do not have very much information on is the zero-age contact systems. These systems are located on the zero-age main sequence or near the lower edge of the main sequence in the H-R diagram. The formation of zero-age contact systems could be a major challenge to theoretical modeling of binary formation. Observational data on these systems are very much needed in our understanding of the nature of these systems and to provide guidance for theoretical modeling.

We believe that these two groups of binary systems deserve special attention in our study of close binary formation and evolution. Thus,

K.S. Cheng et al. (eds.), Stellar Astrophysics, 281–282.

we propose more effort should be devoted to (1) zero-age or near zero-age contact systems, and (2) highly evolved very long-period contact systems.

2. POTENTIAL CANDIDATES

One of the difficult parts of designing an observing program is to develop a method to locate potential candidates. It is suggested that we utilize the spectral-type and period diagram for contact systems. A discussion concerning evolved and zero-age contact system was mentioned in an article of Leung and Schneider (1978). They calculated the loci for critical contact systems for mass ratio of 1.0 and 0.5. A system having a period shorter than the critical period of the same spectral-type (i.e. on the left of the loci) has to be a zero-age system and with over-contact configurations. If a contact system having a period considerably longer than the critical period of the same spectral-type (i.e. on the right of the loci) has to be an evolved contact system. The system is located very far to right of the loci is mostly like the result of mass exchange in the advance shell hydrogen-burning phase of the massive component.

Therefore, the potential zero-age contact systems will likely be found on or near the critical contact line in the spectral-type verses period diagram. The contact systems which are the result of post core hydrogen burning phase will likely be found among the systems that are far to right of the critical contact line in the spectral-type verses period diagram. A word of caution is need. The critical period loci of Leung and Schneider was based on very simple model and spectral-type usually found in the generally binaries or variable stars could have an uncertainty of two or more subtypes. Thus, the method suggested represents a most likelihood approach only.

References

Leung K. C. and Schneider D. P. 1978, ApJ 222, 917.

BALMER DECREMENT AND THE STRUCTURE OF BE-STAR ENVELOPES

T. KOGURE

Bisei Astronomical Observatory Ohkura, Bisei, Oda-Gun, Okayama, 714-1411, Japan

Abstract The observed Balmer decrements $D_{34} = F(H\alpha)/F(H\beta)$ of Be stars are collected in an unified system and compared with theoretical decrements based on the non-LTE solution of the radiation field. It is shown that the Balmer decrements are closely related to the structure of stellar envelope which is characterized by the mean dilution factor, optical thickness of the envelope and spectral subclass of the star. This is in contrast to the explanation of Slettebak et al. (1992) and of Dachs et al. (1990), both of which claimed the decrement to be a function of electron density in the envelope, based on the nebular approach of the decrements.

1. INTRODUCTION

Balmer decrements of Be stars provide important information on the structure of envelopes of Be stars. For deriving the decrements there are two theoretical approaches, one is nebular approach and the other is stellar-envelope approach. Cillié (1932, 1936) originally proposed the former as an application of Menzel & Baker's nebular case. More recently, Drake and Ulrich (1980) extended nebular approach to dense and opaque cases with electron density of $10^8 - 10^{15}\ cm^{-3}$, and optical thickness $\tau(L\alpha)$ of $10^4 - 10^6$. The results of Drake and Ulrich have been applied by Dachs et al. (1990) and by Slettebak et al. (1992) to the interpretation of Balmer decrements of Be stars.

The stellar-envelope approach has been developed originally by Miyamoto (1949) and applied by Kogure (1959) to Be stars. Recently, Kogure et al. (1994) revised the earlier calculation for the radiation field of Be star envelope and derived the Balmer decrement $H\alpha/H\beta$.

In this paper, we first collect the observational data of Balmer decrement $D_{34} = H\alpha/H\beta$ from literature in an unified system in the conversion from equivalent width ratio to the Balmer decrement. Then

K.S. Cheng et al. (eds.), Stellar Astrophysics, 283–290.

we qualitatively discuss on the structure of Be envelopes from the view point of the stellar-envelope approach. In conclusion, we emphasize the importance of dilution factor and the optical thickness of the envelope and of stellar radiation (spectral subclass) as the basic parameters.

2. OBSERVATIONAL MATERIALS

The Balmer decrements of Be stars are defined and measured as follows:

$$D_{34} = \frac{F(H\alpha)}{F(H\beta)} = G_\alpha \frac{W(H\alpha)}{W(H\beta)}, D_{54} = \frac{F(H\gamma)}{F(H\beta)} = G_\gamma \frac{W(H\gamma)}{W(H\beta)}, \quad (1)$$

where F and W are the radiation flux and equivalent width of the emission lines of the respective lines, and G the conversion factor derived by assuming the black-body radiation or energy distribution in some model atmosphere.

We have collected the observed materials from the following sources:

Author(s)	Adopted model	Number of stars	Derived decrement
Rojas (1956)	Black-body	17	D_{34}
Briot (1971)	Black-body	55	D_{34}, D_{54}
Briot(1981)	Black-body	15	D_{34}, D_{54}
Dachs et al.(1990)	Krucz(1971)	26	D_{34}, D_{54}
Slettebak et al.(1992)	Modified Collins(1979)	41	D_{34},D_{54}

In order to compare these decrements homogeneously, we have adopted the Slettebak system using the conversion factor G based on Modified Collins model. Among the decrements D_{34} thus collected are shown the cases of moderate and strong $H\alpha$ emission in Figures 1 and 2, respectively.

3. CALCULATION OF BALMER DECREMENT D_{34}

The present stellar-envelope approach is as follows (Kogure et al. 1994):

1. The envelope of Be stars is simplified by a disk with constant vertical thickness and outwardly decreasing electron density.
2. The envelope is divided into the elements each characterized by the same values of line-of-sight velocity and dilution factor.

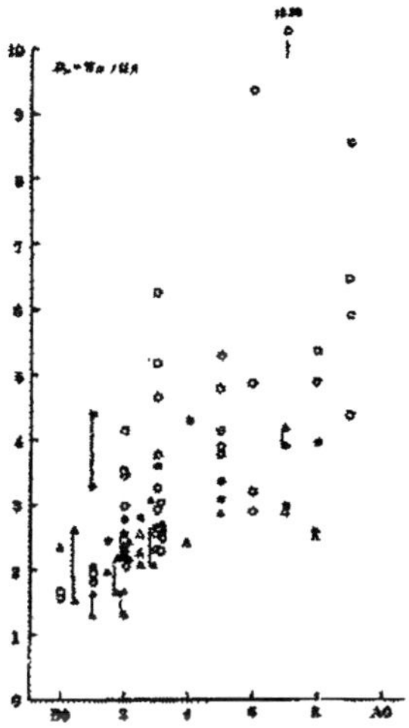

Figure 1 Balmer decrement D_{34} for Be stars with moderate $H\alpha$ emission ($10 < W(H\alpha) < 30$ Å).

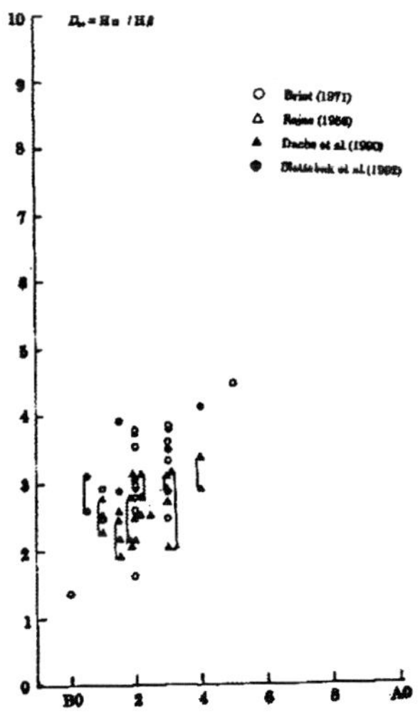

Figure 2 Balmer decrement D_{34} for Be stars with strong $H\alpha$ emission ($W(H\alpha) > 30$ Å).

3. The radiation field of the envelope is solved for the Balmer lines in each envelope-element, based on the non-LTE treatment of Kogure (1959). Stellar parameters are taken from Kurucz's model atmospheres (1979).

4. The Balmer decrements D_{34} and D_{54} are derived as a function of the optical depth $\tau(H\alpha)$, dilution factor W of the envelope-element and of the stellar effective temperature. The dilution factor is defined by

$$W = \frac{1}{2}\{1 - \sqrt{1 - (\frac{R}{r})^2}\} \tag{2}$$

where R denotes the stellar radius, and r the radial distance from the star's center.

Envelope-element is taken as shown in Figure 3 for purely rotating envelope and the derived decrements D_{34} are illustrated in Figures 4 and 5 for a case of optically thick envelope-element with $\tau(H\alpha) = 100$.

4. DISCUSSIONS

The theoretical decrements D_{34} shown in Figures 4 and 5 are those calculated for an envelope-element situated at a dilution factor W. Although the calculation of decrement should be made covering all envelope-elements of a star in order to compare with observed decrements, we shall suppose that the whole envelope is represented by an envelope-element having a mean dilution factor. This may be sufficient for the discussions of the envelope structure derived from the Balmer decrements in two different approaches.

The main features of the decrement derived in the present calculation can be summarized as follows:

1. Figure 4 shows the decrement D_{34} as a function of inverse dilution factor for several cases of stellar effective temperature. The inverse dilution factor is related to the radial distance r/R roughly as follows: where R is the stellar radius.

$1/W$	10	20	40	60	80
r/R	1.7	2.3	3.2	3.9	4.5

It is seen that the decrement D_{34} increases with the extension of the envelope. So that the scatter of observed D_{34} seen in Figures 1 and 2 can be understood as the difference in the extension of the envelopes.

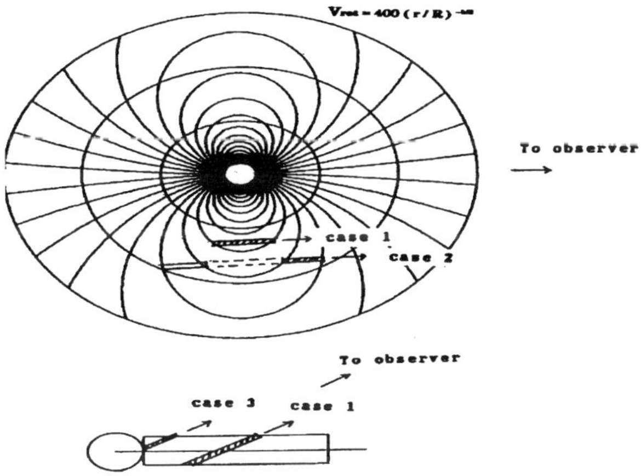

Figure 3 Envelope model : velocity field and envelope-element. Three cases are

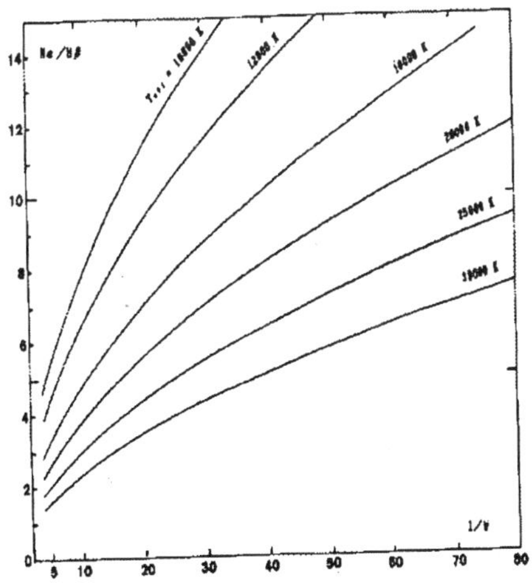

Figure 4 Balmer decrement D_{34} as a function of inverse dilution factor $1/W$. Each curve is characterized by stellar effective temperature.

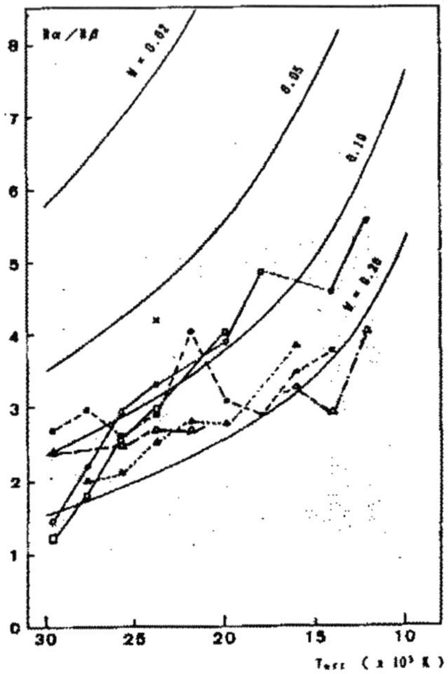

Figure 5 Balmer decrement D_{34} as the function of stellar temperature T_{eff} for several values of W. The observed decrements averaged for each spectral subclass are also plotted.

2. In Figure 5 the decrement D_{34} is shown as the function of stellar temperature for some values of dilution factor W. It is seen that the decrement increases as the decrease of stellar temperature for each value of dilution factor. The observed decrements averaged for each spectral subclass are also plotted in Figure 5. As is seen, the observed decrements D_{34} are distributing mostly in between the curves of $W = 0.2$ and 0.05. This infers that the main contribution for the formation of Balmer emission is confined in the nearer part of the envelope less than around 3 stellar radius, independent of spectral subclass.

3. Theoretical decrements in Figures 4 and 5 show the case of $\tau(H\alpha) = 100$. Though the decrement D_{34} takes smaller values for smaller values of $\tau(H\alpha)$, theoretical curve at $\tau(H\alpha) = 100$ yields an upper asymptotic limit. Namely, the decrements take practically the same values for optically thick envelopes with $\tau(H\alpha) > 100$.

This picture for Be star envelope is markedly different from that of Slettebak et al. (1992) and Dachs et al. (1990), who both adopted the calculation of emission-line spectrum by Drake and Ulrich (1980). They compared the theoretical decrements with observations by taking the electron density as a free parameter and adopting fixed parameters of $Te = 10000\ K$ which is corresponding to $R_{1c} = 3\ s^{-1}$, and $\tau(L\alpha) = 10^5$, where R_{1c} is the ground-state photoionization rate related to the stellar radiation.

Slattebak et al. (1992) have thus argued that the electron density of the envelope is generally in the range of $10^{11} - 10^{13}\ cm^{-3}$, and other features are summarized as follows:

1. Be stars with weaker Balmer emission have envelopes with somewhat higher electron density than those of stars with strong emission.

2. The hotter stars have envelopes with higher electron densities.

3. The envelopes of shell stars have lower electron densities,

4. Shell stars generally have envelopes smaller and compact than the normal Be stars, as suggested from the peak separation in emission-line profiles.

This picture gives somewhat difficult situation in explaining the structure of Be star envelopes. For example, Be stars with stronger emission are expected to have more developed envelope in size and in electron density, and this is contradicting with the above first point. According to Kogure (1990), Be-shell stars are characterized by the existence of sufficiently opaque envelope. The optical thickness of the envelope $\tau(H\alpha)$ should be as high as several thousand. This is a requirement for the formation of strong shell absorption lines in the center of Balmer lines of shell stars. The reason why ordinary Be stars do not show strong shell absorption lines is their small optical thickness less than around $\tau(H\alpha) = 100$. If the arguments (3) and (4) are combined, the envelopes of shell stars should have smaller optical thickness than ordinary Be stars, and this makes contradiction with the existence of shell lines in Be-shell stars.

In conclusion we emphasize the importance of the stellar-envelope approach which takes the dilution effect in the envelopes as the basic parameter rather than electron density.

Acknowledgements

The author expresses his thanks to M. Suzuki of Kanazawa Institute of Technology and M. Mon of Osaka Shoin Women's College for their collaboration both in numerical calculations and in data reduction.

References

Briot, D. (1971) A & A, 11, 57

Briot, D. (1981) A & A, 103, 5

Cillié, G. G. (1932) MN, 92, 820

Cillié, G. G. (1936) MN, 96, 771

Dachs, J., Rohe, D. & Loose, A. S. (1990) A & A, 238, 227

Drake, S.A. & Ulrich, R. K (1980) ApJ Suppl., 42, 351

Kogure, T. (1959) PAS,Japan, 11, 127, 278

Kogure, T. (1990) Ap. Sp. Sci., 163, 7

Kogure, T., Suzuki, M., & Mon, M. (1994) in Pulsation, Rotation and Mass Loss in Early-Type Stars, IAU Symp. No. 162, Balona et al. (eds.), 378

Kurucz, R. L. (1979) ApJ,Suppl. 40, 1

Rojas, H.R. (1956) Thesis (Paris Observatory)

Slettebak, A., Collins, G.W. II, & Truax, R. (1992) ApJ, Suppl.,81, 335.

SYMBIOTIC STARS WITH HIGHLY RESOLVED EMISSION LINE PROFILES

Shin'ichi Tamura & Yuji Ikeda
Astronomical Institute, Graduate School of Science, Tohoku University, Aoba-ku, Sendai, Miyagi 980-8578, Japan
tamura@astr.tohoku.ac.jp

1. INTRODUCTION

This report is concerned with spectroscopic diagnoses on the gaseous region of symbiotic stars based upon our recent work(Ikeda and Tamura, 1999: submitted to P.A.S.Japan). It is now widely accepted that the symbiotic star is a binary system which consists of a cool giant and a hot dwarf surrounded or accompanied with the ionized gas, probably together with dusts. Usually we can not recognize a hot star in the spectrum directly. Therefore, it looks like that a cool giant coexist with the ionized gas which produce rich emission-lines. This is the origin of the designation as symbiotic stars.

2. ANALYSES ON HIGHLY RESOLVED EMISSION-LINES

Recently the deep CCD images are getting to reveal symbiotic nebulae for a part of typical samples in addition to already known objects. While their actual electromagnetic spectrum distributes from the radio to ultraviolet region, even in some cases to the X-ray region, we concentrate to the optical region in our analyses. Our diagnostic tools on the symbiotic stars are H_α, HeIIλ4686, and [OIII]λ5007. A typical sample is displayed in Fig. 1.

Due to the Gaussian de-convolution of the highly resolved emission-line profiles, we can distinguish 1) main, 2) wing, and 3) absorption components from complicated features of H_α. One of our subject is to examine and make clear whether this absorption is real or not(see, Fig. 1a).

In case of HeIIλ4686 line we can see 1) main, and 2) usually blue-shifted additional components(see, Fig. 1b). There is no absorption and

K.S. Cheng et al. (eds.), Stellar Astrophysics, 291–296.

wing components. We will discuss about the FWHM of HeIIλ4686 main component both in the quiescent and the outburst phases.

In order to pursue our subjects, we choosed sample stars which are observable at the northern hemisphere, and obtained emission line data from 34 symbiotic stars. Moreover, ephemerides of 19 stars are already given in the published literatures. Our basic attitude is not to pursue detailed analyses or not to seek each individuality, but to examine these emission-line profiles as a group and try to extract common characteristics of symbiotic stars, by using these ephemerides.

3. RESULTS AND CONCLUSION

First, we present the result of H_α line analyses. For 30 stars among 34 samples we can detect the wing components of H_α which show the FWHM between 200 and 800 km/sec, both in quiescent and outburst phases. The frequency distribution of the FWHM also shows that the significant majority should belong to the range between 300 and 500 km/sec(Fig. 2). It is attributed these FWHM components to the stellar winds of hot stars, while we do not discuss the reason in detail at the moment. Based upon the spectroscopic analyses we can definitely say the apparent absorption features of complicated profiles are produced by real absorption processes in the ionized gas. To analyze these absorption components statistically the radial velocity has to define in uniform way. Hereafter we will use the radial velocities of absorption components relatively to those of hot stars, namely wing components. Then we can consider them only in connection to the binary phases, because the individual characteristics among different binary systems can be excluded. In Fig. 3 the relation of the radial velocities of the absorption component relative to the wing component, $V_{abs} - V_{wing}$ vs. phase, ϕ is displayed. The blue shift of absorption component as a whole can be clearly recognized. Its mean value is -26.5 km/sec. It seems also that the radial velocities of absorption components are independent to phase.

Second, we show the statistical analyses of HeIIλ4686 main components. Fig. 4 is the frequency histogram of their FWHM in both of quiescent and outburst phases. At the outburst phase the FWHM values can be seen only in the relatively wide region. Therefore, it can be considered that the width of HeIIλ4686 main component reflect the activity on the surface of hot stars as predicted by the theory of thermo-nuclear runaway.

Lastly, we mention on the characteristics of forbidden line [OIII]λ5007 as a group. Fig. 5 shows de-convolved each component relative to the hot star, namely the central velocity of H_α wing component, $V_{[OIII} - V_{H_{wing}}$

vs. phase, ϕ. The most conspicuous feature is that the double-line profiles will appear only in the limited phases, 0-0.25 and 0.5-0.75. While samples are not sufficient, this seems to be very important to confirm in near future. According to this figure we can imagine a bipolar or bi-conical flow of [OIII] line region.

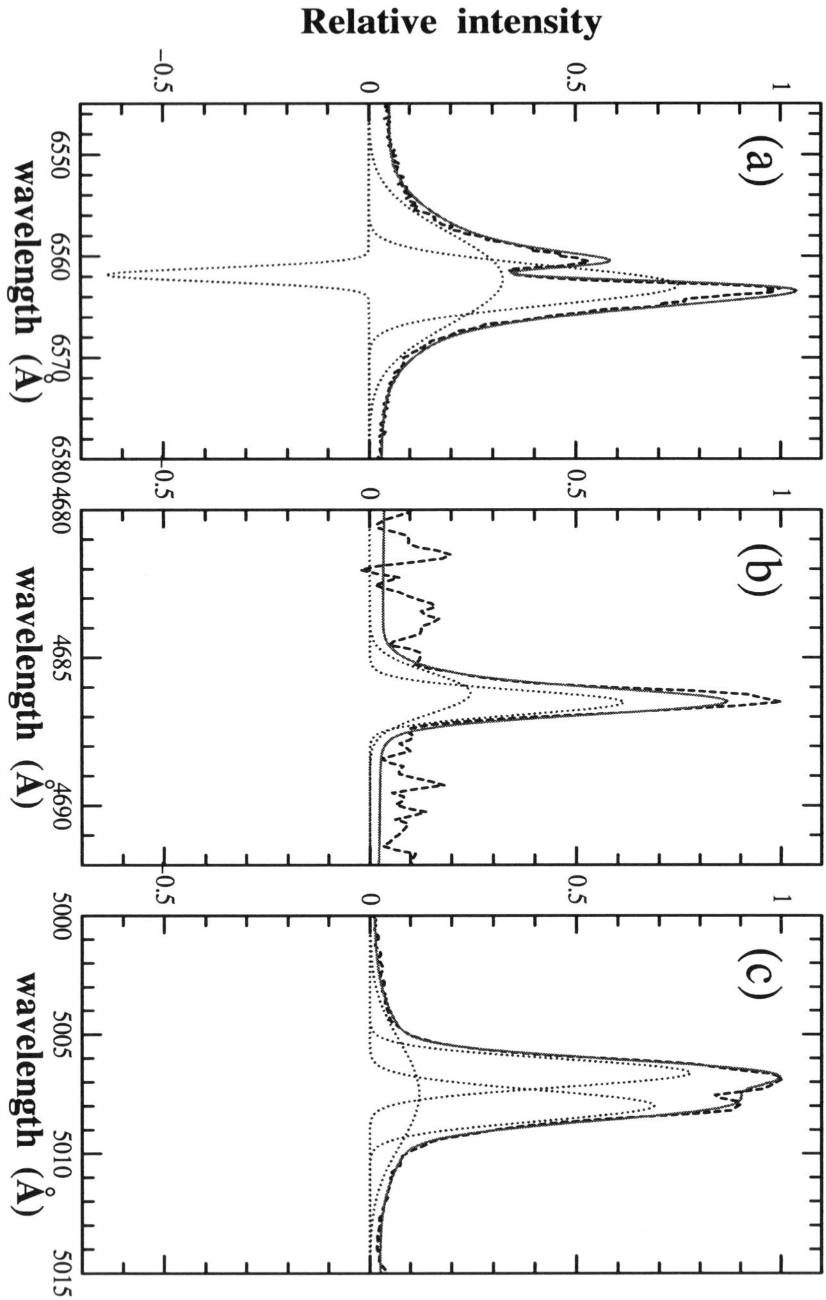

Figure 1 Examples of Gaussian de-convolution. (a) H_α, (b) HeIIλ4686, (c) [OIII]λ5007.

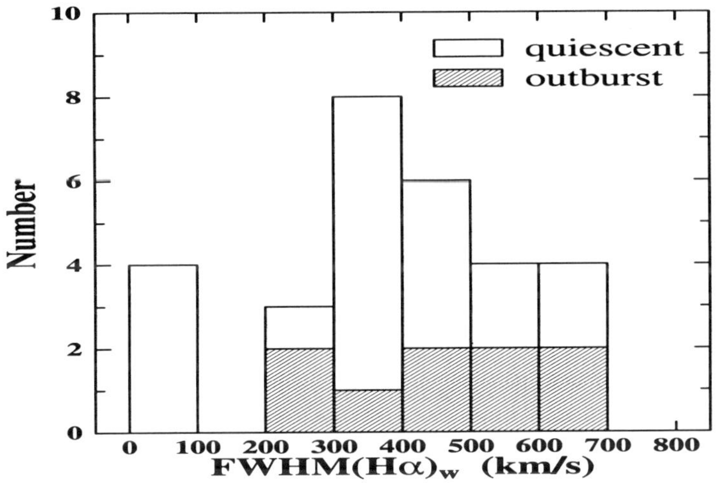

Figure 2 The frequency histogram of the FWHM for the H_α wing components at quiescent and outburst phases.

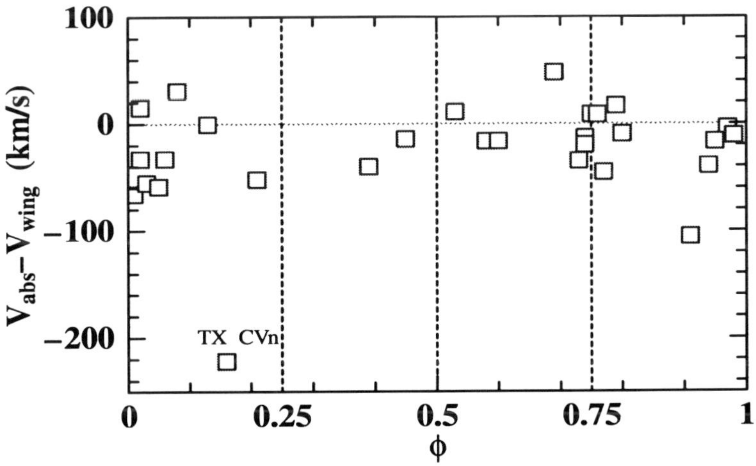

Figure 3 The diagram of $V_{abs} - V_{wing}$ vs. ϕ.

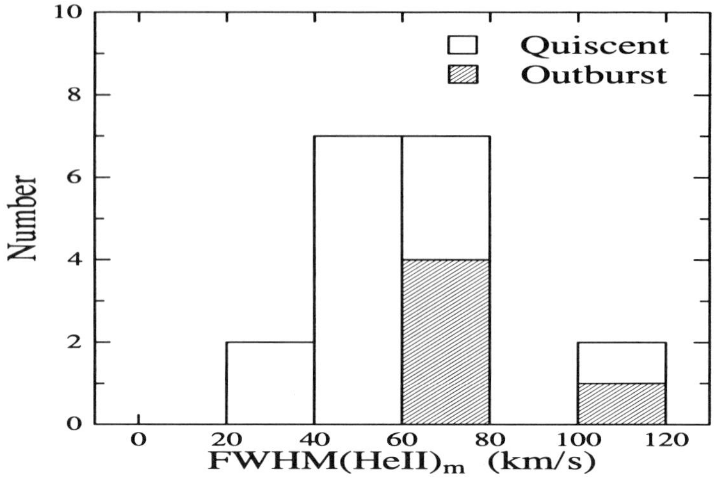

Figure 4 The frequency histogram of the FWHM of the main component of HeIIλ4686 at quiescent and outburst phases.

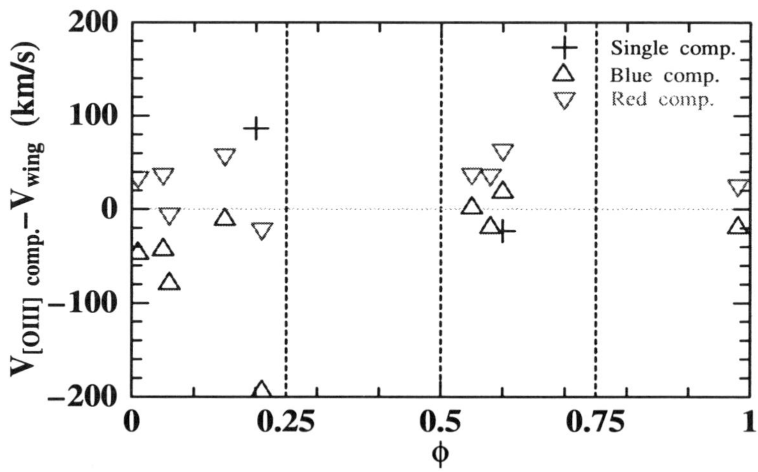

Figure 5 The diagram of $V_{[OIII]} - V_{wing}$ vs. ϕ. V_{wing} means of the central velocity of the wing component of Hα.

LONG-TERM X-RAY VARIABILITY IN X-RAY BINARIES: RXTE/ASM AND BATSE DATA

A. K. H. Kong and P. A. Charles

Department of Astrophysics, Nuclear & Astrophysics Laboratory, Oxford OX1 3RH, United Kingdom

albertk@astro.ox.ac.uk (AKHK), pac@astro.ox.ac.uk (PAC)

Abstract The Rossi X-ray Timing Explorer All Sky Monitor is providing sensitive, long-term measurements of the light curves of many X-ray binaries. The ASM data have already enabled the confirmation of several previously suggested so-called 'super-orbital' periods (in Cyg X–2, SMC X–1 and LMC X–3) and have revealed super-orbital periods in a number of sources (4U 1700–377, GX 354–0 and GX 339–4). Combined with the more extensive database of BATSE (at higher energies), these results provide hints of correlation between the super-orbital and orbital periods. Possible origins for this behaviour will be discussed.

1. INTRODUCTION

Luminous X-ray binaries exhibit two forms of precise periodicities which can be observable to both space and ground-based observatories: these are the binary orbital period (usually in hours to days) and, if the compact object is a neutron star, the spin or pulse period (usually in secs to mins). However, there is a third or 'long-term' periodicity first recognised 20 years ago which is present in many, perhaps all, X-ray binaries and yet is extremely poorly understood. That is because the timescale ($\sim$10s to 100s of days) can be extremely difficult to observe because of satellite constraints and background effects. From the long-term variability survey carried out by Smale & Lochner (1992) using *Vela 5B*, only 3 out of 16 low-mass X-ray binaries (LMXBs; X1820–30, X1916–05, Cyg X–2) were found to have long-term periods. In particular, X1820–30 and X1916–05 have periods of 176 days and 199 days, respectively, while recent observations by *RXTE* ASM confirm the long-term period of $\sim$78 days in Cyg X–2 (Wijnands et al. 1996). These long-term periods are

K.S. Cheng et al. (eds.), Stellar Astrophysics, 297–303.

designated *super-orbital*, as the orbital periods are well-known for all three objects (11.3 mins, 50 mins and 9.8 d, respectively). In high-mass X-ray binaries (HMXBs), super-orbital periods appear to be more common. Cyg X–1, SS433, LMC X–4, Her X–1 and several other sources all have super-orbital periods in the range of 30–300 days (e.g. Priedhorsky & Holt 1987). Table 1 summarizies all the orbital and super-orbital periods of X-ray binaries in which this effect has been seen. The cause of these super-orbital periods is still a subject of great debate, and possible explanations include the precession of a tilted accretion disk, neutron star precession, mass transfer feedback and triple systems (see Priedhorsky & Holt 1987 and Schwarzenberg-Czerny 1992 for more details). More recently, models such as radiation-driven warped accretion discs (e.g. Wijers & Pringle 1999), disc instabilities (Dubus et al. 1999) and magnetically-driven precession (Lai 1999; also in this volume) have been proposed to explain such behaviour. Most of the data in Table 1 comes from low energy X-ray instruments (2–10 keV), but with the BATSE on *CGRO*, we are able to study this effect over a much wider energy range (($>$ 20 keV). In this paper, we report the first combined results of long-term X-ray variability in X-ray binaries from both *RXTE*/ASM and *CGRO*/BATSE.

2. OBSERVATIONS

The All Sky Monitor (ASM) (Levine et al. 1996) on board the *RXTE* (Bradt, Rothschild & Swank 1993) consists of three wide-angle shadow cameras equipped with proportional counters that provide regular and almost continuous intensity measurements of most bright X-ray sources in three different energy bands (1.5–3, 3–5 and 5–12 keV). The ASM covers 80% of the sky every 90 minutes; combined with a 40% duty cycle (Levine et al. 1996), any given source is usually scanned 5–10 times per day. Both individual dwell and one-day average light curves are made available via the WWW by the *RXTE* ASM team.

The *Burst and Transient Source Experiment (BATSE)* on board the *Compton Gamma Ray Observatory (CGRO)* has been operated continuously since its launch in 1991 April. BATSE consists of eight identically configured detector modules with energy channels spanning from 20 to 600 keV (see Fishman et al. 1989). Light curves are made available via the WWW by the BATSE team.

3. PERIOD ANALYSIS AND RESULTS

In order to search for periodic phenomena in the data set, we employed two different methods: the Lomb-Scargle periodogram (Lomb

Table 1 Orbital and super-orbital period of X-ray binaries.

Source	Orbital Period (d)	Super-orbital Period (d)	References
X 1820–30	0.0078	176	1
X 1916–05	0.035	199	1
Cyg X–2	9.8	78	1, 2
GX 339–4	0.62	220	3
GX 354–0	-	72	4
Her X–1	1.7	35	5
LMC X–4	1.408	30.4	5
Cen X–3	2.09	140	5
X 1907+097	8.38	42	5
SMC X–1	3.9	55	5
Cyg X–1	5.6	294	5
SS 433	13	164	5
LMC X–3	1.7	198	6
4U 1700–377	3.41	13.81	7

References: (1) Smale & Lochner 1992; (2) Wijnands et al. 1996; (3) Nowak et al. 1999; (4) Kong et al. 1998; (5) Priedhorsky & Holt 1987; (6) Cowley et al. 1991; (7) König & Maisack 1997

1976; Scargle 1982; hereafter LSP) and phase dispersion minimization (PDM; Stellingwerf 1978). The LSP is a modification of the discrete fourier transform which is generalized to the case of uneven spacing. However, it is most sensitive in those cases where any modulation present is sinusoidal. PDM is suitable even where the data is non-sinusoidal and works by dividing the data into phase bins and minimizing the dispersion within the bins. The deepest minimum relates to the period of the strongest modulation. Finally, a folded light curve is created and used to check the validity of any flagged periods. Low-frequency noise (red noise) can sometimes produce false signals (e.g. Done et al. 1992). We therefore simulated the effect of red noise on our light curves and period-searching technique. We performed period searching as described above for all the X-ray binary light curves available to both *RXTE*/ASM and BATSE. The highest power flagged by the LSP is then tested for significance. This is determined by generating noise data sets with the same time intervals and variance as the true data and then performing the LSP on the resulting data sets. The peak power in each periodogram (which must be purely due to noise) was then recorded. This was then repeated 10,000 times for good statistics. Although there are several reports of long-term X-ray variations in X-ray binaries with *RXTE*/ASM (see e.g. Kong et al. 1998), this is not true with BATSE. From our analysis, only GX 354–0, GX 339–4, Her X–1 and 4U 1700–377 were found to have long-term variations in both databases. In particular, Her X–1 and 4U 1700–377 have been previously reported (Scott & Leahy 1999; König & Maisack 1997).

Fig. 1 displays the power spectra of both *RXTE*/ASM and BATSE data for GX 354–0 and GX 339–4. For GX 354–0, a ~72 d long-term period was found by Kong et al. (1998) with *RXTE*/ASM, and a consistent 68.5±2.8 d with BATSE (see Fig. 1). Furthermore, GX 339–4 shows a 230 ± 3 d in BATSE data, which is consistent with the *RXTE*/ASM ~220 d variation (Nowak et al. 1999; also see Fig. 1).

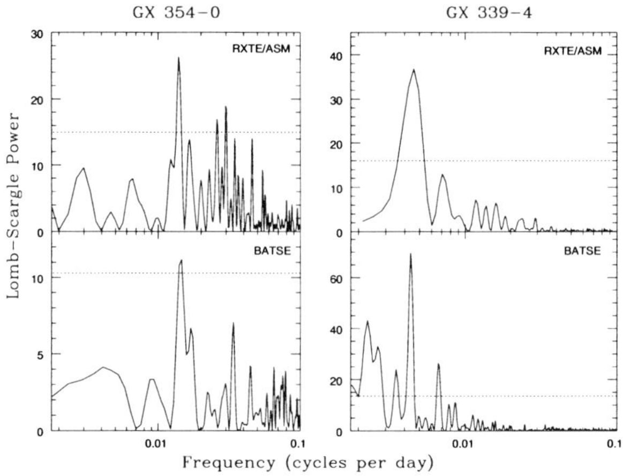

Figure 1 LSP of *RXTE*/ASM (upper panel) and BATSE (lower panel) data of GX 354–0 and GX 339–4. The strongest peak is at ~70 d and ~220 d for GX 354–0 and GX 339–4 respectively. The 99% confidence level is shown by a dashed line.

4. DISCUSSION

We have found a ~70 d and ~220 d long-term X-ray periodicity for GX 354–0 and GX339–4 respectively with *RXTE*/ASM and BATSE. In particular, the *RXTE*/ASM variabilities of GX 354–0 and GX 339–4 have been previously reported by Kong et al. (1998) and Nowak et al. (1999), while we have found similar variations in the BATSE data (see Fig. 1). We note that the above mentioned periodicities also appeared in *Vela 5* and *Ariel 5* data (Kong et al. 1998; Kong et al. 2000, in preparation). This provides strong evidence that such variations are stable and not transitory. Although super-orbital periods have been seen in several X-ray binaries with *RXTE*/ASM and other older missions such as

Vela 5 and *Ariel 5*, they are not common in BATSE data, presumably due to a smaller database. However, such long-term variability is rare in BATSE data even for many HMXBs which show obvious variations at lower energies. Significant long-term variation (in both *RXTE*/ASM and BATSE) have only been found in Her X–1 (Scott & Leahy 1999) in addition to GX 354–0, GX 339–4 (see §3) and 4U 1700–377 (König & Maisack 1997). Due to the limited data at higher energies, it is not clear that the super-orbital period originates mainly in soft X-rays (< 12 keV). From Table 1, we note that super-orbital periods are common in high-mass systems, while they are much more rare in LMXBs. In addition, there is a large spread in values of the ratio of super-orbital to orbital periods for LMXBs. For X1820–30, X1916–05 and Cyg X–2, the ratio is 22100, 5750 and 8 respectively, whereas in massive systems it is in the range of 10 to 100 (Wijers & Pringle 1999). It has therefore been suggested that the origin of the long-term variation in HMXBs and LMXBs may be different (see e.g. Priedhorsky & Holt 1987; Kong et al. 1998; Wijers & Pringle 1999). We plot the orbital periods against the super-

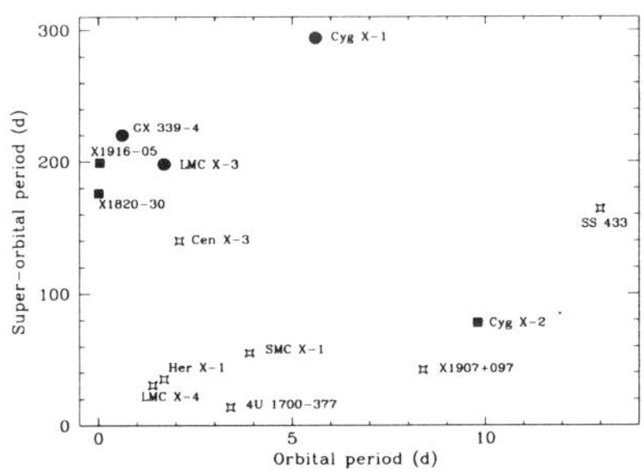

Figure 2 Comparison of orbital periods and super-orbital periods of X-ray binaries (after Wijers & Pringle 1999). HMXB (stars), LMXB (solid squares) and BHC (solid circles)

orbital periods in Fig. 2 and there is no clear correlation between them. However, on further sub-division of this figure into HMXBs, LMXBs and black hole candidates (BHCs), it is clear that they exhibit different values of the orbital/super-orbital ratio. For the LMXBs and BHCs, it is not clear whether they have a correlation due to the limited data available. For HMXBs, there is a slight trend that the orbital period is correlated with the super-orbital period. It is clear that Cyg X–2 is located in the HMXB region and it is no surprise as from the ratio of super-orbital to orbital periods, it is consistent with those of HMXBs. This may reflect the recent proposal for an intermediate-mass companion in Cyg X–2 (Podsiadlowski & Rappaport 1999). Observers might be careful in determining long-term X-ray variations from all-sky surveys since false signals will occur if the observations are not long enough. For example, we fail to identify the 37 d period of X 2127+119 (Corbet et al. 1997) in the *RXTE*/ASM data and the 309 d periodicity of Aql X–1 was shown to be unstable from more extensive data (Kitamoto et al. 1993). In conclusion, long term surveys and detailed pointed observations (in different states of long-term periods) are necessary to reveal the true nature of such variations.

Acknowledgements

We are grateful to Erik Kuulkers and Tariq Shahbaz for many interesting and useful discussions. This paper utilizies results provided by the ASM/RXTE team. The BATSE data is generated by the BATSE team at NASA/MSFC. AKHK is supported by Hong Kong Oxford Scholarship.

References

Bradt, H. V., Rothschild, R. E. and Swank, J. H., 1993, A&AS, 97, 355
Corbet, R., Peele, A. and Smith, D. A., 1997, IAU Circ. 6632
Cowley, A. P., Schmidtke, P. C., Ebisawa, K., Makino, F., Remillard, R. A., Crampton, D., Hutchings, J. B., Kitamoto, S. and Treves, A., 1991, ApJ, 381, 526
Done, C., Madejski, G. M., Mushotzky, R. F., Turner, T. J., Koyama, K. and Kunieda, H., 1992, ApJ, 400, 138
Dubus, G., Lasota, J.-P., Hameury, J.-M. and Charles, P., 1999, MNRAS, 303, 139
Fishman, G.J. et al., 1989, in Proc. of the GRO Science Workshop, ed. W.N. Johnson (Greenbelt: NASA/GSFC), p.2
Lai, D., 1999, ApJ, 524, 1030
Kitamoto, S., Tsunemi, H., Miyamoto, S. and Roussel-Dupre, D., 1993, ApJ, 403, 315
Kong, A. K. H., Charles, P. A. and Kuulkers, E., 1998, NewA, 3, 301
König, M. and Maisack, M., 1997, A&A, 327, L33
Levine, A. M., Bradt, H. V., Cui, W., Jernigan J. G., Morgan, E. H., Remillard, R., Shirley, R. E. and Smith, D. A., 1996, ApJ, 469, L33
Lomb, N. R., 1976, Ap&SS, 39, 447
Nowak, M. A., Wilms, J. and Dove, J. B., 1999, ApJ, 517, 355
Podsiadlowski, P. and Rappaport, S., 1999, in astro-ph/9906045
Priedhorsky W. C. and Holt S. S., 1987, SSRv, 45, 291
Smale A. P. and Lochner J. C., 1992, ApJ, 395, 582
Scargle J. D., 1982, ApJ, 263, 835
Schwarzenberg-Crerny A., 1992, A&A, 260, 268

Scott, D. M. and Leahy, D. A., 1999, ApJ, 510, 974
Stellingwerf R. F., 1978, ApJ, 224, 953
Wijers, R. A. M. J. and Pringle, J. E., 1999, MNRAS, 308, 207
Wijnands R. A. D., Kuulkers E. and Smale A. P., 1996, ApJ, 473, L45

X-RAYS FROM MILLISECOND PULSARS

Werner Becker

Max-Planck-Institut für extraterrestrische Physik, Giessenbachstrasse 1, 85740 Garching, Germany

Abstract The satellite observatories ROSAT, ASCA, RXTE and BeppoSAX have brought important progress in neutron star and pulsar astronomy. With significantly higher sensitivities compared with previous X-ray satellites they allowed for the first time to detect X-ray emission from objects as faint as millisecond (ms) pulsars. In this paper we summarize the current observational status and discuss the empirical constraints on the pulsars' X-ray emission mechanisms.

1. INTRODUCTION

Millisecond pulsars are distinguished from the group of ordinary rotation-powered pulsars by their small spin periods ($P \leq 20$ ms) and high spin stability ($dP/dt \approx 10^{-18} - 10^{-21}$ s/s). Consequently, they are very old objects with spin-down ages of typically $P/2\dot{P} \sim 10^9 - 10^{10}$ years and magnetic dipole components $B_\perp \propto (P\,\dot{P})^{(1/2)}$ of the order of $10^8 - 10^{10}$ G. More than $\sim 75\%$ of the known disk millisecond pulsars are in binaries with a compact companion star, compared with the $\cong 1\%$ of binary pulsars found in the general population. This gives support to the idea that their fast rotation has been acquired by angular momentum transfer during a past mass accretion phase.

Before the launch of ROSAT, nothing was known on the X-ray emission properties of ms-pulsars. According to the standard models of cooling neutron stars, ms-pulsars are too old to expect detectable thermal emission related with the star's heat content at birth (see Tsuruta 1998 and references therein). Thermal X-ray emission, however, may be emitted from polar caps heated up to temperatures of few million degrees by high energy secondary electrons/positrons, streaming back from the outer magnetosphere to the neutron star's polar cap regions. A non-thermal process which may account for X-rays is the emission from rel-

K.S. Cheng et al. (eds.), Stellar Astrophysics, 305–317.

ativistic charged particles accelerated in the pulsar magnetosphere. The interaction of a relativistic pulsar wind with a nearby companion star or with the interstellar medium is another likely scenario. The latter process causes a DC component in the observed emission, and, if the physical conditions are appropriate, will manifest in an extended synchrotron nebula.

Eleven of the 35[1] detected rotation-powered pulsars belong to the small group of millisecond pulsars (see Fig. 9). Although this is almost 1/3 of all X-ray detected rotation-powered pulsars, the origin of their X-ray emission is only known for few of them.

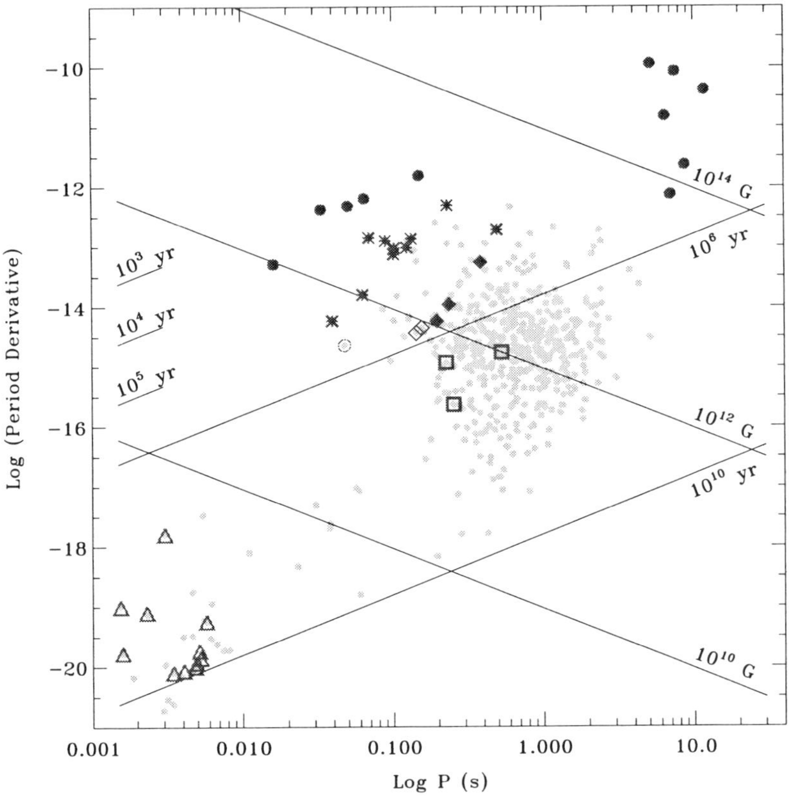

Figure 1 The sample of rotation-powered pulsars plotted with respect to their spin parameters $P, \dot{P}$ (small gray dots). Separate from the majority of ordinary field pulsars are the ms-pulsars in the lower left corner. X-ray detected pulsars are indicated by symbols. Lines representing constant ages $\tau = P/2\dot{P}$ and magnetic field strength $B_\perp = 3.3 \times 10^{19}(P\dot{P})^{1/2}$ are indicated.

[1]An updated version of Table 1 from Becker & Trümper 1997 is available online. See http://www.xray.mpe.mpg.de/~web/bt97/bt97_update.html.

Six of the eleven detected ms-pulsars (PSR B1957+20, J1012+5307, B0751+18, J1744-1134, J1024-0719 and J0030+0451) are identified only by their positional coincidence with the radio pulsar (see also Table 1), and in view of the low number of detected counts do not provide much more than a rough flux estimate. More detailed results are found for the other five ms-pulsars which all provide important empirical information on the X-ray emission mechanisms operating in these objects. The following sections present these five pulsars in greater detail.

2. THE GLOBULAR CLUSTER PULSAR 1821-24 IN M28

The 3 ms-pulsar PSR 1821-24 in the globular cluster M28 shows Crab-like X-ray pulses up to $\sim$ 17 keV (Saito et al. 1997; Kawai & Saito 1999). The two sharp peaks in the pulse profile (see Fig. 4) and the power-law nature of the spectrum argue without doubt for a magnetospheric (non-thermal) origin of the detected emission. The alignment between a radio and X-ray pulse component adds further support to this interpretation and implies a common emission site for the main X-ray and radio pulse component observed at 800 MHz. The photon-index for the *pulsed* emission is found to be $\alpha \sim 1.2$ (Kawai & Saito 1999). Imaging M28 with the ROSAT HRI (Fig. 2) has shown two separate sources: a point source (RX J1824.5-2452P) consistent with the ms-pulsar position and a brighter but extended source RX J1824.5-2452E, whose nature is not yet clear. The spectral results obtained from ASCA for RX J1824.5-2452E favors the model in which RX J1824.5-2452E is a pulsar-powered synchrotron nebula, similar to what is seen for the Crab. An alternative and maybe more likely interpretation, however, is that RX J1824.5-2452E is made of a number of point sources, e.g. accreting binaries containing white dwarfs (cataclysmic variables, CV) or neutron stars (low mass X-ray binaries, LMXB), which have not been spatially resolved by the HRI. An interpretation in terms of low accretion LMXBs is also supported from a long term X-ray luminosity study of M28. Becker & Trümper (1999) found recently that the ROSAT PSPC and HRI data taken in 1991 and 1995 suggest an X-ray flux variability of M28 on time scales of years. The total HRI energy flux taken in March 1995 from M28 is about a factor of 3 higher than the total energy flux deduced from the March 1991 PSPC observation. This behavior is in agreement with the results by Gotthelf & Kulkarni (1997) who discovered an X-ray burst from M28 in the 1995 ASCA data.

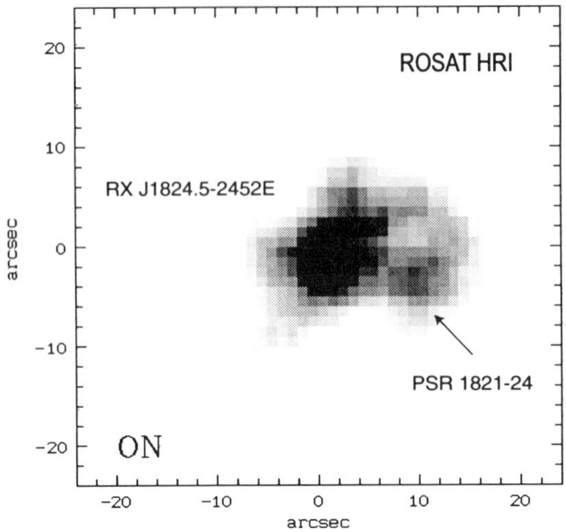

Figure 2 PSR 1821-24 in the globular cluster M28 as seen by the ROSAT HRI. The plot depicts the pulse-on phase of the pulsar. In the pulse-off phase only the extended source RX J1824.5-2452E is visible. (Image from Danner et al. 1997).

3. PSR 1937+21

A recent result based on ASCA observations is the detection of X-ray pulses from PSR 1937+21 (Takahashi et al. 1998), an isolated 1.56 ms pulsar which is the fastest spinning pulsar known so far. The pulsar has a dispersion measure based distance of 3.6 kpc and a rather high column absorption of $N_h \sim 6 \times 10^{21}$ cm^2. The latter prevented the detection of soft X-rays in a $\sim$ 30 ksec ROSAT PSPC observation (Verbunt et al. 1996; Becker & Trümper 1997). Most of the flux detected by ASCA (probably up to 80%) is in the $1.7 - 6.5$ keV band. The X-ray detection is remarkable because the dominating X-ray pulse has a peak width of only $\lesssim 100\mu$s (i.e. ~ 0.08 in phase) and a pulsed fraction of $\sim 75\%$ (see Fig. 5). The narrow peak infers a very small beaming factor which is difficult to explain in terms of thermal emission from hot polar caps. A non-thermal origin is further suggested by the high pulsed fraction and the spectral analysis. The latter, although strongly limited by photon statistics (less than 100 cts were recorded with the ASCA GIS), yields a photon-index of 0.9 ± 1.

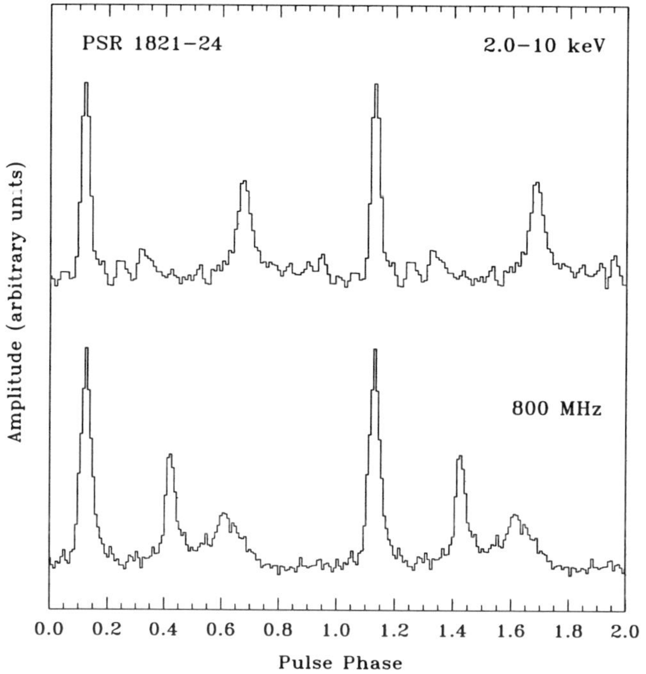

Figure 4 Integrated pulse profiles of the globular cluster pulsar PSR 1821−24 as observed with RXTE (Kawai & Saito 1999) and with the NRAO at 800 MHz (*bottom*) by Backer & Sallmen (1997). Two phase cycles are shown for clarity. The X-ray pulse profile is characterized by a double peak structure with a phase separation of ~ 0.6 between the two peaks. The radio profile at 800 MHz depicts three pulse components. At this frequency, the dominating radio pulse is nearly phase aligned with the primary X-ray pulse.

4. PSR J0218+3242

The 2.3 ms pulsar PSR J0218+4232 is in a two day binary orbit with a low-mass white dwarf companion and shows significant *unpulsed* radio emission throughout the pulse period. The latter has been taken as an indication that the magnetic dipole is almost aligned with the rotation axis.

X-ray pulses from the 2.32 ms pulsar PSR J0218+3242 have been detected by Kuiper et al. (1998) in ROSAT HRI data and subsequently with BeppoSAX by Mineo et al. (2000). The BeppoSAX data are best fitted by a power-law spectrum with photon-index $\sim 1 \pm 0.2$ (phase averaged). The pulse profile shows two peaks which change their relative intensity with increasing photon energy. The flux in the energy band

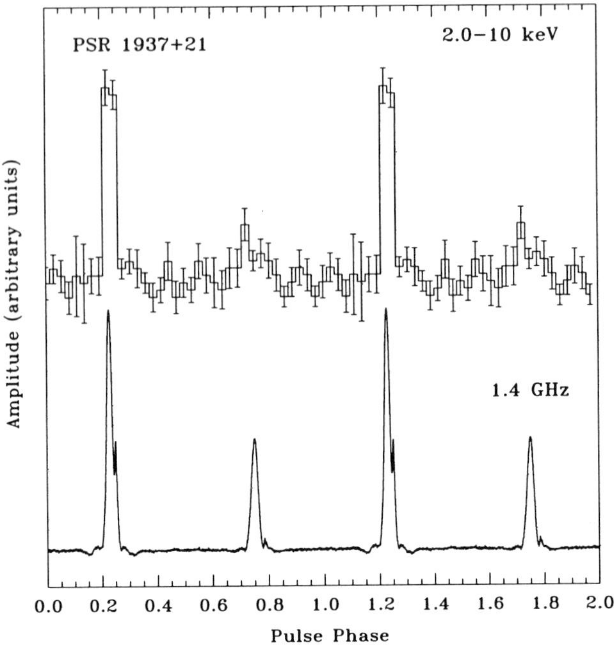

Figure 5 Integrated pulse profiles of the 1.56 ms pulsar PSR 1937+21. The X-ray profile (top) was observed by Takahashi et al. (1998) using ASCA. The radio profile (bottom) was taken with the Arecibo radio telescope at 1.4 GHz (D. Backer, priv. com.). The radio profile is characterized by two narrow peaks, which both have X-ray counterparts (although the counterpart for the second radio peak is only marginal detected). Because the relative phase between the X-ray and radio profile is unknown both profiles have been aligned arbitrarily.

$2 - 10$ keV is $f_x = 4 \times 10^{-13}\,\mathrm{erg}^{-1}\mathrm{s}^{-1}\mathrm{cm}^2$ (Mineo et al. 2000). ROSAT data have revealed the existence of a compact synchrotron nebula surrounding the pulsar (see Fig. 6). The HRI data imply a source extend of about 14 arcsec. Although this is close to the HRI's attitude solution, the BeppoSAX data seem to confirm the existence of a soft nebula component, showing an energy dependent pulsed fraction of $37 \pm 13\%$ ($0.1 - 2.4$ keV), $67 \pm 12\%$ (1.6–4 keV) and $89 \pm 10\%$ ($4 - 10$ keV). A summary of spectral fits based on the BeppoSAX data and pulse profiles for different energy bands are shown in Figure 6.

Noting the spatial coincidence of PSR J0218+4232 with the EGRET source 2EG J0220+4228, Verbunt et al. (1996) tentatively identified the pulsar with the high-energy γ-ray source. Using some additional EGRET observations, applying a combination of spatial and timing analysis, Kuiper et al. (1999) conclude that 2EG J0220+4228 is probably

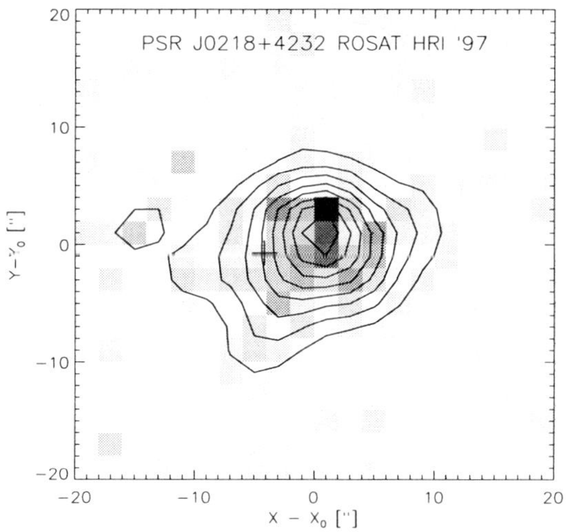

Figure 6 PSR J0218+3242 as observed with the ROSAT HRI. The data imply a source extend of about 14 arcsec, associated with a compact synchrotron nebula surrounding the pulsar. (Image from Kuiper et al. 1998).

multiple: between 0.1 and 1 GeV PSR J0218+4232 is the most likely counterpart, and above 1 GeV the bright AGN 3C 66A is the best candidate counterpart. The third EGRET catalog (Hartman et al. 1999), which is based on more viewing periods than the 2EG catalog, identifies 3EG J0222+4253 (2EG J0220+4228) with 3C 66A, rather than with the ms-pulsar. However, in a note on this source, Hartman et al.(1999) indicate that the identification with 3C 66A stems from the catalog position based on the > 1 GeV map. Furthermore they confirm that for lower energies (100-300 MeV) the EGRET map is consistent with all the source flux coming from the pulsar.

5. PSR 0437-4715 AND J2124−3358

The case for the 5.75 ms binary pulsar PSR J0437-4715 and the isolated 4.93 ms pulsar J2124−3358 is not as certain as for 1821−24, 1937+21 and J0218+3242 for which a non-thermal origin of the radiation is constrained.

There is no common agreement on the origin of the X-rays detected from PSR J0437-4715. Although this ms-pulsar is the closest and brightest of all ms-pulsars detected so far, the small bandwidth of ROSAT and the limited photon-statistic of the ASCA and SAX data did not allow to discriminate between various possible emission scenarios (Fig. 8). Multi-component thermal spectra (thermal polar-cap emission) and non-

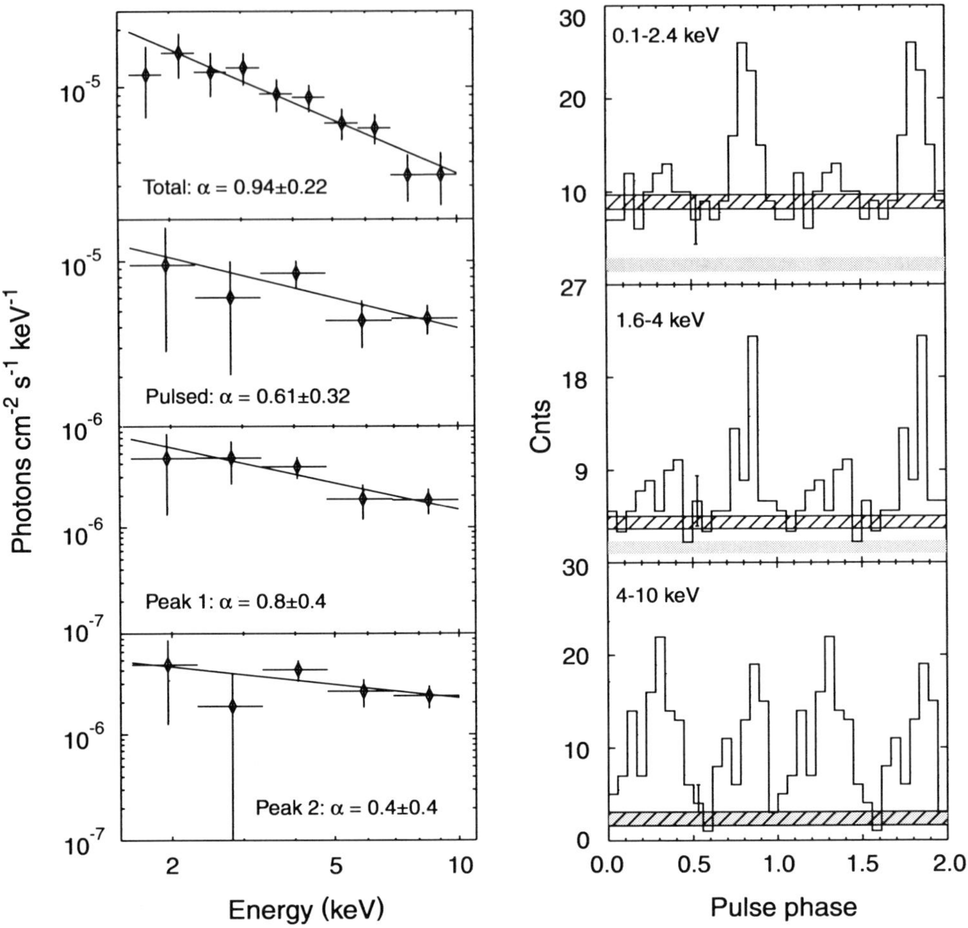

Figure 7 **Left:** Spectral fits based on the BeppoSAX data of PSR J0218+3242 (from Mineo et al. 2000). The upper top panel shows the phase averaged data fitted with a power-law spectrum. Fits to the total pulsed emission and to the separate peaks are shown in the other panels. Although the small photon-statistics results in large uncertainties, the data suggest a spectral hardening of the pulsed emission with increasing energy. **Right:** pulse profiles as observed by ROSAT (top panel) and BeppoSAX. While in the soft band the profile is in agreement with a single pulse, a second peak shows up in the hard band. The gray-shaded area indicates the background level whereas the hatched area indicates the DC-level. The difference between both levels and its decrease with increasing photon energy supports the idea that there is a compact nebula contribution in the soft X-ray band which fades away with increasing photon energy.

thermal spectra (power-law and broken power-law) fit the data equally well (see Becker & Trümper 1999 for a more detailed discussion and further references). The X-ray pulse profile is broad with a single peak stretching almost the entire phase cycle. The pulsed fraction is $\sim 30\%$ in the $0.1 - 2.4$ keV band.

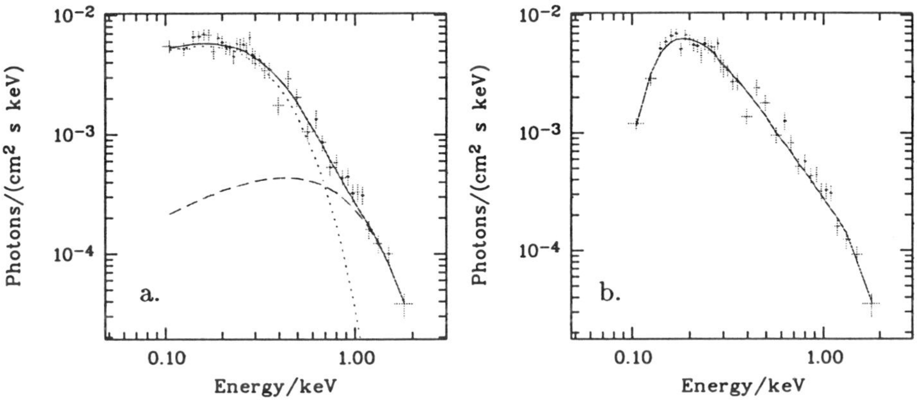

Figure 8 **a.** The ROSAT PSPC data of PSR J0437−4715 fitted with a two component black body model. The harder component (dashed line) represents the black-body spectrum as deduced from spectral fits to ASCA data. The dotted line represents the soft component as obtained from a spectral fit to the ROSAT data. **b.** The PSPC data fitted with a broken power-law which also fits the ASCA data. The break energy in this model is 1.4 keV. The slopes of the first and second power-law components are −2.35 and −4.5, respectively.

PSR J2124−3358 was the first isolated galactic ms-pulsar detected at X-ray energies (Becker and Trümper 1999). Its lightcurve shows indication for a double pulse structure, which in terms of pulsed components and pulse phase separation implies a similarity between the X-ray and radio profile observed at 436 MHz. This similarity points towards a non-thermal origin of the X-ray emission (spectral information from this pulsar is not yet available). However, the shape of a pulse profile itself is not a strong indicator for the origin of the detected emission if the profile (as observed for J2124−3358) is broad: a radiation cone which yields sharp peaks at one aspect angle may well be seen as hardly modulated away from this angle, so that sharp peaks may indicate non-thermal emission processes whereas the reversal – soft modulated emission originates from thermal processes – must not be true. The fraction of pulsed X-ray photons observed from this pulsar is $\sim 33\%$ in the ROSAT band .

6. SUMMARY AND DISCUSSION

The X-ray emission observed from
PSR 1821−24, 1937+21 and J0218+4232
is dominated by non-thermal processes. This is constrained by power-law spectra and pulse profiles with narrow peaks and high pulsed fractions. For the other ms-pulsars like PSR 0437−4715 and J2124−3358 the existing data do not allow to unambiguously identify the emission process. It is not excluded that in these pulsars thermal polar-cap emission dominates. Hot polar caps are predicted by many magnetospheric emission models as a consequence of pair-creation by the vacuum gap discharge, after which a significant amount of highly energetic charged particles is expected to stream back to the neutron star, heating the surface to a few million degrees.

Two of the eleven X-ray detected ms-pulsars are found to show diffuse extended emission: PSR 1821-24 in M28 and PSR J0218+4232. While for 1821-24 the extended emission possibly comes from unresolved globular cluster sources, the faint diffuse emission found around J0218+4232 might be plerionic. Chandra and XMM observations will constrain the existence and origin of such nebulae.

Comparing the stellar parameters of PSR 1821−24, 1937+21 and J0218+4232 with those of the young Crab-like pulsars (for which similar X-ray emission properties are observed), it can be seen that the intensity of the magnetic field at the pulsar's light-cylinder radius[2] $R_L = cP/(2\pi)$ may play a key role in determining the emission properties and the efficiency of $\dot{E} \rightarrow L_X$ conversion (see Fig. ??). All three pulsars have a period in the range $\sim 1.5 - 3$ ms. As a consequence, in spite of its surface magnetic field $B_S \sim 10^8$ G, its magnetic field at the light-cylinder is large ($B_L = B_S(R_S/R_L)^3 \simeq 10^6$). By comparison for the Crab it is $\simeq 2 \times 10^6$, even though B_S is 5 orders of magnitude larger! If this correlation reflects intrinsic dependences of the emission process on P and $\dot{P}$, it implies that the X-ray emission in these pulsars is emitted from a location close to the light-cylinder rather than from a place close to the star's surface.

Another argument in favor of a non-thermal origin for the bulk of the observed X-rays from ms-pulsars is the close correlation between the pulsars spin-down energy $\dot{E}$ and the X-ray luminosity within 0.1-2.4 keV. Becker & Trümper (1997) found that from the young Crab-like pulsars down to the ms-pulsars all X-ray luminosities from ROSAT detected rotation-powered pulsars correlate very closely with the pulsar's spin-

[2]the radius at which the magnetosphere, co-rotating with the neutron star, will have a speed equal to the speed of light

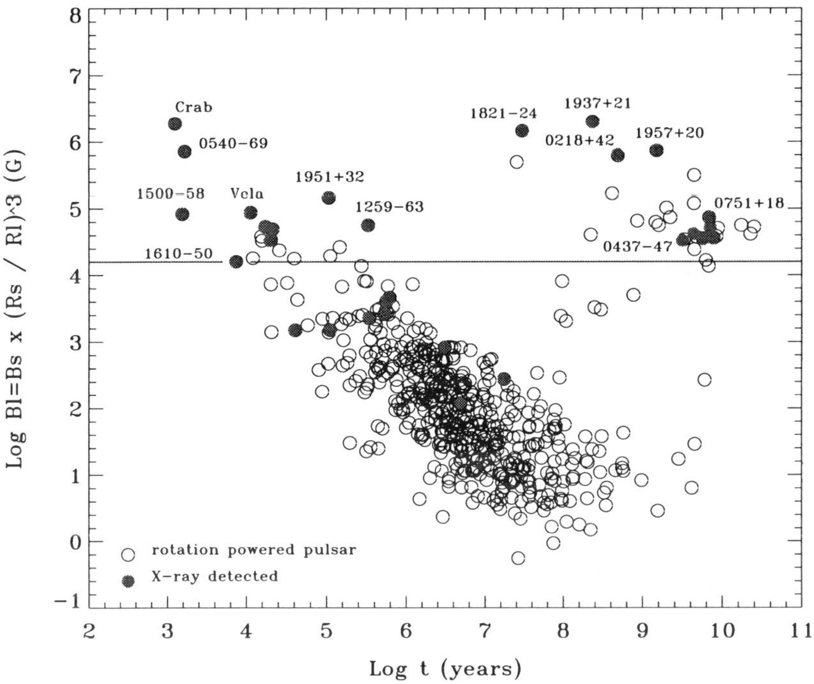

Figure 9 Magnetic field at the pulsar's light-cylinder vs. spin-down age for the radio pulsars of the Princeton pulsar catalog. The 35 X-ray detected pulsars are plotted as filled circles (some are label). The line at Bl=4.2 indicates the level of Bl for the radio pulsar PSR 1610-50 for which (by its young age, i.e. P and $\dot{P}$) Crab- or Vela-like X-ray emission is expected. Because of their small light-cylinder radius, all ms-pulsars have a Bl comparable with those of the young pulsars, for which non-thermal emission is found to be the dominating process.

down energy; implying $L_X = 10^{-3}\dot{E}$. So in particular the ms-pulsars obey the same linear relationship with the same X-ray efficiency as the Crab type pulsars, indicating that their emission is mainly due to magnetospheric processes. It will be a key question for XMM and Chandra to test this correlation and to find out on what level thermal/non-thermal processes contribute. The main progress in respect to this is expected from those ms-pulsars for which no detailed spectral and/or temporal information could be obtained by sensitivity limitations. Table 1 lists the eleven X-ray detected ms-pulsars. All pulsars in the list are promising targets, worth to be studied further with XMM, Chandra and Astro-E in order to better understand the X-ray emission mechanisms working in this kind of pulsars.

Note	Name PSR	P ms	$\dot{P}$	log age yr	$\dot{E}$ erg/s	$\dot{E}/4\pi d^2$ erg/s/cm^2	Nh $10^{21}cm^{-2}$	dist kpc
p	J0437−4715	5.75	5.7e-20	9.50	33.62	1.0e-09	∼ 0.08	0.18
p	B1937+21	1.5	1.1e-19	8.37	36.04	7.1e-10	∼ 2	3.6
p	B1821−24	3.05	1.6e-18	7.48	36.35	6.2e-10	∼ 4	5.5
p	J2124−3358	4.93	1.1e-20	9.86	33.55	4.7e-10	∼ 0.5	0.25
d	J0030+0451	4.86	1.0e-20	9.88	33.53	5.4e-10	∼ 0.13	0.23
d	J1744−1134	4.07	8.6e-19	9.86	33.28	5.5e-10	∼ 0.09	0.17
d	B1957+20	1.60	1.2e-20	9.32	35.06	4.1e-10	∼ 4.5	1.53
d	J1024−0719	5.16	1.8e-20	9.76	33.72	3.6e-10	∼ 0.2	0.35
d	J1012+5307	5.25	1.4e-20	9.76	33.60	1.2e-10	∼ 0.07	0.5
p	J0218+4232	2.32	8.0e-20	8.66	35.29	6.5e-11	∼ 1.8	5.7
d	J0751+1807	3.47	8.0e-21	9.84	33.88	1.5E-11	∼ 0.4	2.0

Table 1 Properties of the X-ray detected millisecond pulsars. The list depicts the pulsars' spin parameters $P, \dot{P}$, their spin-down energy $\dot{E}$ and their characteristic age $P/2\dot{P}$. The column densities N_H and the pulsar distances are estimated from the radio dispersion measure. Pulsars which are identified by positional coincidence with a radio pulsar (d) and by pulsed emission (p) are indicated.

Acknowledgements

We thank D. Backer, Y. Saito, M. Takahashi and S. Shibata for their support with the radio and X-ray pulse profiles used to produce the Figures 3 and 4.

References

Backer, D.C., Sallmen, S., 1997, ApJ, 114, 1539

Becker, W., Trümper, J., 1997, A&A, 326, 682

Becker, W., Trümper, J. 1999, A&A, 341, 803

Danner, R., Kulkarni, S.R., Saito, Y., Kawai, N., 1997, Nat, 388, 751

Gotthelf, E.V., Kulkarni, S.R. 1997, ApJ, 490, L161

Hartman R.C., Bertsch D.L., Bloom S.D., et al. 1999, ApJS, 123,79

Kawai, N., Saito, Y., 1999, 3rd INTEGRAL Workshop "The Extreme Universe", Sept. 1998, Taormina, Italy, in press

Kuiper et al. 1998, A&A, 336, 545

Kuiper L., Hermsen W., Verbunt F., et al. 1999, 3rd INTEGRAL Workshop "The Extreme Universe", Sept. 1998, Taormina, Italy, in press

Mineo, T., Cusumano, G., Kuiper, L., Hermsen, L., Massaro E., Becker, W., Nicastro, L., Sacco, B., Verbunt, F., Lyne, A.G., Stairs, I.H., Shibata, S. 2000, A&A, submitted

Saito, Y., Kawai, N., Kamae, T., et al., 1997, ApJ, 477, L37

Takahashi M., Shibata S., Torii K., Saito Y., Kawai N. 1998, IAU circ. 7030

Tsuruta S., 1998, Physics Reports, 292, 1

Verbunt F., Kuiper L., Belloni T., et al. 1996, A&A, 311, L9

LONG-TERM VARIABILITY PROPERTIES AND PERIODICITY ANALYSIS FOR BLAZARS

J.H. Fan
1. Center for Astrophysics, Guangzhou Normal University, Guangzhou 510400, China
e-mail: jhfan@guangztc.edu.cn
2. Chinese Academy of Sciences-Peking University Joint Beijing Astrophysical Center (CAS-PKU.BAC), Beijing, China

Abstract In this paper, the compiled long-term optical and infrared measurements of some blazars are used to analysis the variation properties and the optical data are used to search for periodicity evidence in the lightcurve by means of the Jurkevich technique and the discrete correlation function (DCF) method. Some explanations have been discussed.

Keywords: Active Galactic Nuclei (AGNs): variability – periodicity

Introduction

The nature of the central regions of quasars and other Active Galactic Nuclei (AGNs) is still an open problem. The study of AGN optical variability can yield valuable information about the mechanisms operating in these sources, with important implications for quasar modeling (see, for instance, Fan et al. 1998). Some particular objects have been claimed to display periodicity in their lightcurves over a variety of timescales (e.g. Jurkevich 1971; Hargen-Thorn et al. 1987; Sillanpaa et al. 1988; Webb et al. 1988; Babadzhanyants & Belokon 1991; Kidger et al. 1992; Liu et al. 1995; Marchenko et al. 1996; Fan et al. 1997, 1998, Fan & Su 1998; Fan 1999a; Zhang et al. 1998; Fan & Lin 1999a; Fan et al. 1999a; Lainela et al. 1999; Lin 1999; Su 2000 and references therein), but, in general, the clear identification of periodic behaviour has been very elusive due to the complexity of the optical lightcurves and the lack of databases large enough as to provide an adequate sampling over large periods.

Blazars are an extremely subclass of AGNs, characterized as rapid and high amplitude variations, high and variable polarization, super-

K.S. Cheng et al. (eds.), Stellar Astrophysics, 319–324.

luminal radio components, nonthermal continuum. Blazars have been monitored for a long time, compilations are available for optical (Fan & Lin 1999b,c) and infrared (Fan & Lin 1999d; Fan 1999b) bands. The infared data are available in http://xxx.lanl.gov/astro-ph/9908104 and -/astro-ph/9910269. The longterm measurements make it possible for one to search for periodicities from the light curves and to discuss the long-term variability properties. Because the characteristic of the astronomical measurements, we adopted the Jurkevich (Jurkevich 1971) and DCF (Discrete Correlation Function) (see Fan & Lin 1999a) methods to the available data in searching for the periods. In the second section we will give the variation properties, in the third section, we present the results and finally in section 4, we give some discussions.

Variation Property

From the complication (Fan et al. 1999b, c, d; Fan 1999b), we found that the largest variations are comparable in both the optical and infrared bands. Largest variations at different wavelengths increase with decreasing wavelength. The variations are also correlated with the highest observed optical polarization, which can be explained using the beaming model. Some objects show that the spectra flatten when the sources brighten while the opposite cases are for some others. For the largest infrared and optical variations, we listed them in Table 1.

Periodicity Analysis Methods

Jurkevich Method. The Jurkevich method (Jurkevich 1971, also see Kidger et al. 1992; Fan et al. 1998a; Fan 1999a) is based on the expected mean square deviation and it is less inclined to generate spurious periodicity than the Fourier analysis used by other authors (e.g. Fan et al. 1997). It tests a run of trial periods around which the data are folded. All data are assigned to m groups according to their phases around each trial period. The variance V_i^2 for each group and the sum V_m^2 of all groups are then computed. If a trial period equals the true one, then V_m^2 reaches its minimum. So, a "good" period will give a much reduced variance relative to those given by other false trial periods and with almost constant values. A further test is the relationship between the depth of the minimum and the noise in the "flat" section of the V_m^2 curve close to the adopted period. If the absolute value of the relative change of the minimum to the "flat" section is large enough as compared with the standard error of this "flat" section (say, five times), the periodicity in the data can be considered as significant and the minimum as highly reliable (Fan et al. 1998a).

DCF Method. The DCF (Discrete Correlation Function) method is intended for analyses of the correlation of two data set. It is described in detail by Edelson & Krolik (1988) (also see Fan et al. 1998b). This method can indicate the correlation of two variable temporal series with a time lag, and can be applied to the periodicity analysis of a unique temporal data set. If there is a period, P, in the lightcurve, then the DCF should show clearly whether the data set is correlated with itself with time lags of $\tau = 0$ and $\tau = P$ (see Fan & Lin 1999a). We have implemented the method as follows.

Firstly, we have calculated the set of unbinned correlation (UDCF) between data points in the two data streams a and b, i.e.

$$UDCF_{ij} = \frac{(a_i - \bar{a}) \times (b_j - \bar{b})}{\sqrt{\sigma_a^2 \times \sigma_b^2}}, \tag{1}$$

where a_i and b_j are points in the data sets, $\bar{a}$ and $\bar{b}$ are the average values of the data sets, and σ_a and σ_b are the corresponding standard deviations. Secondly, we have averaged the points sharing the same time lag by binning the $UDCF_{ij}$ in suitably sized time-bins in order to get the DCF for each time lag τ:

$$DCF(\tau) = \frac{1}{M} \Sigma\, UDCF_{ij}(\tau), \tag{2}$$

where M is the total number of pairs. The standard error for each bin is

$$\sigma(\tau) = \frac{1}{M-1} \{\Sigma\, [UDCF_{ij} - DCF(\tau)]^2\}^{0.5}. \tag{3}$$

Results and Discussion

Very recently, we compiled four data bases for BL Lac objects and OVV/HPQs in the optical and infrared bands (Fan & Lin 1999b,c,d; Fan 1999b). The data bases can be used to discuss the long-term variability properties, the correlated variations and to search for periodicity signatures in the light curves. When the both methods are adopted to those optical data, the both methods give consistent results. The results are presented in Table 2. It is interesting to notice that there is a common period of ~ 1.0 year, which is likely from the effect of the Sun on the measurements.

For the long-term periodicity of variations, there are several explanations: the double black hole model, the hectic jet model, the slim disk model, and the effect of external perturbations to the accretion disk. (e.g. Sillanpaa et al. 1988; Meyer & Meyer-Hofmeister 1984; Horiuchi

& Kato 1990; Abraham & Romero 1999; Fan et al. 1997, 1998a, 1999; Villata & Raiteri 1999; Romero et al. 1999).

Table 1 The largest variations of Blazars

Name (1)	Δm_{opt} (2)	Δm_{IR} (3)	$P_{opt}(\%)$ (4)	Name (1)	Δm_{opt} (2)	Δm_{IR} (3)	$P_{opt}(\%)$ (4)
0048-097	2.7	6.55	27.2	0109+244	3.07	1.58	17.3
0118-272	1.05	0.67	17.	0138-097	1.52	1.69	29.3
0215+015	5.0	2.69	20.	0219+428	2.0	1.61	18.0
0235+164	5.3	5.0	44.	0323+022	1.3	0.97	10.4
0420-014	2.8	2.88	20.	0422+004	2.2	3.25	22.
0521-365	1.4	1.25	11.	0537-441	5.4	3.0	18.8
0716+714	5.0		29.0	0735+178	4.6	2.47	36.
0736+017	1.35	2.71	6.	0754+100	3.16	1.88	26.
0818-128	3.78	2.14	36.	0823-223	1.41	2.32	11.
0829+046	3.58	2.15	12.	0851+202	6.0	3.87	37.2
0912+297	2.25	2.27	13.	1101+384	4.6	2.03	16.0
1144-379	1.92	3.46	8.5	1147+245	1.0	1.34	13.0
1156+295	5.0	4.47	28.	1215+303	3.3	1.05	14.
1219+285	3.13	2.46	10.	1253-055	6.7	4.57	44.
1308+326	4.17	5.55	28.	1418+546	4.8	1.59	24.
1510-089	5.4	1.25	14.	1514-241	3.0	2.64	8.0
1538+149	3.7	0.89	32.8	1641+395	3.0	3.16	35.
1652+398	1.30	2.37	7.0	1727+502	2.1	0.82	6.0
1749+096	2.7	2.21	32.	1807+698	2.0	1.02	12.
1921-293	2.64	3.06	17.	2005-489	0.53	0.31	2.0
2155-304	1.85	1.88	14.2	2200+420	5.31	2.93	23.
2223-052	5.0	3.96	17.3	2240-260		1.66	15.1
2251+158	2.5	1.57	19.	2254+074	3.27	1.36	21.

Notes to the table:
Col. 1 The name of the source;
Col. 2, The largest optical variation;
Col. 3 The largest infrared variation;
Col. 4, The largest optical polarization

Table 2 Investigation Results of Periodicity of 14 Blazars

Name	P_1 (f)(year)	P_2 (f)(year)	P_3 (f)(year)	P_4 (f)(year)
0219+428	0.99	2.14	2.93	4.52
0235+164	1.01	1.56	2.95±0.25	5.87±1.3
0735+178	1.01	14.39±0.51	18.58±0.52	28.62±0.58
0754+100	1.00	17.85±1.3	24.7±0.7	
0851+202	1.01	5.53	11.75±0.5	
1215+303	0.99	2.05±0.1	4.45±0.05	6.89±0.34
1219+285	0.99	1.97	9.0±0.1	14.84±1.55
1226+023	0.98	2.00	13.5±0.2	22.5±.02
1253-055	1.2	7.1±0.4		
1308+326	1.00	2.03	2.95	6.07±0.1
1418+546	0.99	2.99		
1514-241	1.00	2.01	4.09	9.35±0.1
1807+698	0.99	6.0	18.18	
2155-304		4.16	7.0	
2200+420	0.99	13.7±0.4	19.95±0.70	29.6±0.30

Notes to the Table:
P_1, P_2, P_3, P_4 are the possible periods derived from the light curves

Acknowledgements

This work has been supported by the National PanDeng Project of China and the National Natural Scientific Foundation of China. I thank Prof. K.S. Cheng to provide him with the opportunity to present this work and Prof. Lin and Su for their help.

References

Abraham, Z. & Romero, G.E. 1999, A&A, 344, 61, 1999
Angione, R.J. et al., 1981, AJ, 86, 653
Angione, R.J. & Smith, H.J. 1985, AJ, 90, 2474
Babadzhanyants, M.K. & Belokon, E.T., 1991, in Variability of Blazars eds. E. Valtaoja & M. Valtonen, Cambridge University Press, p384
Fan J.H., Xie G.Z., Lin R.G. et al. 1997, A&AS, 125, 525
Fan J.H. & Su C.Y. 1998, Ch.A&A, 23, 22
Fan J.H., Xie G.Z., Pecontal E., et al., 1998a, ApJ, 507, 173
Fan J.H., Adam G. Xie G.Z. et al., 1998b, A&AS, 136, 217
Fan J.H., et al. 1999a, ASP Conf. Ser. Vol. 159, p99
Fan J.H., Romero, G. & Lin R.G. 1999b, submitted

Fan J.H., 1999a, MNRAS, 382, 1032
Fan J.H., 1999b, astro-ph/9910269
Fan J.H. & Lin R.G. 1999a, A&A, (accepted)
Fan J.H. & Lin R.G., 1999b, ApJS, 121, 131, astro-ph/9908104
Fan J.H. & Lin R.G., 1999c, ApJ, (accepted)
Fan J.H. & Lin R.G., 1999d, (preparation)
Hagen-Thorn et al. 1987, Astron. Rep. 41, 1
Honma, F. et al. 1991, PASJ, 43, 147
Horiuchi, T. & Kato, S. 1990, PASJ, 42, 661
Hufnagel, B.R. & Bregman, J.N. 1992, ApJ, 386, 473
Jurkevich, I. 1971, Ap&SS, 13, 154
Kidger, M.R., Takalo, L., Sillanpaa, A. 1992, A&A, 264, 32
Lainela M., et al. 1999, ApJ, 521, 561
Lin R.G. 1999, Acta Astrophysica Sinica, submitted
Liu F.K., Xie G.Z., Bai J.M. 1995, A&A, 295, 1
Marchenko S.G. et al. 1996, ASP Conf. Ser. 110, p105.
Romero, G.E. et al. 1999, in preparation
Sillanpaa, A., Mikkola, S., Valtaoja, L. 1991, A&AS, 88, 225
Sillanpaa, A., Haarala, S., Korhonen, T. 1988, A&AS, 72, 347
Su C.Y. 2000, Acta Astrophysica Sinica, (in press).
Villata M., & Raiteri C.M. 1999, A&A, 347, 30
Webb J.R. et al. 1988, AJ, 95, 374
Zhang X., Xie G. Z., Bai J.M., 1998, A&A, 330, 469

THE OPTICAL, X-RAY AND GAMMA-RAY LIGHT CURVES AND SPECTRA OF PSR B0540-69 AND PSR B1509-58

K.S. Cheng[1], L. Zhang[1,2] and S.F.Ko[1]

[1] *Department of Physics, the University of Hong Kong, P.R. China*

[2] *Department of Physics, Yunnan University, Kunming, P.R. China*

Abstract Based on the three dimensional outer magnetosphere model of pulsars proposed by Cheng, Ruderman and Zhang, we calculate the light curves and spectra of the Crab-like pulsars: PSR B0540-69 and PSR B1509-58. The light curves of optical, X-rays and γ-rays for the Crab-like pulsars are expected to be the same origin, i.e. these high energy photons are produced in a finite region just above outer gaps through synchrotron self-Compton mechanism. For PSR B1509-58, if magnetic inclination and viewing angle are 65° and 75°, and the photon emission region has a mean thickness of $\sim 0.1R_L$, where R_L is the radius of the light cylinder, the expected broad single pulse can compare with the observed one and the model spectrum is consistent with observed spectrum. For PSR B0540-69, a broad single pulse and spectrum similar to the observed data can be produced if magnetic inclination and viewing angle are 50° and 66° respectively and the mean thickness of photon emission region is $\sim 0.04R_L$.

Keywords: X-rays: stars, gamma rays: theory - pulsars: individual (PSR B1509-58, PSR B0540-69)

Introduction

It is commonly accepted that a three dimensional outer gap model is necessary in order to explain the observed light curves and phase-resolved spectra of γ-ray pulsars. Cheng, Ho and Ruderman (1986a, 1986b, hereafter CHR I and CHR II) proposed a outer gap model to explain the observed data of the Crab and Vela pulsars (also see Ho 1989). Their model assumed that the radiation regions are thin in the longitudinal direction. Their double peak γ-ray structure was from two topologically disconnected outer gaps, each of which is associated with

K.S. Cheng et al. (eds.), Stellar Astrophysics, 325–337.

different magnetic poles. However, Romani and co-workers (Chiang & Romani, 1992,1994; Romani and Yadigaroglu 1995; Romani 1996) have shown that only one outer gap with only outgoing current can already produce a broad, irregularly-shaped emission beam of which is particularly dense near the edge, so that two γ-ray peaks would be observed when the line of sight from the Earth crosses these enhanced γ-ray beam regions; the inner region of the beam provided a significant amount of emission between the peaks. Recently, Cheng, Ruderman & Zhang (1999) (hereafter CRZ) have re-considered the three dimensional magnetosphere using various physical processes (including pair production which depends sensitively on the local electric field and the local radius of curvature, surface field structure, reflection of $e^{\pm}$ pairs because of mirroring and resonant scattering) to determine the three-dimensional geometry of the outer gap. They have shown that two outer gaps and both outgoing and incoming currents are in principle allowed, but it turns out that outgoing currents dominate the emitted radiation intensities. The observed features of the Crab pulsar can be explained by this model well.

For the known γ-ray pulsars, two pulsars may be Crab-like pulsars, which are PSR B0540-69 and PSR B1509-58. PSR B0540-69 in the Large Magellanic Cloud (LMC) is one of the youngest and most luminous rotation-powered pulsars, its period and period derivative are 50 ms and 4.8×10^{-13} s s^{-1}, which implies a characteristic age of $\tau = 1.5 \times 10^3$ yr. PSR B1059-58 is a pulsar with a period of 150 ms and period derivative of 1.5×10^{-12} s s^{-1}. Its characteristic age of $\tau = 1.6 \times 10^3$ yr makes it the second youngest known pulsar behind only the Crab pulsar. Here, we describe an outer gap model of Crab-like pulsars and then apply this model to explain the light curves and spectra of PSR B1509-58 and PSR B0540-69.

3D Outer Gap Model of Crab-like Pulsars

The Structure of the Outer Gap

CRZ have proposed a three dimensional outer gap model for rotation-powered pulsars with thin outer gap. They successfully applied this model to the Crab pulsar. In this model, polar cap shape is determined at first because it defines the boundary of the open volume at the stellar surface. Because the outer gaps are within the open volume, the open volume is divided into many parts, in which the shape of each part at the stellar surface is the same as the polar cap shape but the size is smaller. If the coordinate values (x_0, y_0, z_0) represent the values of the last closed field lines at the stellar surface. Then the coordinate values

(x_0', y_0', z_0') for different parts can be determined by using $x_0' = a_1 x_0$, $y_0' = a_1 y_0$ and $z_0' = (1 - (x_0'^2 + y_0'^2))^{1/2}$ and by changing a_1. It should be point out that $a_1 = 1$ corresponds to the last closed field lines and $a_1 < 1$ to open field lines in the magnetosphere. Once the polar cap shape is given, which depends on pulsar parameters such as period, magnetic field and the magnetic inclination angle, the structure of an outer gap can be determined as follows.

According to CRZ, the azimuthal extension of the outer gap ($\Delta\Phi$) is finite and is determined by the local pair production condition. Two mechanisms may be responsible for it. For an oblique dipole, the radius of inner boundary of the outer gap ($r_{in}(\phi)$)(the intersection of the null charge surface with the last closed field lines) changes with azimuthal angle (ϕ). If we define that $\phi = 0$ at the (x, z) plane, then $r_{in}(\phi)$ increases from $\phi = 0°$ to $\phi = 180°$, and decreases from $\phi = 180°$ to $\phi = 360°$. In the ($\mathbf{\Omega}$, μ) plane, generally four gaps (two long gaps labeled gaps 1 and 2 and two short gaps labeled gaps 3 and 4) exist in the outer magnetosphere (CHR I). Because the reflected outgoing $e^{\pm}$ pairs can enter the short gaps, the short gaps can be quenched (CHR I). In this case the extension along azimuthal direction is about 180°, corresponding to the maximum value of the inner boundary of the outer gap is r_{in}' where the null charge surface intercepts the light cylinder. Another mechanism is that the extensions of the outer gap along both azimuthal and radial directions are limited by local pair production processes (see CRZ for details). For the Crab-like pulsars with average size of the outer gap

$$f_0 \approx 5.5 P^{26/21} B_{12}^{-4/7} \left(\frac{\Delta\Phi}{2\pi}\right)^{1/7} \quad , \tag{1}$$

the pair production will take place mainly in the range $r_{in} \leq r \leq r_{lim}'$ where r_{lim}' is estimated as $(r_{lim}'/r_{in})^{-3/8} \sim 1/2$, which gives $r_{lim}' \sim 6 r_{in}$ (CRZ). This limits pair production both along the field lines and in transverse directions, and gives $\Delta\Phi(r_{lim}')$ which depends on the magnetic inclination angle. Therefore, the azimuthal extension of the outer gap is

$$\Delta\Phi \approx min(180°, \Delta\Phi(r_{lim}')) \tag{2}$$

and the pair production region inside the outer gap is in the region from r_{in} to r_{lim}, where

$$r_{lim} = min(r_{in}', r_{lim}') \quad . \tag{3}$$

Within the pair production regions, outgoing and incoming directions for particle flows are allowed. For $r > r_{lim}$ only outgoing current is possible. It should be pointed out that four gaps may be active or inactive for an aligned dipole.

We want to point out that the soft photons in the Crab-like pulsars will be non-thermal photons which are produced by the synchrotron radiation by the secondary pairs (CHR II). CHR II estimated that the fractional size of the outer gap for the crab pulsar is ~ 0.05 by taking into account effect of the synchrotron photons, on the other hand, Eq. (1) gives 0.04. Since this estimate is so close, we will use Eq.(1) as a rough approximation for estimating f_0. In fact, in three dimensional case, such estimate of the fractional height of the outer gap including secondary synchrotron photons becomes very difficult because both the high energy photons and the soft target flux are produced by the gaps themselves (Romani 1996).

Briefly, for Crab-like pulsars, there may be two outer gaps in the outer magnetosphere. The structure of an outer gap is as follow: (i) the fractional height of the outer gap is given by f_0, (ii) the azimuthal extension is about $\sim 160^\circ$ and (iii) the radial extension is from r_{in} to R_L with small region $(r_{lim} - r_{in})$ of photon-photon pair production.

Photon Emission Processes and Patterns

CRZ have argued that most curvature photons emitted from the accelerator will become secondary pairs outside the gap in Crab-like pulsars where the local magnetic field in the outer gap accelerator is sufficiently strong to give KeV synchrotron radiation. These secondary pairs will boost a fraction of the synchrotron photons to high energy γ-rays through inverse Compton scattering. Therefore, for Crab-like pulsars, photon emission comes from the emission region above the outer gap through synchrotron self-Compton (SSC) mechanism. Because those photons emitted into a given pulse phase come from different positions of the outer gap, and curvature radiation, synchrotron radiation and inverse Compton scattering depend on local quantities, e.g. curvature radius, $E_{||}$, B, n_{ph} etc., so it is likely that the radiation spectrum varies with phase. Here we assume that the distribution of secondary electrons/positrons in a steady state is

$$\frac{d^2N}{dE_e} = N_0(r)E_e^{-p} \quad \text{for } E_{min} \leq E_e \leq E_{max} \ , \tag{4}$$

where $N_0(r) \sim (l_{cur}n_{GJ}/E_{cur})$, $l_{cur}(r) = eE_{||}(r)c$ is the local power in curvature radiation produced by a single e^+/e^-, n_{GJ} is the local Goldreich-Julain number density. E_{cur} is the local curvature radiation energy, which is

$$E_{cur}(r) \approx 1.4 \cdot 10^9 f_0^{3/2} B_{12}^{3/4} P^{-7/4} \left(\frac{r}{R_L}\right)^{-1/8} \text{ eV} \ , \tag{5}$$

p is the spectral index. $E_{min} \sim mc^2$ and $E_{max} \approx E_{cur}/2$ are the minimum and maximum energy of the secondary pairs respectively.

As mentioned above, the photon emission comes from the emission region above the outer gap, so it is important to estimate the thickness of the emission region. CRZ have argued that the lower energy photons produced by the synchrotron radiation of the secondary pairs outside the outer gap cannot get into the outer gap, but they can collide with the high energy curvature photons at an angle $\sim f_0$. The number density of these low energy photons is $n_X \sim L_{gap}/ < r >^2 cE_X$, where $L_{gap} \sim f_0^3 L_{sd}$ and $L_{sd} \approx 3.85 \times 10^{31} P^{-4} B^2 12$ ergs/s. Therefore the mean free path which pairs are produced outside the outer gap is $\lambda_{X\gamma} \sim (n_X \sigma_p)^{-1}$, where $\sigma_p \sim \sigma_T/3$ is the pair production cross section and σ_T is Thompson cross section.Using $< r >\sim R_L/2$ and $E_X \approx (mc^2)^2/(f_0 E_{cur})$, we have

$$\lambda_{X\gamma} \sim 7 \times 10^{-3} f_0^{-4} \left(\frac{P}{0.1\text{s}}\right)^5 \left(\frac{E_{cur}}{10\text{GeV}}\right)^{-1} B_{12}^{-2} R_L \ . \tag{6}$$

Corresponding pitch angle is estimated by $\sin\beta(R_L/2) \sim \lambda_{X\gamma}/(R_L/2)$. Then we can calculate the photon spectrum of the synchrotron radiation of these secondary pairs and the spectrum of inverse Compton scattered photons.

Now we describe the photon emission patterns. We assume relativistic charged particles in the open zone radiate along the magnetic field lines in the corotating frame, then we project photon emission on (ζ, Φ) plane, where ζ is the polar angle from the rotation axis and Φ is the phase angle of rotation of the star. In the calculations, we have taken the effects of travel time and aberration into account, and chosen $\Phi = 0$ for emission in the (x, z) plane from the center of the star. Thereforc, for a given magnetic inclination angle, we can calculate the photon emission patterns of a given pulsar on the sky, then we can obtain the light curve for a specific viewing angle. In Appendix I, we describe an interactive animation program which can produce the light curves from outer gap observed at different viewing angle.

Model Results

We apply this three dimensional outer gap model to Crab-like pulsars: PSR B1509-58 and PSR B0540-69, and model the light curves and the spectra of optical, X-rays and γ-rays from these two pulsars. Our calculation procedure is as follow. For a Crab-like pulsar with a given magnetic inclination angle, at first, we trace the magnetic field lines by changing a_1 for a given f_0 (which can be approximated by Eq. 1), and require that the interval (in units of R_L) between the field line and the last closed field line (corresponding $a_1 = 1$) at radius of $R_L/2$ equal to f_0, so we find out a value of a_1 which defines approximately the upper boundary of the outer gap (we label it as a_1^u). Second we trace the magnetic field lines for different a_1 and find a value of a_1 (we label it as a_1^r) which the interval between the field lines with a_1^r and the field lines with $a_1 = a_1^u$ at radius of $R_L/2$ equals to $\sim \lambda_{X\gamma}/R_L$, where $\lambda_{X\gamma}$ is given by Eq.(6). After estimating a_1^u and a_1^r, we calculate the emission pattern of non-thermal photons produced in the emission region above the outer gap and variation of radial distances with pulse phase in this emission region for a given viewing angle. Finally we calculate the light curve and spectrum of this Crab-like pulsar. In our calculation, we assume that the spectral index of the secondary pairs is 2 (i.e. we have approximated the curvature radiation spectrum as a δ function). Although $\sin\beta(R_L)$ can be roughly estimated by using $\sin\beta(R_L) \sim \sqrt{2}\sin\beta(R_L/2)$ with $\sin\beta(R_L/2) \sim \lambda_{X\gamma}/(R_L/2)$, here we use it as a parameter.

PSR B1509-58

PSR B1509-58 is believed to be a Crab-like pulsar because its parameter B/P^2 is very close to that of the Crab pulsar. The observed data indicate that (i) the light curves of hard X-rays, low energy γ-rays have broad single peak, (ii) the X-ray pulse lags the radio by about 0.27 period, with no evidence for any energy dependence in the range 2 - 100 KeV (Rots et al. 1998), (iii) low energy γ-ray pulse lags the radio by 0.32 ± 0.02 (Ulmer et al. 1993) or ~ 0.3 (Rots et al. 1998), and (iv) medium energy γ-ray pulse lags the radio by 0.30 ± 0.06 (Carraminana et al. 1997). Rots et al. (1998) pointed out that the change in phase lags of X-rays and low energy γ-rays would be due to a gradual change in dispersion measure. According to our model, hard X-rays and low energy γ-rays have same origin, i.e. they are produced by synchrotron self-Compton radiation of secondary $e^{\pm}$ pairs of the outer gap. Therefore, the phase offsets of hard X-rays and low energy X-rays with respect to the radio pulse are the same.

In order to determine the emission pattern, we need to know the magnetic inclination angle and viewing angle. Unfortunately, we cannot get any constraints on two angles from radio observation. Therefore, we estimate these two angles by comparing the model pulse features with observed data, i.e. the phase offset with respect to radio pulse is ~ 0.3 and the pulse is a single broad peak. In our calculations, we find that model light curve with a magnetic inclination angle of about 65° and a viewing angle of 75° can compare with the observed light curves. These results are consistent with those estimated by Romani & Yadigaroglu

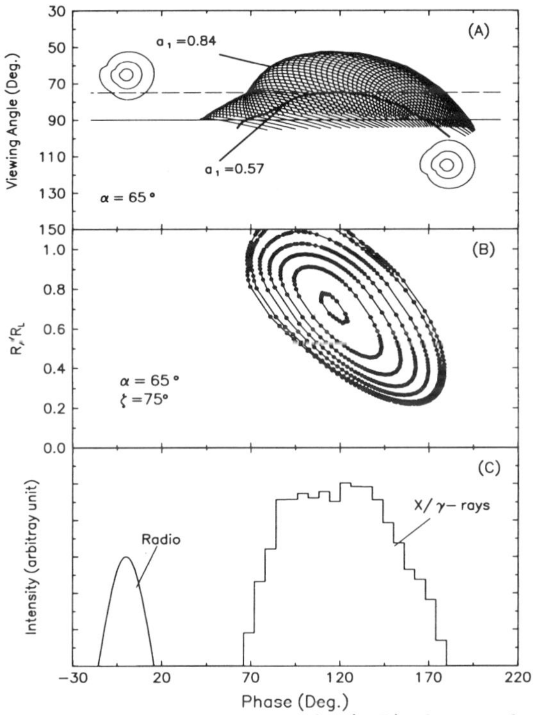

Figure 1 Emission projection onto the (ζ,Φ) plane and pulse profile for PSR B1509-58 ($\alpha = 65°$). Panel A: Photon emission from the pulsar as a function of phase, the solid curves indicate outward photon emission from a single outer gap respectively. The observed lines of sight at $\zeta = 75°$ and $\zeta = 90°$ (dashed and solid lines), and both polar caps (contours) are shown. Panel B: The variation of radial distance with pulse phase for the thickness of the emission region of $0.3R_L$. The curves represent trajectories for $a_1 = 0.84$, 0.82, 0.77, 0.72, 0.67, 0.62, and 0.57. Panel C: Pulse profile for PSR B1509-58. The model X-ray/γ-ray pulse at $\zeta = 75°$ is shown.

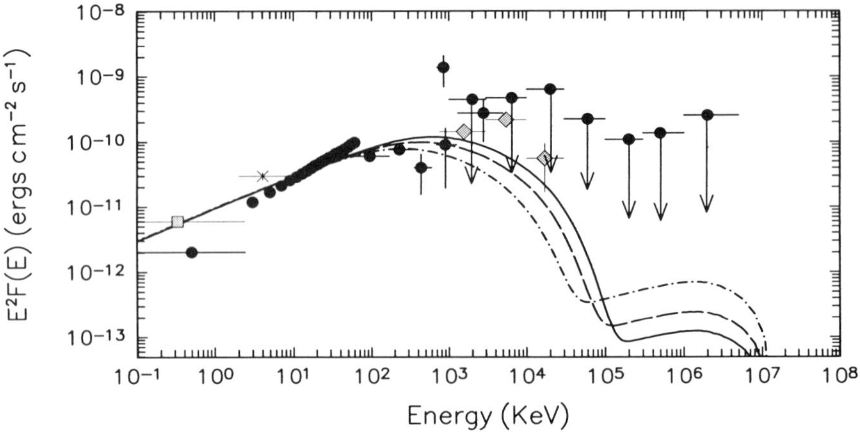

Figure 2 The comparison of model spectra with the observed data for PSR B1509-58. The solid, dashed and dot-dashed curves represent model results for $\sin(\beta(R_L)) = 0.7$, 0.5 and 0.3 respectively. We have used the observed data compiled by Thompson et al. (1999), ROSAT data (solid box) (Becker & Trümper 1997), ASCA data (open box) (Saito 1998) and COMPTEL data (solid diamonds) (Kuiper et al. 1999).

(1995). After determining the inclination and viewing angles, following our calculation procedure described above, we find that $a_1^u \sim 0.84$ for $f_0 \sim 0.135$ and $a_1^r \sim 0.57$ for $\lambda_{X\gamma}(R_L/2) \sim 0.1R_L$. Therefore, the non-thermal radiation is mainly produced in the emission region from $a_1^u = 0.84$ to $a_1^r = 0.57$. In panel A of Fig. 1, the high energy photon emission pattern of the emission region in (ζ, Φ) plane and radio caps are shown. Furthermore, we estimate the radial distances of the emission region which is the function of pulse phase and show them in panel B of Fig. 1. Using Eq. (6), we have $\sin\beta(R_L) \sim \sqrt{2}\sin\beta(R_L/2) \sim 0.3$, then we calculate the intensity of non-thermal photons from the emission region which is function of the pulse phase. Therefore, we obtain the light curves of the Crab-like pulsars once the inclination angle, viewing angle and the thickness of photon emission region are given. The light curves corresponding to the emission pattern of panel A of Fig. 1 are shown in panel C of 1, in which radio pulse is schematic. In this figure, we assume that the radio emission comes from polar cap, so its phase is zero.

Using the parameters given above, we calculate radiation spectrum from PSR B1509-58. The comparison of model spectrum with the observed one is shown in Fig. 2. It can be seen that a large pitch angle is needed in order to explain the observed data, especially in COMPTEL energy range.

PSR B0540-69

PSR B0540-69 is one of the youngest rotation-powered pulsar. Its energetics and age have been noted as similar to those of the Crab pulsar, but detailed comparison of pulse shape reveals some differences. In fact, X-ray/γ-ray emission from the Crab pulsar indicate two peaks with phase separation ~ 0.4. Based on following facts that: (i) it has broad X-ray pulse (the pulse width is about 0.35 phase) which is similar to that of PSR B1509-58, and (ii) both radio and X-ray pulses are observed and radio beam is $6.2^\circ P^{-1/2} \sim 28^\circ$, we expect that the viewing angle is far away from $\xi = 90^\circ$ and the magnetic inclination is not less than $\sim 40^\circ$. However, because the phase offset of X-rays with respect to radio pulse is not clear, it is difficult to estimate the values of the viewing angle and the magnetic inclination. With the constraints of γ-ray data, we choose that the magnetic inclination and the viewing angle are 50° and 66° respectively. Because of large r_{in} for small magnetic inclination, the extension of the outer gap is about 180° and corresponding r_{lim} is determined by the value of r_{in} at $\phi \sim 180^\circ$. For PSR B0540-69, we have $f_0 \sim 0.07$ and $\lambda_{X\gamma} \sim 0.04 R_L$ which give $a_1^u \sim 0.91$ and $a_1^r \sim 0.83$. The high energy photon emission pattern of the emission region from a_1^u to a_1^r in (ζ, Φ) plane and radio caps are shown in panel A of Fig. 3, where outgoing and incoming emission patterns are represents by solid curves and dashed curves respectively. Then we calculate the radial distances of the emission region which vary with the pulse phase. In panel B of of Fig. 3, we show the variation of the radial distance of the emission region with pulse phase. Finally, we calculate photon intensity from the emission region which is the function of the pulse phase and show the light curves corresponding to the emission pattern in panel C of Fig. 3, here we assume that the radio emission comes from polar cap, so its phase is zero and radio pulse is schematic.

We also model the spectrum of PSR B0540-69. The comparison with the observed data is shown in Fig. 4.

Discussion

We have used a three-dimensional outer gap model to calculate the observed light curves and the spectra of the Crab-like pulsars: PSR B1509-58 and PSR B0540-69. For Crab-like pulsars, the non-thermal radiation is from the finite region just above the outer gap through synchrotron self-Compton mechanism. The light curves and spectra depends on the inclination angle (α), viewing angle (ξ) and the thickness of emission region. In our calculation, we have estimated the a_1^u and a_1^r which mean that the intervals between the field lines with a_1^u and the

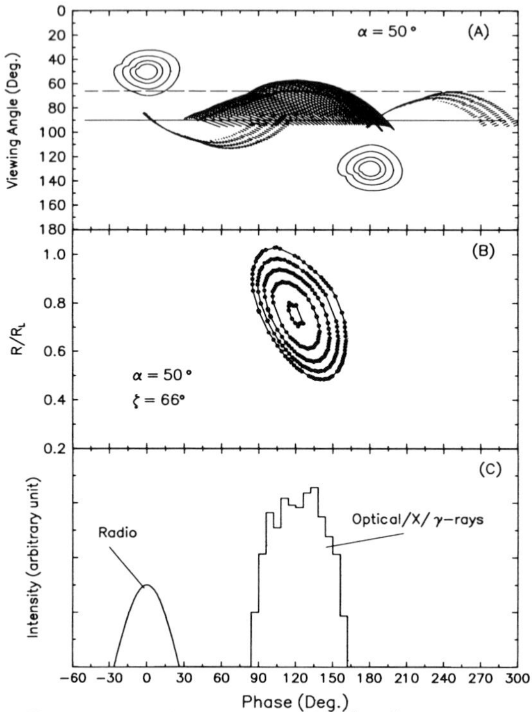

Figure 3 Emission projection onto the (ζ,Φ) plane and pulse profile for PSR B0540-69 ($\alpha = 50°$). Panel A: Photon emission from the pulsar as a function of phase, the solid and dashed curves indicate inward and outward photon emission from a single outer gap respectively. The observed lines of sight at $\zeta = 65°$ and $\zeta = 90°$ (dashed and solid lines), and both polar caps (contours) are shown. Panel B: The variation of radial distance with pulse phase for the thickness of the emission region of $0.1R_L$. The curves represent trajectories for $a_1 = 0.91$, 0.89, 0.87, 0.85, and 0.83. Panel C: Pulse profile for PSR B0540-69. The model X-ray/γ-ray pulse at $\zeta = 65°$ is shown.

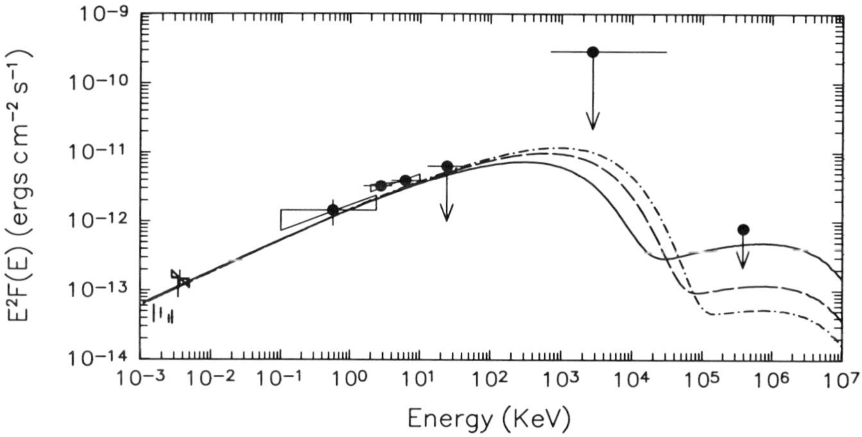

Figure 4 The comparison of model spectra with the observed data for PSR B0540-69. The solid, dashed and dot-dashed curves represent model results for $\sin(\beta(R_L)) =$ 0.05, 0.1 and 0.15 respectively. Observed data at optical waveband are taken from Middleditch et al. (1987), Hill et al. (1997). The data at ROSAT energy range, Beppsax energy range, COMPTEL and EGRET are taken from Finley et al. (1993), Mineo et al. (1999), Hermsen et al. (1994) and Thompson et al. (1994) respectively.

last closed field lines and between the field lines with a_1^u and the last closed field lines at $R_L/2$ are roughly equal to f_0 and the thickness of the emission region. Therefore the emission region is surrounded by the field lines with a_1^u and with a_1^r above the outer gap. In order to produce the light curve and spectrum which are consistent with the observed data, we have found out that the inclination and viewing angles are 65° and 75° respectively and mean thickness of the photon emission region is $\sim 0.1R_L$ for PSR B1509-58, and 50° and 66° respectively and $\sim 0.04R_L$ for PSR B0540-68.

Using CHR model, Cheng & Ding (1994) and Cheng & Wei (1995) have calculated its phase-averaged spectrum of PSR B1509-58 by assuming that it is Vela-like and Crab-like pulsars respectively. Generally, if the mean distance to the outer gap approaches the light cylinder radius, and then the smaller magnetic field makes the Vela-like mechanism more favorable, otherwise the Crab-like mechanism is more favorable. In their calculations, however, they did not consider the offset of the X-ray/γ-rays with respect to the radio. Romani & Yadigaroglu (1995) have considered the observed offset and concluded that $\alpha = 60°$ and $\xi \sim 75°$ based on the geometry of the outer gap, which is consistent with our result. For PSR B0540-69, Cheng & Wei (1995) have calculated the phase-averaged spectrum by assuming that it is a Crab-like pulsar, which is consistent with ours.

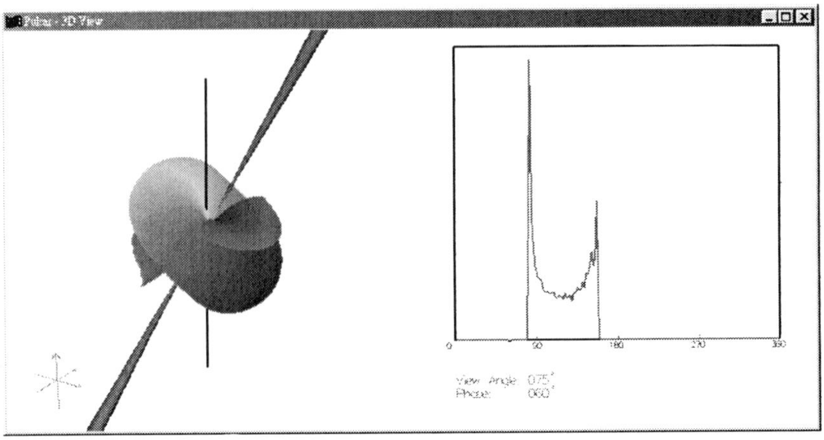

Figure 5 The snapshot of the animation program. The left panal shows the structure of the magnetosphere and the fan-shaped surfaces above the magnetosphere represent the emission regions. The right panel plots the light curve of corresponding parameters. The inclination angle is chosen to be 30o and the viewing angle is 75o. The current phase is 60o.

Acknowledgements

We thank R.W. Romani, M.A. Ruderman for useful discussion and suggestions, D.J. Thompson for providing us the observed data of multiwavelength energy spectrum of PSR B1509-58. This work is partially supported by an RGC grant of the Hong Kong Government and Natural Science Foundation of Yunnan Province.

Appendix I

An Interactive Animation Program for Simulating the Light Curves from the Outer Gap

The graphical library, OpenGL, is used for rendering the three dimensional structure of the pulsar to visualize the structure and the emission morphology of a pulsar. Firstly, the position of the pulsar is defined. The last closed field lines and the upper boundary of the outer gap is traced. Then, the relative position between the camera and the pulsar is defined and the viewport is set. The camera is revolving around the pulsar to simulate the rotation of the pulsar itself.

Fig. 5 is a snapshot of the program screen showing the structure of the magnetosphere and the light curve of the corresponding viewing position. The bean-shape volume represent the last closed field lines. The darker surface above is the upper boundary of the outer magnetospheric gap and the radiations are coming out from there along the magnetic field lines. The cones above the both magnetic poles represent the radio emission beam. Users can define the inclination angle α and the viewing angle

ζ (the polar angle between the rotation axis and the position vector of the camera, and hence defining the position of the observer). The light curve for the corresponding viewing position is immediately shown on the right panel. The small box on the light curve indicate the intensity and phase for a particular time.

The animation is place on the Internet in following address:

http://www.physics.hku.hk/~sfko/pulsar

Both the complete program and the output animated-gif movie are available.

References

Carraminana, A. et al. 1997, in AIP Conf. Proc. 410, Proceedings of the Fourth Compton Symposium, ed. C.D. Dermer, M.S. Strickman, & J.D. Kurfess (New York: AIP), 583
Cheng, K.S., Ho, C. & Ruderman, M.A., 1986a, ApJ, 300, 500 (CHR I)
Cheng, K.S., Ho, C. & Ruderman, M.A., 1986b, ApJ, 300, 522 (CHR II)
Cheng, K.S. & Ding, W.K.Y. 1994, ApJ, 431, 724
Cheng, K.S. & Wei, D.M. 1995, ApJ, 448, 218
Cheng, K.S. & Zhang, L., 1999, ApJ, 515, 337
Cheng, K.S., Ruderman, M. & Zhang, L. 1999, ApJ, accepted (CRZ)
Chiang, J. & Romani, R.W., 1992, ApJ, 400, 724
Chiang, J. & Romani, R.W., 1994, ApJ, 436, 754
Finley, J.P., Ögelman, H., Hasinger, G., & Trümper, J. 1993, ApJ, 410, 323
Hermsen, W. et al. 1994, ApJS, 92, 559
Hill, R.J. et al. 1997, ApJ, 486, L99
Ho, C. 1989,ApJ, 221, 523
Kuiper, L. et al. 1999, A&A, 351, 119
Middleditch, J. & Pennypacker, C.R. 1985, 1985, Nature, 313, 659
Middleditch, J., Pennypacker, C.R., & Burns, M.S. 1987, ApJ, 315, 142
Mineo, T. et al. 1999, A&A, 348, 519
Romani, R.W., 1996, ApJ, 470, 469
Romani, R.W. & Yadigaroglu, I.-A., 1995, ApJ, 438, 314
Rots, A.H. et al. 1998, ApJ, 501, 749
Saito, Y., 1998, Ph.D thesis
Thompson , D.J. et al., 1994, ApJ, 436, 229
Thompson , D.J. et al., 1999, ApJ, 516, 297
Ulmer, M.P. et al. 1993, ApJ, 417, 738
Zhang, L. & Cheng, K.S., 1997, ApJ, 487, 370 (ZC)

ANALYTICAL STUDIES ON THE STRUCTURE AND EMISSION OF THE SS 433 JETS

Noriaki SHIBAZAKI[1], Hajime INOUE[2] and Reiun HOSHI[3]
[1]*Department of Physics, Rikkyo University 3-34-1 Nishi-Ikebukuro, Tokyo 171-8501,*
[2]*The Institute of Space and Astronautical Science3-1-1 Yoshinodai, Sagamihara, Kanagawa 229-8510,*
[3]*Department of Physics, Rikkyo University 3-34-1 Nishi-Ikebukuro, Tokyo 171-8501*

Abstract We study the structure and emission of the SS 433 jets, especially in the inner and hotter portion inside the X-ray emitting region. We find that the jet beams in this portion are of two-temperature and emit a large amount of high energy gamma photons. Our analyses suggests the thick absorbing envelope to exist in the SS 433 system. Based on our results, we discuss the possible acceleration mechanisms for the SS 433 jets.

Keywords: Astrophysical jets, Stars: individual (SS 433), X-rays: binaries

1. INTRODUCTION

The galactic source SS 433 is well-known as an unique object exhibiting a jet phenomenon (for review, see Margon 1984 and Vermeulen 1989). The moving behavior of Balmer and He I lines has been interpreted successfully in terms of the two opposing jets precessing with a period of 164 days. The velocity of matter outflowing in the jet is found to be $0.26c$ and remarkably stable, where c is the speed of light. The SS 433 system is an eclipsing binary with the orbital period of 13 days. The system may consist of an early type star and an accretion disk around a compact object. The compact star may be a neutron star or a black hole, although the definite determination has been still waited.

SS 433 has attracted huge attention from the time of the discovery because of its peculiar behavior. A lot of observations have been conducted in the radio, optical and X-ray bands. Many theoretical studies have also been put forward. In spite of these enormous efforts the most fundamental problems on SS 433 still remain unsolved. Those are the

K.S. Cheng et al. (eds.), Stellar Astrophysics, 339–348.

nature of the compact star and the acceleration and collimation mechanisms of the jet beams.

Recently, the ASCA satellite has detected X-ray lines of various elements, such as Fe, Ni, Mg, Si, S and Ar, from the SS 433 jets (Kotani et al. 1996; Kotani et al. 1997). In the Fe case both the blue-shifted and red-shifted X-ray lines, which are emitted respectively from the approaching and receding jets, have been observed at the same time. Fitting the analytical jet model to these ASCA data, Kotani et al. (1996, 1997) have examined the properties of the X-ray emitting portion of the jet beams. They have derived the important constraints on the jet parameters such as the density, temperature and size of the X-ray emitting region and the mass outflow rate. Their analyses have also implied the X-ray absorbing gas to exist in the system.

We study the structure and emission of jet beams, especially in the inner and hotter portion inside the X-ray emitting region. Based on the results on the jet structure, we discuss the acceleration mechanisms for the SS 433 jets.

We present a simplified model for the SS 433 jets in section 2. We derive the analytic solution for the structure and emission of the X-ray jets in section 3. Deriving the analytic solution, we show that the jet structure is of two-temperature inside the X-ray emitting region in section 4. In the final section we discuss the absorbing envelope and the acceleration mechanisms based on our analytical results on the jet structure.

2. MODEL

The jet may be divided into several characteristic regions depending on the distance from the central engine, the inner region, the X-ray emitting regions, the optical and radio emission regions, and the outer region. Here, we focus on the X-ray emitting region and the hotter region inside of it. For simplicity, we consider the conical jets with a small opening angle of θ.

No observational evidence is found for the velocity variation along the jets in SS 433. This fact implies that the matter entrainment from the ambient medium and other braking processes are umimportant in the region of our interest. Hence, we assume the matter velocity constant, $v = 0.26c$, along the jets.

The mass outflow rate $\dot{M}$ in a jet is written as

$$\dot{M} = r^2 \Omega \rho v \quad , \tag{1}$$

where r is the distance from the apex of the cone or the central engine, ρ the density and Ω the solid angle equal to $\sim \pi(\theta/2)^2$.

The matter in the jets cools, expanding and emitting radiations. Writing the temperature and radiative loss rate respectively as T and Λ, we have for the energy equation

$$\frac{3}{2}\frac{k}{\mu m_H}\frac{dT}{dr} - \frac{kT}{\mu m_H}\frac{1}{\rho}\frac{d\rho}{dr} = -\frac{\Lambda}{v} \quad , \tag{2}$$

where m_H is the mass of hydrogen, k the Boltzmann constant and μ the mean molecular weight of matter. The second term on the left hand side of equation (2) and the term on the right hand side represent the adiabatic and radiative coolings, respectively.

We normalize physical quantities with those at a fiducial point r_0 ; $x = r/r_0$, $y = T/T_0$, $z = \rho/\rho_0$ and $f = \Lambda/\Lambda_0$. Here and in the following the subscript 0 is used to denote the physical quantities at $r = r_0$. Using these parameters, equations (1) and (2) reduce to the dimensionless equations,

$$z = \frac{1}{r^2} \tag{3}$$

and

$$\frac{dy}{dx} + \frac{4}{3}\frac{1}{x}y = -\xi_0 f \quad , \tag{4}$$

respectively. In equation (4) the frec parameter ξ is the ratio of the flow time t_f to the radiative cooling time t_c,

$$\xi = \frac{t_f}{t_c} \quad , \tag{5}$$

where t_f and t_c are defined respectively by

$$t_f = \frac{r}{v} \tag{6}$$

and

$$t_c = \frac{\frac{3}{2}\frac{kT}{\mu m_H}}{\Lambda} \quad . \tag{7}$$

Note that the flow time also expresses the adiabatic cooling time due to the plasma expansion. Equation (4) shows that the free parameter ξ_0 determines the property of solutions and hence the thermal structure of the jet.

3. ONE-TEMPERATURE JETS

We consider the region of the jet beams, where a typical temperature is $\sim$ 10 keV and X-rays are emitted. We assume that the jet is optically thin to X-ray photons, satisfying $\tau_e = \kappa_e \rho R \lesssim 1$, where τ_e is the optical thickness, κ_e is the electron scattering opacity and R is the jet radius. In this region the ion-electron scattering occurs so frequently that the electrons and ions have the same temperature.

3.1 ANALYTIC SOLUTIONS

When the radiative cooling rate is given by the thermal Bremsstrahlung emission,

$$\Lambda = \Lambda_{ff} = 5.7 \times 10^{20} \rho \sqrt{T} \quad \mathrm{ergs\,g^{-1} s^{-1}} \quad , \tag{8}$$

the normalized cooling rate is obtained as

$$f = z\sqrt{y} \quad . \tag{9}$$

Equation (4) together with equations (3) and (9) can be solved analytically to yield

$$\sqrt{y} = \frac{3}{2}\,\xi_0 \frac{1}{x} + \left(1 - \frac{3}{2}\,\xi_0\right) \frac{1}{x^{2/3}} \quad . \tag{10}$$

Note that the solution (10) tends to the adiabatic cooling formula, $y = 1/x^{4/3}$, as ξ_0 approaches to zero. If we intergrate the Bremsstrahlung emission along the jet, we have for the X-ray luminosity emitted from the region outside the distance x,

$$L(x) = r_0^3 \Omega \rho_0 \Lambda_{ff0}\, g(x) \quad \mathrm{ergs\,s^{-1}} \quad , \tag{11}$$

where

$$g(x) = \frac{3}{4}\,\xi_0 \frac{1}{x^2} + \frac{3}{5}\left(1 - \frac{3}{2}\,\xi_0\right) \frac{1}{x^{5/3}} \quad . \tag{12}$$

Adopting this solution at $r/r_0 \geq 1$, Kotani et al. (1996) have calculated the line X-ray emission from Fe, Ni and other elements and compared the numerical results with the ASCA observations for SS 433. A standard set of jet parameters they obtained from comparison are $kT_0 \sim 21$ keV, $\rho_0 \sim 4 \times 10^{-12}$ g cm^{-3}, $\dot{M} \sim 4 \times 10^{-6} M_\odot$ y^{-1}, and $\theta \sim 5°$. Here the fiducial point is chosen to be the base of the X ray emitting region of the jet. At the base of the X-ray emitting region the jet radius R_0 is $\sim 5 \times 10^{10}$ cm and the distance r_0 from the apex of the cone is calculated as $r_0 = R_0/(0.5\theta) \sim 1 \times 10^{12}$ cm. The length of the X-ray emitting region is determined as $\ell_X \sim 10^{13}$ cm. The optical thickness τ_{e0} and the parameter ξ_0 are also calculated as $\tau_{e0} \sim 0.08$ and $\xi_0 \sim 0.09$. Hence, in the X-ray emitting region the jet plasma is transparent and the temperature structure is determined mainly by the adiabatic cooling loss due to the plasma expansion.

3.2 JET INSIDE THE X-RAY EMITTING REGION

The jet radius R_0 of the X-ray emitting region and its distance r_0 from the apex of the cone are at least several orders of magnitude larger than the size of the central engine that drives the jet, a neutron star or a black hole. Although Kotani et al. (1996, 1997) have truncated the inner portion of the jet, it is quite natural to assume that the jet beam extends more down to the vicinity of the central engine. In the following let us examine the property of the jet expected to exist inside the X-ray emitting region.

We apply the solution (10) to the inner region adjacent to the X-ray emitting region. Following equations (3) and (10), the temperature and density of the jet increase with decreasing distance. Hard X and soft gamma rays are emitted from the jet. The energy transfer from ions to electrons through the coulomb scattering becomes less efficient at higher temperatures, whereas electrons lose energy promptly radiating X and gamma rays. The insufficient energy transfer makes the ion and electron temperatures different at $T \gtrsim 2 \times 10^9$ K (see the next section for detail). The jet reaches the critical temperature $T_1 \sim 2 \times 10^9$ K at the distance $r_1 \sim 2.5 \times 10^{11}$ cm. Here and in the following we denote the physical parameters at the innermost region of the one-temperature jet by the subscript 1. The plasma density, the optical thickness and the parameter ξ at $r \sim r_1$ are $\rho_1 \sim 8 \times 10^{-11}$ g cm^{-3}, $\tau_{e1} \sim 0.3$, and $\xi_1 \sim 0.1$, respectively. Hence, the jet is optically thin. The relativistic effect, if included, will enhance the radiative emission rate and the parameter

ξ will become larger. We find that the radiative loss as well as the adiabatic loss also contribute significantly to the plasma cooling near the innermost region of the one-temperature jet. The luminosity of X and gamma rays emitted from the region at $r \gtrsim r_1$ is calculated from equation (12) as $L(x_1) \sim 9 \times 10^{36}$ ergs s^{-1}. We note that this luminosity exceeds the observed X-ray luminosity by approximately an order of magnitude.

4. TWO-TEMPERATURE JETS

We have the two-temperature jet inside the distance r_1, where the ion temperature is higher than the electron temperature. Ions transfer energy to electrons through the ion-electron scattering, which in turn lose energy by the radiative emission. The distribution of ion temperature along the jet beams can also be described by equation (4), taking $\mu = 1$ and substituting the ion energy loss rate for the radiative loss rate. The ion energy loss rate is given approximately by

$$\Lambda = \Lambda_{ie} = \frac{3}{2}\frac{kT}{m_H}\nu_E \quad , \tag{13}$$

where T now expresses the ion temperature and ν_E is the ion-electron collision frequency. The ion-electron collision frequency is written as

$$\nu_E = 6.5 \times 10^{21}\frac{\rho}{T_e^{3/2}} \quad \mathrm{s}^{-1} \quad , \tag{14}$$

where T_e is the electron temperature. We adopt the relativistic Bremsstrahlung emission with the Compton amplification and approximate the emissivity by

$$\Lambda_{ff} = 4.9 \times 10^{16}\eta\rho T_e \quad \mathrm{ergs\ g^{-1}s^{-1}} \quad , \tag{15}$$

where η is the Compton amplification factor. Equation (15) can also be used approximately in the optically thick case unless the optical thickness is substantially larger than one ($\tau_e \gg 1$). The electron temperature is determined from the balance between the energy acquired from ions and the energy lost to radiation and expressed as

$$T_e = 3.8 \times 10^5 \eta^{-2/5} T^{2/5} \quad \mathrm{K} \quad . \tag{16}$$

If we equate the ion and electron temperatures in equation (16), we can derive the critical temperature T_1 above which the plasma is in the two-temperature regime,

$$T_1 \sim 2 \times 10^9 \eta^{-2/3} \quad \text{K} \quad . \tag{17}$$

We choose the position of $r \sim r_1$ as the new fiducial point and normalize the physical quantities with those at $r = r_1$; $x = r/r_1$, $y = T/T_1$, and $z = \rho/\rho_1$. The dimensionless equation (4) with the free parameter ξ_1 replaced for ξ_0 is again derived from the ion energy equation. The parameter ξ is now re-defined as the ratio of the flow time to the ion energy loss time or the ion-electron scattering time given by

$$t_c = \nu_E^{-1} \quad . \tag{18}$$

Furthermore, the ion energy loss rate is written in the dimensionless form as

$$f = zy^{2/5} \quad . \tag{19}$$

We solve equation (4) together with equations (3) and (19) and derive the analytic solution,

$$y^{3/5} = 3\xi_1 \frac{1}{x} + (1 - 3\xi_1)\frac{1}{x^{4/5}} \quad . \tag{20}$$

Using equation (15) and integrating the high energy emission along the jet, we obtain the luminosity of high energy photons emitted outside the distance x, which is approximated by

$$L(x) \sim \frac{3}{5} \, r_1^3 \Omega \rho_1 \Lambda_{ff1} \frac{1}{x^{5/3}} \quad . \tag{21}$$

As we proceed inwards along the jet, the ion and electron temperatures increase following equations (20) and (16). Moreover, the ion temperature increases faster than the electron temperature, rendering the jet of two-temperature. As we move furthermore inwards, we reach the sonic point where the ion sound velocity becomes equal to the flow velocity. Although derived by ignoring the pressure gradient force and by assuming the constant flow velocity, the above solution may be applicable down to the sonic point as long as the approximate estimates of the jet parameters are concerned. Using equation (20) together with $\xi_1 \sim 1$, the location of the sonic point is estimated as $r_s \sim 2 \times 10^{10}$ cm ($x \sim 0.07$), which is very far from the central engine. Here the subscript s denotes the sonic point and the compton enhancement factor is set as $\eta \sim 1$. The ion and electron temperatures, density and jet radius at the sonic point are calculated as $T_s \sim 4 \times 10^{11}$ K, $T_{es} \sim 2 \times 10^{10}$ K, and $R_s \sim 8 \times 10^8$ cm, respectively. The ion temperature is approximately an order of magnitude higher than the electron temperature. The parameter ξ_s is ~ 0.6. As seen from the values of ξ_1 and ξ_s, the radiative cooling as well as the adiabatic cooling contributes significantly to the energy loss of plasma in the two-temperature region of the jet. We find that the luminosity of the high energy photons emitted outside the sonic point amounts to $\sim 4 \times 10^{39}$ ergs s^{-1}, which is approximately comparable to the kinetic luminosity of the outflowing matter in the jet beam.

We note that the electron-positron pairs and appropriate compton amplification in addition to the pressure gradient force should be included in the calculation in order to have more accurate estimates on the jet property around the sonic point. If included, these effects would place the sonic point still farther from the central engine.

5. DISCUSSION AND CONCLUSIONS

The jet beams in SS 433 may originate near the compact object, or more specifically in the inner region of the accretion disk. The X-ray emitting region of the jet is likely to be located at the distance of $\gtrsim 10^{12}$ cm from the compact object. The jet beams must be accelerated and collimated to reach the terminal velocity within the distance of $\sim 10^{12}$ cm.

We show in the previous section that the portions of jets inside the X-ray emitting region are very hot and emit a huge amount of gamma-rays. No observational evidence, however, has not been found for an emission of such high energy photons. Hence, we conclude that the inner and hotter portions of the jets are hidden by the surrounding matter, most probably by the thick envelope and accretion disk. Note that such an envelope, or a dense outflowing atomosphere which envelops the system, is also suggested from the substantial fluctuations of the optical magnitudes and the infrared spectrum (Vermeulen 1989; Band 1987).

The mechanisms of acceleration and collimation for the jet beams in SS 433 are poorly understood. As shown in the section 4, the sonic point is located far from the central and compact object. The terminal velocity of the jet beams exceeds the free fall velocity due to the gravity of the compact object at the sonic point approximately by two orders of magnitude. Hence, the acceleration by the gas pressure such as in the solar wind is ruled out for the case of the SS 433 jets. The radiative acceleration is also raised as a possible mechanism. Our calculations show that the inner and hotter portions of the jets are optically thick along the flow direction. Photons cannot get into the jet and impart momentun to the matter. Hence, the radiative acceleration is also ruled out.

Promising is the magnetically driven jet from the accretion disk (Kudoh & Shibata 1997). The jet is accelerated by the centrifugal force and magnetic pressure. In the magnetically driven jets the terminal velocity of the conical beams is of the order of the Keplerian velocity at the radius of the accretion disk, where the mass outflow begins. Consider a neutron star as a compact object. Then, the terminal velocity of the SS 433 jets indicates that the footpoint of the jet may be within several stellar radii from the star. Meanwhile, the sonic point is quite far at $\sim 10^4$ stellar radii. Note that the ion temperature at the sonic point is close to the maximum reached near the neutron star. If the shape of the jet is conical, the outflowing matter suffers from the sigificant cooling even due only to the adiabatic expansion and hence the temperature is unlikely to be kept so hot up to the sonic point. The same problem also arises even if the compact object is a black hole of the stellar size mass.. Our analyses on the jet temperature favor the columnar rather than conical type for the jet shape inside the sonic point. The magnetic acceleration needs to be studied together with the shape and temperature structure in order to understand the behavior of the SS 433 jet properly.

The collimation of the jet beams is another important issue in the study of the astrophysical jet. The magnetic confinement is one possibility. Our studies show that the magnetic fields, which yield the magnetic pressure comparable to the gas pressure at the sonic point, are of the order of 10^6 G. This field strength may also be used to study and set constraints on the properties of the magnetically driven jets.

References

Band D. L. 1987, Publ. Astron. Soc. Pacific 99, 1269

Kotani T., Kawai N., Matsuoka M., Brinkmann W. 1996, PASJ 48, 619

Kotani et al. 1997, in 4th Compton Symposium on Gamma-Ray Astronomy and Astrophysics

Kudoh T., Shibata K. 1997, ApJ 474, 362

Margon B. 1984, Ann. Rev. Astron. Astrophys. 22, 507

Vermeulen R. C. 1989, Ph. D Thesis, University of Leiden

MAGNETICALLY DRIVEN WARPING, PRECESSION AND RESONANCES IN ACCRETION DISKS

Dong Lai
Center for Radiophysics and Space Research, Cornell University, Ithaca, NY 14853, USA
dong@spacenet.tn.cornell.edu

1. INTRODUCTION

Interaction between an accretion disk and a magnetic star lies at the heart of the physics of a variety of astrophysical systems, including accreting neutron stars, white dwarfs and pre-main-sequence stars (e.g., Frank et al. 1992). The basic picture of disk–magnetosphere interaction was first outlined by Pringle & Rees (1972), following the discovery of accretion-powered X-ray pulsars. These are rotating, highly magnetized ($B \sim 10^{12}$ G) neutron stars (NSs) that accrete material from a companion star, either directly from a stellar wind, or in the form of an accretion disk. The strong magnetic field disrupts the accretion flow at the magnetospheric boundary (typically at a few hundreds NS radii), and channels the plasma onto the polar caps of the NS. The magnetosphere boundary is located where the magnetic and plasma stresses balance, $r_m = \eta\,\mu^{4/7}(GM\dot{M}^2)^{-1/7}$, where M and μ are the mass and magnetic moment of the central star, $\dot{M}$ is the mass accretion rate, η is a dimensionless constant of order unity. In low-mass X-ray binaries containing weakly magnetized ($B \sim 10^8$ G) NSs, the magnetosphere lies close to the stellar surface. In this case, complex field topology and general relativity can affect the determination of the magnetosphere boundary (e.g., Lai 1998). Similar magnetosphere–disk interaction also occurs in T Tauri stars (e.g., Hartmann 1998), where the stellar magnetic field ($B \sim 10^3$ G) strongly affects the accretion flow, as well as in certain class (the so-called DQ Her stars or intermediate polars) of magnetic CVs.

Because of its intrinsic importance for a wide range of astrophysical systems, disk accretion onto magnetic stars has been subjected to many

K.S. Cheng et al. (eds.), Stellar Astrophysics, 349–353.

theoretical studies (see references in Lai 1999). Most previous papers have assumed that the stellar magnetic dipole axis is aligned with the spin axis. Given the complexity of the problem, the aligned model is certainly worth investigating, but it is obviously unrealistic. Moreover, essentially all previous authors have assumed that the spin is aligned with the disk angular momentum — this might appear to be a valid assumption, as it is natural to think that the star attains its spin angular momentum from the accretion disk. However, we have recently shown (Lai 1999) that under quite general conditions, the stellar magnetic field can induce warping in the inner disk and make the disk precess around the stellar spin axis.

2. NEW MAGNETIC EFFECTS IN DISKS

Lai (1999) shows that the inner region of the accretion disk onto a rotating magnetized central star is subjected to magnetic torques which induce warping and precession of the disk. The origin of these torques lies in the interaction between the surface current on the disk and the horizontal magnetic field (parallel to the disk) produced by the inclined magnetic dipole. The key physics can be most easily understood by considering two extreme situations: (1) If the disk is diamagnetic (perfectly conducting) so that the vertical stellar field (perpendicular to the disk) cannot penetrate, an azimuthal screening surface current will be induced on the disk. This current interacts with the radial magnetic field from the star, and a vertical magnetic force results. The uneven distribution of the force induces a net torque acting on the disk, making it precess around the spin axis. This is the *magnetic precessional torque.* (2) If there is a vertical field threading the disk, it will be twisted by the disk rotation to produce a discontinuous azimuthal field ΔB_ϕ across the disk surface, and a radial surface current results. The interaction between this current and the stellar B_ϕ gives rise to a vertical force, and the resulting torque tends to misalign the disk normal vector and the spin axis. This is the *magnetic warping torque.* In more realistic disk models, we expect both the precessional torque and the warping torque to exist.

Since the magnetic torque drives the disk tilt, while the viscosity reduces the tilt, one can derive the criterion for the warping instability. Roughly speaking, the disk warp can grow if the timescale associated with the warping torque is shorter than the viscous time r^2/ν_2. Since the warping torque is a steep function of r, the warping instability occurs only inside a critical radius r_w. Our analysis (Lai 1999) indicates that r_w is typically a few times the canonical Alfv́en radius (the magnetosphere boundary). Thus, as the disk approaches the magnetosphere, its normal vector $\hat{l}$ will tend to be tilted with respect to the stellar spin even if at large radii it is aligned with the spin axis. Also of interest is the "*Magnetic Bardeen-Petterson Effect*". Because of the magnetic precessional torque, the tilted disk will be driven into differential precession (with the precession rate dependent on r). By analogy with the well-known Bardeen-Petterson effect (i.e., the inner region of an accretion disk undergoing Lense-Thirring precession around a rotating black hole tends to align itself with the equatorial plane of the black hole; see Bardeen & Petterson 1975), we expect that the magnetically driven precession also tends to damp the tilt of the inner disk through the action of viscosity. However, since this effect is indirect, the precession-induced alignment will be overwhelmed by the warping instability.

For the general magnetic field–disk geometry, the vertical magnetic force on a disk element varies with the stellar rotation period. This gives rise to a number of *vertical resonances* in the disk. Similarly. there exist *epicyclic resonances* due to the time-dependent radial magnetic force. Although the force expressions are model-dependent, the existence of the resonances appears to be inevitable. These magnetically driven resonances are somewhat similar to the corotation resonance and Lindblad resonances in gravitational systems. The resonances may act as an extra source (in addition to the non-resonant precessional and warping torques discussed above) for generating bending waves and spiral waves in the disk. Near the resonances, fluid elements undergo large out-of-plane and radial excursions, which may lead to thickening of the disk. This may be analogous to the Lorentz resonances (which occur when charged particles move around a rotating magnetic field) in the jovian ring (e.g., Schaffer & Burns 1992). However, because of the fluid nature of the disk, the resonances may not lead to sharp edges in the disk.

3. APPLICATIONS

The magnetically driven warping instability and precession may be related to a number of observational puzzles (for more details, see Lai 1999).

(i) Spin evolution of accreting X-ray pulsars: Recent long-term, continuous monitoring of X-ray pulsars with the BATSE instrument on the Compton Gamma Ray Observatory (CGRO) has revealed a number of puzzling behaviors of the spins of these objects (see Bildsten et al. 1997 and references therein). Several well-measured disk-fed systems (e.g., Cen X-3, GX 1+4 and 4U 1626-67) display sudden transitions between episodes of steady spin-up and spin-down, with the absolute values of spin torques approximately equal (to within a factor of a few).

The transition timescale ranges from days to years. It is likely that the magnetically driven disk warping may be a crucial ingredient in determining the spin behaviors of accreting X-ray pulsars. With the magnetic warping torque, the perpendicular state is an "attractor". The observed sign switching of $\dot{\omega}_s$ (spin derivative) in several X-ray pulsars may be associated with the "wandering" of the inner disk around this "attractor". Rough estimate based on the magnetic torque indicates that the switching timescale (which depends on the stellar field strength, the disk parameters and geometry) ranges from days to years, in agreement with observations.

(ii) Quasi-periodic oscillations in low-mass X-ray binaries: Rapid variability in low-mass X-ray binaries, containing weakly magnetized ($B \sim 10^8$ G) neutron stars, has been studied since the discovery of the so-called horizontal-branch oscillations (HBOs) (see van der Klis 1998). The HBOs are quasi-periodic oscillations (QPOs) (with Q-value $\nu/\Delta\nu$ of order a few) which manifest as broad Lorentzian peaks in the X-ray power spectra with centroid frequencies in the range of 15–60 Hz which are positively correlated with the inferred mass accretion rate. Stella and Vietri (1998) suggested that HBOs (and other low-frequency QPOs) are associated with Lense-Thirring precession of the inner accretion disk around the rotating NS. For this interpretation to be viable, the inner disk must be tilted with respect to the stellar spin axis. Since the Bardeen-Petterson effect tends to keep the inner region of the disk (typically within 100–1000 Schwarzschild radii) co-planar with the star and radiation-driven warping is only effective at large disk radii (Pringle 1996), another mechanism to drive warping in the inner disk is needed. The magnetic warping torque provides a natural source for inducing disk tilt. Moreover, the magnetically driven (retrograde) precession rate is not negligible compared to the Lense-Thirring precession rate, and will contribute to the total precession.

Other possible applications include **(iii) Super-orbital variabilities in X-ray binaries** as a result of warped, precessing disks, and **(iv) Photometric period variations of T Tauri stars**.

Acknowledgements

This work is supported by NASA grant NAG 5-8356 and 5-8484, and by a fellowship from the Alfred P. Sloan foundation.

References

Bardeen, J. M., & Petterson, J. A. 1975, ApJ, 195, L65.
Bildsten, L., et al. 1997, ApJS, 113, 367.

Frank, J., King, A., & Raine, D. 1992, Accretion Power in Astrophysics (Cambridge Univ. Press).
Hartmann, L. 1998, Accretion Processes in Star Formation (Cambridge Univ. Press).
Lai, D. 1998, ApJ, 502, 721.
Lai, D. 1999, ApJ, 524, 1030.
Pringle, J. E. 1996, MNRAS, 281, 857.
Pringle, J. E., & Rees, M. J. 1972, A&A, 21, 1.
Schaffer, L., & Burns, J. A. 1992, ICARUS, 96, 65.
Stella, L., & Vietri, M. 1998, ApJ, 492, L59.
Van der Klis, M. 1998, in "The Many Faces of Neutron Stars" (Proc. NATO ASI).

WHAT DO γ-RAY BURSTS LOOK LIKE?

T. Lu
Department of Astronomy, Nanjing University, Nanjing 210093, China

Abstract There have been great and rapid progresses in the field of γ-ray bursts (denoted as GRBs) since BeppoSAX and other telescopes discovered their afterglows in 1997. Here, we will first give a brief review on the observational facts of GRBs and direct understanding from these facts, which lead to the standard fireball model. The dynamical evolution of the fireball is discussed, especially a generic model is proposed to describe the whole dynamical evolution of GRB remnant from highly radiative to adiabatic, and from ultra-relativistic to non-relativistic phase. Then, Various deviations from the standard model are discussed to give new information about GRBs and their environment. In order to relax the energy crisis, the beaming effects and their possible observational evidences are also discussed in GRB's radiations.

Keywords: γ ray: bursts - shock waves - ISM: jets and outflows

1. INTRODUCTION

γ-ray burst (or shortly, GRB) is an astronomical phenomenon of short duration enhancement of γ-rays from cosmic space. It was discovered by R.W. Klebesadel et al. in 1967 and published in 1973. There have been more than 2000 GRBs discovered now. If there are appropriate satellites in the sky, people could discover one or two GRBs everyday in average. However, they are still among the most mysterious astronomical objects even at present time (Mészáros 1999; Piran 1999a). The difficulty encountered in this field lies in that GRBs occur at random, people can not prepare to observe in advance, they are too short in duration to be studied in detail, they were observed only in γ-ray band which has very low precision in localization and could not be identified or associated with other objects of known distances. Without the knowledge of distances, no serious studies could be made in astrophysics.

In early 1997, the Italian-Dutch Satellite, BeppoSAX, brought the great break-through by rapid and accurate GRB localization and thus

K.S. Cheng et al. (eds.), Stellar Astrophysics, 355–368.

provided a small arcmin (or even smaller) error box. With so small an error box, it identified an X-ray counter-part (now known as X-ray afterglow) of GRB970228 even 8 hours after the γ-ray trigger (Costa et al. 1997). Several hours later, its optical afterglow was also observed (Groot et al. 1997; van Paradijs et al. 1997). Since then, BeppoSAX has observed more than 20 GRBs, of which almost all exhibited X-ray afterglows. Based on the precise localization, many telescopes observed about a dozen optical afterglows and about ten radio afterglows. Up to now, people have observed host galaxies of more than ten GRBs with large red-shifts, showing them definitely at cosmological distances. These great discoveries lead to rapid developments. A lot of questions have now been clarified. However, compared with GRB itself, afterglow appears to be simpler and has been known much better. In contrast, the GRB itself, especially its energy source and origin, still keeps to be mysterious.

The main observational facts of GRBs are as follows (Piran 1999a):

Temporal properties: The GRB duration (T) is very short, usually only a few seconds or tens of seconds, or occasionally as long as a few tens of minutes or as short as a few milli-seconds. Their time profiles are diverse, some may be simply shaped, others complicated, multi-peaked. There seems to be a roughly bimodal distribution of long bursts of $T \geq 2$ s and short bursts of $T \leq 2$ s. The time scales of variability (δT), especially their rising time, may be as short as only milli-seconds or even sub-milli-seconds. Typically, $\delta T \sim 10^{-2}T$.

Spectral properties: The photon energy radiated in GRB is typically in the range of tens keV to a few MeV. However, high energy tail up to GeV or even higher than 10 GeV does exist. The spectra are definitely non-thermal and can usually be fitted by power law $N(E)dE \propto E^{-\alpha}dE$ (or break power law) with index α within about 1.8 to 2.0. The γ-ray fluences are typically in the range of $(0.1 - 10) \times 10^{-6}$ ergs/cm^2.

Spatial distribution: After the launch of the CGRO (Compton Gamma-Ray Observatory) Satellite in 1991, the Burst and Transient Source Experiment (BATSE) showed clearly that the spatial distribution of GRB sources is highly isotropic with almost zero dipole and quadrupole components (Meegan et al. 1992). This distribution favors GRBs at cosmological distances at least statistically. However, GRBs at extended dark halo of our Galaxy could also explain this feature. This led to a great debate between Galactic origin and cosmological origin.

Afterglows: Afterglows are counterparts of GRBs at wave bands other than γ-rays, may be in X-ray, optical, or even radio bands. They are variable, typically decaying according to power laws: $F_\nu \propto t^{-\alpha}$ ($\nu =$ X, optical, ...) with $\alpha = 1.1 - 1.6$ for X-ray, $\alpha = 1.1 - 2.1$ for optical band. X-ray afterglows can last days or even weeks; optical afterglows and radio afterglows months. The most important discovery is that many afterglows show their host galaxies being definitely at cosmological distances (with large red-shifts up to $Z = 3.4$ or even 5). Thus, the debate is settled down, GRBs are at cosmological distances, they should be the most energetic events ever known since the Big Bang.

2. THE STANDARD FIREBALL SHOCK MODEL

Stellar level event: The variability time scale is usually very short. Let $\delta T \sim$ ms, then, the space scale of the initial source, $R_\mathrm{i} < c\delta T \sim 3 \times 10^2$km. Hence, even for black hole, considering $R = 2\mathrm{G}M/c^2$, we have $M \leq \frac{\mathrm{c}^3 \delta T}{2\mathrm{G}} \sim 10^2\ \mathrm{M}_\odot$. If the burster is not a black hole, its mass should be much smaller. Thus, we can conclude that the GRB should be **a stellar phenomenon** and the burster should be **a compact stellar object** which may be related with neutron star (or strange star) or stellar black hole.

Fireball: From the measured fluence F and the measured distance D, if emission is isotropic, we can calculate the total radiated energy to be $E_0 = F(4\pi D^2) \approx 10^{51}(F/(10^{-6}\mathrm{ergs/cm}^2))(D/(3\mathrm{Gpc}))^2$. Thus, very large energy (10^{51} ergs) is initially contained in a small volume of $(4/3)\pi R_i^3 \sim 1 \times 10^{23}$ cm^3. This should be inevitably a fireball, of which the optical depth for $\gamma\gamma \longrightarrow \mathrm{e}^+\mathrm{e}^-$, $\tau_{\gamma\gamma}$, is very large. Consider a typical burst with an observed fluence F, at a distance D, with a temporal variability time scale δT, its average optical depth can be written as:

$$\tau_{\gamma\gamma} = \frac{f_\mathrm{p}\sigma_\mathrm{T} F D^2}{R_\mathrm{i}^2 m_\mathrm{e} c^2} \sim 10^{17} f_\mathrm{p} \left(\frac{F}{10^{-6}\mathrm{ergs/cm}^2}\right)\left(\frac{D}{3\ \mathrm{Gpc}}\right)^2\left(\frac{\delta T}{1\ \mathrm{ms}}\right)^{-2}, \quad (1)$$

where f_p denotes the fraction of photon pairs satisfying $\sqrt{E_1 E_2} > \mathrm{m_e\ c^2}$.

For so large an optical depth, there seem to appear two serious difficulties. First, the radiation in an optically thick case should be thermal, while the observed radiation is definitely non-thermal. Second, high energy photons should be easily converted into $\mathrm{e}^+\mathrm{e}^-$ pairs, while the observed high energy tail indicates that this convertion has not happened. However, it is very interesting to note that just such a large optical depth paves the way to solve both of them.

Compactness problem: In fact, the luminosity of the thermal radiation, according to the Stefan-Boltzmann law, should be proportional to the surface of the fireball which is initially so small that the thermal radiation can not be observed. However, just due to the large optical depth, the radiation pressure should be very high and could accelerate the fireball expansion to become ultra-relativistic with a large Lorentz factor γ. After expanding to a large enough distance, it may be getting optically thin. At this time, the non-thermal γ-ray bursts can be observed. Does such a large distance contradict the compactness relation $R_{\rm i} \leq c\delta T$ compared with the milli-second variabilities? To answer this question, let us first note that this relation holds only for non-relativistic (rest) object with $R_{\rm i}$ denoting its space scale. For an ultra-relativistically expanding fireball, the compactness relation should be relaxed to

$$R_{\rm e} \leq \gamma^2 c\delta T, \tag{2}$$

here $R_{\rm e}$ is the space scale of the expanding fireball with Lorentz factor γ. Considering two photons we observed at two different times apart by δT, as the emitting region is moving towards the observer with a Lorentz factor $\gamma \gg 1$, the second photon should be emitted at a far nearer place than the first one. This gives effectively short time variabilities and leads to the additional factor γ^2 appearing in the above compactness relation.

The factor $f_{\rm p}$ in the optical depth $\tau_{\gamma\gamma}$ also sensitively depends on the ultra-relativistic expansion of the fireball. As for this case, the observed photons are blue-shifted, in the comoving frame, their energy should be lower by a factor of γ, and fewer photons will have sufficient energy to produce pairs. This gives a factor depending on spectral index α, namely a factor of $\gamma^{2\alpha}$ in $\tau_{\gamma\gamma}$.

Ultra-relativistic expansion: Therefore, the optical depth $\tau_{\gamma\gamma}$ will decrease by a factor of $\gamma^{4+2\alpha}$ for the ultra-relativistically expanding fireball (Goodman 1986; Paczyński 1986; Piran 1999a; Krolik & Pier 1991):

$$\tau_{\gamma\gamma} = \frac{f_{\rm p}}{\gamma^{2\alpha}} \frac{\sigma_{\rm T} F D^2}{R_{\rm e}^2 m_{\rm e} c^2} \approx \frac{10^{17}}{\gamma^{(4+2\alpha)}} f_{\rm p} \left(\frac{F}{10^{-6}{\rm ergs/cm^2}}\right) \left(\frac{D}{3\ {\rm Gpc}}\right)^2 \left(\frac{\delta T}{1\ {\rm ms}}\right)^{-2}. \tag{3}$$

Note, the spectral index α is approximately 2, we will have $\tau_{\gamma\gamma} < 1$ for $\gamma > 10^{17/(4+2\alpha)} \sim 10^2$. Thus, in order for the fireball to become optically thin, as required by the observed non-thermal spectra of γ-ray bursts, its expanding speed should be ultra-relativistic with Lorentz factor

$$\gamma >\sim 10^2.$$

This is a very important character for GRBs, which limits the baryonic mass contained in the fireball seriously. If the initial energy is E_0, then the baryonic mass M should be less than

$$E_0/(c^2\gamma) \leq 10^{-5} {\rm M}_\odot (E_0/(2 \times 10^{51}{\rm ergs})), \tag{4}$$

otherwise, the initial energy can not be converted to the kinetic energy of the bulk motion of baryons with such a high Lorentz factor. Most models related with neutron stars contain baryonic mass much higher than this limit. This is the famous problem named as "baryon contamination".

It is worthwhile to note that this very condition $\gamma >\sim 10^2$ can also explain the existence of the high energy tail in the GRB spectra, as the observed high energy photons should be only low energy photons in the frame of emitting region, they are not energetic enough to be converted into e^+e^- pairs.

Internal-external shock: What is the radiation mechanism in the fireball model? The fireball expansion has successfully made a conversion of the initial internal energy into the bulk kinetic energy of the expanding ejecta. However, this is the kinetic energy of the associated protons, not the photons. We should have another mechanism to produce radiation, otherwise, even after the fireball becoming optically thin, the γ-ray bursts can not be observed. Fortunately, the shocks described below can do such a job.

The fireball can be regarded as roughly homogeneous in its local rest frame, but due to the Lorentz contraction, it looks like a shell (ejecta) with width of the initial size of the fireball. As the shell collides with inter-stellar medium (ISM), shocks will be produced (Rees & Mészáros 1992; Katz, J., 1994; Sari & Piran 1995; Mitra 1998). This is usually called as external shocks. Relativistic electrons that have been accelerated in the relativistic shocks will usually emit synchrotron radiation. As the amount of swept-up interstellar matter getting larger and larger, the shell will be decelerated and radiation of longer wave length will be emitted. Thus, an external shock can produce only smoothly varying time-dependent emission, not the spiky multi-peaked structure found in many GRBs. If the central energy source is not completely impulsive, but works intermittently, it can produce many shells (or many fireballs) with different Lorentz factors. Late but faster shells can catch up and collide with early slower ones, and then, shocks (internal shocks) thus produced will lead to the observed bursting γ-ray emission (Rees & Mészáros 1994; Paczyński & Xu 1994). This is the so called external-internal shock model, internal shocks give rise to γ-ray bursts and external shocks to afterglows. The internal shocks can only convert a part of their energies to the γ-ray bursts, other part remains later to interact with the interstellar medium and lead to afterglows. Typically, the GRB is produced at a large distance of about 10^{13} cm to the center, such a large distance is allowed according to the relaxed compactness relation $R_e \leq \gamma^2 c\delta T$, while its afterglows are produced at about 10^{16} cm or even much farther. This internal-external shock scenario, under the simplified assumptions of uniform environment with typical ISM number density of $n \sim 1\mathrm{cm}^{-3}$, isotropic emission of synchrotron radiation and only impulsive energy injection, is known as the standard model.

Spectra of afterglows: The instantaneous spectra of afterglows, according to this model, can be written as $F_\nu \propto \nu^\beta$, with different β for different range of frequency ν (Sari et al. 1998; Piran 1999b). Let ν_{sa} be the self absorption frequency, for which the optical depth $\tau(\nu_{sa}) = 1$. For $\nu < \nu_{sa}$, we have the Wien's law: $\beta = 2$. For $\nu_{sa} < \nu < \min\ (\nu_\mathrm{m}, \nu_\mathrm{c})$, we can use the low energy synchrotron tail, $\beta = -1/3$. Here ν_m is the synchrotron frequency of an electron with characteristic energy, ν_c is the cooling frequency, namely the synchrotron frequency of an electron that cools during the local hydrodynamic time scale. For frequency within ν_m and ν_c, we have $\beta = -1/2$ for fast cooling ($\nu_\mathrm{c} < \nu_\mathrm{m}$) and $\beta = -(p-1)/2$ for slow cooling ($\nu_\mathrm{m} < \nu_\mathrm{c}$). For $\nu > \max(\nu_\mathrm{m}, \nu_\mathrm{c})$, we have $-p/2$. Here, p is the spectral index of the emitting electrons: $N(E) \propto E^{-p}$.

3. DYNAMICAL EVOLUTION OF THE FIREBALL

During the γ-ray bursting phase and the early stage of afterglows, the fireball expansion is initially ultra-relativistic and highly radiative, but finally it would be getting into non-relativistic and adiabatic, a unified dynamical evolution should match all these phases. In fact, the initial ultra-relativistic phase has been well described by some simple scaling laws (Mészáros & Rees 1997a; Vietri 1997; Waxman 1997; Wijers et al. 1997), while the final non-relativistic and adiabatic phase should obey the Sedov (1969) rule, which has well been studied in Newtonian approximation. The key equation (Blandford & McKee 1976; Chiang & Dermer 1999) is

$$\frac{\mathrm{d}\gamma}{\mathrm{d}m} = -\frac{\gamma^2 - 1}{M}, \tag{5}$$

here m denotes the rest mass of the swept-up medium, γ the bulk Lorentz factor, and M the total mass in the co-moving frame including internal energy U. This equation was originally derived under the ultra-relativistic condition. The widely accepted results derived under this equation are correct for ultra-relativistic expansion. Accidentally, these results are also suitable for the non-relativistic and radiative case. However, for the non-relativistic and adiabatic case, they will lead to wrong result "$v \propto R^{-3}$" (v is the velocity), while the correct Sedov result should be "$v \propto R^{-3/2}$", as first pointed out by Huang, Dai and Lu (1999a,b).

It has been proved (Huang, Dai & Lu 1999a,b) that in the general case, the above equation should be replaced by

$$\frac{d\gamma}{dm} = -\frac{\gamma^2 - 1}{M_{ej} + \epsilon m + 2(1-\epsilon)\gamma m}, \tag{6}$$

here M_{ej} is the mass ejected from GRB central engine, ϵ is the radiated fraction of the shock generated thermal energy in the co-moving frame. The above equation will lead to correct results for all cases including the Sedov limit. This generic model is suitable for both ultra-relativistic and non-relativistic, and both radiative and adiabatic fireballs. As proved by Huang et al. (1998a,b), Wei & Lu (1998a) and Dai et al. (1999a), only several days after the burst, a fireball will usually become non-relativistic and adiabatic, while the afterglows can last some months, the above generic model is really useful and important.

4. COMPARISON AND ASSOCIATION OF GRB WITH SN

Supernova was known as the most energetic phenomenon at the stellar level. SN explosion is the final violent event in the stellar evolution. Dynamically, it can also be described as a fireball, which however expands non-relativistically. After the SN explosion, there is usually a remnant which can shine for more than thousands of years and be well described dynamically by Sedov model (Sedov 1969).

GRB is also a phenomenon at the stellar level. However, it is much more energetic and much more violent than SN explosion! It has been proved to be described as a fireball, which expands ultra-relativistically. The GRB may also leave a remnant which shines for months now known as afterglow.

Their comparison is given in Table I: In April 1998, a SN 1998bw was

found to be in the 8' error circle of the X-ray afterglow of GRB 980425 (Galama et al. 1998; Kulkarni et al. 1998). However, its host galaxy is at a red-shift z=0.0085 (Tinney et al. 1998), indicating a distance of 38 Mpc (for $H_0 = 65$ km s^{-1} Mpc^{-1}), which leads the energy of the GRB to be too low, only about 5×10^{47} ergs, 4 orders of magnitude lower than normal GRB.

Later, in the light curves of GRB 980326 (Bloom et al. 1999; Castro-Tirado & Gorosabel 1999b) and GRB 970228 (Reichart 1999; Galama et al. 1999b), some evidence related with SN was found. This is a very important question worth while to study further (see e.g. Wheeler 1999). These two violent phenomena, GRB and SN, might be closely related. They might be just two steps of one single event (Woosley et al. 1999;

Table I

	GRBs	SNs
Burst	**Bursting γ-rays**	**SN explosion**
Energy up to	10^{54} ergs	10^{51} ergs
Time Scale	10 sec	Months
Profile	irregular	smooth
Wave Band	γ-ray	Optical
Relic	**Afterglow**	**Remnant**
Time Scale	Months	10^3 Years
Wave Band	Multi-band	Multi-band
Understanding		
Fireball Expansion	Ultra-relativistic	Non-relativistic
Mechanism	???	Stellar Core Collapse
Key Process	???	Neutrino process

Cheng & Dai 1999; Wang et al. 1999b; Dai 1999d). It is interesting to note that the first step might provide a low baryon environment for the second step to produce GRB. Such a kind of models can give a way to avoid the baryon contamination.

5. INNER ENGINE AND ENERGETICS

There have been a lot of models proposed to explain the central engine of GRBs (see e.g. Castro-Tirado 1999a; Piran 1999a; Cheng & Dai 1996; Dai & Lu 1998b). All these objects are related with compact stars such as neutron star (NS), strange star (SS), black hole (BH) etc. For example, binary mergers (NS-NS, NS-BH, ...), massive star collapsing, phase transitions (NS to SS) and others have been proposed. To build a successful model for central engine, the most difficult task is to solve the baryon contamination. There seem to be three kinds of ways: 1) based on BH, which can swallow baryons; 2) based on SS, of which baryons are only contained in its crust with mass less than 10^{-5} $M_{\odot}$; 3) based on the two-step process pointed out in above section.

A system of a central BH with a debris torus rotating round it may form after compact star merging or massive star collapsing. In this system, two kinds of energies can be used: the rotational energy of the BH and the gravitational energy of the torus. The rotational energy of the

BH can be extracted via the B-Z (Blandford & Znajek 1977) mechanism (Mészáros & Rees 1997b; Paczyński 1998). For a maximally rotating BH, its rotational energy can be extracted up to 29% of the BH rest mass, while the gravitational binding energy of the torus can be extracted up to 42% of the torus rest mass. Lee, Wijers and Brown (1999) recently studied the possibility to use these mechanisms in producing GRB.

The phase transition from neutron star to strange star can release huge energy to account for GRB. As an estimate, we can reasonably assume that about 20-30 MeV is released per baryon during the phase transition. Total energy released this way can be up to about $(4-6) \times 10^{52}$ ergs. Strange star is the stellar object in the quark level. Whether it exists or not is a fundamental physical/astrophysical problem. Its main part is a quark core with large strangeness (known as strange core). There could be a thin crust with mass of only about $\sim (10^{-6} - 10^{-5})\mathrm{M}_\odot$ (Alcock et al. 1986; Huang & Lu 1997a,b; Lu 1997; Cheng et al. 1998), all baryons are contained in the crust. It is interesting to note that this baryonic mass is low enough to avoid the baryon contamination. Kluźniak and Ruderman (1998) proposed differentially rotating neutron stars as an origin of GRBs. Dai and Lu (1998b) used this mechanism to the case of differentially rotating strange stars and proposed a possible model for GRB without baryon contamination.

6. NEW INFORMATION IMPLIED BY THE DEVIATIONS FROM THE STANDARD MODEL

The standard model described above is rather successful in that its physical picture is very clear, it gives results very simple, and observations on GRB afterglows support it at least qualitatively but generally. However, various quantitative deviations have been found. They indicate that the simplifications made in the standard model should be improved. These deviations may reveal important new information, such as non-uniform environment, additional energy injection, beaming effects of radiation and others.

Wind environment effects: Dai and Lu (1998c) analysed the afterglows of GRB970616 and others. They studied the general case of $n \propto R^{-k}$ for the non-uniform environment density (here n is the number density of the environment medium) and found that the X-ray afterglow of GRB970616 can well be fitted by $k = 2$, and pointed out that this non-uniformity may be due to the existence of a stellar wind. After the detailed studies by Chevalier & Li (1999a,b), the stellar wind model has become widely interested. People are aware that it may contain many implications about the pregenitors of GRBs and provide strong support to their massive star origin.

Additional energy injection: The optical afterglows of GRB 970228 and GRB 970508 had some complexities, showing down-up-down variation in their light curves. These features can be explained by long time scale energy injection from their central engines (Dai & Lu 1998a,b; Rees & Mészáros 1998; Panaitescu et al. 1998). For example, a millisecond pulsar with super-strong magnetic field can be produced at birth of GRB. As the fireball expands, the central pulsar can continuously supply energy through magnetic dipole radiation. Initially, the energy supply is small enough, the afterglow shows declining. As it becomes important, the afterglow shows rising. However, the magnetic dipole radiation should itself attenuate later. Thus, the down-up-down shape would appear naturally. Dai and Lu (1999c) analysed GRB 980519, 990510 and 980326, and obtained the results agreeing well with observations.

Additional radiations: Though the synchrotron radiation is usually thought to be the main radiation mechanism, however, under some circumstances, the inverse Compton scattering may play an important role in the emission spectrum, and this may influence the temporal properties of GRB afterglows (Wei & Lu 1998a,b).

Later data in the afterglows of GRB 970228 (Reichart 1999; Galama et al. 1999b) and 980326 (Bloom et al. 1999; Castro-Tirado & Gorosabel 1999b) may show the deviations as additional contributions from supernovae.

Beaming effects: GRB 990123 has been found very strong in its γ-ray emission, and the red shift of its host galaxy is very large (z=1.6) (Kulkarni et al. 1999a; Galama et al. 1999a; Akerlof et al. 1999; Castro-Tirado, et al. 1999a; Hjorth, et al. 1999; Andersen, et al. 1999). If its radiation is isotropic, the radiation energy only in γ-rays is already as high as $E_\gamma \sim 3.4 \times 10^{54}$ ergs, closely equals two solar rest energy ($E_\gamma \approx 2\mathrm{M}_\odot \mathrm{c}^2$)! As the typical mass of the stellar object is in the order of $\sim 1\mathrm{M}_\odot$, while the radiation efficiency for the total energy converting into the γ-ray emission is usually very low, such a high emission energy is very difficult to understand (Wang, Dai & Lu 1999a).

A natural way to relax the energy crisis is to assume that the radiation of GRB is beaming, rather than isotropic. Denote the jet angle as Ω, then the radiation energy E will be reduced to $E\Omega/4\pi$. At the same time, the estimated burst rate should increase by a factor of $4\pi/\Omega$. Are there any observational evidences for the jet in GRB and its afterglow? Pugliese et al. (1999), Rhoads (1997, 1999), Sari et al. (1999) and Wei

& Lu (1999a,b) have discussed this question. Kulkarni et al. (1999b) observed that two days after the burst, the decaying was getting more steepening, appearing as a break in the light curve of GRB 990123, and they regarded this as the evidence for jet. Recently, Huang et al. (1999c) calculated the influences of various parameters on the jetted emission of GRB, showed that a break in the light curve may appear in the case of narrow jet and for small electron energy fraction and small magnetic energy fraction.

Dense environment effects: However, Dai and Lu (1999b) pointed out that a shock undergoing the transition from a relativistic phase to a non-relativistic phase may also show a break in the light curve, if there are dense media and/or clouds in the way, this break may happen earlier. This model could also give an explanation for the observed steepening. Recently, Wang et al. (1999c) proved that the dense environment model can also explain well the radio afterglow of GRB 980519 (Frail et al. 1999). Thus, we should study the break further to take both beaming and dense environment effects into account.

7. CONCLUSION

The standard internal-external shock model, which is built under many simplifications, has been proved to be well fitted by observations qualitatively but generally. Based on the success of this model, it should be very important to study the deviations from the standard model, which indicate that the simplifications should be relaxed in some aspects. Hence, the deviations contain important new information and have been a fruitful research area.

In contrast to the rapid progress in understanding the nature of afterglows, GRB itself has not yet been clear. However, this is a very important problem. The solution about the origin of GRB is related with most fundamental questions in physics and astrophysics, such as black holes, stellar objects from quark level, physics and properties of the farthest stellar phenomena and others. It may also give new and important information for cosmology. Within five or ten years, there should still be a lot of further surprising achievements about these most violent events.

Acknowledgements

This work is supported by the National Natural Science Foundation of China.

References

Akerlof, C., et al., 1999, Nature, 398, 400.
Alcock, C., Farhi, E., Olinto, A., 1986, ApJ 310, 261.
Andersen, M.I., et al., 1999, Science, 283, 2075.
Blandford, R.D., MaKee, C.F., 1976, Phys. Fluids, 19, 1130.
Blandford, R.D., Znajek, R.L., 1977, MNRAS, 179: 433.
Bloom, J.S. et al., 1999, Nature, 401, 453.
Castro-Tirado, A.J., et al., 1999a, Science, 283, 2069.
Castro-Tirado, A., Gorosabel, J., 1999b, astro-ph/9906031.
Cheng, K.S., Dai, Z.G., 1996, Phys. Rev. Lett., 77, 1210.
Cheng, K.S., Dai, Z.G., Lu, T., 1998, Int. J. Mod. Phys. D, 7, 139.
Cheng, K.S., Dai, Z.G., 1999, astro-ph/9908248.
Chevalier, R.A., Li, Z.-Y., 1999a, ApJ, 520, L29.
Chevalier, R.A., Li, Z.-Y., 1999b, astro-ph/9908272.
Chiang, J., Dermer, C.D., 1999, ApJ, 512, 699.
Costa, E., et al., 1997, Nature, 387, 783.
Dai, Z.G., Lu, T., 1998a, A&A, 333, L87.
Dai, Z.G., Lu, T., 1998b, Phys. Rev. Lett., 81, 4301.
Dai, Z.G., Lu, T., 1998c, MNRAS, 298, 87.
Dai, Z.G., Huang, Y.F., Lu, T., 1999a, ApJ, 520, 634.
Dai, Z.G., Lu, T., 1999b, ApJ, 519, L155.
Dai, Z.G., Lu, T., 1999c, astro-ph/9906109.
Dai, Z.G., 1999d, in this Proceedings.
Frail, D.A., et al., 1999, astro-ph/9910060.
Galama, T.J., et al., 1998, Nature, 395, 670.
Galama, T.J., et al., 1999a, Nature, 398, 394.
Galama, T.J., et al., 1999b, ApJ, astro-ph/9907264.
Goodman, J., 1986, ApJ, 308, L47.
Groot, P.J. et al., 1997, IAUC, No.6584.
Hjorth, J., et al., 1999, Science, 283, 2073.
Huang, Y.F., Lu, T., 1997a, Chin. Phys. Lett., 14, 314.
Huang, Y.F., Lu, T., 1997b, A&A, 325, 189-194.
Huang, Y.F., Dai, Z.G., Lu, T., 1998a, A&A, 336, L69.
Huang, Y.F., Dai, Z.G., Wei, D.M., Lu, T., 1998b, MNRAS, 298, 459.
Huang, Y.F., Dai, Z.G., Lu, T., 1999a, MNRAS, 309, 513.
Huang, Y.F., Dai, Z.G., Lu, T., 1999b, Chin. Phys. Lett., 16, 775, astro-ph/9906404.
Huang, Y.F., Gou, L.J., Dai, Z.G., Lu, T., 1999c, astro-ph/9910493.
Katz, J., 1994, ApJ, 422, 248.
Klebesadel, R.W., Strong, I.B., Olson, R.A., 1973, ApJ, 182, L85.
Kluźniak, W., Ruderman, M., 1998, ApJ, 505, L113.

Krolik, J.H., Pier, E.A., 1991, ApJ, 373, 277.
Kulkarni, S. R., et al., 1998, Nature, 395, 663.
Kulkarni, S. R., et al., 1999a, Nature, 398, 389.
Kulkarni, S. R., et al., 1999b, astro-ph/9903441.
Lee, H.K., Wijers, R.A.M.J., Brown, G.E., 1999, astro-ph/9906213.
Lu, T., 1997, Pacific Rim Conference on Stellar Astrophysics, A.S.P. Conf. Ser. Vol. 138, eds. K.L. Chan, K.S. Cheng, H.P. Singh, 1998, 215.
Meegan, C. A., et al., 1992, Nature, 355, 143.
Mészáros, P., Rees, M.J., 1992, MNRAS, 257, 29.
Mészáros, P., Rees, M.J., 1997a, ApJ, 476, 232.
Mészáros, P., Rees, M.J., 1997b, ApJ, 482, L29.
Mészáros, P., 1999, astro-ph/9904038.
Mitra, A., 1998, ApJ, 492, 677.
Paczyński, B., 1986, ApJ, 308, L51.
Paczyński, B., Xu, G., 1994, ApJ, 427, 708.
Paczyński, B., 1998, ApJ, 494, L45.
Panaitescu, A., Mészáros, P., Rees, M.J., 1998, ApJ, 503, 314.
Piran, T., 1999a, Phys. Rep., 314, 575.
Piran, T., 1999b, astro-ph/9907392.
Pugliese, G., Falcke, H., Biermann, P.L., 1999, astro-ph/9903036.
Reichart, D.E., 1999, ApJ, 521, L111.
Rees, M.J., Mészáros, P., 1992, MNRAS, 258, 41.
Rees, M.J., Mészáros, P., 1994, ApJ, 430, L93.
Rees, M.J., Mészáros, P., 1998, ApJ, 496, L1.
Rhoads, J., 1997, ApJ, 487, L1.
Rhoads, J., 1999, ApJ, 525, 737, (astro-ph/9903399).
Sari, R., Piran, T., 1995, ApJ, 455, L143.
Sari, R., Piran, T., Narayan, R., 1998, ApJ, 497, L17.
Sari, R., Piran, T., Halpern, J.P., 1999, ApJ, 519, L17.
Sedov, L., 1969, Similarity and Dimensional Methods in Mechanics (Academic, New York), Chap.IV.
Tinney, C. et al., 1998, IAU Circ. 6896.
Van Paradijs, J., et al., 1997, Nature, 386, 686.
Vietri, M., 1997, ApJ, 488, L105.
Wang, X.Y., Dai, Z.G., Lu, T., 1999a, astro-ph/9906062.
Wang, X.Y., Dai, Z.G., Lu, T., Wei, D.M., Huang, Y.F., 1999b, astro-ph/9910029.
Wang, X.Y., Dai, Z.G., Lu, T., 1999c, astro-ph/9912492.
Waxman, E., 1997, ApJ, 485, L5.
Wei, D.M., Lu, T., 1998a, ApJ, 499, 754.
Wei, D.M., Lu, T., 1998b, ApJ, 505, 252.

Wei, D.M., Lu, T., 1999a, astro-ph/9908273.
Wei, D.M., Lu, T., 1999b, astro-ph/9912063.
Wheeler, J.Craig, 1999, astro-ph/9909096.
Wijers, R.A.M.J., Rees, M.J., Mészáros, P., 1997, MNRAS, 288, L51.
Woosley, S.E., Macfadyen, A.I., Heger, A., 1999, astro-ph/9909034.

ENERGY AND RADIATION MECHANISMS OF GAMMA-RAY BURSTS

Z. G. Dai[1] and K. S. Cheng[2]
[1]*Department of Astronomy, Nanjing University, Nanjing 210093, China*
[2]*Department of Physics, University of Hong Kong, Hong Kong, China*

Abstract For recent three years major advances have been made in the field of gamma-ray bursts (GRBs). These advances include the discovery of some GRBs associated with supernovae and the detection of steepened light curves of a few optical afterglows. In order to understand these observational results, we here discuss the existing models for GRB energy sources and radiation mechanisms for afterglows. We suggest that the models for delayed formation of black holes in supernovae (including our two-step model) may be a plausible mechanism because such models can avoid the so-called baryon contamination problem. The steepened afterglow light curves may provide evidence for the existence of dense media in the vicinity of GRBs.

Keywords: gamma-ray: bursts - shock waves - supernovae: general

1. INTRODUCTION

Any energy model for cosmological GRBs must satisfy two essential requirements: (1) The model should produce an extremely relativistic fireball, which should subsequently emit an amount of gamma-ray isotropic energy $E_\gamma \sim 10^{55}$ ergs implied by the observed fluences, the cosmological distance scale the radiative efficiency. If anisotropic emission with a beaming factor of $\Delta\Omega/4\pi \sim 0.01$ is assumed, this energy can be reduced to $E_{\rm jet} \sim$ a few times 10^{53} ergs. (2) The rapid variability of GRBs and their nonthermal spectra (Woods & Loeb 1995) requires that the Lorentz factor of the fireball be 100-1000. This implies that the fraction of contaminating baryons must be less than 1%. If the emission is anisotropic, the mass of loading matter $\Delta M \leq 0.01 E_{\rm jet}/c^2 \sim 10^{-3} M_\odot$.

Two currently popular models of GRB progenitors are the merger of two compact objects (neutron stars and black holes) and the collapse of massive stars. The latter model has become more favorable since the

K.S. Cheng et al. (eds.), Stellar Astrophysics, 369–377.

discovery of the Type Ic supernova SN 1998bw in the error box of GRB 980425 (Galama et al. 1998). The high energy inferred for the optical supernova, $(2-3)\times10^{52}$ ergs (Iwamoto et al. 1998; Woosley, Eastman & Schmidt 1999), and the high expansion velocity inferred for the radio supernova (Kulkarni et al. 1998) strengthen the GRB-SN connection. Recently, this connection was further confirmed by Bloom et al. (1999), Reichart (1999) and Galama et al. (1999), who found the dramatical brightening and extreme reddening of the optical afterglows of GRB 980326 and GRB 970228 at late times, respectively. However, it is quite difficult to understand such a connection. From the above second essential requirement, too many baryons possibly exist in the vicinity of the collapsing core in the hypernova models (Woosley 1993; Paczyński 1998) so that an ultrarelativistic jet forming during the collapse of the core rapidly become nonrelativistic. This conclusion is consistent with the numerical studies (MacFadyen & Woosley 1999): an ultrarelativistic jet converts to a nonrelativistic large bubble. A simple reason for this may be that a large amount of radiative energy is released impulsively but the mass of contaminating baryons is in fact of the order of $M_*(\Delta\Omega/4\pi)\sim0.1M_\odot(M_*/10M_\odot)(10^2\Delta\Omega/4\pi)$, where M_* is the mass of the matter around the core.

Recently Cheng & Dai (1999) have proposed a scenario, in which a supernova explosion first may produce a massive neutron star, and about two hours later the star will start to accrete the fall-back supernova ejecta at a hypercritical rate, while an accretion disk will form near the stellar surface due to its large angular momentum. In particular neutrino annihilation and/or Poynting flux along the rotation axis of the star may lead to a jet which will pushes out of its front matter. The mass of the accreting neutron star will eventually reach the maximum mass about several hours after the supernova explosion and thus will promptly collapse to a rapidly rotating black hole surrounded by a torus. The gravitational binding energy of the torus, which is of the order of several 10^{52} ergs, may be dissipated into the supernova ejecta, which may in turn give rise to an energetic hypernova. Another jet with energy of a few times 10^{53} ergs and with low-mass baryon contamination will occur along the rotation axis of the black hole by extracting its rotational energy via the Blandford-Znajek's (1977) mechanism. We expext that the second jet can produce a GRB and its afterglow.

Delayed formation of black holes in supernovae has been widely discussed (also see a review of Woosley et al. 1999). We emphasize that our two-step model is an analytical one, in which hypercritical accretion of a newborn neutron star and delayed formation of a black hole could produce two discrete jets. The first jet will push out of its front matter and leave an exit for the second jet, which could thus be relatively free of baryons.

2. OUR TWO-STEP MODEL

In the first step, the core collapse of a ($\geq 19M_\odot$) massive star leads to a newborn neutron star and a successful supernova explosion, and then the neutron star hypercritically accretes its surrounding fall-back supernova ejecta. The supernova explosion scenarios involve an outgoing shock wave. When this shock enters the hydrogen envelope, the deceleration of matter occurs. This deceleration sharpens into a reverse shock. When the reverse shock reaches the neutron star surface, the star starts to accrete the fall-back supernova ejecta at a hypercritical rate. The time scale for the reverse shock to pass through the core is about the same as the time that the outgoing shock front takes to reach the stellar surface, which is $t_0 \sim 2\,\mathrm{hr}(R_*/3\times10^{12}\mathrm{cm})(M_{\rm ej}/10M_\odot)^{1/2}(E_{\rm sn}/10^{51}\mathrm{ergs})^{-1/2}$, where R_* is the presupernova stellar radius, $M_{\rm ej}$ is the mass of the ejected matter and $E_{\rm sn}$ is the supernova explosive energy (Shigeyama, Nomoto & Hashimoto 1988). For SN1987A, t_0 is about 2 hr (Shigeyama et al. 1988). In order to derive the rate of mass accretion onto the neutron star after t_0, we follow Brown & Weingartner (1994). By assuming the neutron star to be at rest with respect to its ambient matter and using the Bondi's (1955) spherical accretion theory, we obtain the accretion rate

$$\dot{M} = 5.63\times10^{-4}\left(\frac{M}{1.8M_\odot}\right)^2\left(\frac{v_f}{10^8\mathrm{cm\,s^{-1}}}\right)^{-15/8}\left(\frac{t}{1\mathrm{yr}}\right)^{-15/8} M_\odot\,\mathrm{yr}^{-1}, \quad (1)$$

where M is the neutron star mass ($\sim 1.8M_\odot$) and v_f is the final velocity (after being slowed down by the reverse shock) of the carbon-oxygen core. For SN1987A, $v_f \sim 6\times10^7\ \mathrm{cm\,s^{-1}}$ (Woosley 1988). In deriving equation (1), we have assumed that the ambient matter of the neutron star is radiation-dominated due to the effect of the outgoing shock. Owing to this assumption, our accretion rate is slightly larger than that of Brown & Weingartner (1994). The time at which radiation significantly affects accretion can be estimated based on $\dot{M}(t_{\rm cr}) = \dot{M}_{\rm cr}$. Hence, we obtain this timescale $t_{\rm cr} = 0.7v_{f,8}^{-1}\,\mathrm{yr}$, where $v_{f,8} = v_f/10^8\mathrm{cm\,s^{-1}}$. Below, we take $v_{f,8} = 1$. When $t < t_{\rm cr}$, the accretion is hypercritical and the total accreted baryon mass is given by

$$\Delta M_{\rm acc} = \int_{t_0}^{t}\dot{M}dt = 1.81M_\odot v_{f,8}^{-15/8}[(t_0/1\mathrm{hr})^{-7/8} - (t/1\mathrm{hr})^{-7/8}]. \quad (2)$$

For the modern realistic equation of state for neutron matter chosen in the next section, an accreting neutron star with initial mass of $1.76M_\odot$ will collapse to a rapidly rotating black hole when $\Delta M_{\rm acc} = 0.55M_\odot$.

Before the collapse, the accreted matter forms a disk near the neutron star because the accreted angular momentum may be up to $3^{1/2}R_s c$ per gram, where R_s is the Schwarzchild radius (Woosley & Chevalier 1989). The temperature of the accretion disk near the neutron star can be estimated by the following equation: $\eta \dot{M} c^2 = \dot{\epsilon}_\nu \Delta\Omega_d R_{ns}^3/4$, where η is the efficiency for the conversion of the gravitational energy to heat (~ 0.1), $R_{\rm ns}$ is the neutron star radius ($\sim 10^6$ cm), $\Delta\Omega_d$ is the solid angle of the disk, and $\dot{\epsilon}_\nu = 1.0 \times 10^{25}(T/{\rm MeV})^9\,{\rm erg\,s^{-1}\,cm^{-3}}$ is the neutrino pair energy production rate per unit volume (Dicus 1972). For typical values of these parameters (e.g., $\Delta\Omega_d \sim 3$), we obtain $T \sim 6$ MeV. The total energy for neutrino losses is approximated by $E_\nu = GM\Delta M_{\rm acc}/R_{\rm ns} \sim 3 \times 10^{53}$ ergs. Since anisotropic neutrino emission takes place due to the effect of the disk (Kluzniak 1998), neutrino annihilation along the rotation axis of the neutron star leads to a jet. By using the efficiency of neutrino annihilation, $\chi \sim 0.3\%$ (Goodman, Dar & Nussinov 1987; Kluzniak 1998), we obtain the energy of the jet

$$E_{\rm jet,1} \sim E_\nu \chi \sim 10^{51}\ {\rm ergs}. \tag{3}$$

It should be pointed out that this efficiency is much larger than that of Popham, Woosley & Fryer (1999), in which an accretion disk surrounding a black hole is advection-dominated. In the present case, however, matter is accreted onto the neutron star surface where the conversion of the gravitational energy to heat (η) is much more efficient. Since this thermal energy is released via neutrino emission, the efficiency of neutrino annihilation should increase substantially. In addition to the neutrino annihilation mechanism, there is another possible mechanism to produce a jet proposed by Katz (1997), in which the magnetic field amplified by the differential rotation of the disk may result in a strong Poynting flux. The jet produced by these mechanisms will push its front baryonic matter whose velocity is given by $v_{\rm jet} = (2E_{\rm jet,1}/\Delta M')^{1/2} \sim 3\times 10^9(M_*/10M_\odot)^{-1/2}(10^2\Delta\Omega/4\pi)^{-1/2}\,{\rm cm\,s^{-1}}$, where $\Delta M' = M_*(\Delta\Omega/4\pi)$ is the mass of baryons loading with the jet. It should be noted that this velocity is much larger than that of the outgoing shock. As numerically studied by MacFadyen & Woosley (1999), this nonrelativistic jet will expand laterally, producing a bubble. When $t \sim 5$ hr, at which the collapse of the neutron star may take place as argued in the next section, the radius of the bubble is estimated as $R_b \sim v_{\rm jet}(t - t_0) \sim 3 \times 10^{13}(M_*/10M_\odot)^{-1/2}(10^2\Delta\Omega/4\pi)^{-1/2}$ cm, which is much larger than the presupernova stellar radius R_*.

In the second step, the neutron star promptly implodes to a rapidly rotating black hole surrounded by a torus when the mass of the star increases to the maximum mass and meanwhile its rotation frequency increases to the upper limit due to the accreted angular momentum. It has been argued by Brown & Bethe (1994) and Bethe & Brown (1995) that hypercritical accretion of neutron stars in supernovae (e.g., SN1987A) may lead to collapse of the stars to low-mass black holes. These authors considered a soft equation of state like kaon condensation for neutron matter as the starting point of their discussions. Here we adopt a more realistic modern equation of state named UV14+TNI (Wiringa, Fiks & Fabrocini 1988). This equation of state gives the property of a non-rotating neutron star at the maximum mass: maximum gravitational mass $M_{max} = 1.84M_{\odot}$, corresponding baryon mass $M_b = 2.17M_{\odot}$, radius $R_{\rm ns} = 9.51$ km and moment of initia $I = 1.5 \times 10^{45}\,{\rm g\,cm^2}$ (Wiringa et al. 1988). According to this property, the maximum frequency of rotation for this neutron star equation of state is given by $\Omega_{max} = 7.84 \times 10^3 (M_{max}/M_{\odot})^{1/2} (R_{\rm ns}/10\,{\rm km})^{-3/2} {\rm s}^{-1} = 1.17 \times 10^4\,{\rm s}^{-1}$ (Cook, Shapiro & Teukolsky 1994). The property of a rotating neutron star at this maximum frequency and at the allowable maximum mass is as follows (Cook et al. 1994): maximum gravitational mass $M_{max} = 2.19M_{\odot}$, corresponding baryon mass $M_b = 2.55M_{\odot}$, radius $R_{\rm ns} = 12.7$ km and angular momentum $J = 2.85 \times 10^{49}$ erg s or the Kerr rotation parameter $a = cJ/GM_{max}^2 = 0.67$.

After having these properties, we discuss implications of hypercritical accretion of a newborn neutron star with gravitational mass of $1.76M_{\odot}$. First, hypercritical accretion may produce and maintain a large bubble, as argued in the last section.

Second, hypercritically accreted matter may rapidly submerge the magnetic field of the neutron star (Muslimov & Page 1995). Once the accreted mass reaches $0.01M_{\odot}$, the buried magnetic field ohmically diffuses out after $\sim 10^8$ yr (Geppert, Page & Zannias 1999). This implies that the magnetic field of the neutron star could be always weak in the accretion timescale.

Third, hypercritical accretion may spin up the neutron star (Woosley & Chevalier 1989). A $1.76M_{\odot}$ neutron star has the baryon mass of $2.0M_{\odot}$ and thus this star needs to accrete matter with mass of $0.55M_{\odot}$ to become a maximum rotating neutron star with maximum mass. As suggested by Woosley & Chevalier (1989), the accreted angular momentum from the mixed mantle and helium core of the ejecta may be as large as $3^{1/2}R_s c$ per gram corresponding to 5.54×10^{48} erg s per $0.1M_{\odot}$ accreted mass. Thus, after accreting $0.55M_{\odot}$ mass, the neutron star can obtain the angular momentum of 3×10^{49} erg s. But, the maximum angular momentum of a rapidly rotating neutron star at maximum mass is only 2.85×10^{49} erg s. The remaining angular momentum is $\Delta J = 0.15 \times 10^{49}$ erg s. How is ΔJ dissipated? Fortunately, this remaining accreted angular momentum can be easily carried away by gravitational radiation (Wagoner 1984). Owing to hypercritical accretion, therefore, the mass of the neutron star not only reaches the maximum mass but it is spun up to the maximum rotation frequency.

Finally, once the mass of the accreted matter reaches $0.55M_\odot$, the neutron star will promptly collapse to a rapidly rotating black hole. From equation (2), we can estimate the time at which the collapse will occur: $t \sim 5.1$ hr. The resulting black hole has the rotational energy

$$E_{\rm rot} = f(a)M_{\rm BH}c^2 \approx 3 \times 10^{53}\,{\rm ergs}, \tag{4}$$

where $f(a) = 1 - \sqrt{(1+\sqrt{1-a^2})/2} \approx 0.067$ at $a \approx 0.67$, and the mass of the black hole has been taken to be $2M_\odot$. After the collapse, not all mass will be immediately accreted; the outmost layers with a small fraction (a few percent) of the total mass, in fact, have centrifugal accelerations beyong the local gravitational attraction (Vietri & Stella 1998), leading to a torus. In addition, a small amount of accreted ejecta still stagnate in the torus. Thus, the total mass of the torus may be $M_t \geq 0.1M_\odot$. The presence of the torus will give rise to the following two effects: (i) The binding energy of the torus is $E_b = GM_{\rm BH}M_t/R_t = 3 \times 10^{52}\,{\rm ergs}(M_t/0.1M_\odot)(R_t/20\,{\rm km})^{-1}$, where R_t is the typical radius of the torus. Because the magnetic field of the torus may be amplified to 10^{15} G due to differential rotation in the torus, this binding energy can be dissipated into a magnetized relativistic wind (Usov 1994; Mészáros & Rees 1997; Katz 1997). Because such a wind may be easily absorbed by the outgoing supernova ejecta as argued by Dai & Lu (1998a, 1998b), almost all of the binding energy may convert to the expansion energy of the ejecta, which may further give rise to a supernova with a much brighter optical luminosity. (ii) In the presence of the torus, the rotational energy of the hole can be extracted by the Blandford-Znajek (1977) mechanism (Mészáros, Rees & Wijers 1998). The power for this mechanism is $P_{\rm BZ} = 1.7 \times 10^{51}a^2 f(a)(M_{\rm BH}/M_\odot)^2(B/3 \times 10^{15}{\rm G})^2{\rm erg\,s^{-1}} = 2.5 \times 10^{50}(B/3 \times 10^{15}{\rm G})^2{\rm erg\,s^{-1}}$ (Lee, Wijers & Brown 1999). The rotational energy of the black hole will be dissipated in the timescale of $\sim 10^3$ s if the magnetic field strength is of the order of 3×10^{15} G. We believe that this mechanism will produce an ultrarelativistic jet because the mass of matter loading with this jet can be estimated by

$$\Delta M = \Delta M'(\Delta\Omega/4\pi) = 10^{-3}M_\odot(M_*/10M_\odot)(10^2\Delta\Omega/4\pi)^2. \tag{5}$$

In fact, since the angular momentum of the presupernova star may be rather large, the centrifugal force should reduce the density of matter along the rotation axis and thus the above estimate is an upper limit. After a short acceleration phase, the Lorentz factor of the jet should be

$$\Gamma = E_{\rm rot}/(\Delta M c^2) \geq 150(M_*/10M_\odot)^{-1}(10^2\Delta\Omega/4\pi)^{-2}, \qquad (6)$$

which is consistent with the constraints from the rapid variability of the light curves and the nonthermal spectra of GRBs (Woods & Loeb 1995). The collisions among the shells in such a jet will produce a GRB and subsequently the deceleration of the jet in its ambient medium will result in an afterglow.

3. SOME OPTICAL AFTERGLOWS AND DENSE MEDIA

Recent observations show that the temporal decay of the R-band afterglow from GRB 990123 steepened about 2.5 days after the burst (Kulkarni et al. 1999; Castro-Tirado et al. 1999; Fruchter et al. 1999). Dai & Lu (1999a) have proposed a possible explanation for such a steepening: a shock expanding in a dense medium has undergone the transition from a relativistic phase to a nonrelativistic phase. We have found that this model is consistent with the observations if the medium density is about $3\times10^6\,{\rm cm}^{-3}$. By fitting our model to the observed optical and X-ray afterglow quantitatively, we have further inferred the electron and magnetic energy fractions of the shocked medium and find these two parameters are about 0.1 and 2×10^{-8} respectively. The former parameter is near the equipartition value while the latter is about six orders of magnitude smaller than inferred from the GRB 970508 afterglow. We have also discussed possibilities that the dense medium can be produced.

Dai & Lu (1999b) have discussed this model in more details. In particular, we have investigated the effects of synchrotron self absorption and energy injection. A shock in a dense medium becomes nonrelativistic rapidly after a short relativistic phase. The afterglow from the shock at the nonrelativistic stage decays more rapidly than at the relativistic stage. Since some models for GRB energy sources predict that a strongly magnetic millisecond pulsar may be born during the formation of GRB, we have discussed the effect of such a pulsar on the evolution of the nonrelativistic shock through magnetic dipole radiation. We have found that after the energy which the shock obtains from the pulsar is much more than the initial energy of the shock, the afterglow decay will flatten significantly. When the stellar effect disappears, the decay will steepen again. These features are in excellent agreement with the afterglows of GRB 980519, GRB 990510 and GRB 980326. Furthermore, our model fits very well all the observational data of GRB 980519 including the last two detections. Wang, Dai & Lu (1999) have numerically calculated the trans-relativistic evolution of a shock in a dense medium and given an excellent fit to the radio afterglow data of GRB 980519.

References

Bethe, H.A., & Brown, G.E. 1995, ApJ, 445, L129

Blandford, R.D., & Znajek, R.L. 1977, MNRAS, 179, 433

Bloom, J. et al. 1999, Nature, 401, 453

Brown, G.E., & Weingartner, J. C. 1994, ApJ, 436, 843

Brown, G.E., & Bethe, H. A. 1994, ApJ, 423, 659

Bondi, H. 1952, MNRAS, 112, 195

Castro-Tirado, A.J., et al., 1999, Science, 283, 2069

Cheng, K.S. & Dai, Z.G. 1999, astro-ph/9908248

Cook, G.B., Shapiro, S.L., & Teukolsky, S.A. 1994, ApJ, 424, 823

Dai, Z.G., & Lu, T. 1998a, A&A, 333, L87

Dai, Z.G., & Lu, T. 1998b, Phys. Rev. Lett., 81, 4301

Dai, Z.G., & Lu, T. 1999a, ApJ, 519, L155

Dai, Z.G., & Lu, T. 1999b, astro-ph/9906109

Dicus, D.A. 1972, Phys. Rev. D, 6, 941

Fruchter, A.S. et al. 1999, astro-ph/9902236

Galama, T.J. et al. 1998, Nature, 395, 670

Galama, T.J. et al. 1999, ApJ, submitted (astro-ph/9907264)

Geppert, U., Page, D., & Zannias, T. A&A, 345, 847

Goodman, J., Dar, A., & Nussinov, S. 1987, ApJ, 314, L7

Iwamoto, K. et al. 1998, Nature, 395, 672

Lee, H.K., Wijers, R.A.M.J., & Brown, G.E. 1999, astro-ph/9905373

Katz, J.I. 1997, ApJ, 490, 633

Kluzniak, W. 1998, ApJ, 508, L29

Kulkarni, S.R. et al. 1998, Nature, 395, 663

MacFadyen, A., & Woosley, S.E. 1999, ApJ, 524, 262

Mészáros, P., & Rees, M.J. 1997, ApJ, 482, L29

Mészáros, P., Rees, M.J., & Wijers, R.A.M.J. 1998, ApJ, 499, 301

Muslimov, A. & Page, D. 1995, ApJ, 440, L77

Paczyński, B. 1998, ApJ, 494, L45

Popham, R., Woosley, S.E., & Fryer, C. 1999, ApJ, 518, 356

Reichart, D.E. 1999, ApJ, 521, L111

Shigeyama, T. Nomoto, K., & Hashimoto, M. 1988, A&A, 196, 141

Usov, V.V. 1994, MNRAS, 267, 1035

Vietri, M., & Stella, L. 1998, ApJ, 507, L45

Wagoner, R.V. 1984, ApJ, 278, 345

Wang, X.Y., Dai, Z.G., Lu, T., 1999a, astro-ph/9906062.
Wiringa, R.B., Fiks, V., & Fabrocini, A. 1988, Phys. Rev. C, 38, 1010
Woods, E., & Loeb, A. 1995, ApJ, 453, 583
Woosley, S.E. 1988, ApJ, 330, 218
Woosley, S.E. 1993, ApJ, 405, 273
Woosley, S.E., & Chevalier, R.A. 1989, Nature, 338, 321
Woosley, S.E., Eastman, R.G., & Schmidt, B. P. 1999, ApJ, 516, 788
Woosley, S.E., MacFadyen, A., & Heger, A. 1999, astro-ph/9909034
Wang, X.Y., Dai, Z.G., & Lu, T. 1999, MNRAS, submitted

For The Best View

ON THE INVERSE COMPTON SCATTERING MODEL OF RADIO PULSARS

G. J. Qiao[1,2], R. X. Xu[1,2], J. F. Liu[1,2], J. L. Han[1,3], B. Zhang[1,2]
[1] *Beijing Astrophysics Center, CAS-PKU, Beijing 100871, China*
[2] *Astronomy Department, Peking University, Beijing 100871, China*
[3] *National Astronomical Observatories, CAS, Beijing 100012, China*

Abstract Some characteristics of the inverse Compton scattering (ICS) model are reviewed. At least the following properties of radio pulsars can be reproduced in the model: core or central emission beam, one or two hollow emission cones, different emission heights of these components, diverse pulse profiles at various frequencies, linear and circular polarization features of core and cones.

1. INTRODUCTION

"The crisis today, over twenty years later, is that theory has produced nothing more useful to compare observations with in order to interpret them" (Radhakrishnan 1992). On observational aspect, emission beam of a radio pulsar can be divided into two (core, inner conical) or three (plus an outer conical) components through careful studies of the observed pulse profiles and polarization characteristics (Rankin 1983; Lyne & Manchester 1988).

Many pulsar profiles at meter wavelength are dominated by core components. In the usual models of radio pulsars which are related to polar cap, it is difficult to get a central or "core" emission beam. Considering such obvious discrepancy between observations and theory, several authors (e.g., Beskin et al. 1988; Qiao 1988a,b; Wang et al. 1988, Lyutikov et al. 1999) presented their models for the core emission. In the following, we discuss some characteristics of the inverse Compton scattering (ICS) model (Qiao 1992; Qiao and Lin 1998, hereafter as QL98; Liu et al. 1999; Qiao et al. 1999; Xu et al. 2000).

K.S. Cheng et al. (eds.), Stellar Astrophysics, 379–384.

2. ON EMISSION BEAMS

2.1 BASIC IDEA OF THE ICS MODEL

The basic idea of the ICS model (Qiao 1988a,b; QL98) can be briefly described as following. Low-frequency electromagnetic waves are produced in the polar cap due to gap sparking and afterwards propagate out freely; Such low energy photons are then inverse-Compton-scattered by the secondary particles from the pair cascades, and the up-scattered waves are the radio emission observed from pulsars.

With the conditions $B \ll B_q = 4.414 \times 10^{13}$ Guass and $\gamma\hbar\omega_0 \ll m_e c^2$, the frequency formula of the ICS mechanism (Qiao 1988a,b; QL98) is

$$\omega \simeq 2\gamma^2\omega_0(1 - \beta\cos\theta_i), \tag{1}$$

where ω_0 and ω are the frequencies of incident and scattered waves, respectively. The Lorentz factor of the secondary particles is $\gamma = 1/\sqrt{1-\beta^2}$, and θ_i is the incident angle (the angle between the direction of motion of a particle and an incoming photon). Using the formula above, we have obtained so called "beam-frequency figure" (see QL98, Qiao et al. 1999), which is the plot of the beaming angle θ_μ (the angle between the emission direction and the magnetic axis) versus observing frequency ν. The shape of emission beams, emission heights and other emission properties can be derived from the "beam-frequency figure".

2.2 THE CENTRAL EMISSION BEAM AND HOLLOW CONES

Two distinct types of emission components were identified from observations, namely, 'core' or central emission beam near the center and one or two hollow cone(s) surrounding the core. The ICS model can reproduce one core and two cones at the same time. The emission beam in the model could consist of (1) Core + inner cone emission, for pulsars with shorter rotational periods; (2) Core + inner and outer cones, for pulsars with longer periods. Furthermore, the core emission may be a small hollow cone in fact, i.e., not be fully filled. This can be identified from some pulsar profiles through "Gaussian de-composition" (Qiao et al. 1999).

In the model, we found that these emission components are emitted at different heights. The 'core' emission is emitted at the place closed to the surface of neutron star, the 'inner cone' at a higher place, and the 'outer cone' at the highest place.

We also found that the sizes of the emission beams should change with frequency. As observing frequency increases, the 'core'emission beam becomes narrow; the 'inner cone' size increases or at least keeps constant; but the 'outer cone' size decreases. For given magnetic inclination and impact angles, we can figure out how the shape of pulse profiles vary with frequency (see Qiao et al.1999). This is the theoretical base for the classification of radio pulsars.

2.3 CLASSIFICATION FOR RADIO PULSARS

The pulse profile shapes, especially the variation of profiles with the observing frequencies are very important for understanding the emission mechanism of pulsars. In the ICS model, various pulse profiles and its evolution over a wide frequency range can be well simulated. As the impact angle gradually increases, pulsars can be grouped into two types (and further sub-types) according to the ICS model (Qiao et al. 1999).

Type I – Pulsars with only core and inner conical emissions. Such pulsars usually have shorter periods (thus have larger polar caps). There are two sub-types. Type Ia: Pulsars of this type have very small impact angle, and normally show single pulse profiles at low frequencies, but triple profiles at high frequencies. Prototype in this form are PSR B1933+16, PSR B1913+16. Type Ib: Pulsars of this type have larger impact angle than that of Type Ia. Though at low frequencies the pulsars also show single profiles, they will evolve to double profiles at higher frequencies, since the lines-of-sight only cut the inner conical beam. An example of this type is PSR B1845-01.

Type II – Pulsars with all three emission beams. Pulsars with average period often have such a feature. As the impact angle gradually increases, these pulsars can be grouped into two or three sub-types (Qiao et al. 1999). IIa: Pulsars of this type have five components at most observing frequencies, since the small impact angle makes the line-of-sight cut all the three beams (the core, the inner cone and the outer cone). One important point is that in the ICS model, the five pulse components will evolve to three or even one component at very low frequencies (see QL98, Fig. 6a, line A). Such a feature have been observed from PSR B1237+25 (Kuzmin et al. 1998). IIb: For this type, the impact angle is larger than IIa, so that at higher frequencies, the line-of-sight does not cut the core beam. Thus the pulsars show three components at low observing frequencies, but four components when frequency is higher, and finally just double profiles at highest frequencies. PSR B2045−16 is an good example. IIc: Pulsars of this type have the largest impact angle, so that only the outer conical beam is cut by the line-of-sight. The pulse profiles of this type are double at all observing frequencies, with the separation between the two peaks decreasing at higher frequencies. This is just the traditional radius-to-frequency mapping which has also been calculated in the curvature radiation (CR) picture. The prototype is PSR B0525+21. Alternative situation of this type is that pulsars have single profiles at most of observing frequencies, but become double components at very low observing frequencies. An example is PSR B0950+08.

3. LINEAR AND CIRCULAR POLARIZATION

Pulsar radio emission is generally found to be linearly polarized over all longitudes of profiles, some times as high as up to 100%. The position angle sweep is in an 'S' shape, which can be well understood within the rotating vector model. However, depolarization and position angle jumping are often found in the integrate profiles of some pulsars. Considering the retardation effect due to relative phase shift between pulsar beam components (the core and conal emission components are retarded from different heights in ICS model), we find that the phase shift of beam centers of the different components could cause the depolarization and position angle jump(s) in integrated profiles (Xu et al. 1997).

3.1 BASIC IDEA FOR POLARIZATION IN ICS MODEL

We have calculated the polarization feature of the ICS model. Now we would like to present one key point about how the circular polarization is produced in the ICS model.

As many authors (Ruderman & Sutherland, 1975; QL98) argued, there may exist the inner gap above the pulsar polar cap. The continuous gap formation and breakdown provide both low frequency waves with $\omega_0 \sim 10^5$ s^{-1} and out-streaming relativistic particles with $\gamma \sim 10^3$. The low frequency waves will be up-scattered by relativistic particles (moving direction $\mathbf{n}_\mathrm{e}$) to observed radio waves. The frequency (ω) of out-going waves is determined by eq.(1), while their complex amplitudes $\mathbf{E}$ are determined by (Liu et al. 1999)

$$\mathbf{E} = C\frac{\sin\theta'}{\gamma^2(1-\beta\cos\theta')^2}e^{i(\frac{\omega_0}{c}R-\frac{\omega}{c}\mathbf{R}\cdot\mathbf{n}+\phi_0)}\mathbf{e}_\mathrm{s}, \tag{2}$$

here $\mathbf{n}$ is the observing direction, $\mathbf{R}$ is the low frequency wave vector, θ' is the angle between $\mathbf{n}_\mathrm{e}$ and $\mathbf{n}$, C is a constant, $\mathbf{e}_\mathrm{s}$ is a electric unit-vector in the co-plan of $\mathbf{n}_\mathrm{e}$ and $\mathbf{n}$, ϕ_0 is the initial phase determined by incident wave.

Moving out and losing energy via inverse Compton scattering, the particle undergoes a decay in γ, which is assumed to be (QL98)

$$\gamma = \gamma_0[1 - \xi(r - R_0)/R_e]. \tag{3}$$

The electric field of the total scattered wave at a direction $\mathbf{n}$ is the sum of $\mathbf{E}$ of each electron if such electrons are scattered coherently.

The emission region for a certain ω can be obtained by eq.(1). As pointed out in QL98, there are generally three possible emission zones, corresponding to core, inner and outer cones. The scattered electromagnetic waves can be superposed coherently if the low-frequency waves are from *same* sparking and the emission region is *smaller* than $2\pi c/\omega_0$. The coherent superposition of scattered waves will result in circular polarization as well as linear polarization (Xu et al. 2000).

3.2 FURTHER CONSIDERATIONS AND NUMERICAL RESULTS

'Subpulse' circular polarization patterns

When the line of sight sweeps across a mini-beam, we can see a *transient* "subpulse". If the line of sight sweeps across the center of a core or inner conal mini-beam, the circular polarization will experience a central sense reversal, or else it will be dominated by one sense, either left hand or right hand according to its traversal relative to the mini-beam.

Subpulse Position Angles

Subpulse position angles show diverse values, which are generally centered at the projection of the magnetic field. The variation range is quite small for sub-pulses from outer cone emission zones, and becomes larger when emission height decreases, that is, larger for inner cone and core component. The position angles values are scattered around the value of the projection of the magnetic field lines. When emission of all the sub-pulses are summed up, the mean position angle will have the mean value. So naturally the mean position angle is related to the projection of magnetic field lines, exactly as assumed by the rotation vector model.

Circular and linear polarization of mean profiles

An observer can see bunches of particles all around his line of sight. The polarization he receives from individual bunches is different. When the magnetic axis is inclined from the rotation axis, the probability of the sparkings was assumed to decrease exponentially with the azimuthal angle from the projection of the rotational axis in our simulation. We then get significant (but *smaller* than in subpulse) circular polarization for core components of mean pulses. But this is not the case for the cone components.

4. CONCLUSION

The ICS model can reproduce many observational properties of radio pulsars. There are emission beams of core, inner and outer cones. These beams are emitted from different heights. The pulse profiles changes with frequencies, similar to what observed from many pulsars. The transient 'sub-pulses' have very high circular polarization (sometime as strong as 100%). In mean pulse profile, circular polarization is much higher in core than that in inner cone and oute cone.

Acknowledgements

This work is supported by National Nature Sciences Foundation of China, by the Climbing project of China, the Doctoral Program Foundation of Institution of Higher Education in China and by the Youth Foundation of Peking University.

References

Beskin, V.S., Gurevich, A.V., Istomin, Y.N., 1988, Ap&SS, 146, 205
Liu, J.F., Qiao, G.J., Xu, R.X., 1999, Chin. Phys. Lett., 16, 541
Lyne A.G., Manchester R.N., 1988, MNRAS, 234, 477
Lyutikov, M., Blandford, R.D., Machabeli, G., 1999, MNRAS, 305, 338
Qiao, G.J., 1988a, Vistas in Astronomy, 31, 393
Qiao, G.J., 1988b, in: High Energy Astrophysics, ed. G.B. Borner, p.88, Berlin: Springer-Verag.
Qiao, G.J., 1992, in: IAU Colloq. 128, eds. T.H. Hankins et al., p. 239, Poland: Pedagogical Univ. Press
Qiao, G.J. Lin, W.P., 1998, A&A, 333, 172 (QL98)
Qiao, G.J. Liu, J.F.,Zhang, B. Han, J.L.,1999, ApJ, submitted
Radhakrishnan V. 1992, in: IAU Colloq. 128, eds. T.H. Hankins et al., p. 267, Poland: Pedagogical Univ. Press
Rankin J.M. 1983, ApJ, 274, 333
Ruderman, M.A. Sutherland, P.G., 1975, ApJ, 196, 51
Wang,D.Y., Wu X.J., Chen H., 1988, Vistas in Astronomy, 31, 399
Xu, R.X., Liu, J.F., Han, J.L., Qiao, G.J., 2000, ApJ, in press
Xu, R.X., Qiao, G.J. Han J.L. , 1997, A&A, 323, 395

THE EFFECT OF ICS ON GRB AFTERGLOWS

D.M. Wei
Purple Mountain Observatory, Academia Sinica, Nanjing 210008, P.R.China

Abstract The detection of delayed X-ray, optical and radio emission, "afterglow", associated with γ-ray bursts (GRBs) is consistent with fireball models, where the emission are produced by relativistic expanding blast wave, driven by expanding fireball at cosmogical distances. The emission mechanisms of GRB afterglow have been discussed by many authors and synchrotron radiation is believed to be the main mechanism. The observations show that the optical light curves of two observed gamma-ray bursts, GRB970228 and GRB GRB970508, can be described by a simple power law, which seems to support the synchrotron radiation explanation. However, here we shall show that under some circumstances, the inverse Compton scattering (ICS) may play an important role in emission spectrum and this may influence the temporal properties of GRB afterglow. We expect that the light curves of GRB afterglow may consist of multi-components, which depends on the fireball parameters.

Keywords: γ ray: bursts - shock waves

1. INTRODUCTION

The origin of γ-ray bursts has been one of the great unsolved mysteries in high energy astrophysics for about 30 years. The recent discovery of fading sources at X-ray and optical wavelengths coincident with the locations of some γ-ray bursts thus provides a good opportunity to probe the nature of these high energy events ((Costa et al. 1997; van Paradijs et al. 1997; Bond et al. 1997)). The detection of absorption lines in the optical afterglow of GRB970508 provides the first direct estimate of source distance, constraining the redshift of GRB970508 to the range of $0.8 \leq z \leq 2.3$ ((Metzger et al. 1997)).

The observed properties of GRB afterglow are broadly consistent with models based on relativistic blast waves at cosmological distances ((Mészáros & Rees 1997; Vietri 1997; Waxman 1997a; 1997b; Wijers Rees & Mészáros 1997; Wei & Lu 1998)). In fireball models of GRB after-

K.S. Cheng et al. (eds.), Stellar Astrophysics, 385–391.

glow, the huge energy released by an explosion ($\sim 10^{52}$ $ergs$) is converted into kinetic energy of a shell expanding at ultra-relativistic speed. After the main GRB event occurred, the fireball continues to propagate into the surrounding gas, driving an ultra-relativistic shock into the ambient medium. The expanding shock continuesly heats fresh gas and accelerates relativistic electrons to very high energy, which produce the delayed emission on timescale of days to months.

There are only two γ-ray bursts, GRB970508 and GRB970228, related with which afterglows in optical band have been observed. In the fireball model, it is widely believed that afterglow is produced through synchrotron radiation of relativistic electrons, and the effect of inverse Compton scattering can be neglected, which semms to be supported by the observations that the light curves of above two gamma-ray bursts can be described by a simple power law ((Galama et al. 1997)). However, here we show that under some circumstances, the ICS may have important effect on the emission spectrum and this will affect the light curve of GRB afterglow.

In next section, we use two different methods to calculate the specific intensity of ICS. One is to calculate the emission spectrum directly, the other is to estimate the relative importance of ICS and synchrotron radiation by comparing the energy density of magnetic field and synchrotron radiation. Of course, these two methods give almost the same result. In section 3, we discuss the effect of ICS on the temporal behavior of GRB afterglow, we find that, in general cases, the light curve may be diverse. Finally some discussions and conclusions are given in section 4.

2. THE INTENSITY OF INVERSE COMPTON SCATTERING

The process of inverse Compton scattering has also been discussed by several authors (e.g. (Waxman 1997a)), but they have not considered the influence of ICS on GRB afterglow, and they think ICS to be not important in GRB afterglow. Here we present a detailed calculation of the intensity of ICS using two different methods.

First, we calculate the emission spectrum of ICS directly. Assuming that after shock acceleration, the electrons have a power law energy distribution $N(\gamma) = k_e \gamma^{-p}$ $(\gamma_{min} \le \gamma \le \gamma_{max})$, then the photon spectrum of ICS can be expressed as ((Blumenthal & Gould 1970))

$$\frac{dN}{dtdE} = \frac{3}{8}\sigma_T c k_e f(p) E^{-(p+1)/2} \int \epsilon^{(p-1)/2} n(\epsilon) d\epsilon \tag{1}$$

where σ_T is the Thomson cross section, $f(p) = 2^{p+3}\frac{p^2+4p+11}{(p+3)^2(p+1)(p+5)}$, and c is the speed of light. We assume that the soft photon spectrum also obeys a power law, $n(\epsilon) = n_0\epsilon^{-\alpha}$ $(\epsilon_1 \le \epsilon \le \epsilon_2)$. Note that, in our case, the soft photons are produced by synchrotron radiation from these same electrons, so $\alpha = (p+1)/2$. Then from eq.(1) we have

$$\frac{dN}{dtdE} = \frac{3}{8}\sigma_T c k_e f(p) n_0 E^{-(p+1)/2}\ln(\frac{\epsilon_2}{\epsilon_1}) \tag{2}$$

In the fireball model, the electron density in the comoving frame is $n_e = \int N(\gamma)d\gamma = \Gamma n_1$, where n_1 is the density of surrounding gas and Γ is the bulk Lorentz factor, then we have $k_e = (p-1)\Gamma n_1 \gamma_{min}^{p-1}$. Thus we can obtain the ratio of the specific intensity of ICS at peak energy ($E = \epsilon_n = \gamma_{min}^2\epsilon_m$) to the synchrotron radiation at peak energy ($E = \epsilon_m$)

$$y = \frac{I_{ICS}}{I_{syn}} = \frac{3}{8}\sigma_T(p-1)n_1 f(p) r \ln(\frac{\epsilon_2}{\epsilon_1}) \tag{3}$$

where $\ln(\frac{\epsilon_2}{\epsilon_1}) = 2\ln(\frac{\gamma_{max}}{\gamma_{min}})$. The minimum and maximum electron Lorentz factors can be estimated by physical processes. For a power law energy distribution, the average electron Lorentz factor $\bar{\gamma} \sim \gamma_{min}$ for $p > 2$, and it is widely believed that after the shock acceleration, the average electron energy should be $\bar{\gamma} \sim \xi_e(m_p/m_e)\Gamma$, where m_p (m_e) is the mass of proton (electron). The maximum electron energy may be limited by shock acceleration and radiative energy loss. It has been shown that $\gamma_{max} \sim 5 \times 10^7 B^{-1/2}$ ((Cheng & Wei 1996)), where B is the magnetic field strength, which is usually written as $B = (\xi_B 8\pi\Gamma^2 n_1 m_p c^2)^{1/2}$, so $\ln(\frac{\gamma_{max}}{\gamma_{min}}) \sim 10$. In addition, after the shock acceleration the spectral index of electron distribution p is typically between 2 and 3 (e.g. (Blandford & Eichler 1987)), therefore we see that $y \simeq 10^{-24}n_1 r$.

There is another way to estimate the intensity of ICS, i.e. to calculate the ratio of the synchrotron radiation energy density (u_{syn}) to the magnetic energy density (u_B), $R = u_{syn}/u_B$, $u_B = B^2/8\pi$ and $u_{syn} \sim n_e P_{syn} r/\Gamma c$, where P_{syn} is the synchrotron radiation power. This method is simpler and more important, since the value of R indicates which process is more efficient for electron energy loss, inverse Compton scattering or synchrotron radiation, so it is necessary to calculate the value of R carefully. It is easy to show that $R \simeq 10^{-24}\bar{\gamma}^2 n_1 r$. In the afterglow model, if the fireball expands outward adiabatically, we have

$$r/r_0 = (t/t_0)^{1/4} \qquad \Gamma/\Gamma_0 = (t/t_0)^{-3/8} \tag{4}$$

where the deceleration radius $r_0 = r_d = 5\times 10^{16}(E_{51}/n_1)^{1/3}(\Gamma_0/100)^{-2/3}$ (cm), and the characteristic time $t_0 \simeq r_0/2\Gamma_0^2 c \simeq 10^2(E_{51}/n_1)^{1/3}(\Gamma_0/100)^{-8/3}$ (s)

((Mészáros & Rees 1997)) with E_{51} being the burst energy in units of 10^{51} *ergs* and Γ_0 the initial Lorentz factor. Then we obtain

$$R = 17\xi_e^2 n_1^{1/2} E_{51}^{1/2} t_{day}^{-1/2} \tag{5}$$

Obviously, the emission power of ICS is not much smaller than that of synchrotron radiation even for several days after the GRB event, thus the effect of ICS should not be neglected. The ratio of intensities of ICS to synchrotron radiation at peak energy is $I_{ICS}/I_{syn} \sim R\epsilon_{syn}/\epsilon_{ICS} \sim R\bar{\gamma}^{-2} \sim 10^{-24} n_1 r$, which is in agreement with the previous result.

3. THE EFFECT OF ICS ON GRB AFTERGLOW

In the case where the effect of ICS cannot be neglected, the emission spectrum should consist of two components. There exists a critical energy ϵ_c, below which the spectrum is dominated by synchrotron radiation and above which the spectrum is dominated by ICS, our purpose is to calculate the value of ϵ_c.

We assume that, in the comoving frame, the synchrotron radiation intensity has the form $I_\epsilon \propto \epsilon^{-\alpha}$ for $\epsilon < \epsilon_m$ and $I_\epsilon \propto \epsilon^{-\beta}$ for $\epsilon > \epsilon_m$. Since in our situation the soft photons produced through synchrotron radiation are scattered by the same electrons, so the Compton scattered spectrum should have nearly the same form as that of synchrotron radiation, i.e. $I_\epsilon \propto \epsilon^{-\alpha}$ for $\epsilon < \epsilon_n$ and $I_\epsilon \propto \epsilon^{-\beta}$ for $\epsilon > \epsilon_n$, therefore the total intensity is $I_\epsilon \propto \epsilon^{-\alpha}$ for $\epsilon < \epsilon_m$ or $\epsilon_c < \epsilon < \epsilon_n$, and $I_\epsilon \propto \epsilon^{-\beta}$ for $\epsilon_m < \epsilon < \epsilon_c$ or $\epsilon > \epsilon_n$. Then from eq.(4) and the relation $I_{\epsilon_m}(\frac{\epsilon_c}{\epsilon_m})^{-\beta} = I_{\epsilon_n}(\frac{\epsilon_c}{\epsilon_n})^{-\alpha}$ we can obtain the value of ϵ_c

$$\epsilon_c = f(\alpha,\beta)\xi_B^{1/2}\xi_e^{2-\frac{2\alpha}{\beta-\alpha}} n_1^{-\frac{3-\alpha}{4(\beta-\alpha)}} E_{51}^{\frac{1}{2}-\frac{1+\alpha}{4(\beta-\alpha)}} t_{day}^{-\frac{3}{2}-\frac{1-3\alpha}{4(\beta-\alpha)}} \quad eV \tag{6}$$

where $f(\alpha,\beta) = 20 \times 17^{-\frac{1}{\beta-\alpha}} 10^{\frac{6(1-\alpha)}{\beta-\alpha}} 8^{\frac{2(1-\alpha)}{\beta-\alpha}}$.

It has been shown that, in the adiabatic case, the electron Lorentz factor $\gamma_e \propto t^{-3/8}$, the typical energy $\epsilon_m \propto t^{-3/2}$, and the comoving specific intensity of synchrotron radiation at peak energy is $I'_{\epsilon_m} \propto t^{-1/8}$ ((Mészáros & Rees 1997)). From the relation $I_{ICS}/I_{syn} \sim R\gamma_e^{-2}$ it is easy to show that the intensity of ICS at peak energy $I'_{\epsilon_n} \propto t^{1/8}$ and $\epsilon_n \propto t^{-9/4}$, then the observed peak flux $F_{\epsilon_m} \propto t^2\gamma^5 I'_{\epsilon_m} \propto t^0 \sim$ constant, and $F_{\epsilon_n} \propto t^2\gamma^5 I'_{\epsilon_n} \propto t^{1/4}$. Therefore we can conclude that, if our observation is fixed at energy ϵ, then the observed flux $F_\epsilon \propto F_{\epsilon_m}(\frac{\epsilon}{\epsilon_m})^{-\alpha} \propto t^{-\frac{3}{2}\alpha}$ for $\epsilon < \epsilon_m$, $F_\epsilon \propto F_{\epsilon_m}(\frac{\epsilon}{\epsilon_m})^{-\beta} \propto t^{-\frac{3}{2}\beta}$ for $\epsilon_m < \epsilon < \epsilon_c$, $F_\epsilon \propto F_{\epsilon_n}(\frac{\epsilon}{\epsilon_n})^{-\alpha} \propto t^{\frac{1}{4}-\frac{9}{4}\alpha}$ for $\epsilon_c < \epsilon < \epsilon_n$, and $F_\epsilon \propto F_{\epsilon_n}(\frac{\epsilon}{\epsilon_n})^{-\beta} \propto t^{\frac{1}{4}-\frac{9}{4}\beta}$ for $\epsilon > \epsilon_n$.

Here the most interesting quantity is the critical energy ϵ_c, which is dependent on the fireball parameters, i.e. the fireball energy, surrounding gas density, energy fractions in electrons and magnetic field, and the spectral index of synchrotron radiation. In particular, it is easy to show that the effect of inverse Compton scattering is important only for large values of spectral index. As an example, we take $\alpha = 0.25$, $\beta = 1.4$ (these values are consistent with the observed γ-ray burst spectra), then the value of ϵ_c

$$\epsilon_c = 1.85(\frac{\epsilon_{syn}}{100KeV})(\frac{\gamma_0}{600})^{-4}\xi_e^{-0.43}n_1^{-1.1}E_{51}^{0.23}(\frac{t}{6\text{day}})^{-1.55} \quad eV \quad (7)$$

where we have used the relation $\epsilon_{syn} = 5.2\times10^{-3}\gamma_0^4\xi_B^{1/2}\xi_e^2 n_1^{1/2}\ eV$, which is the peak energy of γ-ray burst spectrum. So we can see that, for the typical values of GRB events, i.e. $E \sim 10^{51}\ ergs$, $n \sim 1\,cm^{-3}$, $\xi_e \sim 0.3$, the critical energy ϵ_c crosses the optical band for about six days after the burst. Furthermore, we can calculate the peak energy of ICS

$$\epsilon_n = 1.87(\frac{\epsilon_{syn}}{100KeV})(\frac{\gamma_0}{600})^{-4}(\frac{\xi_e^2}{0.1})n_1^{-3/4}E_{51}^{3/4}(\frac{t}{60\text{day}})^{9/4} \quad eV \quad (8)$$

Obviously, if we take the same parameters as above, then ϵ_n should cross the optical band for about two months, so we expect that the time during which the optical flux varies as $F \propto t^{\frac{1}{4}-\frac{4}{9}\alpha}$ rather than $F \propto t^{-\frac{3}{2}\beta}$, and according to the values of α and β, the brightness may decrease very slowly than before.

4. DISCUSSION AND CONCLUSION

The detection of γ-ray burst in the optical and radio bands has greatly furthered our understanding these objects, especially the shape of the light curves of GRB afterglow provide important information on exploring their emission mechanisms. Here we present a detailed calculation to show that the temporal properties of GRB afterglow may be quite diverse due to the effect of inverse Compton scattering.

We have shown that the inverse Compton scattering may play an important role in the GRB afterglow. From eq.(5) one sees that $R \propto t^{-1/2}$, it seems that ICS is more efficient than synchrotron radiation in the early time of GRB afterglow (especially in γ-ray burst epoch, it seems that ICS dominates the electron energy loss), however, in fact this is not true, since in the early time, the photon energy of synchrotron radiation is so high that in the electron rest frame the photon energy is greatly larger than m_ec^2, so the Compton scattering occurs in the extreme Klein-Nishina limit, and thus the scattering is not efficient. It can easily be shown that Compton scattering is efficient only when about $t > 10^3\ s$ after GRB event.

Here we point out that the light curves of GRB afterglow may consist of four components, which depend on the characteristic energies ϵ_m, ϵ_c, ϵ_n and the detector frequency ϵ. However, it should be noted that not all the GRB afterglows could contain so many components, the necessary condition for having four components is $\epsilon_c < \epsilon_n$. It is obvious that if $\epsilon_c > \epsilon_n$, the light curves should have three components. In addition, if the ICS efficiency is low enough that there is no intersection between the spectrum of ICS and synchrotron radiation, then the light curves only have two components.

It should be noted that the effect of inverse Compton scattering is strongly dependent on the model parameters, especially on the spectral index of synchrotron radiation, the contribution from inverse Compton scattering is important only for those bursts with large spectral indices, and the spectral indices may be estimated by the ratio of X-ray flux to the optical flux.

The observations from BATSE show that the spectra of GRBs are rather diverse, varying from burst to burst. For most bursts, their spectra can be described by a broken power law, and the distribution of the spectral indices below and above the peak energy is rather wide. In general, below peak energy the spectral indices (α) are mainly between 0 and 0.5, while above peak energy the slopes (β) are typically between 1 and 1.5 ((Band et al. 1993)), so we expect that the temporal properties of GRB afterglows (which strongly depend on the parameters α and β) would be very diverse, and contain multi-components.

Acknowledgements

This work is supported partly by the National Natural Science Foundation of China.

References

Band, D., et al., 1993, ApJ, 413, 281
Blandford, R., & Eichler, D., 1987, Phys. Report, 154, 1
Blumenthal, G., & Gould, R., 1970, Rev. Mod. Phys., 42, 237
Bond, H.E., et al., 1997, IAUC. 6654
Cheng, K.S., & Wei, D.M., 1996, MNRAS, 283, L133
Costa, E., et al., 1997, Nature, 387, 783
Galama, T., et al., 1997, astro-ph/9712322
Mészáros, P., & Rees, M.J., 1997, ApJ, 476, 232
Metzger, M.R., et al., 1997, Nature, 387, 878

van Paradijs, J., et al., 1997, Nature, 386, 686
Vietri, M., 1997, ApJ, 478, L9
Waxman, E., 1997a, ApJ, 485, L5
Waxman, E., 1997b, ApJ, 489, L33
Wei, D.M., & Lu, T., 1998, ApJ, 499,754
Wijers, A.M.J., Rees, & Mészáros, P., 1997,MNRAS, 288, L51

A POSSIBLE MECHANISM FOR THE METALLICITY OF THE FIRST GENERATION STAR

Y. Lu[1,2] G. Zhao[1] J.R. Shi[1]

[1] *Beijing Astronomical Observatory, Chinese Academy of Sciences, Beijing, 100012, China*

[2] *Department of Physics, Normal University of Huazhong, Wuhan, 430079, China*

ly@yac.bao.ac.cn

zg@orion.bao.ac.cn

sjr@yac.bao.ac.cn

Abstract We propose that the absence of the metal zero star, the first generation star (FGS), may be explained by the mechanism for metallicity enrichment of a star in the Galaxy based on the host Galaxy's disk-star interactions. The mechanism involves an accreting of interstellar medium (ISM). The zero metal star may accrete a small amount of metal rich material of the ISM and polluted the star's composition. We suggest here that the expected degree of the polluted relative average metallicity $[Fe/H]$ of a star may be $[Fe] \sim -4.5$.

Keywords: Galactic chemical evolution, solar neighborhood, G-dwarf, supernova

1. INTRODUCTION

Up to a decade ago, searches for the first generation star (FGS) whose chemical composition is strictly left by the Big Bang had led to the result that the observational limit towards the lowest metallicities $[Fe/H]$ (Beers, Preston & Shectman 1985) is now about $[Fe/H] = -4$. No a star with zero metallicity has been found, or even no one star near $[Fe/H] = -5$. The idea that a FGS could be born from a gas having the chemical composition inherited from the Big Bang was inescapable, and even if such stars were not observed. Many scenarios have been invoked to explain why no FGS have been found (e.g. Truran & Cameron 1971; Yoshii 1981; Cayrel 1986; De Murchi & Paresce 1995). We here propose that the absence of FGS may be explained by the mechanism for metallicity enrichment of a star in our Galaxy based on the host

K.S. Cheng et al. (eds.), Stellar Astrophysics, 393–397.

Galaxy's disk-star interactions. The mechanism involves accreting of interstellar medium (ISM) by a passing star through the host Galaxy's disk.

2. THE CHEMICAL EVOLUTION OF THE ISM

We adopt a multizone model of the Galactic chemical enrichment proposed by Portinari et al.(PCB98). In each ring the gas is assumed to be chemically homogeneous, and the normalized gas density for each chemical species E is $G(E,r,t)$. The equations for the evolution of $G(E,r,t)$ at a given Galactocentric radius r and time t is

$$\begin{aligned}
\frac{dG(E,r,t)}{dt} &= -Z_{ISM}(E,r,t)\wp(r,t) \\
&+ \int_{M_l}^{M_{b1}} \wp[r,t-\tau(M)]\phi(M) \times \left[\frac{M_{ej}(E,M,Z)}{M_{av}}\right] dM \\
&+ \beta \left[\frac{M_{ej}^I(E)}{M_{av}}\right] \int_{M_{b1}}^{M_{b2}} \phi(M_b) \int_{\mu_m}^{1/2} f(\mu)\wp[r,t-\tau(\mu M_b)]d\mu dM_b \\
&+ (1-\beta) \int_{M_{b1}}^{M_{b2}} \wp[r,t-\tau(M)]\phi(M) \times \left[\frac{M_{ej}(E,M,Z)}{M_{av}}\right] dM \\
&+ \int_{M_w}^{M_{b2}} \wp[r,t-\tau(M)]\phi(M) \times \left[\frac{M_{ej}(E,M,Z)}{M_{av}}\right] dM \\
&+ \left[\frac{dG(E,r,t)}{dt}\right]_{inf} \qquad (1)
\end{aligned}$$

Each term of Eq.(1) is the same meaning as that of PCB98. The contribution of the infalling gas to the evolution of the normalized gas fraction $G(r,t)$ is (Arnett 1971; Dwek, 1998; Timmes et al. 1995)

$$\left[\frac{dG(r,t)}{dt}\right]_{inf} = \frac{1}{\sigma(t_G)} \left[\frac{d\sigma_{tot}(r,t)}{dt}\right]. \qquad (2)$$

Where $\sigma(t_G) = \sigma_\odot = 50 M_\odot pc^{-2}$ (PCB98) is the total surface density at the present age t_G of the Galaxy . With these assumptions, we could derive the evolution of the gas fraction and the mass fraction $[Fe/H]$ in the ISM of Galaxy as a function of time, which is the same as the result of PCB98's Model B.

3. ACCRETION OF THE ISM (GAS) BY FGS IN THE GALAXY

When a star passes through the Galaxy, it obtains one ample opportunity to accrete the ISM following subsonically in its vicinity in the Galaxy disk. The rate of accretion of a star strongly depends on the relative velocity between the star and interstellar gas of the Galaxy. If we assume that an interstellar gas of the Galaxy is unperturbed and its velocity vector $\vec{V_g}$ is assumed to be the circular velocity $\vec{V_{rot}}$ ($V_{rot} = \sqrt{\frac{GM_{tot}}{r}}$) in the Galactic disk. Then the relative velocity between a star and a gas of the ISM is

$$\vec{V_{rel}} = \vec{V_s} - \vec{V_{rot}}; \tag{3}$$

We treat the star as a test particle. The star with velocity vector $\vec{V_s}$ moves in the gravitational field of the Galaxy with a box orbit bounded by the apo and the peri-galactic distances, r_{ap},r_{pr}, respectively. The maxium box angle is ξ_{max} (Yoshii & Saio , 1979). The accretion process of a solar-type star ($m \sim 0.9 M_\odot$)is similar to the three-dimensional problem considered by Bondi & Hoyle (1944). Then the accretion rate onto a star is

$$\dot{m} = 2.5\pi G^2 m^2 \rho_{ISM}(r,t) v_{rel}(t)^{-3}, \tag{4}$$

$\rho_{ISM}(r,t)$ is the gas density of ISM of the Galaxy and

$$\rho(r,t) \quad = \quad \sigma_\odot G(r,t)/2H \tag{5}$$

The average accretion rate $\dot{m}_a$ is obtainted by multiplying equation $\dot{m}$ by the fraction of time that stars spend within the interstellar gas of the Galaxy disk. This fraction can be estimated from the accretion of the time τ_{acc} during which the star is passing through the Galaxy disk: If a star vertical velocity is assumed to $v_z(t) \sim v_{rel}(t)/\sqrt{3}$, and if this mass is accreted twice per orbital period ($P = 2\pi R/v_{rot}$), then the average accretion rate is $\dot{m}_a \sim (\tau_{acc}\dot{m})/P$. Although many elaborate analyses have been made, the masses of FGS have not been well converged (Carlberg 1981; Palla et al. 1983; Yoshii & Saio 1986; Uehara et al.1996; Nakamura & Umemura 1999). In this paper, we assume that the mass of the star is about $0.9 M_\odot$, and the age of FGS is $t_{age} \sim 10$ Gyr. We adopt the thickness of the disk is $H = 10^{-2} R_\odot$ (Van der Kruit & Searle 1982), and the maxium box angle is $\xi_{max} = 0.2$ (Yoshii & Saio, 1979). The average accreted mass m_a through the ISM with a typical star of a mass $0.9 M_\odot$ is

$$m_a = \int_0^{t_{age}} 2.5\pi G^2 m^2 \rho_{ISM}(r,t) v_{rel}(t)^{-3} dt \ . \tag{6}$$

This accreted matter, although minuscule, can affect the elemental abundances in the surface convective zones for stars.

4. METAL ENRICHMENT FOR THE ZERO METAL FGS

The interstellar medium contains heavy elements which are synthesized in stars and are dispersed through supernovae explosions. If a sufficient amount of material which contains heavy elements is accreted on a star, the stellar surface would be increasely contaminated with time. The degree of enrichment depends on the depth to which the accreted material is mixed. Here we assume that it is continuously mixed only in the convective envelope of a star with the mass of m_c. For the star with the mass $m \sim 0.9M_\odot$, the typical values of m_c is $m_c \sim 10^{-4}M_\odot$ (Pinsonneault et al. 1992). The enriched abundance of FGS is

$$Z_*(Fe) = m_a(Fe)/m_c \ , \tag{7}$$

and

$$[Fe/H] = lg\left(\frac{Z_g(Fe)}{Z_\odot(Fe)}\right); \tag{8}$$

Then

$$m_a(Fe) \quad = \quad m_a \frac{Z_g(Fe)}{Z_\odot} * \left(\frac{m(Fe)}{m}\right)_\odot ; \tag{9}$$

Where $Z_g(Fe)$ is the present enrichment abundance of ISM of the Galaxy. For the present illustration, we take the star moves in the gravitational field of the Galaxy with box angle $\xi_{max} = 0.2$. Then, from Eq.(1.8) to Eq.(1.10), we obtain

$$[Fe/H]_* \quad = \quad [Fe/H] + lg\left(\frac{m_a}{m_c}\right) \sim -4.5 \tag{10}$$

5. CONCLUSIONS

From the above investigation, Our main results are:

The metal abundance $[Fe/H]$ of the FGS (with mass $0.9M_\odot$) polluted by accreting the ISM with the box angle $\xi_{max} = 0.2$ amounts to $[Fe/H] \sim -4.5$. Then the absence of FGS which has the chemical composition inherited from the Big-Bang could be explained by the hypothesis of enrichment of the FGS surface metal abundance due to accretion of metal-rich material through encounters with the ISM of our Galaxy.

The mass accreted on a FGS strongly depends on the relative velocity between a FGS and the ISM as well as the path length of the orbit in the region occupied by the ISM. If ξ_{max} is smaller, then the relative velocity v_{rel} remains smaller, and the probability increase the accreting the gas of ISM because of $\dot{m} \propto v_{rel}^{-3}$. However, the stars with large ξ_{max} have so large relative velocity that they could escape the Galaxy. Since most of the metal poor stars are in eccentric orbits much inclined (large ξ) to a Galactic plane (Eggen et al. 1962), this is why few very metal-poor stars could be observed.

Acknowledgements

The research work is supported by the National Natural Science Foundation of China.

References

Arnett, W.A. (1971). APJ, 219: 1008–1013
Beers, T.C., Preston, G.W., Shectman, S.A. (1985). AJ, 90: 2089–2092
Bondi, H., & Hoyle, F. (1944). MNRAS, 104: 273
Carlberg, R.G. (1981). MNRAS, 197: 1021
Cayrel, R. (1986). A&A, 168: 81
De Marchi, G., & Paresce, F. (1995a). A&A, 304: 202
Dwek, E. (1998). APJ, 501: 643
Eggen, O.J., Lynden-Bell, D. & Sandage, A.R. (1962). APJ, 136: 748
kamura, F. & Umemura, M. (1999). APJ, 515: 239-248
Palla, F., Salpeter E.E., & Stahler, S.W. (1983). APJ, 271:632
Pinsonneault, M.H., Deliyannis, C.P. & Demarque, P. (1992). APJS, 78: 179
Portinari L., Chiosi, C. & Bressan, A. (1998). A&A, 334: 505 (PCB98)
Timmes, F.X., Woosley, S.E., & Weaver, T.A. (1995). APJS, 98: 617 (TWW95)
Truran, J.W., & Cameron, A.G.W. (1971). APSS, 14: 179
Uehara,H., et al. (1996). APJ, 473: L95
Van der Kruit, P.C. & Searle, L. (1982). A&A, 95: 105
Yoshii, Y. (1981). A&A, 97: 280
Yoshii, Y. & saio H. (1979). PASJ, 31: 339

THE INFLUENCE OF THE STELLAR MASS RATIO ON SPIRAL SHOCKS IN ACCRETION DISKS AROUND COMPACT OBJECTS

Giuseppe Lanzafame[1],Fabio Maravigna[2],Gaetano Belvedere[3]

[1] glanzafame@alpha4.ct.astro.it

[2] fma@sunct.ct.astro.it

[3] gbelvedere@alpha4.ct.astro.it

[1] *Osservatorio Astrofisico di Catania, Viale A. Doria, 6, Città Universitaria, I-95125, Catania, Italy*

[23] *Institute of Astronomy, University of Catania, Viale A. Doria 6, 95125 Catania ITALY*

Abstract We investigated, in the Smooth Particle Hydrodynamics (SPH) framework, the development of spiral structures and shock fronts in the radial flow of accretion discs in close binary systems. These shock waves take place when the initially radial flow penetrating the disc bulk, reduces substantially its speed becoming suddenly subsonic. To this purpose, keeping constant the mass of the compact primary ($M_1 = 1M_\odot$), the separation between the two components and the injection speed at the inner lagrangian point L_1 (close to the local sound speed), we carried out four 2D SPH simulations for four values of the stellar mass ratio M_2/M_1.

The results show that spiral structures and shock fronts develop for low values of M_2/M_1, whilst they become less evident for higher values. This behaviour is interpreted in terms of initial specific angular momentum at L_1, whose position, relative to the companions, depends on the stellar mass ratio.

Keywords: spiral shocks - accretion disks - secondary to primary mass ratio

1. INTRODUCTION

In the present work we performed a systematic investigation on accretion discs around a compact primary ($1M_\odot$), in a close binary system, considering different secondary masses ($0.5M_\odot$, $1M_\odot$, $2.5M_\odot$, $5M_\odot$), and taking a constant injection flux speed of the order of the local sound

K.S. Cheng et al. (eds.), Stellar Astrophysics, 399–403.

speed ($\simeq 12Km/s$). Our purpose was to study the conditions under which spiral structures and spiral shock waves in the radial flux may develop as a function of the stellar mass ratio M_2/M_1.
Indeed, our results showed that M_2/M_1 is a dominant factor in spiral structure and shock formation. In fact, these structures become more and more evident for smaller values of M_2/M_1. We found that the development of spiral structures and shocks in the radial flow at the outer edge of accretion discs, is well correlated to the angular momentum of the injected gas at L_1, since the relative position of the inner lagrangian point, with respect to the primary, is a function of M_2/M_1.
These simulations were carried out in 2D, in order to limit the damping effect of the artificial viscosity and to avoid too long computational times. In fact, in 3D, the artificial viscosity (Monaghan, 1985, Monaghan and Lattanzio, 1985), seems to prevent spiral structures formation at least for $\gamma = 1.01$ (Lanzafame and Belvedere, 1997).

2. THE MODEL

The binary system characteristics are determined by the masses of the companions and the separation between their centres: $d_{12} = 3 \cdot 10^{11}$ cm, the latter being fixed for all four models.
For the model equations and the physical meaning of our parametrization in terms of the adiabatic index γ, we refer the reader to Lanzafame et al. (1994a).
The adopted value, $\gamma = 1.01$, simulates a physical situation in which there is a local equilibrium between energy inputs and outputs. In order to make the equations dimensionless, we adopted the following normalization factors: $M = M_1 + M_2$ for the masses, $v_0 = \sqrt{G\ (M_1 + M_2)/d_{12}}$ for velocities (so that the orbital period is normalized to 2π), $\rho_0 = 10^{-9}$ g/cm^3 for the density, $P_0 = \rho_0 v_0^2\ dyn/cm^2$ for the pressure, $T_0 = (\gamma - 1) \cdot v_0^2\ m/k_B$ K for the temperature and v_0^2 for the thermal energy per unit mass.
The adopted Kernel width, or resolution length, was $h = 0.005$, and the integration (or interpolation) domain is a sphere of radius 0.6, centered on the primary.
As to the physical conditions at the inner and outer disc edges, we refer the reader to Lanzafame et al. (1994a).

3. RESULTS AND CONCLUSIONS

In all four simulations we got a stationary state after ≈ 75 orbital periods for the lowest mass ratio model and after ≈ 35 orbital periods

for the highest mass ratio model. As in previous papers (Molteni et al. 1991, Lanzafame et al. 1992, 1993, 1994ab) we get a "stationary state" when, statistically, the total number of particles does no longer change: $dN/dt = 0$.
Fig. **??** shows the accretion discs for the four models considered. It is clear that the value of the stellar mass ratio M_2/M_1 affects disc formation, structure and dynamics, and determines its dimension and total mass.
It is evident the presence of well defined spiral structures and radial shock fronts for low values of the mass ratio such as $M_2/M_1 = 0.5$ and $M_2/M_1 = 1$, whilst these structures are less remarkable for $M_2/M_1 = 2.5$ and $M_2/M_1 = 5$. Consequently,we suggest a mass ratio $M_2/M_1 \simeq 2.5 \,\text{div}\, 5$ to be an upper limit for spiral structure formation.
The resulting temperature radial distribution ($T \propto r^{-3/4}$) agrees with the standard analytical models which include radiative cooling in the disk bulk. This behaviour indicates once more that a disc with a small γ-value represents a reasonable approximation to a real disc, except in the outermost regions.
The resulting specific angular momentum $J = r^2\Omega \propto r^{1/2}$ is statistically in good agreement with the keplerian law, this confirming again the validity of our approach for all four models.
The main results of our simulations can be interpreted as follows.
It is clear that the stellar mass ratio determines the position of L_1 relative to the two companions, and the Roche-lobes geometry and size.
Now, increasing M_2/M_1 implies L_1 to get closer to the primary, this resulting in an initial injection flow with smaller and smaller angular momentum and higher and higher radial acceleration. Thus, the direction of the flow tends to be more and more radial, that means that the injection flow penetrates deeper and deeper into the disc bulk. In this physical situation, the particle flow tends to be accreted directly onto the primary and is less effective to convert mechanical energy into thermal energy by collisional interaction. This explains why, increasing M_2/M_1, discs are smaller, less massive, colder and show a higher $\dot{M}_{acc}$. Moreover, the pressure forces in the disc bulk are clearly smaller for higher M_2/M_1 since, the particle density being approximately equal as a function of the radial distance, the temperature lowers on average. This reduces the braking effect on the supersonic flow and thus the possibility of formation of spiral structures and radial shock fronts. On the other hand, notice that, comparing the two extreme models ($M_2/M_1 = 0.5$ and $M_2/M_1 = 5$), a temperature lowering implies a decrease in the sound speed c_s in the disc bulk ($c_s \sim T^{1/2}$), this also disfavouring a stationary transition from a supersonic regime to a subsonic one at the outer disc

edge. Instead, spiral structures and radial shock fronts at the outer disc edge are prominent when the azimuthal motion are relevant. This occurs when the initial angular momentum of the injection flow at the L_1 point is large, namely for low M_2/M_1, for nearly sonic injection speed.
Notice that the fading out of spiral structures for increasing M_2/M_1 is not a consequence of the lack of resolution due to the decreasing number of particles. Indeed, no consistent differences were found in a supplementary simulation for $M_2/M_1 = 5$, in which a smaller value $h = 0.003$ (that means more resolution) was adopted.

References

Lanzafame G., Belvedere G., Molteni D. 3D simulations of the SS Cyg accretion disc in the quiescient phase. *MNRAS*, 258:152–158, 1992.

Lanzafame G., Belvedere G., Molteni D. A three dimensional smoothed particle hydrodynamics simulation of the active phase of the SS Cyg-type discs and its implications for the mass transfer burst model. *MNRAS*, 263:839–850, 1993.

Lanzafame G., Belvedere G., Molteni D.: The 35-d modulation of the X-ray emission of Her X-1 in the framework of the SOD model: results of a three dimensional SPH simulation. *MNRAS*, 267:312–322, 1994a.

Lanzafame G., Belvedere G., Molteni D. SPH models of the tilted accretion disc of Her X-1. *Mem. Soc. Astron. It.*, 65(4):1193–1198, 1994b.

Lanzafame G., Belvedere G. Wind accretion in close binary systems: a 3D smoothed particle hydrodynamics simulation of disc structure and dynamics. *MNRAS*, 284:957–966, 1997.

Molteni D., Belvedere G., Lanzafame G. Three-dimensional simulation of polytropyc accretion discs. *MNRAS*, 249:748–754, 1991.

Monaghan J. J., Lattanzio J. C. A refined particle method for astrophysical problems. *A&A*, 149:135–143, 1985.

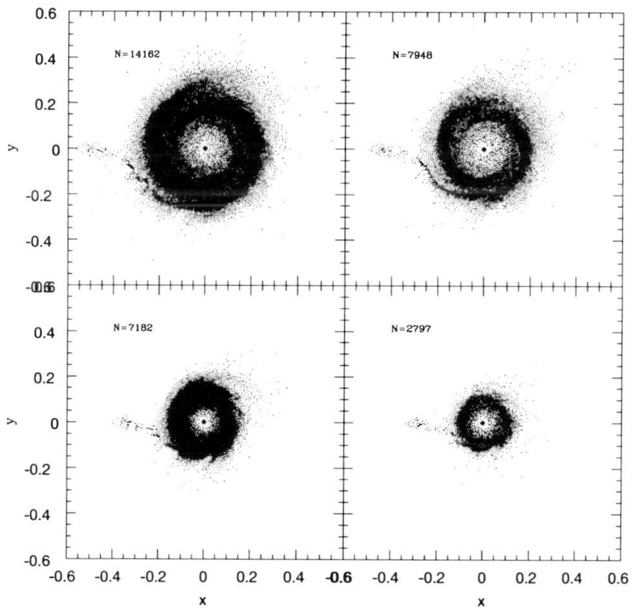

Figure 1 Plots of accretion discs for the four models. Top left $M_2 = 0.5M_\odot$; top right $M_2 = 1M_\odot$; bottom left $M_2 = 2.5M_\odot$; bottom right $M_2 = 5M_\odot$.

48V

ROTATING NEUTRON STARS FOR A NEW MICROSCOPIC EQUATION OF STATE

I. BOMBACI,* A.V. THAMPAN,◇ and B. DATTA ♭

*Dipartimento di Fisica, Università di Pisa, and INFN Sez. di Pisa, via Buonarroti, 2, I-56127 Pisa, Italy

◇Inter-University Centre for Astronomy and Astrophysics (IUCAA), Pune 411 007, India

♭Indian Institute of Astrophysics, Bangalore 560 034, India

Abstract Using a recent microscopic equation of state (EOS) for neutron star matter, we construct equilibrium sequences of rapidly rotating neutron stars in general relativity. The sequences are the normal and supramassive evolutionary sequences of constant rest mass. Also calculated is the radius of the marginally stable orbit and its dependence on the rotation rate Ω, relevant for modeling of kilo–Hertz quasi–periodic oscillations in X–ray binaries.

1. INTRODUCTION

The structural properties of rapidly rotating neutron stars are important in modeling a variety of phenomena of astrophysical interest, such as millisecond pulsars, low–mass X–ray binaries (LMXBs) and Quasi Periodic Oscillators (QPOs). Models of rapidly rotating neutron stars in general relativity must be constructed numerically. In the present contribution we report recent calculations by Datta, Thampan & Bombaci (1998), of equilibrium sequences of rapidly rotating neutron stars in general relativity. The various equilibrium sequences that we construct are the normal and the supramassive evolutionary sequences of constant rest mass.

A key input in determining the structure of neutron stars is the equation of state (EOS) of high density matter. The *core* of a neutron star consists of a quantum liquid composed of an uncharged mixture of neutrons, protons, electrons and muons in equilibrium with respect to the

K.S. Cheng et al. (eds.), Stellar Astrophysics, 405–409.

weak interaction (β–stable nuclear matter) [1]. Baldo, Bombaci & Burgio (1997) developed two models for the EOS of β – stable matter; the basic input of this microscopic approach being the bare nucleon–nucleon (NN) interaction that reproduces the experimental features of NN scattering experiments and the deuteron properties. For our presentation here, we use one of the EOS models (referred to as BBB1 EOS in Baldo *et al.* 1997), namely the one based on the Argonne V14 two–body nuclear force, implemented by the Urbana VII model for the three–body nuclear force. This microscopic EOS satisfies the basic requirements of reproducing the empirical saturation point for symmetric nuclear matter, the symmetry energy and the incompressibility parameter at the saturation density (Baldo *et al.* 1997). Moreover the BBB1 model has the desirable physical feature that the velocity of sound in the medium does not violate the causality condition. Therefore, it can be taken to be *realistic* EOS.

2. RAPIDLY AND RIGIDLY ROTATING RELATIVISTC STARS

We use the methodology described in detail in Datta, Thampan & Bombaci (1998) to calculate the structure of rapidly rotating neutron stars. For completeness, we briefly describe the method here. For a general axisymmetric and stationary space–time, assuming a perfect fluid configuration, the Einstein field equations reduce to ordinary integrals (using Green's function approach). These integrals may be self consistently (numerically and iteratively) solved to yield the value of metric coefficients in all space. Using these metric coefficients, one may then compute the structure parameters, moment of inertia and angular momentum corresponding to initially assumed central density and polar to equatorial radius ratio. The values of the structure parameters and the metric coefficients, so computed, may then be used (as described in Thampan & Datta 1998) to calculate parameters connected with stable circular orbits (like the innermost stable orbit and the Keplerian angular velocities) around the configuration in question.

3. RESULTS

The equilibrium sequences of rotating neutron stars depend on two parameters: the central density (ρ) and the rotation rate (Ω). For purpose

[1] At ultra–high density, matter might undergo a transition to other exotic hadronic components like hyperons, a K^- condensate or a deconfined phase of quark matter. Here, we do not consider such a possibility.

Table 1 Neutron star properties. The listed quantities are: angular velocity (Ω) in units of 10^4 $rad\ s^{-1}$ of the neutron star as measured by an observer at infinity, central density (ρ) in units of $g\ cm^{-3}$, equatorial radius (R) in km, gravitational mass (M_G) and baryonic mass (M_B) in solar units, eccentricity (e), moment of inertia (I) in units of 10^{45} $g\ cm^2$, ratio of the rotational energy to the total gravitational energy (T/W), radius (r_{orb}) of the innermost stable orbit in km.

Ω	ρ	M_G	M_B	R	e	I	T/W	r_{orb}
0.000	3.09E+15	1.788	2.082	9.646	0.000	1.428	0.000	15.845
1.095	2.56E+15	2.135	2.471	13.129	0.703	2.428	0.120	13.490
1.079	2.44E+15	2.133	2.468	13.264	0.706	2.465	0.120	13.558
0.403	1.40E+15	1.590	1.799	11.201	0.361	1.522	0.021	12.478

of illustration, we choose three limits in this parameter space. These are: (i) the static or non–rotating limit, (ii) the limit at which instability to quasi–radial mode sets in and (iii) the centrifugal mass shed limit. The last limit corresponds to the maximum Ω for which centrifugal forces are able to balance the inward gravitational force.

In Table 1 we list some neutron star properties corresponding to: the non-rotating maximum mass configuration (first row in Tab. 1); the maximum mass rotating configuration (second row); the maximum angular momentum configuration (third row); a star with $M_G = 1.59 M_\odot$, having a value of angular velocity equal to that of the fastest known pulsar: PSR 1937+21 (fourth row).

In Fig. 1 we show (a) the functional dependence of the gravitational mass with central density and (b) the gravitational mass–equatorial radius relation. The bold solid curve represents the non–rotating or static limit, and the bold dashed curve the centrifugal mass shed limit. The long dashed curve is the constant–Ω sequence corresponding to the period $P = 1.558$ ms. The thin solid curves that are roughly horizontal are the constant rest mass evolutionary sequences. The evolutionary sequences above the maximum stable non-rotating mass configuration are the supramassive evolutionary sequences, and those that are below this limit are the normal evolutionary sequences. The almost vertical thin dashed line is the limit for instability against quasi–radial modes. The supramassive evolutionary sequences beyond the quasi–radial mode instability limit are represented by dotted lines. The maximum mass sequence for this EOS corresponds to a rest mass value of 2.471 $M_\odot$. The supramassive sequences lie in the rest mass range of

2.356 $M_\odot < M_0 < 2.471\ M_\odot$. The gravitational mass of the maximum stable rotating configuration is 2.135 $M_\odot$and its radius is 13.129 km.

From Fig. 1(b) we see that EOS model BBB1 predicts the millisecond pulsar PSR 1937 + 21, with an assumed mass value of 1.4 $M_\odot$, to have an equatorial radius of 11 km.

References

Baldo, M., Bombaci, I., & Burgio, G.F. 1997, Astron. Astrophys. , 328, 274

Datta, B., Thampan, A.V., & Bombaci, I. 1998, Astron. Astrophys. 334, 943

Thampan, A.V., & Datta, B. 1999, MNRAS, 297, 570

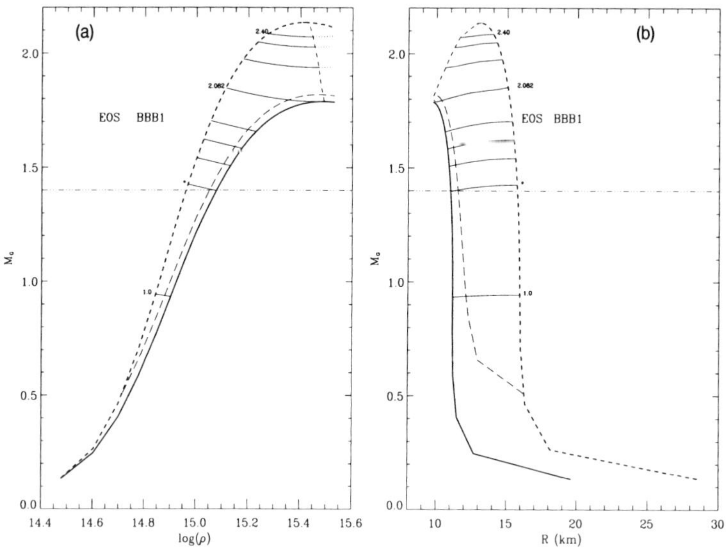

Figure 1 (a) The functional dependence of the gravitational mass (M_G) with central density (ρ) and (b) the gravitational mass–equatorial radius (R) relation. M_G is in solar units ($M_\odot$), ρ in units of $g\ cm^{-3}$ and R in kms. The bold solid curve represents the non–rotating or static limit, and the bold dashed curve the centrifugal mass shed limit. The long dashed curve is the constant–Ω sequence corresponding to the period $P = 1.558$ ms. The thin solid curves that are roughly horizontal are the constant rest mass evolutionary sequences. The evolutionary sequences above the maximum stable non-rotating mass configuration, are the supramassive evolutionary sequences and those that are below this limit, are the normal evolutionary sequences. The almost vertical thin dashed line is the limit for instability against quasi–radial modes. The supramassive evolutionary sequences beyond the quasi–radial mode instability limit are represented by dotted lines. The numbers against some of the curves represent the baryonic mass for the corresponding sequence; the equilibrium sequence that has 1.4 $M_\odot$configuration at the static limit is represented by an asterisk against it.

THE RADIAL, VERTICAL AND AZIMUTHAL INSTABILITY OF AN ISOTHERMAL DISK WITH THREE-DIMENSIONAL MAGNETIC FIELD

S.X.Ding[1,3], L.T. Yang[1] and Y.Lu [1,2]
[1]*Department of Physics, Huazhong Normal University, Wuhan, 430079, China*
[2]*Beijing Astronomical Observatory, Chinese Academy of Sciences, Beijing, 100012, China*
[3]*Department of physics, Xiang Fan University. Xiang Fan, 441053, China*

Abstract The radial, vertical and azimuthal instability of an isothermal magnetized accretion disk is examined in this paper. We find that the magneto-acoustic modes and non-axisymmetric mode are always unstable and the viscouse mode is always stable. The azimuthal perturbation affects the instability of the magneto-acoustic modes. But doesn't affect the instability of non-axisymmetric mode, in pure magnetized disk, but there are opposite effects in viscosity and magnetized disk. Our results are useful to explain the periodic variability of astronomical objects.

Keywords: Accretion disk – MHD – instability

1. INTRODUCTION

As a powerful and attractive theoretical mode, accretion disks are widely believed to be sources of high-energy radiation for X-ray binaries and active galatic nucleus (AGNs). Since the standard thin disk model was constructed, in the early 1970s (Shakura & Sunyaev 1973), the stability of the disk has become an important area in the theory of accretion disks. This is because a lot of quasi-periodic, or periodic luminosity variations of various astronomical objects such as Dwarf Novae, X-ray binaries, AGNs and Quasars have been successfully explained by instabilities in accretion disks. Magnetic field can play an important role in accretion disks. Recently, Balbus & Hawley (1991) showed that accretion disks are dynamically unstable to an axisymmetric shear instability whenever a weak vertical field is present. Wu et al. (1995) &

K.S. Cheng et al. (eds.), Stellar Astrophysics, 411–415.

Yang et al. (1995) studied the magnetized isothermal accretion disks and explained the luminosity variations of some astronomical objects successfully. More recently. Yang et al. (1997) examined the instability of an isothermal disk with three-dimensional magnetic fields. They used their model to explain the variability of FU Orions and T Tauri stars successfully.In most research mentioned above, only the radial pertturbations were considered. However, a more realistic analysis of the stability of a geometrically thin accretion disk should include both the radial and azimuthal perturbations because of the strong shear of viscous flow. In this paper, our aim is to investigate the radial-azimuthal instability of an isothermal accretion disk with three dimensional magnetic fields.

2. BASIC EQUATION AND DISPERSION RELATION

For simplicity, we assume an isothermal, geometrically thin accretion disk and adopt the pseudo-Newtonian potential (Paczynski & Wiita 1980):

$$\Phi = \frac{-GM}{R - R_g} \tag{1}$$

Here M is the mass of central object and is the Schwazchild radius, R_g is the distance from the accretion flow to the centre of the disk, and$R = (r^2 + z^2)^{1/2}$, where cylindrical coodinates (r, ϕ, z) are adopted. The magneto-hydrodynamical equations for the magnetized disk are as follows:

$$\frac{\partial \Sigma}{\partial t} + \frac{1}{r}\frac{\partial (rv_r)}{\partial r} + \frac{1}{r}\frac{\partial (\Sigma v_\phi)}{\partial \phi} + \frac{\partial (|sigmav_z)}{\partial z} = 0 \tag{2}$$

$$\frac{\partial v_r}{\partial t} + v_r\frac{\partial v_r}{\partial r} - \frac{v_\phi^2}{r} + \Omega\frac{\partial v_r}{\partial \phi} + v_z\frac{\partial v_r}{\partial z} + r\Omega_k^2 + \frac{1}{\Sigma}\frac{\partial W}{\partial r} + \frac{W}{\Sigma}\frac{\partial ln\Omega_k}{\partial lnr}$$
$$-\frac{B_z}{4\pi\rho}\frac{\partial B_r}{\partial z} - \frac{B_\phi}{4\pi\rho r}\frac{\partial B_r}{\partial \phi} + \frac{B_\phi^2}{4\phi\rho r} + \frac{B_z}{4\pi\rho}\frac{\partial B_z}{\partial r} + \frac{B_\phi}{4\pi\rho}\frac{\partial B_\phi}{\partial r} = 0 \tag{3}$$

$$\frac{\partial v_{\phi r}}{\partial t} + v_r\frac{\partial v_\phi}{\partial r} + \frac{v_\phi v_r}{r} + v_z\frac{\partial v_\phi}{\partial z} + \Omega\frac{\partial v_\phi}{\partial \phi}$$
$$-\frac{\alpha c_s^2}{\Sigma r^2}\frac{\partial}{\partial r}(\Sigma\Omega_k^{-1}r^3\frac{\partial \Omega}{\partial r}) + \frac{2}{\Sigma r}\frac{\partial (Hp)}{\partial \phi} - \frac{1}{4\pi\rho}B_z\frac{\partial B_\phi}{\partial z} - \frac{1}{4\pi\rho}B_r\frac{\partial B_\phi}{\partial r}$$
$$-\frac{1}{4\pi\rho}B_z\frac{B_\phi B_r}{r} + \frac{B_z}{4\pi\rho r}\frac{\partial B_z}{\partial \phi} + \frac{B_r}{4\pi\rho r}\frac{\partial B_r}{\partial \phi} = 0 \tag{4}$$

$$\frac{\partial B_z}{\partial z} + \frac{1}{r}\frac{\partial}{\partial r}(B_r r) + \frac{1}{r}\frac{\partial B_\phi}{\partial \phi} = 0 \tag{5}$$

$$\frac{\partial B_r}{\partial t} - \frac{1}{r}\frac{\partial(v_r B_\phi - v_\phi B_r)}{\partial \phi} + \frac{\partial(v_z B_r - v_r B_z)}{\partial z} = 0 \qquad (6)$$

$$\frac{\partial B_\phi}{\partial t} + \frac{\partial(v_r B_\phi - v_\phi B_r)}{\partial z} + \frac{\partial(v_z B_r - v_r B_z)}{\partial z} = 0 \qquad (7)$$

$$\frac{\partial B_z}{\partial t} + \frac{1}{r}\frac{\partial}{\partial r}[r(v_z B_r - v_r B_z)] + \frac{1}{r}\frac{\partial}{\partial \phi}[r(v_\phi B_z - v_z B_\phi)] = 0 \qquad (8)$$

$$W = c_s^2 \Sigma \qquad (9)$$

We assume perturbations in the form $\delta Q = Q_0 exp[i(\omega t - k_r r - K_z z + N\phi)]$ for the quantities $\rho, P, v_r, v_\phi, B_r, B_\phi, B_z$ in the perturbations. Taking the perturbed quantities to the basic equations and setting the determinants of the coefficients in the perturbed equation to zero. We get a dispersion relation: $a_1\tilde{\sigma}^4 + a_2\tilde{\sigma}^3 + a_3\tilde{\sigma}^2 + a_4\tilde{\sigma} + a_5 = 0$. Where $\tilde{\sigma} = i(\omega - k_r v_r - k_z v_z + N\Omega)/\Omega$, and $a_i (i = 1...5)$ is the coefficients given by functions of some parameters of disk structure and perturbations.

3. NUMERICAL SOLUTION AND DISCUSSION

In this section, we solve the dispersion relation numerically for different parameters. We consider only the outer region of the disk, and assume $\tilde{k} \approx 1, \Omega \approx 1$ for the outer region (Wu et al.1995). In Fig. 1, the solid and dashed lines correspond to the case with N=1 and 20. The curves (1,1'), (2,2'), (3,3'), (4,4') represent the non-axisymmetric, outside and inside magneto-acoustic and viscous modes respectively. We don't deal with the viscosity in the left panel of Fig. 1. With the increase of azimuthal wavenumber N, the magneto-acoustic modes have become more unstable and the viscous mode become more stable, but there is no effect on the non-axisymmetric mode. This result is opposite to Wu et al's paper (1995I). In Wu et al's paper, they pointed out that the acoustic modes are always stable, and the increase of N enhances the growth rate of non-axisymmetric mode. The differences between W et al. And this paper come from the fact that we consider not only the radial, vertical and azimuthal perturbations but also the magnetic fields, so even though in an inviscid disk, the magnetic viscosity is instead of viscosity. The right panel of Fig. 1 is the case with magnetic and viscosity for the outer disk. From the right panel of Fig. 1, we know that the growth rates of magnetic-acoustic modes decrease when N become large. This result is opposite to the left panel of Fig. 1. The growth rate of non-axisymmetric mode increase rapidly in the long wavelenth region. The reason is that the effect of viscosity damps the magneto-acoustic modes but the magnetic field enhance the instability of these modes. Finally, Mckee (1990,1991) concluded that the coupling of radial and azimuthal perturbations can eliminate the viscous instability. However, Wu et al. (1995I) pointed out that the viscosity instability still exists, whether for standard disk or for slim disks. Now, in our work, the viscous mode is always stable. This result is the same as the work by Mckee (1990, 1991),and different from Wu et al.(1995I). The reason is that we consider the isothermal magnetized accretion disk with the coupling of radial, vertical and azimuthal perturbation. Mckee considered the radiation pressure dominant disk without the magnetic field. So the magnetic pressure affects the stability of viscosity mode instead of the radiation pressure. In summary,the disk becomes more unstable because of the coupling of radial, vertical and azimuthal perturbations and the effect of magnetic field in the disk. Our results are useful to explain the periodic variability of astronomical objects.

Acknowledgements

This project was supported by the National Natural Science Foundation of China

References

Balbus S.A., Hawley J.F. (1991). APJ, 376: 214
Mckee M.R. (1990). A&A, 235: 521
Mckee M.R. (1991). A&A, 251: 689
Paczynaski B., Wiita P.J. (1980). A&A, 88: 23
Shakura N.I., Sunyaev R.A. (1973). A&A, 24: 317
Wu X.B., Yang L.T., Xie G.Z., Yang P.B. (1995). A&A, 295: 833
Wu X.B., Li Q.B., Zhao Y.H., Yang L.T. (1995I). APJ, 442: 736
Yang L.T., Henning Th., Lu Y., Wu S.P. (1997). MNRAS, 288: 965
Yang L.T., Wang D.X., Wu S.P., Liu Z.D. (1995) A&A, 303: 635

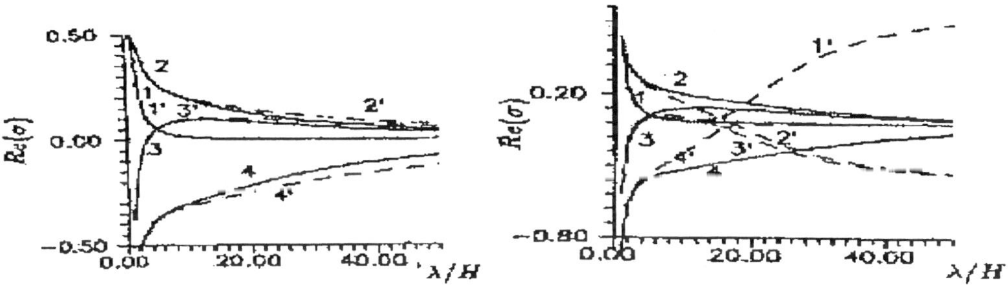

Figure 1 For the left panel, the parameters $(\beta,\alpha,f,\frac{\partial lnB_r}{\partial lnz},\frac{k_z}{k_r},\frac{k_r^2}{k^2},\tilde{H},\ \xi)$ equal to (0.5,0.0,0.5,1.5,10,20,0.6,0.003,0.1); for the right panel, the parameters $(\beta,\alpha,f,\frac{\partial lnB_r}{\partial lnz},\frac{k_z}{k_r},\frac{k_r^2}{k^2},\tilde{H},\ \xi)$ equal to (0.5,0.0,0.5,1.5,10,20,0.6,0.003,0.1)

9A

ORBITAL PERIOD OF THE LOW-INCLINATION SW SEXTANTIS STAR V442 OPHIUCHI

D. W. Hoard[1], John R. Thorstensen[2] and Paula Szkody[3]
[1] *Cerro Tololo Inter-American Observatory,*
[2] *Dartmouth College,*
[3] *University of Washington*

1. INTRODUCTION

V442 Ophiuchi was identified as a cataclysmic variable (CV) by Szkody & Wade (1980) based on its photometric and spectroscopic properties. They noted He II λ4686 emission comparable in strength to Hβ. Extensive time-series photometry by Szkody & Shafter (1983) did not reveal an eclipse, implying $i \lesssim 65°$. Hoard (1998; also Hoard & Szkody 1999) suggested that V442 Oph is a low-inclination SW Sextantis star (e.g. Hoard 1998, Thorstensen et al. 1991) based on its single-peaked emission lines, strong He II emission, and a weak transient absorption feature observed in its Balmer emission lines during 1995.

2. ORBITAL PERIOD

Szkody & Shafter (1983) reported $P_{\rm orb} = 0.1403$ d from Hα and He II λ4686 radial velocity curves, while Patterson (1998, private communication) suggested $P_{\rm orb} = 0.1243$ d based on an ≈ 0.02 mag modulation in the light curve of V442 Oph. Clearly, these periods are 1-day aliases. In 1995 and 1998, we obtained time-resolved optical spectra of V442 Oph on the 3.5-m telescope at Apache Point Observatory (APO). Radial velocity curves constructed from these spectra were unable to discriminate between the period aliases. In June 1999, we obtained additional optical spectra, on the 2.4-m telescope at the Michigan-Dartmouth-MIT Observatory (MDM), spanning a large range of hour angle (6.35 h) to break the period alias. We measured velocities from the Hα line wings using the double-Gaussian method (Schneider & Young 1980) and fit a sine function with the best period yielded by a periodogram of the velocities (as in Thorstensen et al. 1996). The spectroscopic observations and ephemerides are summarized in Table 1. We remeasured the veloc-

K.S. Cheng et al. (eds.), Stellar Astrophysics, 417–421.

Table 1 Observation Log and Ephemerides

Run	*UT Date*	*Site*	*N*	$T_0{}^a$	*P* *(d)*
1	1995 Aug 28–29	APO	13	49957.7161(19)	0.1250(4)
2	1998 Apr 8–10	APO	71	50912.9475(14)	0.1233(3)
3	1999 Jun 4, 9–10	MDM	14	51338.6566(17)	0.12435(7)

[a]Apparent emission line inferior conjunction (blue-to-red crossing), HJD −2400000.

ities from the APO spectra using the same technique, and re-fit them using the best period from the 1999 data as a starting value. Table 2 lists the parameters of sinusoidal fits to the velocity data obtained for each observing run; Figure 1 shows all of the velocity points and the sinusoidal function that best fits the combined data set. A Monte Carlo test (as in Thorstensen & Freed 1985) verified that we had resolved the 1-day period alias ambiguity. *The orbital period of V442 Oph is 0.1243 d.* Interestingly, this places V442 Oph at the upper end of the so-called CV "period gap."

Table 2 Fits to Hα Radial Velocities

Run	*K* *(km s^{-1})*	γ *(km s^{-1})*	σ *(km s^{-1})*
1	180(17)	−56(12)	33
2	149(9)	7(6)	45
3	170(19)	−95(11)	34

3. DOPPLER TOMOGRAPHY

We constructed Doppler tomograms from the 1998 Apr 09 UT spectra (our best-sampled data set) using the fourier-filtered back-projection algorithm (Kaitchuck et al. 1994 and references therein). Figure 2 shows tomograms of Hα, Hβ, He II λ4686, and He I λ4471. The spectra were phased according to the appropriate ephemeris from Table 1. Since V442 Oph does not eclipse, the phase zero point was set by rotating the He II tomogram until the emission was situated on the $-V_y$-axis, at the expected location of the white dwarf (WD). This implies $K_{\rm WD} \sin i \approx$ 100–150 km s^{-1} and a phase lag of ≈ 0.08 in the Balmer emission line velocities (the latter is typical of the SW Sex stars). The Hα and Hβ tomograms (as well as He II) are very similar to those found for high-inclination examples of SW Sex stars (e.g., BH Lyn, Hoard & Szkody 1997; PX And, Hellier & Robinson 1994). The He I line underwent a pronounced transient absorption event in this data set (see Fig. 3). This is reflected in its unusual tomogram, which shows a strong absorbing region near the velocity origin as well as weaker absorption regions in the ($-V_x$, $+V_y$) and ($+V_x$, $-V_y$) quadrants. The physical origin of this absorption is unclear since in this low-inclination system we should be able to see most or all of the disk face and avoid vertically-extended obscuring structures. The presence of prominent emission in the ($+V_x$, $+V_y$) quadrant in the Hβ and He I tomograms is also unusual and defies a simple physical explanation.

References

Hellier, C., Robinson, E. L. 1994, ApJL, 431, L107

Hoard, D. W. 1998, PhD dissertation (University of Washington)

Hoard, D. W., Szkody, P. 1999, to appear in New Astron. Rev. – also see `http://www.ctio.noao.edu/~hoard/warner/v442.html`

Hoard, D. W., Szkody, P. 1997, ApJ, 481, 433

Kaitchuck, R. H., et al. 1994, ApJS, 93, 519

Schneider, D. P., Young, P. 1980, ApJ, 238, 946

Szkody, P., Shafter, A. W. 1983, PASP, 95, 509

Szkody, P., Wade, R. A. 1980, PASP, 92, 806

Thorstensen, J. R., Freed, I. B. 1985, AJ, 90, 2082

Thorstensen, J. R., Patterson, J., Shambrook, A., Thomas, G. 1996, PASP, 108, 73

Thorstensen, J. R., et al. 1991, AJ, 102, 272

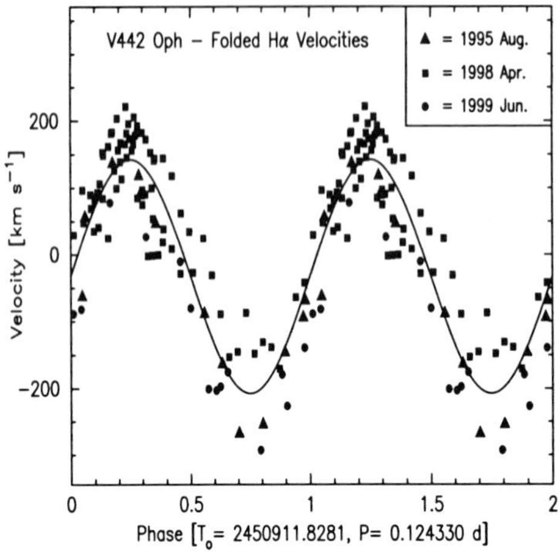

Figure 1 All Hα velocity points with the best fit sine function, $v(t) = \gamma + K \sin[2\pi(t - T_0)/P_{\rm orb}]$.

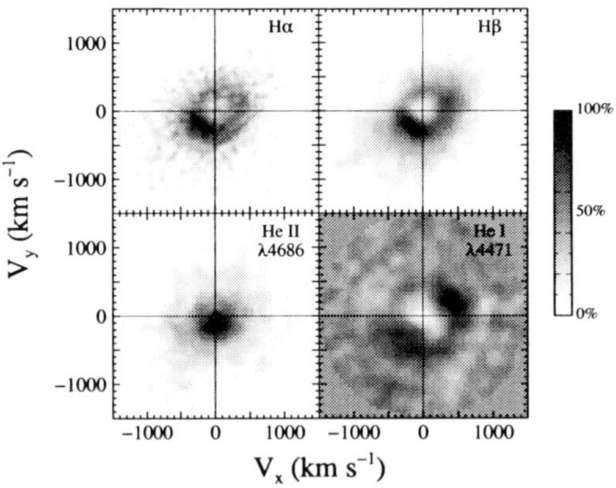

Figure 2 Doppler tomograms of V442 Oph emission lines. The intensity scale extends from the spectral continuum (0%) to the emission peak (100%), except in the He I tomogram, where it starts at the deepest absorption below the continuum (0%) and extends through the continuum (≈45%) and emission peak (100%).

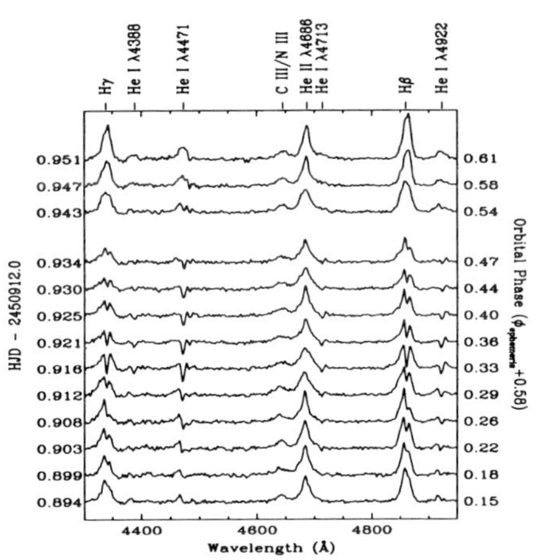

Figure 3 A sub-set of the spectra from 1998 Apr 09 UT. Note the deep transient absorption feature in the Balmer and especially He I lines, and the strong, single-peaked He II line.

A NEW MASS FUNCTION OF A SYMBIOTIC STAR, HBV475

Yuji Ikeda & Shin'ichi Tamura

Astronomical Institute, Graduate School of Science, Tohoku University, Aoba-ku, Sendai, Miyagi 980-8578, Japan

ikeda@astr.tohoku.ac.jp

Abstract We present highly resolved spectroscopic data of HBV475 (=V1329 Cyg) in the optical region, which have been obtained during the past 10 years at the Okayama Astrophysical Observatory. We analyzed line profiles of Hα, Hγ, HeIIλ4686, [OIII]λ5007 and [FeVII]λ6086 by using Gaussian de-convolution method and investigated variation of individual components of the profiles as a function of the orbital phase. We revised the mass function obtained from orbital elements of the hot star as $f_{\rm h}(M) \sim 1.2$ M$_\odot$. It differs considerably from previously published values, which were larger than 20 M$_\odot$. However, our new mass function is more reasonable for a symbiotic system consisting of a red giant and a hot star, which is currently presumed to be a white dwarf.

Keywords: symbiotic, spectroscopy, line profiles, HBV475

1. INTRODUCTION

Hamburg-Bergedorf Variable 475 (= HBV475, V1329 Cyg) belongs to the group of symbiotic novae. Since its sudden brightness in 1964, eyes at all wavelengths, including from X-ray to radio, have turned toward HBV475. Even though a number of observational data on HBV475 have been obtained for the last thirty years, there are some unresolved problems, which prevent us from constructing a reasonable model. At present, it seems that the most important point of unsolved problems is to determine a realistic mass function. Both Iijima et al. (1981) and Nussbaumer et al. (1986) obtained $f_{\rm h}(M) > 20$M$_\odot$ from radial velocity analysis of the nebular emission lines. However, this very large value of the mass function is not consistent with a basic model – red giant + white dwarf – supported by other observations.

K.S. Cheng et al. (eds.), Stellar Astrophysics, 423–427.

High dispersion emission spectra can give a key to solve this problem. Generally, the complex emission line profiles of the HBV475 spectrum reflect a complicated structure of the ionized gas, which probably consists of more components differing in kinematics. Radial velocity analysis based on low or intermediate dispersion spectra couldn't provide the correct mass function due to contamination by gas components which do not correlate with the stellar motion. In this paper, we de-convolute our obtained five emission lines (Hα, Hγ, HeIIλ4686, [OIII]λ5007 and [FeVII]λ6086) into a few components and reanalyze their radial velocities. Then, we provide a new and reliable mass function of HBV475.

2. OBSERVATIONS AND ANALYSIS

We have observed highly resolved emission lines of HBV475 at several epochs between October 1983 and December 1994 at the Okayama Astrophysical Observatory (OAO) by using the coudè spectrograph of the 74-inch reflector, at the region of Hα, Hγ, HeIIλ4686, [OIII]λ5007, and [FeVII]λ6086. We also use highly resolved spectra previously published by Wallerstein et al. (1989) and Ivison et al. (1994) added to our data. We have 13 spectra of Hα, 3 of Hγ, 5 of HeIIλ4686, 15 of [OIII]λ5007, and 2 of [FeVII]λ6086 in total.

All line profiles of HBV475 form very complicated profiles. We can presume that all emission lines consist of a few components. Further we will assume that each component is approximated by a Gaussian function. Thus, to get characteristics of individual components, we can use a method of Gaussian de-convolution, which we described in Ikeda & Tamura (2000a) in more detail. Each component (k) is determined by its position (v_k) - heliocentric radial velocity, broadening (σ_k) defining its FWHM$_k$, and the intensity (A_k) which gives its height above the local continuum.

For example, the Hα line was de-convolved into three Gaussians, which represent the main emission component, the absorption component and the broad wing of the observed profile. The HeIIλ4686 line was fitted by two components, namely the main component and the satellite component. The [OIII]λ5007 line was matched by three individual emission profiles, which are red, central, and blue components, respectively (see Ikeda & Tamura 2000b).

3. RESULTS

In our attempt to derive a new, more realistic mass function, we pay our attention to the wing component of Hα, the main component of HeII, and the central component of [OIII]. Their heliocentric radial velocities are plotted in the phase diagram of figure 1. The reason why we select

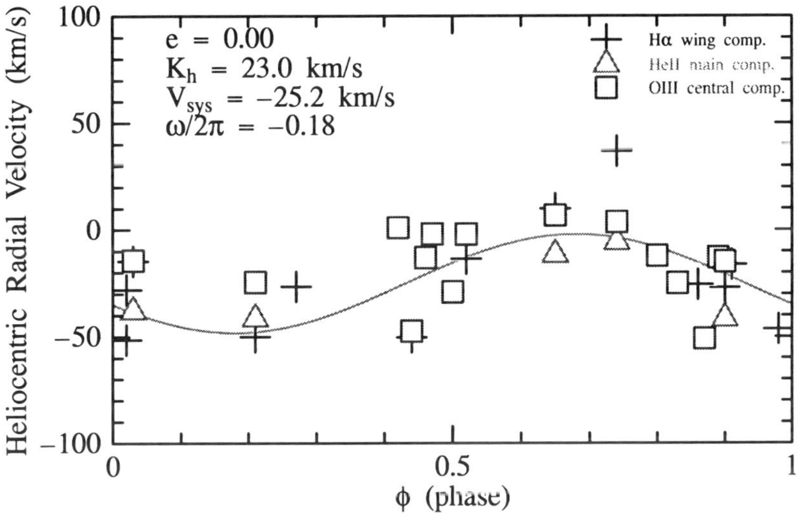

Figure 1 Phase diagram of radial velocities of the Hα wing components, the HeIIλ4686 main components, and the [OIII]λ5007 central components. A solid curve represents our solution of the orbital motion of the hot star.

these components is as follows. Among three components producing the Hα line (i.e. the main emission, absorption, and the wing component), only the radial velocity of the wing component shows a cyclic variation with the orbital phase, while radial velocities of neither the main nor absorption component show such variations. Ikeda & Tamura (2000a) mentioned that the wing component of the Hα line in symbiotic stars could originate in the hot star wind and its central velocity reflects the orbital motion of the hot component. In addition, we found also that main component of HeII and the central component of [OIII] behave similarly. In spite of a large scatter in figure 1, one can recognize that radial velocities of all lines follow the same type of variation, namely at $\phi \sim 0.75$, and they all have the required attributes of the radial velocity curve of the hot component in HBV475. This fact suggests that these emissions occupy approximately the same region around the hot star. Thus, we can search for an orbital solution by the least square fit with following the equations,

$$V_{\text{radial}} = V_{\text{sys}} + K[e\cos\omega + \cos(\omega + \psi)],$$

$$K = \frac{2\pi}{P}\frac{a\sin i}{(1-e^2)^{1/2}},$$

where $V_{\rm radial}$ is the radial velocity of the line, $V_{\rm sys}$ is the space velocity, e is the orbital eccentricity, ω is the longitude of the periastron, ψ is the true anomaly, K is the semiamplitude of the radial velocity curve, P is the orbital period, a is the semi-major axis of the orbit, i is the orbital inclination. Setting $P = 956.5$ dy (Shild & Schmid 1997), best solution for the hot component orbit is defined by $e = 0.00$, $V_{\rm sys} = -25.2 \pm 1.5$ km s^{-1}, $K_{\rm h} = 23.0 \pm 2.0$ km s^{-1} and $\omega/2\pi = -0.18 \pm 0.01$. Thus, the mass function obtained from such orbital elements for the hot component $f_{\rm h}(M)$ becomes,

$$\begin{aligned} f_{\rm h}(M) &= (M_{\rm cool}\sin i)^3/(M_{\rm hot}+M_{\rm cool})^2, \\ &= 1.036\times10^{-7}K_{\rm h}^3P(1-e^2)^{3/2} \quad {\rm M}_\odot, \\ &\sim 1.2\pm0.3 \quad {\rm M}_\odot, \end{aligned}$$

where P is in days and $K_{\rm h}$ in km s^{-1}. This result looks reasonable for a typical symbiotic binary. For example, Mikolajewska & Kenyon's (1992) orbital elements for AX Per suggest $f_{\rm h}(M) = 0.4$ M$_\odot$, Thackeray & Hutchings (1974) suggested the mass function in AR Pav to be 0.14 M$_\odot$, Skopal et al. (1997) derived $f_{\rm h}(M) = 1.4$ M$_\odot$ for BF Cyg and Paczynski (1965) suggested $f_{\rm h}(M) = 3.6$ M$_\odot$ for T CrB. Assuming the mass of the hot star in HBV475 to be 0.65 M$_\odot$ (Iben & Tutukov 1996), obtained from mass-luminosity relationship for degenerate dwarfs with hydrogen burning shells, then $i = 87°$ (Schild & Schmid 1997), the mass of the cool star and the total mass are 2.1 M$_\odot$ and 2.8 M$_\odot$, respectively.

Therefore, we conclude *HBV475 is not a massive system, but it can be a normal mass symbiotic system consisting of a normal cool giant and white dwarf.*

References

Iijima T. Mammano A. Margoni R. 1981, Ap&SS. 75, 237
Ikeda Y. Tamura S. 2000, PASJ. submitted. (=2000a)
Ikeda Y. Tamura S. 2000, PASJ. submitted. (=2000b)
Ivison R.J. Bode M.F. Meaburn J. 1994, A&AS. 103, 201
Mikolajewska J. Kenyon S.J. 1992, AJ. 103, 579
Munari U. Margoni R. Mammano A. 1988, A&A. 202, 83
Paczyński B. 1965, Acta Astronomica. 15, 197
Schild H. Schmid H.M. 1997, A&A. 324, 606
Skopal A. Vittone A. Errico L. Bode M.F. Lloyd H.M. Tamura S. 1997, MNRAS 292, 703

Thackeray A.D. Hutchings J.B. 1974, MNRAS 167, 319
Wallerstein G. Solf J. Mikolajewska J. Crampton D. Brugel E. 1989, PASP. 101, 189

NEW CATACLYSMIC VARIABLES FROM RASS

X.J. Jiang[1,2], D. Engels[3], J.Y. Wei[1,2], and J.Y. Hu[1,2]

1. Beijing Astronomical Observatory, Chinese Academy of Sciences Beijing 100012, P.R. China; Email: xjjiang@class1.bao.ac.cn

2. National Astronomical Observatories, Chinese Academy of Sciences

3. Hamburger Sternwarte, Gojenbergsweg 112, D-21029 Hamburg, Germany

Keywords: surveys, X-rays: general, X-rays: galaxies, galaxies: active

1. INTRODUCTION

During the ROSAT All-Sky Survey (RASS) about 80,000 X-ray sources with a detection likelihood $\geq$ 10 were detected, from which 18,811 sources having a PSPC count rate larger than 0.05 cts s^{-1} and detection likelihood $\geq$ 15 were compiled in the RASS-BSC (Voges et al. 1996a). More than 65% RASS sources remain unidentified (Voges et al. 1996b). The optical identification of this large numbers of X-ray sources is a challenge. Since often more than one candidate is present in the error circle and the follow-up optical spectroscopy of all the counterparts down to a given optical limit is very time-consuming, we pre-identify ROSAT sources on objective prism plates from the Hamburg Quasar Survey (HQS, Hagen et al. 1995). The low-dispersion spectra allow to draw conclusions on the nature of the optical candidates and provide unambiguous identifications of the X-ray sources in many cases.

Cataclysmic variables (CVs) consist of a white dwarf accreting matter from a low mass companion star in a semi-detached system. Depending on the mass transfer rate and on the magnetic field of the white dwarf, the matter leaving the inner Lagrangian point may either from an accretion disc or flow directly toward the white dwarf pole following the magnetic field lines. A large variety of behavior, optical appearance and X-ray luminosities are observed from these systems. We selected 15 candidates of new CV which are not contained in the SIMBAD database as of July 1998 from Hamburg/RASS database for optical identifications.

K.S. Cheng et al. (eds.), Stellar Astrophysics, 429–432.

2. OBSERVATIONS AND RESULTS

All 15 sources were observed with the 2.16 m reflector at Xinglong Station, Beijing Astronomical Observatory from July 1998 through June 1999. The low resolution spectra were taken with the OMR spectrograph, using a Tektronix 1024 CCD as a detector. Two gratings of 300 g mm^{-1} and 150 g mm^{-1} were employed in order to get large wavelength coverage. The gratings give the dispersion of about 4.8Å$pixel^{-1}$ and 9.6Å$pixel^{-1}$ respectively. All observations were obtained at airmass less than 1.5. The raw two-dimensional data were reduced following standard procedures using the IRAF program package.

10 new CVs were discovered from the 15 sources, the other 5 sources are AGNs, dMe, and main sequence stars. Figure 1 shows the reduced spectra of the 10 CVs, table 1 lists the detailed information of the CVs, where column (1) is ROSAT source designation; column (2) & (3) are the optical right ascension and declination of the CVs (epoch 2000.0); column (4) is the ROSAT total count rate as given in RASS-BSC; column (5) is the hardness ratio HR1=(B-A)/(B+A), where A means the count rate in the energy range 0.1-0.4keV and B in the energy range 0.5-2.0keV; column (6) is the hardness ratio HR2=(D-C)/(D+C), where C means the count rate in the energy range 0.5-0.9keV and D in the energy range 0.9-2.0keV; column (7) is the B magnitude from the individual HQS plates.

Acknowledgements

The ROSAT project is supported by the Ministerium für Bildung, Wissenschaft, Forschung und Technologie (BMBF/DARA) and by the Max-Planck-Gesellschaft (MPG).

References

Hagen H.J., Groote D., Engles D., Reimers D., 1995, A&AS 111, 195
Voges W., Aschenbach B., Boller Th., *et al.*, 1996a, IAU Circ. 6420
Voges W., Boller Th., Dennerl K., *et al.*, 1996b, MPE-Report 263, 637

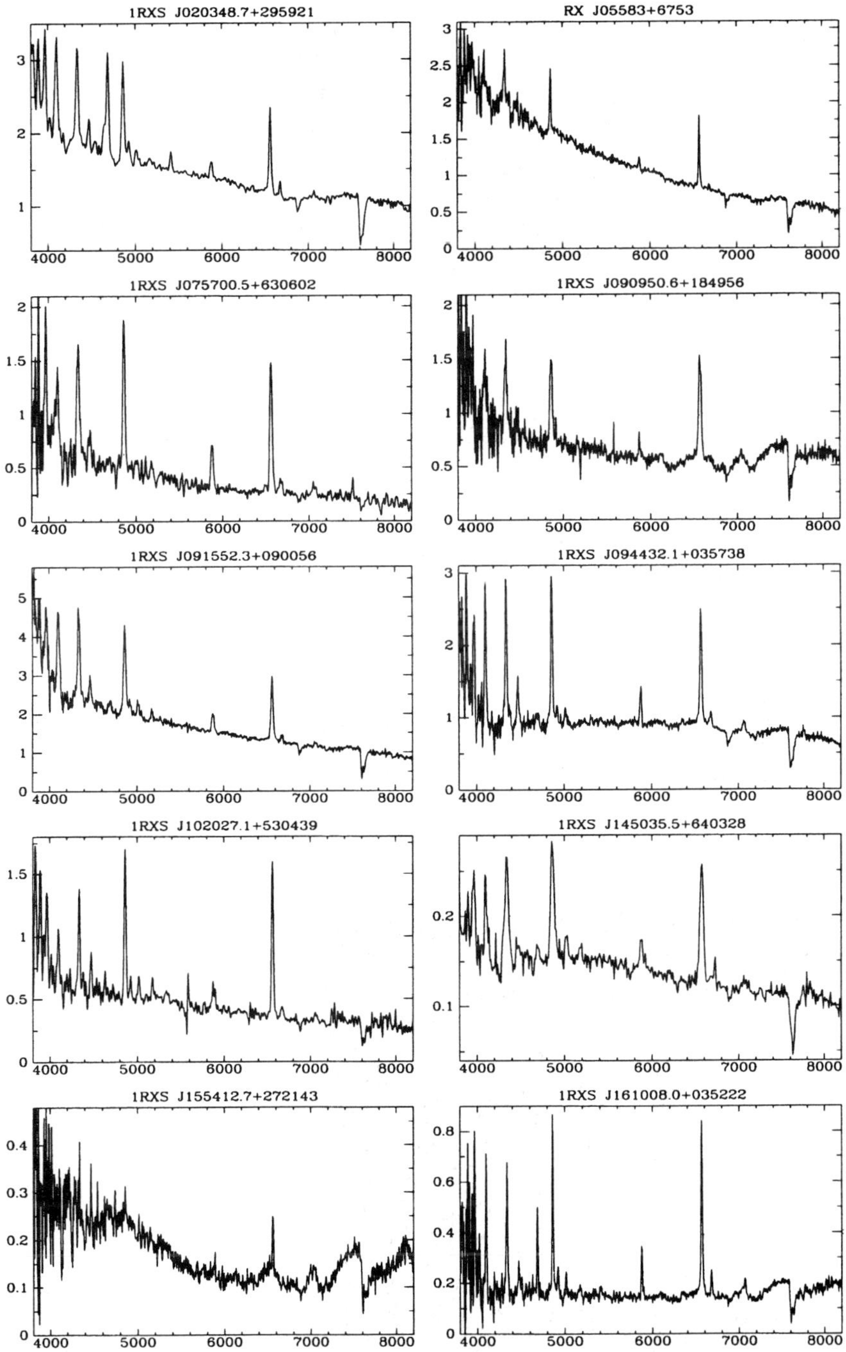

Figure 1 Optical spectra of the counterparts to the X-ray sources. f_λ in units of 10^{-15} erg cm^{-2} s^{-1} Å^{-1} is plotted against wavelength in Å

Table 1 Properties of the new CVs

ROSAT name	RA (2000)	Dec (2000)	CR (cts/s)	HR1	HR2	Bmag
1RXS J013126.2+360241	013125.9	360230	0.402	0.24	0.18	16.5
1RXS J020348.7+295921	020348.5	295925	0.539	−0.96	−1.00	15.9
RX J05583+6753	055817.9	675346				
1RXS J075700.5+630602	075701.4	630601	0.152	0.79	0.43	18.2
1RXS J090950.6+184956	090950.5	184947	0.084	−0.33	−0.12	16.4
1RXS J091552.3+090056	091551.7	090048	0.181	0.53	0.15	15.2
1RXS J094432.1+035738	094431.8	035806	0.089	0.81	0.37	16.4
1RXS J102027.1+530439	102026.6	530432	0.119	0.35	0.28	16.9
1RXS J145035.5+640328	145038.4	640329	0.071	0.06	0.66	17.6
1RXS J155412.7+272143	155412.3	272151	0.113	−0.78	−0.13	16.7
1RXS J161008.0+035222	161007.4	035233	0.358	−0.41	0.01	18.1

OXYGEN ABUNDANCE EVOLUTION IN THE SOLAR NEIGHBORHOOD

Y.C. Liang[1,2] G. Zhao[1,2] J.R. Shi[1,2]
[1] *Beijing Astronomical Observatory, Chinese Academy of Sciences, 100012 Beijing, China*
[2] *National Astronamical Observatories, Chinese Academy of Sciences, 100012 Beijing, China*
lyc@yac.bao.ac.cn, zg@orion.bao.ac.cn, sjr@yac.bao.ac.cn

Abstract According to standard infall-model of Galactic chemical evolution models, adopting the two sets of recently published nucleosynthesis yields of Type II supernova explosion (as parameters of our model A and model B respectively), we calculated the abundance evolution of oxygen in the solar neighborhood. Then we compared the predicted age-metallicity relationship and metallicity distribution of G-dwarfs with the observational data. Our results show that the yields from Nomoto et al. (1997a) is more suitable to explain the observed trend of oxygen in the solar neighborhood than that of Woosley & Weaver (1995).

Keywords: Galactic chemical evolution, solar neighborhood, oxygen, supernova

1. INTRODUCTION

Oxygen is produced mainly in short-lived, massive stars through Type II supernova explosion (SN II). The abundance evolution of oxygen can reflect the enhancements of interstellar medium due to SN II explosion, which can produce main fraction of some other elements, such as Mg, Ca etc. While a substantial contribution to iron comes from Type Ia supernova (SN Ia) with longer lived progenitors. Recently, two literatures calculated the SN II explosion yields (Woosley & Weaver 1995, hereafter WW95; Nomoto et al. 1997a), and give different yields of O and Fe. In this paper, according to standard infall-model of Galactic chemical evolution models (Matteucci & François 1989, hereafter MF89; Timmes et al. 1995, hereafter TWW95 etc.), we make the two sets of SN II yields be parameters of our model A and model B respectively to calculate the abundance evolution of oxygen in the solar neighborhood. Our Galactic chemical evolution model and the corresponding parame-

K.S. Cheng et al. (eds.), Stellar Astrophysics, 433–437.

ters are described in section 2. The calculation results and conclusion are given in section 3.

2. GALACTIC CHEMICAL EVOLUTION MODEL

The rate of change of the surface mass density of isotope i in the gas at a given Galactocentric radius r and time t is

$$\frac{d\sigma_{g,i}(r,t)}{dt} = death - birth + infall \ . \tag{1}$$

The first term describes the enrichment of isotope i due to stellar processing, while the second term describes the disappearance of isotope i due to new star formation. "Infall" represents the infall of primordial material. More details of every term can be found in MF89 and TWW95. We adopt the same infall rate as that of TWW95 but with $\sigma_\odot = 55 M_\odot pc^{-2}$ (Sackett 1997). The Star Formation Rate (SFR) function is given by: $\Psi(r,t) = \nu\sigma_{tot}(r,t)\left[\frac{\sigma_{gas}(r,t)}{\sigma_{tot}(r,t)}\right]^\kappa M_\odot pc^{-2} Gyr^{-1}$, where ν is the efficiency factor, which can be adjusted freely to give the best fit to the solar abundance of Grevesse & Sauval (1998). κ is a constant between 1 and 2. A Salpeter-like Initial Mass Function (IMF) is adopted: $\Phi(m) \propto m^{-(1+x)}$, which is normalized from $\int_{m_l}^{m_u} m\Phi(m)dm = 1$ $(m_l = 0.1M_\odot, m_u = 40M_\odot)$. The yields of intermediate mass and low mass single stars are from Marigo et al. (1996) (1-4$M_\odot$) and Renzini & Voli (1981) (4-8$M_\odot$, with $\alpha = 1.5$ and $\eta = 0.33$), which is same as Prantzos & Silk (1998). The yields of Type Ia SN is taken from W7-model of Nomoto et al. (1997b). β is an amplitude factor of SN Ia. The SN II yields in our model A and model B are from Nomoto et al. (1997a) and WW95 (their model B) respectively. The stellar main-sequence lifetimes are from Schaller et al. (1992). The input parameters in our two models are given in table 1.

Table 1 Input parameters for the solar neighborhood in our model A and model B

model	*model A*	*model B*
SFR	$\nu = 5.0, \kappa = 1.4$	$\nu = 2.4, \kappa = 1.5$
IMF	x=1.35	x=1.31
SN II yields	Nomoto et al. (1997a)	WW95
β	0.035	0.030

3. RESULTS AND CONCLUSION

Using our Galactic chemical evolution models, namely, model A and model B, we calculated the abundance evolution of oxygen in the solar neighborhood, which is given in Fig. 1 . The calculation results can be compared with the observational data of dwarfs which represent perfect record of the chemical composition of the interstellar medium at the time the stars formed. Our results show that the model A is more suitable to explain the observed trend of oxygen in the solar neighborhood than model B. Comparing the yields of oxygen of Nomoto et al. (1997a) and WW95, we know that the former produce more oxygen in the higher mass range, which can be understood in terms of the higher $^{12}C(\alpha,n)^{16}O$ rate. However, WW95 produce significantly more iron than Nomoto et al. (1997a) in stars of $M > 25M_{\odot}$. So the predicted [O/Fe] with WW95 yields is too low, especially at low metallicity. Also, our results can fit other important observational constraints in the solar neighborhood, namely, the age-metallicity relationship, which reflects the evolution of [Fe/H] in the gas as a function of time, and metallicity distribution of G-dwarfs, which can represent a sample of stars that has almost never been removed by stellar evolution. The corresponding results are shown in Fig. 2 and 3 respectively.

References

Barbuy, B. (1988). A&A, 191: 121–127

Barbuy, B. & Erdelyi-Mendes, M. (1989). A&A, 214: 239–248

Carlberg, R. G., Dawson, P.C., Hsu, T. et al. (1985). ApJ, 294: 674-681

Edvardsson, B., Andersen, J., Gustafsson, B. et al. (1993). A&A, 275: 101–152

Gratton, R. G. & Ortolani, S. (1986). A&A, 169: 201–207

Grevesse, N. & Sauval, A. J. (1998). Space Sci. Rew., 85: 161–174

Israelian, G., Garcia Lopez, R. J. & Rebolo, R. (1998). ApJ, 507: 805–817

Marigo, P., Bressan, A. & Chiosi, C. (1996). A&A, 313: 545–564

Matteucci, F. & François, P. (1989). MNRAS, 239: 885–904 (MF89)

Meusinger, H., Reimann, H. -G., & Stecklum, B. (1991). A&A, 245: 57–74

Nomoto, K., Hashimoto, M. et al. (1997a). Nucl. Phys., A616: 79c–90c

Nomoto, K., Iwamoto, K. et al. (1997b). Nucl. Phys., A621: 467c–476c

Prantzos, N. & Silk, J. (1998). ApJ, 507: 229-240

Renzini, A. & Voli, M. (1981). A&A, 94: 175–193

Rocha-Pinto H. J. & Maciel W. J. (1996). MNRAS, 279: 447–458

Sackett, P. D. (1997). ApJ, 483: 103–110

Schaller, G., Schaerer, D., Meynet, G. et al. (1992). A&AS, 96: 269–331

Sneden, C., Kraft, R.P., Prosser, C.F. et al. (1991). AJ, 102, 2001–2021
Spite, M. & Spite, F. (1991). A&A, 252: 689–692
Timmes, F. X., Woosley, S. E. & Weaver, T. A. (1995). ApJS, 98: 617–658 (TWW95)
Woosley, S. E., & Weaver, T. A. (1995). ApJS, 101: 181–235 (WW95)

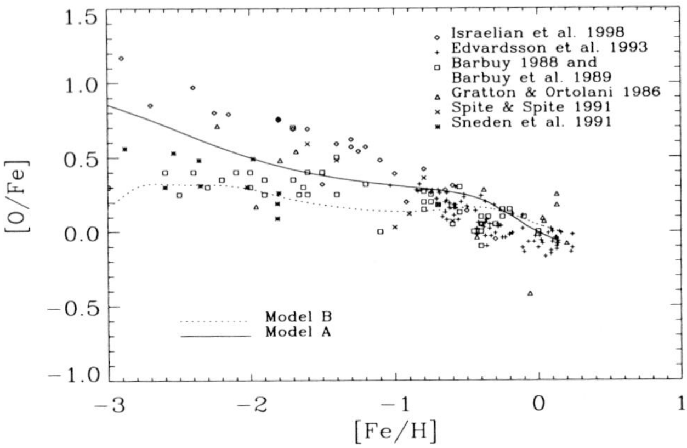

Figure 1 The predicted abundance ratio [O/Fe] as a function of [Fe/H] by our model A (solid line) and model B (dotted line).

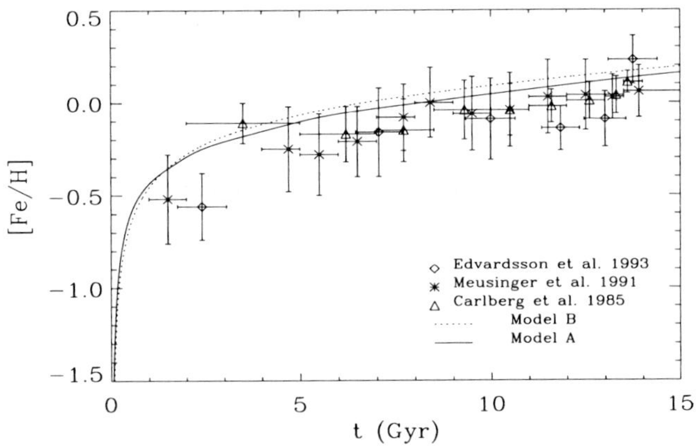

Figure 2 The predicted age-metallicity relationship in the solar neighborhood by our model A (solid line) and model B (dotted line).

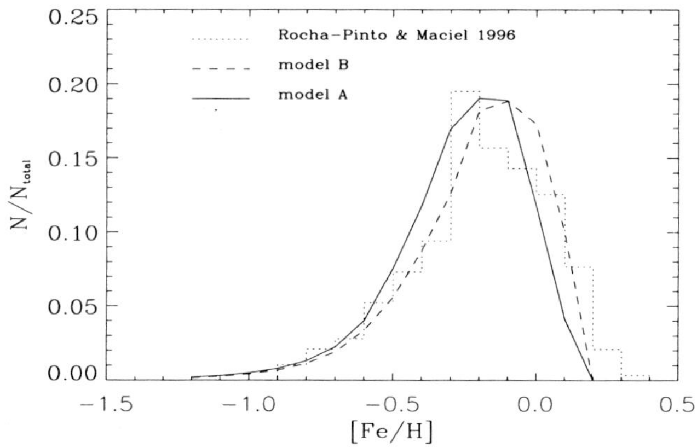

Figure 3 The predicted metallicity distribution of G-dwarfs in the solar neighborhood by our model A (solid line) and model B (dashed line).

THREE VERY METAL-POOR, EXTREMELY WIDE BINARIES

Christine Allen, Arcadio Poveda and M.A. Herrera
Instituto de Astronomía, Universidad Nacional Autónoma de México

Abstract We have searched for wide binaries among the metal-poor, large proper-motion stars from the catalogues of Schuster et al. Out of 48 stars with [Fe/H] < −2.00 we have identified three common-proper-motion companions which turn out to have extreme characteristics. Of particular interest are the very large expected major semiaxes, all larger than 20 000 AU, and up to 125 000 AU. We argue that the persistence of such old wide systems is the result of the small fraction of their lifetimes that the binaries spend within the galactic disk, as well as to their large space velocities. These binaries may be remnants of long-ago disrupted globular clusters or dwarf galaxies.

1. INTRODUCTION

In a recent investigation (Allen et al. 2000) we have identified over a hundred wide, common proper-motion binaries belonging to the old disk, the thick disk or the halo of the Galaxy. These systems were found by searching for common proper-motion companions to the more than 1200 high-velocity, metal-poor primaries of the lists of Schuster et al. (Schuster & Nissen 1988, Schuster et al. 1993). For these stars, Strongren photometry has allowed the determination of metallicities, distances and, in many cases, individual ages. The values obtained for the metallicities and ages of a number of these stars place them among the oldest, most metal-poor stars of the Galaxy. For the purposes of this work, we concentrate our attention on those stars with [Fe/H] < −2.00 as representatives of the very early stellar population of our Galaxy.

Among 48 stars with [Fe.H] < −2.00 present in the lists of Schuster et al. we have found only three wide binaries. Table 1 lists the main properties of these systems.

K.S. Cheng et al. (eds.), Stellar Astrophysics, 439–443.

2. A SEARCH FOR WIDE BINARIES AMONG HIGH-VELOCITY, METAL-POOR STARS.

For each star of the Schuster et al. catalogues a machine search for common proper-motion companions was made among the NLTT stars within 12 minutes of arc. We found a total of 123 wide binaries, that is, binaries with expected major semiaxes larger than 25 AU (Allen et al. 1999).

Many of these systems were observed by Hipparcos. Hipparcos distances were adopted for those stars with relative errors in their trigonometrical parallaxxes of less than 15%. For stars with larger errors, the photometric distances derived by Schuster et al. were preferred.

Absolute visual magnitudes for the primaries were calculated from the adopted distances and the apparent visual magnitudes of Schuster et al. Absolute visual magnitudes for the secondaries were calculated from the apparent red and photographic (blue) magnitude differences between the components, $\delta(m_R)$, $\delta(m_B)$, as estimated by Luyten.

Linear separations between the components were calculated from the angular separations and the adopted distances. The linear separations were transformed into expected values for the major semiaxes by means of the statistical relation (Couteau 1960): $E(\log\ a) - E(\log\ s) = 0.146$. Ages and metallicities were taken from Schuster et al.

3. THE BINARIES WITH [Fe/H] < -2.00

4. DISCUSSION

In the first place, we note the remarkable separations of these three binaries. The fraction of extremely wide binaries ($< a > > 20\ 000$ AU) among the group with [Fe/H] < -2.00 is 3/3, whereas the incidence of such binaries among the group with [Fe/H] > -2.00 is 3/120 (about 2.5%). For comparison, we calculate the fraction of extremely wide binaries ($< a > > 20\ 000$ AU) among a quasi-complete sample extracted from the wide binaries ($< a > > 25$ AU) of the solar vicinity (Poveda et al. 1994) as 4/168 (about 2.4%). The probability of randomly obtaining in three trials three extremely wide binaries out of a population with a typical frequency of 2.5% is about 10^{-5}.

Also remarkable is that two of the galactic orbits are retrograde, while the third is only marginally direct.

The persistence of the duplicity for very old systems, like the stars in Table 1 is an intriguing problem, whose explanation could be a result of the characteristics of their galactic orbits. These systems spend most of

Table 1 Principal properties of the three binaries

	W1828 (Hip 15396)	G015-010 (Hip 74199)	G040-014 (-)
$V_p(km\ s^{-1})$	417.5	306.7	413.1
V_{gc}	229.4	248.4	256.0
$E(100\ km_2\ s^{-2})$	-1163.1	-1131.5	-1095.8
$h(10\ kpc\ km\ s^{-1})$	-143.8	30.4	113.9
$R_{min}(kpc)$	7.6	1.1	2.9
$R_{max}(kpc)$	9.6	15.6	16.4
$\|z_{max}\|(kpc)$	5.2	7.1	1.4
Eccentricity	0.12	0.87	0.70
$[Fe/H]$	-2.05	-2.14	-2.34
Age(Gy)	20.9	-	20.3
$< a >$(AU)	55410	125073	32232

Note. Kinematic and orbital parameters were calculated with the distances determined by Schuster et al., because for these stars Hipparcos parallaxes have large errors; proper motions were taken from Hipparcos.

the time far from the galactic plane, which is where most of the massive perturbers are located. THeir high space velocities also contribute to make the rare encounters relatively ineffective for their disruption.

The chemical abundances, kinematics, galactic orbital parameters, and separations of these binaries are marked different from those of the great majority of wide binaries among high-velocity and metal-poor stars. This leads us to suggest that they may be members of halo moving clusters (Poveda et al. 1992) which in turn, may result from the dissolution of globular clusters or accreted dwarf galaxies.

5. SUMMARY AND CONCLUSIONS

We stress that the binaries were selected from our list of 123 systems only by the criterion that their primaries should have [Fe/H] < -2.00. The three binaries found by this selection criterion turned out to have extreme properties, thich we now list:

1. Two of the binaries have strongly retrograde orbbits, thile the third is only marginally direct.

2. The separations, and hence the expected major semiaxes, of all three binaries are extremely large.

3. Two systems have individual age determinations; both turn out to be very old ($\approx$ 20 Gy), among the oldest in Schuster et al.'s lists.

4. The peculiar velocities are quite large ($V_p > 300\ km\ s^{-1}$).

5. Although the galactocentric velocities are moderate, the distances from the galactic plane reached by these systems, z_{max}, are large, which clearly places them among the halo population.

Acknowledgements

We acknowledge partial financial support for this work from Conacyt, through Grant Nr. 25502-E.

References

Allen, C. & Santillán, A., 1991, Rev. Mex. Astron. Astrof., 22, 255

Allen, C., Poveda, A., adn Herrera, M.A., 2000, Astron. Astroph., in press.

Couteau, P., 1960, J. des Observateurs, 43, No. 3.

Poveda, A. Allen, C. & Schuster, W.J., 1992, IAU Symp., 149, "Stellar Populations of Galaxies", B. Barbuy adn A. Renzini, Eds. (Reidel, Dordrecht), p.471.

Poveda, A. Herrera, MA., Allen, C., Cordero, G. & Lavalley, C. 1994, Rev. Mex. Astron. Astrof. 28,43,1994.

Schuster, W.J. and Nissen, P.E., 1988, Astron. Astrop. Suppl, 73, 225.

Schuster, W.J. Parrao, L. and Contreras, M.E., 1993, Astron. Astrop. Suppl., 97,951.

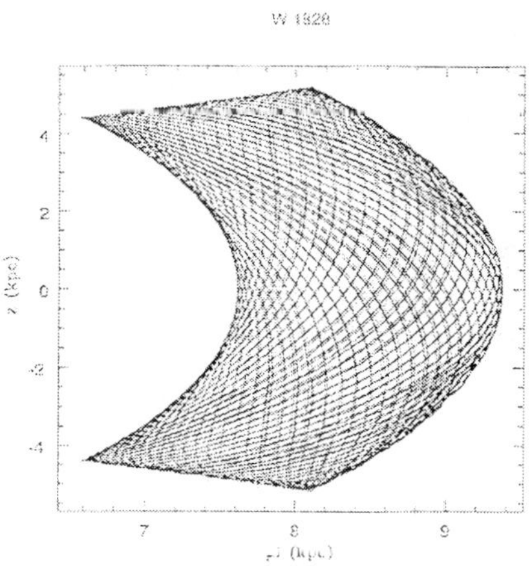

Figure 1. The meridional galactic orbit of W1828. calculated with the galactic potential of Allen & Santillán (1991). The orbit was computed backwards in time for 18 Gy. Note that the binary spends only a short time within the galactic disk.

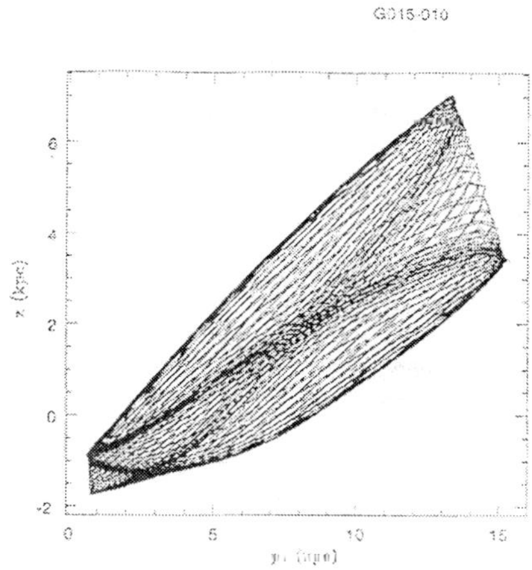

Figure 2. The meridional galactic orbit of G015-010, computed as in Fig. 1. The orbit is of tube type. This binary also spends only a short time within the galactic disk.

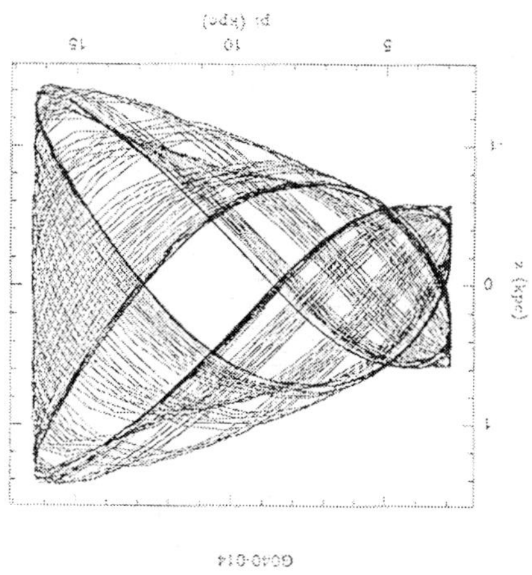

Figure 3. The meridional galactic orbit of G040-014, computed as in Fig. 1.
Of the three binaries, G040-014 spends the longest fraction of its lifetime within the galactic disk; it also has the smallest separation.

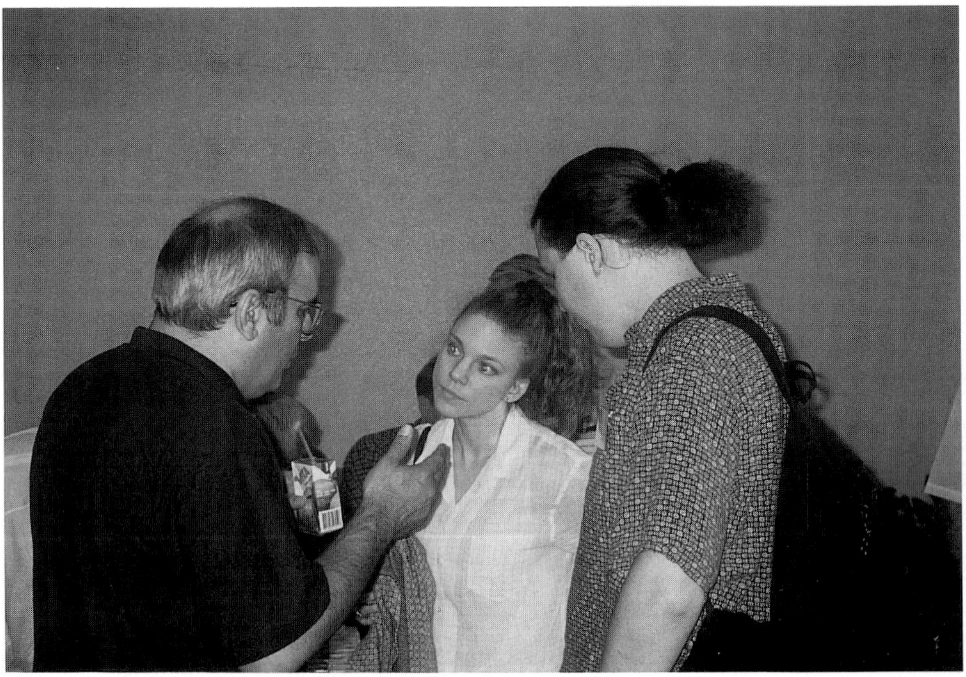

LITHIUM IN G DWARFS

H.M. Qiu[123], G. Zhao[13], Z.W. Li[2]
[1] *Beijing Astronomical Obseratory, Chinese Academy of Sciences*
[2] *Department of Astronomy, Beijing Normal University*
[3] *National Astronomical Observatories, Chinese Academy of Sciences*

Abstract We present determinations of fundamental parameters and lithium abundances of 10 G dwarfs. The analysis of our data shows that the maximum lithium abundance decreasess with decreasing T_{eff}, but with a large scatter. One of the sample stars shows a high lithium abundance as much as 2.34 dex. The correlation between the lithium abundance and absolute visual magnitude or stellar age, does not seem very good. The scatter cannot be uniquely attributed to different initial compositions or to pure age effect.

Keywords: stars: fundamental parameters-stars: abundances

1. INTRODUCTION

Lithium is one of the fragile elements, which can be easily destroyed by thermonuclear reactions at fairly low temperature ($\sim 2.5 \times 10^6$), the abundance of which is of importance as diagnostic of cosmological models. Relative to other PopI dwarfs, lithium in the sun is depleted which first interpreted as due to convection of surface material to the solar interior regions. The first compressive set of observations of main-sequence stars was made by Herbig (1965). He found an exponential decay of lithium abundance in solar-mass stars with a time constant of $\sim 10^9$ yrs and a correlation between the strength of the emission core of the Ca II H & K lines and lithium abundance. Thus, lithium abundance in main-sequence stars will be decreasing rapidly with stellar age. In recent years, large-scale studies on such stars with larger samples have been made (e.g. Pasquini et al. 1994). They reconfirmed the scatter of the distribution of lithium abundance vs. Ca II emission or [Fe/H], and pointed out that lithium was not a good age proxy for G dwarfs.

K.S. Cheng et al. (eds.), Stellar Astrophysics, 445–449.

2. OBSERVATIONS AND REDUCTIONS

The sample stars have been selected from Preliminary Version of the Third Catalogue of Nearby Stars. The observations were carried out on October 9-12, 1998 with the Coudé Echelle Spectrograph and a 1024 $\times$ 1024 Tek CCD attached to the 2.16 m telescope at Beijing Astronomical Observatory (Xingiong, China). The entire wavelength coverage was 5680 $\sim$ 9000$\AA$. The signal-to-noise ratio is greater than 150 over the whole region.

The reductions were carried out with standard MIDAS packages in a similiar procedure used by Zhao et al. (2000). The equivalent widths are weighted averages of those obtained by the direct integration of the line profile and Gaussian fitting.

3. ANALYSIS

3.1 STELLAR PARAMETERS

The effective temperature was derived from the Strömgren color index using the recent calibration by the infrared-flux method (Alonso et al. 1996). The initial metallicities are based on the calibration of Schuster et al. (1989). The microturbulence veloicty for each star was determined by forcing all the Fe I lines to give the same abundance. By using available hipparcos parallaxes, we obtained the surface gravity, absolute magnitude, stellar mass and age based on Nissen et al. (1997). For comparison, we also derived the surface gravity, absolute magnitude and microturbulence parameter from photometric indices by Edvardsson (1993) calibration. Particularly, for three binaries, we follow the usual practice to determine the surface gravities by forcing Fe I and Fe II to indicate consistent abundances due to the relative large uncertainties of their photometric indices.

3.2 ABUNDANCE DETERMINATIONS

We have selected unblended absorption lines with reliable transition probabilities for abundance analysis. By using the corresponding iron abundance well determined by the other iron lines, the contribution of the neighbouring Fe I $\lambda 6707.441\AA$ to the Li+Fe blend is deducted. The gf value and wavelength of Fe and Li are taken from Fuhr et al. (1988) and Andersen et al. (1984) respectively.

Model atmospheres are generated by the ATLAS9 code (Kurucz 1994). The final abundances (in the usual scale $\log N(H) = 12$) together with the parameters are presented in Table 1. The derived [Fe/H] are in good agreement with that derived from photometry.

Table 1 Stellar Parameters and final abundances

HD	T_{eff}	$\log g$	[M/H]	[Fe/H]	ξ_t	M_v	Mass	Age	EW	A(Li)
4307	5742	4.00	-0.27	-0.18	1.28	3.641	1.15	6.11	56.3	2.34
13043	5791	4.15	0.05	0.13	1.18	4.038	1.06	7.99	21.0	1.89
20630	5611	4.00^{α}	0.06	0.08	1.15	5.037	1.00	6.12	30.4	1.90
21019	5564	3.83^{α}	-0.22	-0.28	1.28	3.368	1.17	5.23	12.2	1.44
34411	5805	4.10^{α}	0.11	0.13	1.38	4.200	1.10	7.21	24.8	1.98
38858	5631	4.40	-0.26	-0.17	1.05	5.014	0.90	13.10	7.8	1.29
157214	5713	4.26	-0.22	-0.25	0.95	4.594	0.89	13.10	...	...
165401	5659	4.32	-0.51	-0.38	1.45	4.865	0.84	16.06	<1.3	<0.59
168009	5743	4.29	0.04	0.10	1.15	4.518	0.98	9.57	<4.3	<1.14
176377	5735	4.46	-0.30	-0.21	1.30	4.955	1.00	4.06	30.5	2.01

Table 1 reports the effective temperature (col. 2), gravities log g from hipparcos parallax (col. 3), metallicitics[M/H] (col. 4) from Nissen et al. (1989)and [Fe/H] based on the present analysis (col. 5), microturbulence velocity (col. 6),absolute visual magnitude M_v (col. 7), stellar mass (col. 8), stellar age (col. 9), lithium equivalent widths (col. 10) and lithium abundance (col. 11). $^{\alpha}$ indicates log g determined from spectra.

For four sample stars, lithium measurements are also reported in the literature. Based on several independent measurements and comparison with existing measurements of the lithium line in each star, we estimate an uncertainty of about 10% in our lithium equivalent width which may cause an error of 0.04 dex in log $N(Li)$. In order to evaluate the errors introduced by the model parameters, we computed abundances by changing the stellar parameters seperately. The results show that changing 100 K in T_{eff} would induce an error of 0.09 dex in log $N(Li)$ and changing 0.5 dex in log g would induce an error of 0.01 dex, while the effect introduced by uncertainty in the adopted ξ_t or [M/H] is negligibly small.

4. CONCLUDING REMARKS

At the top of Fig. 1 the lithium abundances for the sample stars are plotted as a function of T_{eff} and [Fe/H] respectively. No obvious trend can be seen from the figure. But we can clearly see that the maximum lithium abundances decrease with decreasing T_{eff}, and a very large scatter is present. The Sun seems representative of the low-Li stars, and one of the sample stars shows a high lithium abundance as much as 2.34 dex. All the sample stars suffered lithium depletion to varying degrees relative to meterorites. Plotting the target stars in a diagram with stellar age and M_v as abscissa and lithium abundances as ordinates (Fig. 1), we can see, in general, the maximum lithium abundance decreases with increasing age and large scatter exists around the solar age, although more data are needed to draw firm conclusions. Thus, the assumption that the dispersion of lithium abundance presented at each color is just an age effect or different initial composition seems not work, more complex mechanisms maybe exist to provide different amounts of lithium depletion for G dwarfs.

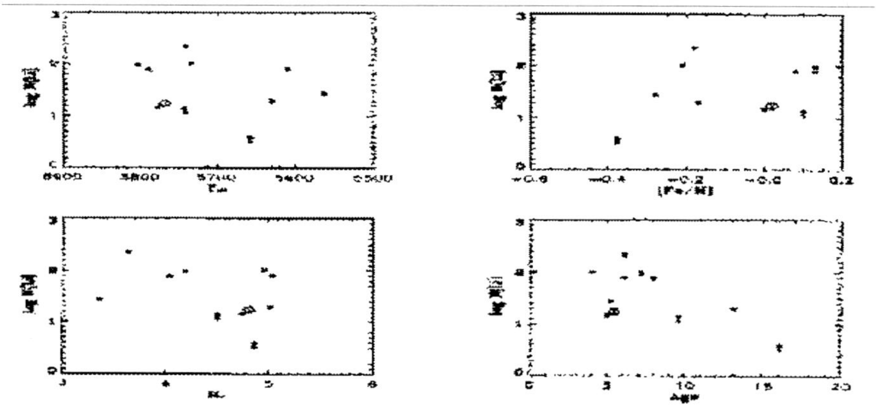

Figure 1 Lithium abundance as a function of effective temperature, stellar metallicities, absolute visual magnitude and stellar age. Symbol with arrow pointing downward signify upper limits to lithium abundance. The sun is denoted $\odot$.

Acknowledgements

We gratefully acknowledge the help of Y.Q. Chen in computing the model parameters for us. This work was funded by the National Natural Science Foundation of China.

References

Alonso A., Arribas S.,Martinez-Roger C., 1996, A&A, 313, 873
Andersen J.,Lambert D.L., 1984, A&A, 136, 65
Edvardsson B., Andersen J., Gustafsson B. et al., 1993, A&A, 275, 101
Fuhr J.R. et al., 1988, J. Phys. Chem. Ref. Data, 17, Suppl. 4
Herbig G.H., 1965, ApJ, 141, 588
Nissen P.E. et al., 1997, ESA SP-402, Hipparcos Venice '97, Ed. B. Battrick, p. 225
Olsen E.H., 1994, A&AS, 106, 257
Pasquini L., Liu Q., Pallavicini R., 1994, A&A, 287, 191
Schuster W.J., Nissen P.E., 1989, A&A, 221, 65
Zhao G., Qiu H.M., Chen Y.Q. & Li Z.W.,2000, ApJS,127

TH EAK TRAM
Except for
access
前往此區者
不在此限

OPTICAL OBSERVATIONS OF THE X-RAY TRANSIENTS X1658−298 AND X1608−522

Stefanie Wachter (Cerro Tololo Inter-American Observatory)
and Alan P. Smale (NASA/GSFC)

1. X1658−298

The X-ray transient (XT) X1658−298 was discovered in 1976 by Lewin, Hoffmann & Doty (1976). The detection of type I bursts indicates that the compact object in the system is a neutron star (NS). Observations during a temporary brightening of the source in 1978 showed dips in the X-ray lightcurve. Detailed analysis of the combined 1976–1978 data set by Cominsky & Wood (1984, 1989) revealed that X1658−298 is one of the rare low mass X-ray binaries that exhibits eclipses of the central X-ray source by the mass donating star. The dipping activity lasts for about 25% of the 7.1 hour orbital cycle followed by an eclipse of $\sim$ 15 min duration.

The optical counterpart of X1658−298 was identified during the 1978 X-ray outburst with a faint ($V = 18.3$), blue star (V2134 Oph). X1658−298 entered an X-ray off–state in 1979 and the counterpart became undetectable with a magnitude limit of $V > 23$ (Cominsky, Ossmann, & Lewin 1983). Renewed X-ray activity from X1658−298 was detected by BeppoSAX on April 2–3 1999 (In't Zand et al. 1999). Follow-up observations were quickly scheduled with RXTE between 1999 April 5–15 and with the CTIO 1.5 m and YALO telescope from UT 1999 April 29 to May 3.

Apart from the discovery observations during the 1978 outburst, optical data of V2134 Oph are sparse. We previously conducted the first photometric study of V2134 Oph in 1997 April/May and found the source only $\sim$ 1 mag fainter than the brightness reported when the system is X-ray active (Wachter & Smale 1998). Our lightcurve surprisingly also did not show any modulation across the binary cycle. X-ray transients in quiescence generally display photometric variability due to ellipsoidal variations. A comparison between our quiescent and outburst frames from 1997 and 1999 (Figure 1) resolves this puzzling behavior: the star

K.S. Cheng et al. (eds.), Stellar Astrophysics, 451–455.

we observed ("A") is in fact an unrelated close companion to the actual optical counterpart.

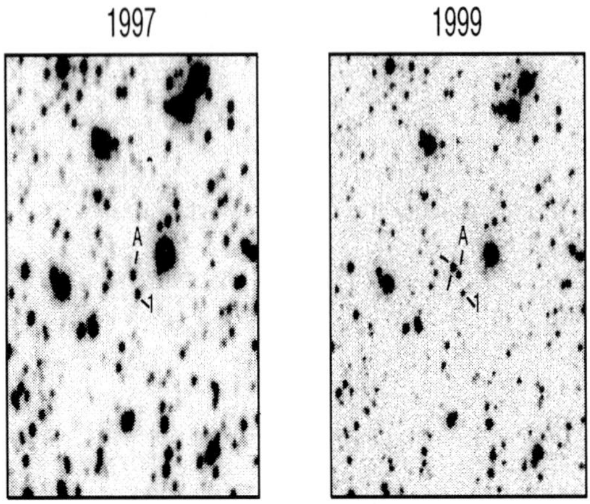

Figure 1 50" ×50" I band exposures of the X1658−298 field obtained with the CTIO 1.5m telescope while the source was in quiescence (left, 1997 May 5) and during outburst (right, 1999 May 2). The actual counterpart (flagged) is 0.8" east and 1.0" north of A.

The X-ray observations were scheduled to occur centered on the expected times of eclipse, as extrapolated from the ephemeris of Cominsky & Wood (1989). We find a spread of eclipse ingress/egress times of 6–13 sec, and a mean eclipse duration of 901.9±0.8 sec. The eclipse centers occur an average of 407.4 sec earlier than predicted by the ephemeris of Cominsky & Wood (1989). We have combined our eclipse timings with the eclipse centers from the observations of Cominsky & Wood (1984, 1989), to produce an updated ephemeris. A parabolic ephemeris is required; the $\dot{P}/P$ term implies that the orbital period of the system is decreasing on a timescale of 10^7 yr.

Figure 2 shows our optical outburst data folded according to our updated X-ray ephemeris. The time of the X-ray mid-eclipse is defined as phase 0. The deepest point of the optical lightcurves on each night was chosen as a reference point for the brightness of the system and the data (vertically) shifted accordingly. The folded V band lightcurve clearly shows a distinct central drop in brightness within $\sim$ 0.2 mag of the faintest observed magnitude. We determined the optical eclipse center to occur at phase 0.004 ± 0.003. A close-up of the central region is shown on the bottom left hand side of Figure 2 together with the fit to the 1999 Apr 5 X-ray eclipse (dotted line). Our data indicate that there is no significant offset between the time of mid-eclipse in the X-ray and optical and that the narrow component of the optical eclipse is of the same duration as the X-ray eclipse. This implies a distinct optical emission region associated with the X-ray emitting area.

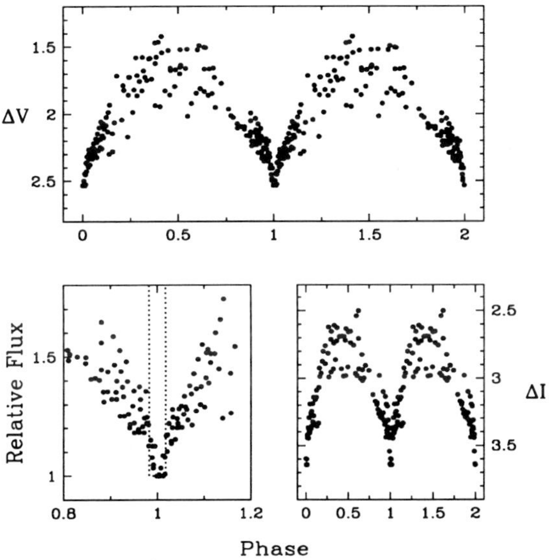

Figure 2 Our V and I band data of V2134 Oph folded according to our updated X-ray ephemeris.

2. X1608−522

X1608−52 is another member of the rare group of XTs with NS primaries (based on the detection of type I X-ray bursts). It exhibits large X-ray outbursts on a semi-regular timescale of about 600-700 days (Lochner & Roussel-Dupre 1994). The optical counterpart of X1608−52, QX Nor, was discovered by Grindlay & Liller (1978) after an outburst in 1977. During quiescence, the counterpart fades to as yet undetected faint optical magnitudes. Wachter (1997) detected QX Nor in the IR in quiescence and in the IR and optical a few months after the 1996 outburst. The next outburst of X1608−52, after the most recent one in 1998 Jan, would be expected at the beginning of 2000 if the source adheres to the 600-700 day recurrence pattern. We obtained optical images of the field in 1999 May, June and July and found the source as bright as or brighter than shortly after the outburst in 1996. This is very surprising since the optical brightness of XTs is thought to be intimately linked to their X-ray activity. The RXTE All Sky Monitor data confirm that X1608−52 remains in quiescence and consequently QX Nor should be at the unobservable optical limit previously reported by Grindlay & Liller (1978) and Wachter (1997). It is unclear whether those faint states are in fact the more unusual state of the source and whether QX Nor is routinely at the currently observed brightness level, which would make the system accessible to orbital period and even radial velocity measurements. We searched the individual and combined data sets of QX Nor for a periodic variability component. A strong signal for a period between 0.52 - 0.55 days is found in each individual data set, however, this period is not apparent in the combined data set. While the same periodicity is present in the data from each month, a phase shift seems to occur from one run to the next. The present data set is insufficient to explore whether there is any systematic trend to the phase shifts.

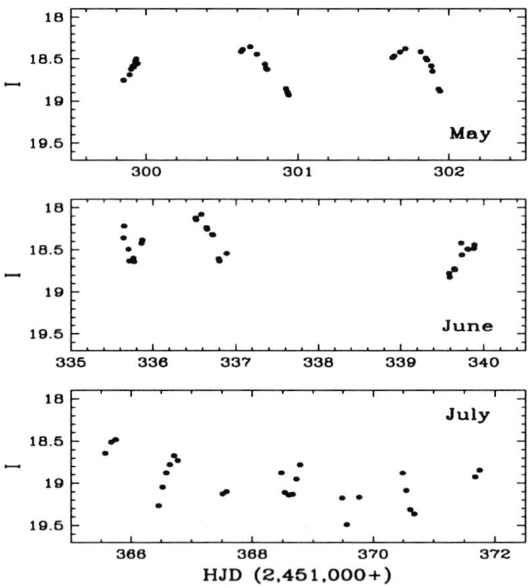

Figure 3 I band lightcurve of QX Nor, the optical counterpart of X1608−52, obtained during three separate observing runs in 1999 with the CTIO 1.5m and 0.9m telescopes.

References

Cominsky, L., Ossmann, W., & Lewin, W. H. G. 1983, ApJ, 270, 226
Cominsky, L. R. & Wood, K. S. 1984, ApJ, 283, 765
Cominsky, L. R. & Wood, K. S. 1989, ApJ, 337, 485
Grindlay, J. E. & Liller, W. 1978, ApJL, 220, L127
In't Zand, J. & Heise, J. 1999, IAUCirc. 7138
Lewin, W. H. G., Hoffmann, J. A. & Doty, J. 1976, IAUCirc. 2994
Lochner, J. C. & Roussel-Dupre, D. 1994, ApJ, 435, 840
Wachter, S. 1997, ApJ, 485, 839
Wachter, S. & Smale, A. P. 1998, ApJL, 496, L21

STUDY OF REDDENING IN THE YOUNG OPEN CLUSTER NGC 6913

Jun-Jie Wang[1] and Jing-Yao Hu[2]
[1]*Beijing Astronomical Observatory, Chinese Academy of Sciences, Beijing 100012, P.R. China*
National Astronomical Observatory, Chinese Academy of Sciences, Beijing 100012, P.R. China,
[2]*Beijing Astronomical Observatory, Chinese Academy of Sciences, Beijing 100012, P.R. China*
[1]wangjj@class1.bao.ac.cn, [2]Hujy@class1.bao.ac.cn

Abstract An extreme variation of extinction is found across the young open cluster NGC 6913. The distribution of extinction in the region is correlated with the effective temperature of the stars for stars earlier than F8. It is found that extinction in the cluster center is relatively homogeneous, but very large. The north and south of this cluster have the largest extinction . The eastern and western parts seem to be low reddening regions. A mean distance modulus of 10.17 ± 0.14 mag. of the cluster is derived, which corresponds to a distance of 1.08 kpc.

Keywords: Open cluster, interstellar reddening

1. INTRODUCTION

The young open cluster NGC 6913(20h23m.9, +38.32(2000)) is located in the Cygnus OB1 association. In 1973, Sanders computed the probabilities of cluster membership for 228 stars in the field of NGC 6913, based on relative proper motions (Sanders, 1973). He found there are 105 stars with probabilities greater than 50% out of 228 stars in the field. Before 1980, only small part of the stars have UBV photoelectric values. Much more complete UBV photoelectric photometry was made by Joshi et al. in 1983, using the 104-cm Sampurnanand reflector of the Uttar Pradesh State Observatory. The standard deviations of their observations were better than $\pm$ 0.025 mag. in (U-B), (B-V) and V magnitudes (Joshi et al., 1983). From their photometric study, the distance

K.S. Cheng et al. (eds.), Stellar Astrophysics, 457–461.

modulus to the cluster is obtained to be 10.85, which corresponds to a cluster distance 1.5kpc. It was also found that the cluster stars have ages between 0.3 to 1.75 Myr with a few even younger than 0.3 Myr. This means NGC 6913 is a very young open cluster.

Since the cluster is so young, it should have many stars in the pre-main-sequence phase. The study of intestellar matter inside young open cluster is quite important for a thorough understanding of the star formation processes

2. OBSERVATIONS

The spectroscopic observations were carried out between Jun. 7 and Jun. 11 in 1994 with the 2.16 m telescope at Xinglong station of Beijing Astronomical Observatory, using a thin Tektronix 512×512 CCD and an UNIVASAL spectrograph. A grating with dispersion of 195Å/mm was used. The spectral resolution is 5.26Å/pixel, and the wavelength coverage is from 4205 to 6898Å.

100 stars were selected from Sanders' list of open cluster NGC 6913. All the stars in the sample have membership probability greater than 50%. A membership probability of 50% guarantees that most of the non-cluster members have been removed. A cross identification between Sanders' finding chart and ST Atlas was made so as to get the accurate coordinates of the stars.

3. DISCUSSION

We classified the total 100 stars in NGC 6913 spectroscopically. Many E(B-V) values derived from our spectral classification are different from those based on photometry (Ep(B-V)). Some show a very large difference. 32 stars have absolute values of (E(B-V)-Ep(B-V)) greater than 0.3 mag., the largest one is as much as 1.48 mag. (#219, hereafter refers Sander's number)! We find that the stars with large absolute values of (E(B-V)-Ep(B-V)) are mainly some early B type stars and most of the F type stars, in which the values of Ep(B-V) of the B type stars are far lower than that of E(B-V), and the values of Ep(B-V) of the F type stars are higher than that of E(B-V). The results are mainly due to the effect of reddening and the uncertainty of photometric classification, moreover they have different R values. The difference shows the importance of the spectroscopic observations in studying young star clusters which include many pre-main sequence stars.

In order to construct the H-R diagram for NGC 6913, the distance modulus must be known accurately. The distance of NGC 6913 estimated by various authors ranges from 0.8 kpc to 2.8 kpc. Up to now

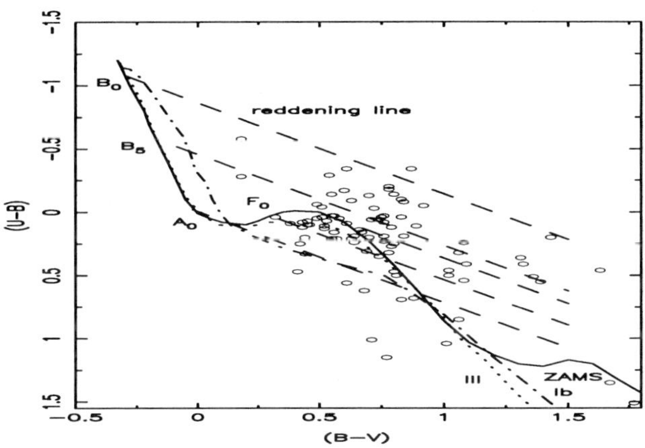

Figure 1 The $(U\text{-}B)_0$ versus $(B\text{-}V)_0$ and the (U-B) versus (B-V) colour-colour diagrams of NGC 6913, in which the reddening line is plotted with a slope 0.72. The zero-age main sequence line, III and Ib type luminosity line are drawn in.

a relatively dependable value of the distance is (1.5 ± 0.12) kpc, corresponding to a distance modulus 10.85 ± 0.15 mag (Joshi et al.,1983). This is obtained by fitting the ZAMS given by Allen onto the unevolved part of the cluster main sequence present in the (V_0, $(B\text{-}V)_0$) and (V_0, $(U\text{-}B)_0$) diagrams respectively. .

We obtained the cluster distance by means of the ZAMS fitting method. Three main-sequence stars(#124, 126 and 130) satisfied the strict conditions, i.e. membership probability around 80 % (or more), OB type stars , location within the central region , $0.70 < E(U\text{-}B)/E(B\text{-}V) < 0.80$ (normal reddening law) and a value of E(B-V) not too high compared with other OB stars in this cluster. After fitting to the ZAMS locus, a mean distance modulus of 10.17 ± 0.14 mag. is derived, which corresponds to a distance of 1.08 kpc. The ((U-B), (B-V)) colour-colour diagrams of NGC 6913 are plotted in Fig. 1, in which the zero-age main sequence line, III and Ib type luminosity line are drawn. In this diagram, the reddening line is also plotted with a slope 0.72.

Fig. 2 shows the spatial distribution of extinction across the NGC 6913 field, in which the diameter of the circle is 12', corresponding to twice the possible smallest cluster diameter given by Lynga (1987). From this figure, we find that most of the nine stars with E(B-V) greater than 1.2 locate on the edge of the field. Most of the stars with E(B-V) between 0.9 and 1.2 are concentrated in the central area. All of the stars except one with E(B-V) between 0.6 and 0.9 lie in the western part (right part of the figure) of the cluster. Those with relatively small values of E(B-V) are distributed randomly over the entire cluster region. It is clear

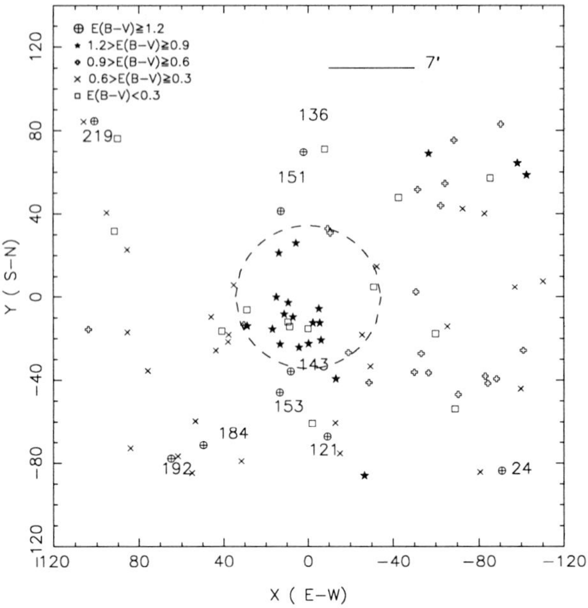

Figure 2 The spatial distribution of extinction across NGC 6913, in which the diameter of the circle is two times Lynga diameter of 12'.

that the region with the largest mean value of extinction is in the central part. The western region comes second.

On the same scale as Fig. 2, another figure (not presented) show the spatial distribution of those stars with high mass and relatively low mass respectively. Surprisingly, it was found that the distribution of OB stars coincide very well with the distribution of large colour excess across the field. All nine stars with E(B-V) greater than 1.2 are B-type stars. Clearly, the surroundings of the massive stars seem to be more obscure than those of the low-mass stars.

The plot $(B\text{-}V)_0$ versus E(B-V) (not presented) clearly show a dependence of E(B-V) on spectral type. For those stars earlier than F8 (i.e. $(B\text{-}V)_0$ less than 0.6), the colour excess increases together with the effective temperature of the stars. But the stars with spectral type later than F8 seem to show reversed characteristics. Because the number is too small, we can not confirm the latter point. A similar result for the whole field was obtained by Sagar (1987). The reason causing this trend is still a puzzle.

References

Joshi U. C., Sanwal B. B., Sagar R.,1983, PASJ 35, 405

Lynga G., 1987, Catalogue of open cluster data, computer based catalogue available through CDS, Strasbourg, France
Sagar R., 1987, MNRAS 228, 483
Sanders W. L., 1973, A&AS 9, 221

LUMINOSITY OF TYPE IA SUPERNOVAE AND HUBBLE CONSTANT

Wang Xiaofeng & Li Zongewei

Department of Astronomy, Beijing Normal University, Beijing, 100875, P.R., China

lizw@bnu.edu.cn

Abstract Luminosity of 41 Type Ia supernovae (SNe Ia) have been studied solely from the observational angle. The connections between the peak luminosity of SNe Ia and distance-independent observables (DIOs) could be established. The variation in the luminosity of the SN mainly depends on two separate DIOs: SN Ia's $B - V$ color at maximum light and $\triangle m_{15}(B)$ (or other measurements of the light curve shape). For most of SNe Ia, the main parts of the observed $B - V$ colors at maximum light are intrinsic. The correlation of $(B - V)_{max}$ with the luminosity is more efficient in reducing the scatter in the Hubble diagrams than other DIOs. When SNe Ia in nearby galaxies with cepheid distances are tied to 41 Hubble-flow SNe Ia, the value of present Hubble constant is obtained, which is 61±5 (statistical) $kms^{-1}Mpc^{-1}$.

1. INTRODUCTION

It has long been recognized that SNe Ia are better distance indicators (see Branch 1998). Cepheid calibrated SNe Ia and Hubble- flow SNe Ia ($0.01 < Z < 0.1$) are combined to determine the Hubble constant (Hamuy et al. 1996a); high redshift SNe Ia ($0.3 < Z < 1.0$) could be used to examine the nature of the redshift using the time dilation (Riess et al. 1997) as well as to measure the expansion history of the universe (Riess et al. 1998, Perlmutter et al. 1999); SNe Ia could also be used to reveal the peculiar motions of distant galaxies when combining with their positions on the sky (Tammann 1998). All of these applications rely on a common assumption that SNe Ia are "standard candles" or nearly standard candles. However, with the improvement of photometric techniques and the accumulation of SN Ia's data, the diverse properties involved with SNe Ia have taken on in many respects, including spectra, light curve shape, color of SNe Ia, and type (or color) of the host galaxies. These DIOs are thought to be able to account for

K.S. Cheng et al. (eds.), Stellar Astrophysics, 463–473.

the variation in the peak luminosity SNe Ia more or less. Some other physical parameters, such as expansion velocity, spectra features, et al., also show correlation with the peak luminosity, however, the data available is rather limited. Among these DIOs, $\Delta m_{15}(B)$ (see Phillips 1993) has long been considered to be the dominant factor influencing the luminosity of SNe Ia. Most authors solely used $\Delta m_{15}(B)$ to standardize the peak luminosity of SNe Ia, nevertheless the correlation of SN Ia's $B-V$ color at maximum light with the peak luminosity is stronger than $\Delta m_{15}(B)$ (see Fig.2 of Hamuy 1996a and Fig.2.of Hamuy 1996b). On the other hand, Riess et al. (1996a) used multicolor light curve shape (MLCS) method to estimate the intrinsic luminosity and color of a Type Ia supernova from the shapes of the observed light and color curves, which is efficient in reducing the scatter in the Hubble diagrams.

In this paper, a simple two-step procedure are used to correct first for SN Ia's color and then for$\Delta m_{15}(B)$ with respect to the observed peak luminosity of SNe Ia. Section 2 briefly describes the data set. Section 3 is concerned with the luminosity of SNe Ia in function of second parameters. The discussion on Hubble constant is given in section 4. Section 5 states conclusions.

2. DATA SET

The source of our data consist of 29 CTIO SNe Ia (see Table 8 in Hamuy 1996c) and 22 CFA SNe Ia (see Table 3 in Riess 1999, hereafter R99). For the 22 CFA SNe Ia, the apparent peak magnitudes and $\Delta m_{15}(B)$ are recalculated by using χ^2 fitting technique similar to Hamuy et al. (1995). Our results show some difference from what R99 calculated. In this paper, we adopted our fitting results except that $log(CZ)_{helio}$ and $(B-V)^0_T$ are taken from R99. Our final sample of SNe Ia contains 41 SNe Ia, which conforms to the criteria that the heliocentric velocities of the host galaxy are greater than $4,000\ kms^{-1}Mpc^{-1}$ and the spectra near maximum light are known to be Branch-normal.

The following parameters are used in our study: (1) $log(CZ)_{helio}$: the heliocentric velocity of the host galaxy; (2) B_{max}, V_{max}: the apparent B and V peak magnitudes, both corrected for the foreground extinction and K terms; (3) $(B-V)_{max}$: the $B-V$ color of SNe Ia at maximum light, represented by B_{max} V_{max}; (4)$\Delta m_{15}(B)$: defined by Phillips(1993) as the amount in magnitudes that the B light curve decays in the first 15 days after maximum; (5)M^B_{max} , M^V_{max}: the absolute B and V peak magnitudes of the SN calculated from the apparent magnitudes of (2), the redshift from (1), and assumed value of the Hubble constant of $H_0 = 60\ kms^{-1}Mpc^{-1}$; (6) $(B-V)^0_T$: the integrated color of the host galaxy corrected for foreground extinction as well as for K terms.

3. LUMINOSITY OF SNE IA

The peak luminosity of SNe Ia is indicated by the absolute B or V peak magnitudes. As far as our 41 sample are concerned, the peak luminosity vary from -18.04mag to -19.86mag in B magnitudes, and from -18.63mag to -19.78mag in V band. There are a lot of literatures on that DIOs govern the luminosity of SNe Ia. Of them, SN Ia's color, $\Delta m_{15}(B)$, host galaxy's color are relatively important parameters.

The absolute B and V peak magnitudes of the SN are plotted versus their color at maximum light in Figure 1. Generally speaking, the dimmer the SN are, the redder they appear to be. For the 41 SNe Ia, $(B-V)_{max}$ covers a range from -0.1 to 0.59. In principle, the variation in color can be intrinsic to the present sample of the SN or can be caused by absorption in the host galaxy. In practice, it is difficult for us to distinguish carefully between these two factors. However, if the conclusion that most of the distant galaxies share the same extinction law with the Milky way (see Riess et al.1996b) is admitted, we could estimate which of these two factors dominates by comparing the phenomenological extinction coefficients R(B), R(V) in the form $R(B-V)$ we get by SN Ia's sample with standard Galactic extinction coefficients (R_B 4.1, R_V 3.1).

$(B-V)_{max}$ show stronger correlation with the absolute peak magnitudes. The linear regression fitting is given below,

$$M_B = -19.38 + 2.09(B-V)_{max} \tag{1}$$

$$M_V = -19.36 + 1.03(B-V)_{max} \tag{2}$$

Where $(B-V)_{max}$ contains $E(B-V)$ and $(B-V)_0$ intrinsic to SN Ia. The phenomenological extinction coefficients are much smaller than the standard Galactic extinction coefficients, and this result is in agreement with what Tripp (1997) got by 29 CTIO sample. It might imply that most of our present SNe Ia do not suffer significant absorption in the host galaxy, hence the main parts of the observed SN Ia's colors are intrinsic. Another way to validate our conclusion is to compare the phenomenological coefficients in spirals and E/S0. Table1 contains relevant information for this comparison. No significant difference is found between spirals and E/S0.

The connection between the luminosity of SNe Ia (corrected for $(B-V)_{max}$) and $\Delta m_{15}(B)$ could also be established (see Figure 2), which is given by,

$$M_B^{\dagger} = -19.45 + 0.60(\Delta m_{15}(B) - 1.1) \tag{3}$$

$$M_V^{\dagger} = -19.44 + 0.56(\triangle m_{15}(B) - 1.1) \quad (4)$$

Where the slope of the correlation of decay rate of the light curve with $M^{\star}_{B,V}$ is much flatter than what Phillips(1993) got and is also flatter than what Hamuy et al.(1996c) obtained, indicating a relatively weaker dependence of the peak luminosity on the parameter, $\triangle m_{15}(B)$.

The peak luminosity of SNe Ia (corrected for the galactic absorption) are also plotted as a function of the color of the host galaxy, $(B-V)^0_T$, in Figure 3. A weaker correlation is found between them as what Branch et al (1996) claimed. However, when the sample is divided at $(B-V)^0_T = 0.75$ in Figure 3, clear difference shows between these two populations. The peak luminosity of SNe Ia in bluer galaxies appears to have relatively smaller scatter, while a wider scatter shows for the SN in redder galaxies. Moreover, the connection between the peak luminosity and $(B-V)^0_T$ is more closely in redder galaxies than in bluers.

4. HUBBLE CONSTANT

The connections between the peak luminosity and some DIOs have been established. We could say, at least from the observational angle, the peak luminosity of the SN mainly depends on $(B-V)_{max}$, $\triangle m_{15}(B)$, and even $(B-V)^0_T$. These DIOs could be used to standardize the peak luminosity of the SN, which is the same as using the periods of Cepheid variables as DIOs to standardize their absolute magnitudes. Of them, $(B-V)_{max}$ and$\triangle m_{15}(B)$ are independent each other, while $(B-V)^0_T$ is correlated with $\triangle m_{15}(B)$ (see Figure 4), so the former two parameters are used to standardize the peak luminosity of the SN. We list the results of the two-step correction in Table 2. The dispersion of the Hubble diagram is reduced significantly after the color correction, and the scatter is further reduced by means of $\triangle m_{15}(B)$ correction. The final dispersion of the Hubble diagram decreases to 0.15mag, which could be attributed to the observational error. Figure 5 shows the Hubble diagram after two-step correction. The Hubble linear relations are given below,

$$M_B^c = 5\log(cz)_{helio} - 3.348(\pm 0.024)\sigma_B = 0.15, N = 41 \quad (5)$$

$$M_V^c = 5\log(cz)_{helio} - 3.339(\pm 0.024)\sigma_B = 0.16, N = 41 \quad (6)$$

So far we know cepheid distances of 8 galaxies which have produced 9 SNe Ia. They are IC4182 (SN 1937C), NGC5253 (SNe 1895B, 1972E), NGC4496A (SN 1960F), NGC4414 (SN 1974G), NGC 4536 (SN 1981B), NGC3627 (SN 1989B), NGC4639 (SN 1990N), NGC 3368 (SN 1998bu). The mean luminosity is $M_B(max) = -19.47 \pm 0.07$, $M_V(max) = -19.48 \pm 0.07$(see Tammann & Reindl 1999). However, $(B-V)_{max}$ or $\triangle m_{15}(B)$

of SNe 1895B, 1960F is unknown, so only 7 calibrated SNe Ia can be used if the analogue corrections are applied to the calibrators. That is,

$$M_B^c(max) = -19.43 \pm 0.04\ mag \tag{7}$$

$$M_V^c(max) = -19.42 \pm 0.04\ mag \tag{8}$$

When we match Eqs. (7) and (8) to (5) and (6) respectively, the values of the Hubble constant can be calculated. The values of H_0 without $(B-V)_{max}$ or $\Delta m_{15}(B)$ correction are also listed in Table 2 for the purpose of comparison. The Hubble constant will be underestimated by $\sim$ 13% without these corrections. The second parameter corrections are necessary since the local calibrators are somewhat bluer and somewhat slower decliners than the distant SNe Ia.

In the procedure of determining the Hubble constant, the possible system errors have not been considered. These errors include: the reliability of cepheid distance. The issue is concerned with the possibility of a significant metallicity dependence (see Beaulieu 1997); the selection bias of the SNe Ia in the Hubble flow. Many of the CTIO events peaked not far above the detection limit, and we might overestimate the peak luminosity of the Hubble-flow SNe Ia. Therefore, the current estimate of H0 are probably too high; the accuracy of the calculations of the theoretical light curve. The main issue here is the value of the ejected nickel mass.

5. CONCLUSION

Many DIOs appear to govern the luminosity of the SN. However, the variation of the peak luminosity is mainly attributed to two separate observables: $(B-V)_{max}$ and $\Delta m_{15}(B)$. Most of the SNe Ia do not suffer significant absorption in the host galaxy. The dispersion of the Hubble diagram could be reduced to 0.15mag after $(B-V)_{max}$ and $\Delta m_{15}(B)$ corrections. Combined with present cepheid calibrated SNe Ia , the value of the Hubble constant of $H_0 = 61 \pm 5\ kms^{-1}Mpc^{-1}$ is obtained.

Acknowledgements

The work reported here was funded by the NSF of China(19733002).

References

Beaulieu, J. P., et al. A & A, 1997, 318, L47

Branch, D. Ann. Rev. Astron. Astrophys.,1998, 36, 17

Branch, D., Romanishin, W., Baron, E. ApJ., 465, 73

Table 1 The phenomenological extinction coefficients in B and V bands

Type	Number	R(B)	R(V)
Spirals	25	2.44(0.27)	1.43(0.30)
E/S0	16	2.04(0.62)	0.86(0.59)
All	41	2.09(0.25)	1.03(0.27)

Table 2 The zero points and dispersions of the Hubble diagram

Correct for	ZP_B	σ_B	ZP_V	σ_V
No correction	-3.125(0.055)	0.35	-3.197(0.041)	0.26
$(B-V)_{max}$	-3.276(0.033)	0.21	-3.272(0.033)	0.21
$(B-V)_{max}$,$\triangle m_{15}(B)$	-3.348(0.024)	0.15	-3.339(0.024)	0.16

Hamuy, M., et al. A.J., 1995, 109,1

Hamuy, M., et al. A.J., 1996a, 112, 2398

Hamuy, M., et al. A.J., 1996b, 112, 2391

Hamuy, M., et al. A.J., 1996c, 112, 2408

Perlmutter, S. ApJ., 1999, 517, 565

Hoflich, P., & Khokhlov, A. ApJ., 1996, 457, 500

Phillips, M.M. ApJ., 1993, 413, L105

Riess, A.G., Press, W.H., & Kirshner, R.P. ApJ., 1996a, 473, 388

Riess, A.G. et al. ApJ., 1996b, 473, 588

Riess, A.G. A.J., 1997, 114, 722

Riess, A.G. A.J., 1998, 116, 1009

Riess, A.G., et al. A.J., 1999, 117, 707

Tammann, G.A. astro-ph/9805013

Tammann G.A., & Reindl, B. astro-ph/9903220

Tripp, R. A & A, 1998, 331, 815

Table 3 The values of the Hubble constant

Correct for	$H_0(B)$	$H_0(V)$
No correction	53.8	55.4
$(B-V)_{max}$	58.2	58.4
$(B-V)_{max}$,$\triangle m_{15}(B)$	60.8	60.8

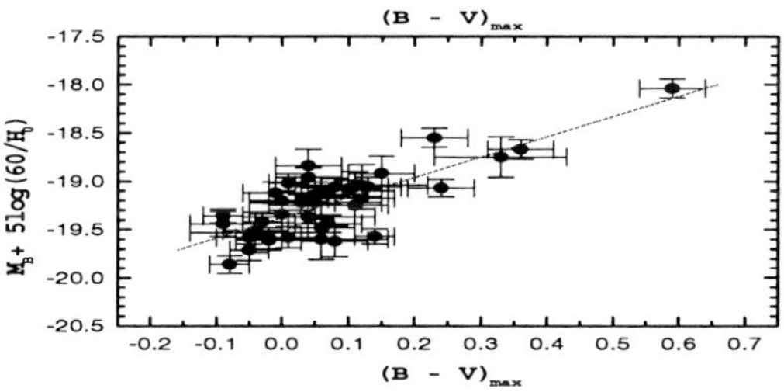

Figure 1 The absolute V, B magnitudes of 41 SNe Ia, plotted as a function of their $(B-V)_{max}$ colors.

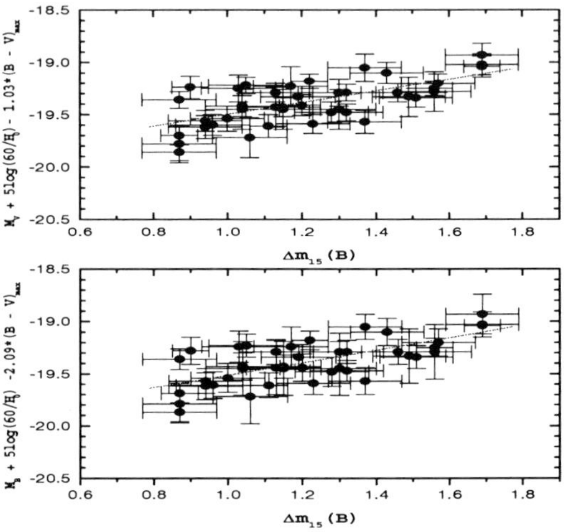

Figure 2 The absolute V, B magnitudes after $(B - V)_{max}$ correction, plotted vs. $\Delta m_{15}(B)$.

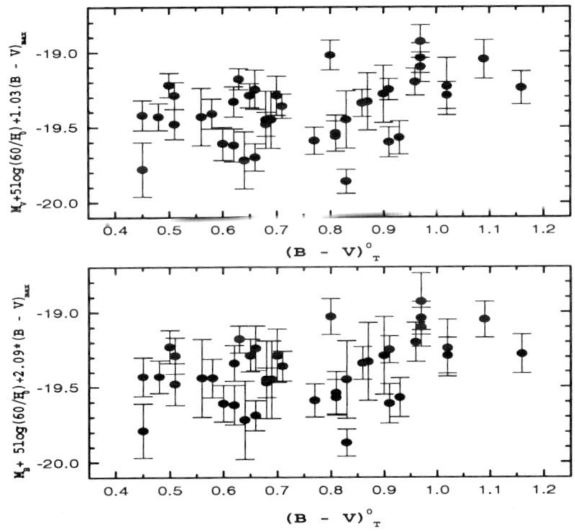

Figure 3 The absolute V, B magnitudes after $(B-V)_{max}$ correction, plotted vs. the color of the host galaxy, $(B-V)^0_T$.

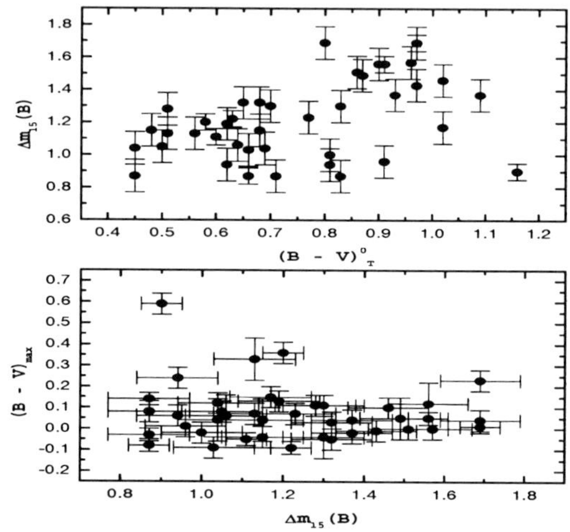

Figure 4 (top) $\Delta m_{15}(B)$ plotted as a function of the color of host galxy, $(B-V)_T^0$; (bottom) SN Ia's $(B-V)$ color plotted as a function of $\Delta m_{15}(B)$.

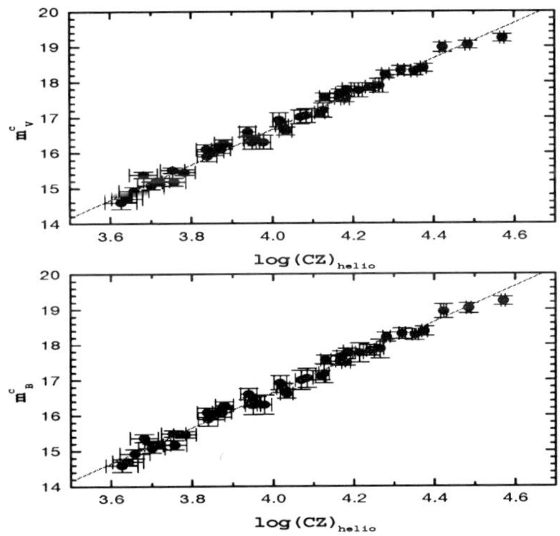

Figure 5 The Hubble diagram after $(B-V)_{max}$ and $\triangle m_{15}(B)$ corrections.

THE EXPLANATION FOR THE STRANGE SHAPE OF MEAN PULSE OF PULSAR PSRJ2019+2425

X.J. Wu., J. Pan., & X.B. Xu

Chinese Academic of Science-Peking University joint Beijing Astrophysical Center and Department of Geophysics, Peking University, Beijing ,100871

Abstract The mean pulse of PSRJ2019+2425 is very wide, which consist of three asymmetric separate pulses. The five components were found via fitting the mean profile with multi- gaussian components. There are relative phase shifts, resulted from spin of pulsar and the different emission altitude of components. With very short period and small magnetic inclination angle, the components emitted at outermost place will possibly shift apart from the components at lower altitudes. The two components of outer cone can shift about 180 degree in the longitude, and the shape become very wide and asymmetric.

1. INTRODUCTION

Two basic geometry pictures are considered in understanding the interpulse of pulsar, one is "orthogonal rotator" and the other is "aligned rotator", which is named for its almost aligned rotation axes and magnetic axes. At very small magnetic inclination angle, while with adequate emission beam width and appropriate seeing angle with respect of sight line and rotation axes, a single emission beam could be double observed through which interpulse is produced. Which model should be applied in understanding interpulse? It is a hot point after the first discussion of Manchester & Lyne (1977).

Presently both geometry pictures have marks on explaining canonical pulsars with interpulse. But preference of explaining interpulse of millisecond pulsar, is "orthogonal rotator" (Chen, Ruderman & Zhu, 1998).

However, the core, inner cone and outer cone (Rankin, 1993) of which the emission beam consists are emitted at different altitudes, which will result in relative phase shifts at receiving between the pulses components

K.S. Cheng et al. (eds.), Stellar Astrophysics, 475–479.

contributed by these elements of beam as the joint effects of retardation, aberration and magnetic field sweep back (Pan & Wu, 1999). Once period short enough, and magnetic inclination angle small adequately, the mean pulses components contributed by the outermost emission cone will shift greatly away from the components emitted at very low places with large chances and act as interpulse. Or in another words, millisecond pulsars as aligned rotators are highly possible to develop interpulses.

We find that the strange mean pulse profile of PSR J2019+2425 at 430MHz can be well understood as an aligned rotator within the theory of pulse components phase shifting. It is a new model of developing interpulse of aligned rotator but different with the classical "aligned rotator" model.

2. PHASE SHIFTING OF PULSE COMPONENTS

Radio emission beam consists of the center core and two concentric loops, inner cone and outer cone with increasing emission altitudes (Rankin, 1993). Under the combined operation of retardation, aberration and magnetic filed sweep back that result from the spin of pulsar, pulses emitted at different altitudes have different arrival time and result in phase shifts on the mean pulse profile. The distribution of the positions of pulse components will be asymmetric but not symmetric as observed in the co-rotating reference (Pan & Wu, 1999). Define the star center as zero altitude point, the phase shift estimation at altitude r is

$$\delta\phi(r) = \delta\phi_a + \delta\phi_r - \delta\phi_{mfs}. \tag{1}$$

The phase shift $\delta\phi > 0$ expresses that the emission of the component will received earlier than what RVM predicts, while $\delta\phi < 0$ indicates delay. The $\delta\phi_a$ is aberration of the center of the beam. At altitude r the shift of phase by aberration is

$$\delta\phi_a = tan^{-1}(\gamma r/r_{LC}) \tag{2}$$

In which $\beta = \Omega r(sin\alpha)/c$ and $\gamma = (1-\beta^2)^{-1/2}$. If $r << r_{LC}$, $\gamma \simeq 1$. The $\delta\phi_r$ is retardation, which is purely a propagation effect. Suppose emission of a frequency is emitted at altitude r, then there will be a longitude shift relative to $r = 0$, $\delta\phi_r = r/r_{LC}$. The MFS is caused by pulsar rotational energy loss in the form of magnetic dipole radiation. The electromagnetic torque on the neutron star results in a toroidal bending of the magnetic field opposite to the sense of the star's rotation (S83, S85). At the altitude r the phase shift caused by MFS is

$$\delta\phi_{mfs} \simeq 1.2\sin^2\alpha(r/r_{LC})^3 \tag{3}$$

$$\delta\Phi = \delta\phi(r_2) - \delta\phi(r_1) \tag{4}$$

There are a group of relations between between total longitude shift ($\delta\phi$) and emission height for the different magnetic inclination angle (α) of $5°, 7°$, and $10°$. A turning point cxist on these three curves at altitude of r_p. The r_p is an important new parameter. Curves of $\delta\phi$ as function of at ordinary magnetic inclination angles ($30° \sim 70°$) are calculated before (Pan & Wu, 1999), which indicate that the maximum values of $\delta\phi$ are not obviously greater than $90°$ and further mean that $\delta\Phi$ won't be large. But as observing the curves of at magnetic inclination angles $\alpha = 10°, 7°, 5°$, we find that $\delta\phi$ can be surprising large, for example $\delta\phi \sim 180°$ at $r/r_{LC} \sim 2$, which reveal a way leading to interpulse in that one emission cell is near star surface while another one's altitude is close to $r/r_{LC} \approx 2$.
The problem is if emission region can get such a high place. A theoretical limit is that emission altitude must satisfy $r \sin\alpha < r_{LC}$, which implies that r/r_{LC} can be much greater than unit because of the existence of α, especially a very small α. So, what is the practical qualification for such possibility? Assuming a emission cell is near star surface where its $\delta\phi(r_1) \approx 0$, with $\alpha \approx 10°$ for estimation, the altitude r_2 of the emission cell at higher place should have $r_2/r_{LC} \approx 1.2$ to 2.0 for $\delta\Phi \approx 140° - 180°$, or as function of pulsar's spin period P.

$$r_2 \approx (0.76 - 0.96) \times 10^5 P(km) \tag{5}$$

The outermost cell of emission beam is its outer cone which emission altitude is about several hundred kilometers (Blaskiewicz, Cordes & Wasserman, 1991; Rankin, 1993; Wu, Gao, et al, 1998). From equation (5) one can see that period should satisfy

$$P < 10^{-2}s \tag{6}$$

It appears that millisecond pulsar is most possibly the target.

3. MEAN PULSE AND AN EXPLANATION FOR THE MEAN PULSE OF PSR J2019+2425

J2019+2425 is in a binary system. Its period is 3.935ms (Nice & Taylor, 1993). What arouse interest are the three separate pulses of its mean pulse profile at 430MHz while ordinarily only two separate pulses can be seen in the mean pulses profile with interpulses. The positions (P) of maximum intensity and the full half power widths (w_{50}) of pulse components are via fitting the mean profile with multi-Gaussian components (the fitting model can be seen in the paper of Wu, Gao et al, 1998). For convenience, name the components in the order of phase increasing as A, B, C, D, E. The parameters of the pulse components of 430MHz mean Pulse Profile of J2019+2425 can be calculated shown in Table 1.

Table 1. parameters of the pulse components of PSRJ2019+2425

component	A	B	C	D	E
P (deg.)	63.7	202.9	279.0	313.2	326.0
W_{50} (deg.)	19.2	18.3	21.9	23.2	11.9
S_{peak}	8.9	6.7	33.0	100.0	58

S_{peak} is the relative peak value of the component. Obviously the main pulse includes C, D and E three components, but which one of A and B should be counted as interpulse? Taking the point of maximum intensity of D as the center of main pulse, the separation between A and the main pulse is about 262° while $\sim$ 110° between B and D. Both values are hard to explain as the separation of interpulse and main pulse. And if choosing one as interpulse, what is the status of the other?

The sound opinion is that A and B are two components contributed by the same emission cell (outer cone). Then the center of interpulse lies between A and B. Assuming the outer edge of emission cone is delimited by the half power width, the center of interpulse is at point of 133.1° and the separation between interpulse and main pulse is 180.1°.

How to understand this kind of interpulse configuration? The mean pulse shows the five components, it is consists with the "core, inner and outer cone " model suggested by Rankin, but the distribution of the five components is not agree with this model.

Analyze the case of phase shifting model. The way of solving the difficulty of taking J2019+2425 as aligned rotator lies in considering the relative phase shifts between pulse components. On this explanation, the Rankin's model also adopted. It is easy to give a rough estimation of the emission altitude of outer cone. Assuming D is the core component, which emitted at $r_1 \sim 10km$ and the magnetic inclination angle $\alpha = 10°$, since the separation between interpulse and main pulse $\delta\Phi = 180.1°$, according to Eq.(1) and (2), the emission altitude of A and B satisfies $r_2/r_{LC} \approx 2.47$. Having $r_{LC} = 187.9km$, $r_2 = 464km$ and $r_2 \sin\alpha/r_{LC} = 0.43$, which shows that the emission altitude is far less than the light cylinder radius.

Acknowledgements

This work is supported by NSF of China, the Climbing Program from the state Science and Technology Commission of China, and by RFDP Of China.

References

Blaskiewicz, M., Cords, J. M. & Wasserman, I., 1991, Astrophys. J., 370, 643

Chen, K., Ruderman, M. & Zhu, T., 1998, Astrophys. J., 493, 397

Manchester, R. N. & Lyne, A. G., 1977, Mon. Not. R. Astr. Soc., 181, 761

Nice, D. J. & Taylor, J. H., 1993, Astrophys. J. Lett., 402, L49

Pan, J. & Wu, X., 1999, Chin. Phys. Lett., 16, No.4, 307

Rankin, J. M., 1993, Astrophys. J., 405, 285

Shitov,Yu.P., 1983, Sov. Astron., 27, 314

Shitov,Yu.P., 1985, Sov. Astron., 29, 33

Wu, X., Gao, X., Rankin, J. M., Xu, W. & Malofeev, V. M., 1998, Astron. J., 116, 1984

A EXPLANATION OF THE CHANGE IN THE ORTITAL PERIOD OF THE CONTACT BINARY V781 TAURI

Yulan Yang and Qingyao Liu

Yunnan Observatory, Chinese Academy of Sciences, Kunming, Yunnan, China

United Laboratory of Optical Astronomy, Chinese Academy of Sciences, China

National Astronomical Observatories, Chinese Academy of Sciences, China

bily@public.km.yn.cn

Abstract New times of minimum light determined from the observations of the W UMa type contact binary V781 Tau are presented. According to these times of minima and those collected from the references, the change in the orbital period of the system is analyzed. The result reveals that the orbital period of V781 Tau decreased continually from 1949 to 1998 and the change rate in the period is $\delta p/p = -5.0 \times 10^{-11}$ .The decrease in the period of the binary is explained by a model of the contracting secondary component of the system. The shrinking velocity of $6.77 \times 10^{-5} cms^{-1}$ as responsible for the change in the orbital period of the system is obtained, which is in agreement with the results shown in the study of the thermal relaxation oscillation of contact binaries by Wang (1994).

Keywords: star, contact binary, period change

1. INTRODUCTION

One of many interesting problems in the study of W UMa contact binaries is to explain the change in the orbital period of the systems. The variety of the period changes of W UMa binaries makes it difficult to understand their real physical mechanism. However, it is clear that the change in the period of a W UMa contact binary should result from their unstable structure and evolution. According to Lucy (1976), Flannery (1976) and Robertson & Eggleton (1977), a zero-age W UMa binary is obliged to the undergo periodic thermal relaxation oscillation about a state of marginal contact. It can be expected that some observable changes in the orbital period of the systems should arise because of such

K.S. Cheng et al. (eds.), Stellar Astrophysics, 481–489.

oscillations. Unfortunately, the time scale of 10^7 years of the oscillations is too long to be observed completely for the short astronomical history, but it is noticed that the results obtained by Wang (1994) could provide a way to do something for this test. Wang's study suggests that the two subtypes of W UMa type binaries are in two different thermal relaxation oscillation states: the secondary components of W-subtypes of W UMa type binaries are shrinking whereas the ones of A-subtypes are swelling. Therefore, the orbital period of the W-subtypes should be continually decreasing whereas that of the A-subtypes should be continually increasing. For acquiring the observational evidence to test the results of Wang's study (1994) the authors are interested in the observation and analysis of the change in the period of a W UMa type binary. The W UMa type binary V781 Tau is the first of our program stars.

The variability of V781 Tau was discovered by Harris (1979), who classed it as a new W UMa type binary with the orbital period of $0^d.3994$. Berthold (1981) derived 14 times of minimum light of the system from 232 photographic plates taken from 1958 to 1977 and showed an orbital period with a sudden increase in 1.5 seconds in 1971. After two years he presented a revision of his previous results (Berthold, 1983). The photoelectric times of minimum light of V781 Tau were published by Diethelm (1981a, 1981b), Pohl (1987) and Keskin (1989). Cereda et al. (1988) made extensive photoelectric observations in B and V bands and published very good light curves. They found a photometric solution of their light curves and obtained rough parameters of the binary. The observations of the radial velocity and absolute dimensions were published by Lu (1993), who re-analyzed the Cereda et al.'s light curves by using the Wilson-Devinney code, with the results revealing that V781 Tau is a W-subtype W UMa binary and the primary component resembles a main-sequence star and the secondary is very close to the ZAMS in the HR diagram and in the mass-radius diagram.

2. OBSERVATIONS

The observations of V781 Tau in B and V bands were carried out on December 31, 1997 and on January 4, 1998 with the PI1024TKB CCD photometric system attached to the 100-cm reflector telescope at the Yunnan Observatory in China. The effective field of view of the photometric system is 6.5 square arc minute at the Cassegrain focus and its BV color system (Yang and Li, 1999) approximates the standard Johnson BV photometric system. The three brighter stars except for the variable in the CCD field are used as the comparison and check stars, which are different from those used by other observers, so their coordinates together with those of the variable are given in Table 1. Integration time for each image is 100 seconds. In each of the two bands, 45 images during the primary eclipse and 38 images during the secondary eclipse have been obtained to determine times of minima of the system.

Table 1 The coordinates of the comparison and check stars

star	R.A. (1998.0)	Dec. (1998.0)
V781 Tau	$05^h50^m06^s$	$26^o57'41''$
comparison 1	05 50 15	26 59 47
comparison 2	05 50 18	26 56 42
check	05 50 13	26 56 34

The photometric program of the TRAF software is used to reduce the images. The results of the reductions show that the difference between the magnitudes of the check star and the comparison star 1 is constant within probable errors of ±0.008 mag. for the V band and ±0.011 mag. for the B band. However, the difference between the magnitudes of the check star and the comparison star 2 is ±0.018 mag. for the V band and ±0.022 mag. for the B band. So the star 1 is used as a comparison star to reduce our observations of V781 Tau. Extinction corrections are not made because the comparison star 1 is very close to the variable.

From the observation made during the primary and the secondary eclipses, a mean epoch of the primary minimum light in the B and V bands and that of the secondary minimum light in the B and V bands are derived by means of the parabola fitting. The present times of minima together with those collected from the references are listed in Table 2, where the second column entry denotes the observation method: V refers to the visible observation, PG to the photographic one and PHE to the photoelectric one.

Table 2 Times of minimum light for V781 Tau

JD(hel.) 2400000+	Meth.	E	$(O-C)_1$	$(O-C)_2$	Ref.
32881.460	PG	-31812.5	-.0193	0.001	Berthold,1983
33950.515	PG	-28713.0	-.0110	0.005	ditto
34775.368	PG	-26321.5	-.0088	0.004	ditto
35540.371	PG	-24103.5	-.0149	-.004	ditto
36610.285	PG	-21001.5	-.0098	-.003	ditto
36957.442	PG	-19995.0	-.0041	0.002	ditto
38088.397	PG	-16716.0	-.0071	-.003	ditto
38440.378	PG	-15695.5	-.0061	-.003	ditto
39536.327	PG	-12518.0	-.0067	-.006	ditto
40981.332	PG	-8328.5	0.0003	-.000	ditto
41329.345	PG	-7319.5	-.0002	-.001	ditto
41330.381	PG	-7316.5	0.0010	0.000	ditto
41337.279	PG	-7296.5	0.0008	-.000	ditto
42839.363	PG	-2941.5	0.0044	0.003	ditto
43853.9088	PHE	0	-.0008	-.0022	Cereda etal.,1988
43874.9482	PHE	61.0	-.0009	-.0023	ditto
44636.339	PHE	2268.5	0.0024	0.0014	ditto
44637.371	PHE	2271.5	-.0003	-.0013	ditto
46113.4117	PHE	6551.	0.0006	0.0010	Pohl et al.,1987
46115.309	PHE	6556.5	0.0009	0.0013	ditto
46775.4651	PHE	8470.5	0.0004	0.0018	Cereda et al.,1988
46788.3981	PHE	8508.0	-.0007	0.0007	ditto
46798.4005	PHE	8537.0	-.0006	0.0007	ditto
46802.3682	PHE	8548.5	0.0006	0.0020	ditto
46802.5396	PHE	8549.0	-.0005	0.0009	ditto
48268.2285	PHE	12798.5	-.0041	0.0000	Keskin etal.,1989
48268.4014	PHE	12799.0	-.0036	0.0005	ditto
48607.2728	PHE	13781.5	-.0057	-.0008	ditto
48607.4454	PHE	13782.0	-.0056	-.0007	ditto
50814.1688	PHE	20180.0	-.0125	-.0015	this paper
50818.1359	PHE	20191.5	-.0119	-.0008	this paper

3. CHANGE IN THE PERIOD AND EXPLANATION

The (O-C)1 residuals in Table 2 are calculated by using the ephemeris given by Cereda et al. (1988), i.e.

$$Min.I = HJD2443853.9096 + 0.3449094E \tag{1}$$

The (O-C) diagram of the change in the orbital period of V781 Tau is shown in Fig 1, where (O-C)1 residuals listed in Table 2 are used.

It is clear that the period of this binary star decreased continually from 1949 to 1998. All epochs in Table 2 are used for a least square solution to obtain a quadratic ephemeris. The improved ephemeris is described as follows:

$$Min.I = HJD2443853.9110(15) + 0.34490929(18)E - 2.5(2) \times 10^{-11}E^2 \tag{2}$$

which is used to calculate the (O-C)2 in Table 2. The period change rate of V781 Tau is $\delta p/p = -5.0 \times 10^{-11}$.

The secular instability of the secondary component of a W UMa type binary arises because they obtain the luminosity transferred from the primary component (Lucy & Wilson 1979; Hazlehurst 1985; Wang 1994). So it could be suggested that decrease in the period of V781 Tau is caused by the contraction of the secondary component, then its shrinking velocity can be calculated from the decrease rate in the orbital period.

The Kepler's third law can be written as

$$A^3 = 74.5p^2M \tag{3}$$

where A is the separation between the two components in solar radii, p represents the orbital period in days and M indicates the total mass of the two components in solar mass. From the definition of the relative radius of one of the two components, one may have

$$A = \frac{R_1 + R_2}{r_1 + r_2}. \tag{4}$$

According to Binnendijk (1970) and Lacy (1977), one may acquire

$$r_1 + r_2 = 0.76 \tag{5}$$

and

$$\frac{R_2}{R_1} = q^{0.92}. \tag{6}$$

Inserting the equations (6), (5) and (4) into the equation (3), one may have

$$R_2^3(1+\frac{1}{q^{0.92}})^3 = 32.7p^2M. \tag{7}$$

Assuming conservation of total mass of the system, from the equation (7), one can obtain

$$\frac{dR_2}{dt} = \frac{21.8pMq^{2.76}}{R_2^2(1+q^{0.92})^3}\frac{dp}{dt}. \tag{8}$$

Since the period of V781 Tau decreases, from the equation (8) and the parameters published by Lu (1993), one may find a contracting velocity of the secondary component, $dR_2/dt = -6.77 \times 10^{-5} cms^{-1}$, namely the contraction of 21.4 m/year, which is in agreement with the results shown in the study of the thermal relaxation oscillation of contact binaries by Wang (1994).

4. DISCUSSION

The decrease in the orbital period of V781 Tau has presented the evidence to test the interesting results presented by Wang (1994), who suggested that the two subtypes of W UMa binaries should be in two different thermal relaxation oscillation states: the secondary components of the W-subtypes of W UMa type binaries are shrinking whereas the ones of A-subtypes are swelling. According to Wang(1994), a W-subtype W UMa binary has a contracting secondary component, which releases some gravitational energy to support its surface effective temperature higher than that of the primary component. The luminosity radiated by the secondary component consists of the three parts, the first is from the nuclear reaction in the secondary, the second is from the transferred luminosity from the primary and the third is from the released gravitational luminosity by the secondary component. Suppose the secondary component with a mass of m_2 and a radius of R_2 spherically symmetric, then its gravitational energy E_g is written (Kippenhahn & Weigert 1990) as follows:

$$E_g = \frac{3Gm_2^2}{(n-5)R_2}, \tag{9}$$

where n=3 for the main-sequence stars. For the low mass secondary components of W-subtype W UMa binaries, $n = 1.5$ adopted by Wang (1994). ¿From the equation (9), the following equation can be derived

$$\frac{dE_g}{dt} = \frac{3Gm_2^2}{(5-n)R_2^2}\frac{dR_2}{dt}. \tag{10}$$

Lu (1993) gave the physical parameters of V781 Tau and pointed out that the components of this system are main-sequence stars. A luminosity generated from the nuclear reactions of the secondary component with a mass of $0.501m_\odot$ is about $0.048L_\odot$. Assuming that the transferred luminosity from the primary component could only make the surface effective temperature of the secondary component the same as that of the primary component, then a maximum value of the transferred luminosity was about 0.65 $L_\odot$. The observed luminosity of the secondary component is $0.71L_\odot$ and it can be supposed that the over-luminosity of $0.012L_\odot$ was the gravitational luminosity released by the secondary component. From the equation (10), one may find a minimum contracting velocity of the secondary component, $dR_2/dt = -2.45 \times 10^{-6} cms^{-1}$. This minimum shrinking velocity is less than the contracting velocity of $6.77 \times 10^{-5} cms^{-1}$ calculated from the decrease in the period, because the superior limit of the transferred luminosity from the primary component is adopted. In fact, adopting the contracting velocity of $6.77 \times 10^{-5} cms^{-1}$ calculated from the decrease in the period and from the equation (10), one may find the gravitational luminosity released by the secondary component, $dE_g/dt = 0.330L_\odot$. Thus, the transferred luminosity from the primary component is only about $0.33L_\odot$ rather than $0.65L_\odot$.

The paradox of the over-luminous secondary component of the W UMa binary was explained by the introduction of large-scale energy transfer from the primary to the secondary (Lucy 1968). In the present discussion for V781 Tau, the secondary component of the system is continuously shrinking at the velocity of $6.77 \times 10^{-5} cms^{-1}$ and releasing the gravitational luminosity of $0.330L_\odot$, while the orbital period of the system was continuously decreasing with the change ratio of $\delta p/p = -5.0 \times 10^{-11}$. Then the transferred luminosity from the primary component may be only a part of the over-luminosity, the other part may be from the gravitational luminosity released by the secondary component when it shrinks. According to Mochnacki (1981), the ratio of energy transfer of V781 Tau should be of $\Delta L/L_1 = 0.45$, but in the present discussion, that is only of $\Delta L/L_1 = 0.22$.

It is interesting to notice that the W-phenomena of W UMa binaries arise also because of the contracting secondary components rather than only because the energy transfer from the primary to the secondary. The transferred luminosity from the primary should never cause the temperature of the secondary component to be higher than that of the primary if there were non-gravitational luminosity released by the secondary component.

In order to test the contracting model of the secondary components of the W-subtype W UMa Binaries, observations and analyses in the orbital period of more systems will be needed.

Acknowledgements

The authors would like to thank Dr. F. Li for his assistance in the observation and Dr. P. G. Niarchos for his useful advice. The authors would also like to express their gratitude for the support from the Chinese National Science Foundation Committee and the Chinese Academy of Sciences.

References

Berthold, T. 1981, IBVS, No.1942
Berthold, T. 1983, IBVS, NO.2443
Binnendijk, L. 1970, Vistas Atron., 12, 217
Cereda, L., Misto, A., Niarchos, P. G., & Poretti, E. 1988, A&AS, 76, 255
Diethelm, R. 1981a, Bed. Beo. Schweizer. Astron. Ges. 52, 7
Diethelm, R. 1981b, Bed. Beo. Schweizer. Astron. Ges. 53, 4
Flannery, B. P. 1976, ApJ, 205, 317
Harris, A. W. 1979, IBVS, No.1556
Hazlehurst, J. 1985, A&A, 145, 25
Keskin, V. & Pohl, E. 1989, IBVS, No.3355
Kippenhahn, R. & Weigert, A. 1990, Stellar Structure and Evolution, Berlin Springer, 184-185
Lacy, C. H. 1977, ApJ. Suppl. Series, 34, 479
Lu, W. X. 1993, AJ, 105, 646
Lucy, L. B. 1968, ApJ, 151, 1123
Lucy, L. B. 1976, ApJ, 205, 208
Lucy, L. B. & Wilson, R. E. 1979, ApJ, 231, 502
Mochnacki, R. 1981, ApJ, 245, 650
Pohl, E., Akan, M. C., Ibanoglu, C., Sezer, C., & Gudur, N. 1987, IBVS, No.3078
Robertson, J. A., & Eggleton, P. P. 1977, MNRAS, 179, 359
Wang, J. M. 1994, Apj, 434, 277

Yang, Y. L., & Li, L. F. 1999, Publication of Yunnan Observatory, No.1, p32

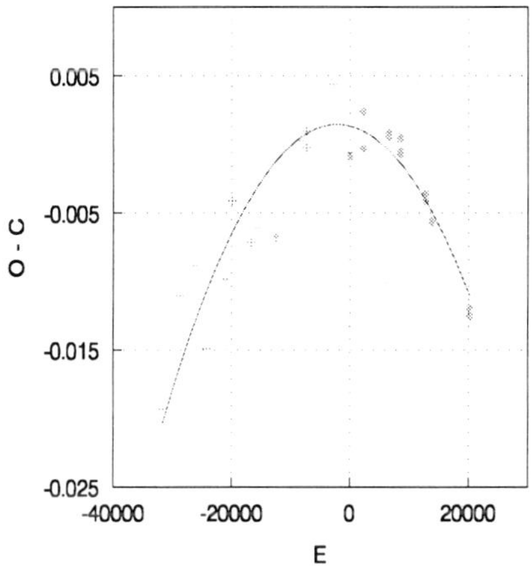

Fig.1.The O-C diagram of the period change for V781 Tau. The solid circles indicate the photoelectric observations and the crosses express the photographic or the visual one.

Figure 1 The (O-C) diagram of the change in the orbital period of V781 Tau

EXIT

THE CIRCUMSTELLAR ENVELOPE OF THE POST-AGB STAR

Yu Zhi-yao[1,2] and J. Chapman[3]
[1] *Shanghai Astronomical Observatory, Shabghai 200030, China*
[2] *National Astronomical Observatories, Chinese Academy of Sciences*
[3] *Division of Radiophysics, CSIRO, Australia*

Abstract OH17.7-2.0 is a typical POST-AGB star. We have observed the circumstellar OH maser radiation of OH17.7-2.0 using Parkes 64m radiotelescope, Australia. We have obtained four peaks in its spectra: the both peaks are from weaker- sources found in our observation run, and the other two are from the stronger-source observed previously. We used a two-dimensional Gaussian fitting procedure to obtain best-fitting positions for each of the four OH peaks. The best-fitting positions show that the position of the stronger OH maser source is clearly consistent with that of a IRAS point source. According to the best-fitting positions the characters of the both weaker-sources have been analysed. Meanwhile the time variation of the weaker-source has been studied.

Keywords: POST-AGB star, circumstellar envelope, OH maser

1. INTRODUCTION

The masing species OH, H_2O and SiO are known to exist in the circumstellar envelopes surrounding post-AGB stars. Each molecule samples a different velocity and temperature regime within envelope. The OH masers are found in the cool, outer regions of the envelopes. And observations of the OH can provide us with valuable information on the conditions and structure of these regions.

OH17.7-2.0 is a atracting post-AGB star. There is the typical circumstellar maser in its circumstellar envelope. Heske et al.(1990) observed OH17.7-2.0 and obtained its emission spectrum using CO (J=2-1). The mass loss rate has been got from its emission spectrum using the model i.e. collision and near-infrared pump exciting CO. Bowers, Johnston, and Spencer(1983) resolved all source using VLA. Diamond et al. (1985) resolved the 12 features of the circumstellar OH maser using MERLIN.

K.S. Cheng et al. (eds.), Stellar Astrophysics, 491–494.

Their results suggested an expanding shell model of the circumstellar envelope. But, the resolution of MERLIN is lower. The VLBI observation by Bower et al.(1980) suggested that the circumstellar OH maser consisted of some maser components, which angularsize is less than 40 mas. The maser componenets distributed in scale about 0."4.

OH17.7-2.0 is a post-AGB star which is of great interest. The 1612-MHz OH emission from OH17.7-2.0 is characterized by a double-peaked spectrum with the two peaks. Such a line profile has been assumed to represent maser emission from an expanding circumstellar shell. We observed the circumstellar OH maser using Parkes 64-m radio telescope in Australia. By Gaussian fitting the best fitting position of the peaks of the OH maser sources has been obtained.

2. OBSERVATION

The observations for the circumstellar OH maser radiation of OH17.7-2.0 were made with the Parkes 64m radiotelescope in Australia. The syatem temperature at 18cm of the radiotescope is 40k. The HPBW of the antenna at 18cm is 12.'5. The observations were made with the both left and right circular polarizations. The observations for the OH 1612 MHz maser radiation were made on the both January 9, and 10 of 1993. The rest center frequency is 1612.231 MHz. The channel number is 1024. The waveband is 4MHz. The velocity resolution is 1.44km/s. The observations for the circumstellar OH1612MHz maser radiation of OH17.7-2.0 were made with the follow positions in the Table 1. According to the ob-

Table 1 Observational position

observational position	RA(1950)(h m s)	DEC(1950)(° ' ")
1.IRAS position	18 27 39.8	-14 31 04.0
2.Central position	18 27 31.5	-14 31 35.0
3.North position	18 27 31.5	-14 25 35.0
4.South position	18 27 31.5	-14 37 35.0
5.East position	18 27 56.3	-14 31 35.0
6.West position	18 27 06.7	-14 31 35.0

servational results the parameters including the flux density and velocity of the peaks are shown in the Table 2. According to the observational results it is found that there is a weak source entetled the weak source 1 in the OH1612MHz spectrum. The peak velocity is 32.7km/s. The

Table 2 The parameters including flux density(Jy) and velocity(km/s) of the peaks

position	weak source 1		strong source			
	peak 1		peak 2		peak 3	
	flux density	velocity	flux density	velocity	flux density	velocity
1.IRAS	0.20	32.7	61.2	48.7	64.1	73.4
2.Center	0.55	32.7	59.7	48.7	62.6	73.4
3.North	0.09	32.7	33.4	48.7	35.5	73.4
4.south	0.67	32.7	28.0	48.7	29.4	73.4
5.East	-	-	44.4	48.7	45.9	73.4
6.West	1.52	32.7	19.2	48.7	20.6	73.4

strong source shows double-peaked structure. The best-fitting positions of the three OH peaks are shown in the Table 3 with a two-dimonsional Gaussian fitting procedure. On March 31 of 1993 the weak source 1

Table 3 The best-fitting position

		RA(1950)(h m s)	DEC(1950)(° ' ")
Strong source	peak 2	18 27 38.72	-14 31 08.00
	peak 3	18 27 39.04	-14 31 10.00
Weak source	peak 1	18 26 58.36	-14 35 48.40

of circumstellar OH 1612MHz maserr radiation of OH17.7-2.0 was observed again with the Parkes 64m radio telescope. The observations for the weak source 1 were made with the follow positions in the Table 4. Carefully analysing the second observational results it is found that the root of the peak 2 of the strong source is superimposed by a small peak entitled peak 4. The peak velocity of the peak 4 is 52.5km/s. We suggest that the peak 4 is a new weak OH source. According to the second observational results it is found that the flux density of the peak 4 is 2.3 Jy. The best-fitting positions of the three OH peaks in the second observational results are shown in the Table 5 with a two-dimensional Gassian fitting procedure. According to our observations and Diamond et al.(1985)'s results it is found that the source parameters of OH17.7-2.0

Table 4 The observational location for the weak source

observation position	RA(1950)(h m s)	DEC(1950)(° ' ")
Central position	18 26 58.0	-14 35 48.4
North position	18 26 58.0	-14 29 48.4
South position	18 26 58.0	-14 41 48.4
East position	18 27 28.0	-14 35 48.4
West position	18 26 34.0	-14 35 48.4

Table 5 The best-fitting position of the three peaks

source	peak	V(km/s)	RA(1950)(h m s)	DEC(1950)(° ' ")
Weak source 1	peak 1	32.5	18 27 01.05	-14 35 29.1
Weak source 2	peak 4	52.5	18 27 08.80	-14 35 02.7
Strong source	peak 2	48.7	18 27 44.27	-14 30 42.1

is as the velocity of the central star V_0=61km/s, the distance D=5.4kpc. From our observational results it is obtained that the offset of the weak source 1 approximately is 10', and the offset of the weak source 2 approximately is 4'23'. Diamond et al. (1985) have obtained that the OH shell approximately is 1'15" using MERLIN. Thus the weak source 1 and the weak source 2 are new OH maser sources, respectively.

The monitoring observations for the OH1612MHz maser of the weak source 1 were made. According to the curve of its flux density it is found that the time scale of the variation is 520.5 day.

This work was supported by the National Foundation of Natural Science of China, the Astronomy Committee, and the Union Lab of Radio astronomy, Chinese Academemy of Science.

References

Bowers,P.F., Johnston,K.J., and Spencer, J.H., Ap.J., 1983, 274,733.
Bowers,P.F., Reid,M.J., Johnston,K.J. et al., Ap.J., 1980, 242, 1088.
Diamond,P.J., Norris,R.P., Rowland P.R. et al., MNRAS, 1985, 212, 1.
Heske,A., Forveille,T., Omont,A. et al., A.Ap., 1990, 239.173.

THE STATISTICS OF X-RAY EMISSION FROM GALACTIC YOUNG PULSARS

L. Zhang[1,2] and K.S. Cheng[1]

[1] *Department of Physics, University of Hong Kong, Pokfulam Road, Hong Kong, China*

[2] *Department of Physics, Yunnan University, Kunming, China*

e-mail addresses: lzhang@star.physics.hku.hk, hrspksc@hkucc.hku.hk

Abstract We study statistical properties of the Galactic population of X-ray pulsars with ages less than 10^6 years using the Monte Carlo method. Combining the initial distributions of neutron stars at birth together with either the X-ray pulsar model proposed by Cheng & Zhang (1999) or the expected X-ray flux obtained by the empirical formula proposed by Becker & Trümper (1997), we obtain that (i) the distance, period, period derivative and X-ray energy flux distributions of simulated X-ray pulsars in both models are consistent with those of the known X-ray pulsars observed by ROSAT if the minimum detectable X-ray energy flux is 2×10^{-14} ergs cm^{-2} s^{-1} and (ii) the ratio of expected X-ray pulsars in our model to those in Becker & Trümper's model is about 0.6. Moreover, as the minimum detectable X-ray energy flux decreases, the simulated distributions from the two models are obviously different except for the distance distributions.

Keywords: X-rays: theory - pulsars: general - stars: neutron - stars: statistics

Introduction

Recently, ROSAT and ASCA observations of pulsars have detected 33 X-ray emitting pulsars ((Becker & Trümper 1997), (Becker & Trümper 1999)), of which 31 are Galactic and the pulsars PSR B0540-69 and PSR J0537-6909. Among these Galactic X-ray pulsars, there are 21 normal pulsars , 9 millisecond pulsars and 1 pulsar in a binary system (see the updated version of table 1 from Becker & Trümper (1997)). Furthermore, 18 of the 21 normal pulsars have ages less than 10^6 years (we refer to them as young X-ray pulsars). This information allows us to perform useful statistical studies of the observed X-ray pulsar parameters such as their distance, period, magnetic field and X-ray flux distributions. In

K.S. Cheng et al. (eds.), Stellar Astrophysics, 495–499.

order to do so, we need to know the initial distributions of pulsars which can be obtained from the statistical analysis of radio pulsars. In this paper, we will study the statistical properties of young pulsars which emit X-rays in our Galaxy. Obviously, the initial distributions of the pulsars provide the basic information for the study of the statistical properties of X-ray pulsars. Furthermore, we need a model which describes X-ray emission from rotation-powered pulsars. Here, combining both the model presented by Cheng & Zhang (1999) (hereafter CZ model) and the empirical formula obtained by Becker & Trümper (1997) (hereafter BT model) with the initial distribtions of the pulsars, we study the statistical properties of X-ray emission from the pulsars.

Monte Carlo Simulation of Galactic Pulsars

Now we consider the statistic properties of young X-ray pulsars in our Galaxy. The initial distributions of pulsar parameters such as position, velocity, period and magnetic field strength at birth used here are given by Zhang & Cheng (1999). Using these initial distributions, We evolve its period and follow its trajectory until the present time. At any time t, the evolution of the pulsar's period and period derivative are determined. The pulsar's location at any time t is determined by following its motion in the Galactic gravitational potential. Using the equations given by Paczynski (1990), the orbit integrations are performed using the 4th order Runge Kutta method with a variable time step (Press et al. 1992) for the variables R, V_R, z, V_z and ϕ. Once the coordinates (R(t),z(t),ϕ) are determined, we can calculate the Galactic longitude l, latitude b and the distance to the pulsar d. Whether or not the pulsar will be observed in the radio band depends on the radio detectability of pulsars: (i) radio energy flux must be greater then the. minimum detectable average flux density in a radio survey of pulsars and (ii) the radio beam is point to the Earth (the radio beaming fraction satisfies the expression given by Emmering & Chevalier (1989).

In order to model the Galactic young X-ray pulsars, we use CZ model to determine the energy flux of X-ray emission from a pulsar. In their model, in principle, the X-ray emission from a pulsar in the ROSAT energy range mainly consists of non-thermal and soft thermal components. Here, we assume that the magnetic inclination angles of the X-ray pulsars have a random distribution. For comparison, we also use BT modelto calculate the energy flux of an X-ray pulsar. Furthermore, we also need to know the minimum detectable energy flux of ROSAT. Generally, the minimum detectable energy flux in ROSAT depends strongly on the observation time, i.e. it will decreases with longer exposure time. However, for the typical neutron star observation done with ROSAT, the minimum detectable energy flux is in the range of a few times 10^{-14} ergs cm^{-2} s^{-1} (Becker, private communication) Therefore, we set the minimum detectable energy flux in ROSAT energy region to be $S_{Xmin} \sim 2 \times 10^{-14}$ erg cm^{-2} s^{-1}. Those pulsars whose X-ray energy fluxes are greater than S_{Xmin} will be considered as detectable X-ray pulsars.

Simulation Results and Discussions

We will use the Monte Carlo simulation code described in Cheng & Zhang (1998) to calculate the distributions of distance, period, period derivative and X-ray flux of the Galactic X-rays pulsars with ages less than 10^6 years. In our simulations, we choose an initial period $P_0 = 30$ ms, and use BT model and CZ model to determine the expected X-ray energy fluxes from the young pulsars respectively.

In Fig. 1, the normalized cumulative distributions of distance, period, period derivative and X-ray energy flux of the pulsars in both our model (long dashed curves) and Becker & Trümper's model (short dashed curves) are shown, where $L_X = 3.6 \times 10^{-5} L_{sd}^{1.04}$ is used for Becker & Trümper's model. For comparison, the corresponding observed distributions are also shown. A KS test indicates that in our model the maximum deviations of distance, period, period derivative and flux distributions from the observed distributions are 0.22, 0.14, 0.21 and 0.19 respectively. It can be seen that all cumulative distributions cannot be rejected at the $> 80\%$ confidence level. From Fig. 1, one can see that the simulated distributions for both models are consistent with the observed data.

Finally, we would like to estimate the number of possible Galactic young X-ray pulsars from our statistical analysis. Using 2.0×10^{-14} erg s^{-1} cm^{-2} as the minimum detectable energy flux and taking the radio pulsars with ages $\leq 10^6$ years in the Princeton pulsar archive, BT model predicts 46 Galactic young X-ray pulsars. From our simulations, however, the ratio of simulated Galactic young X-ray pulsar numbers in our model to those obtained in BT model is about 0.6. Therefore, our model predicts that about 28 Galactic pulsars with ages $\leq 10^6$ years should be X-rays emitters. It should be pointed out that the minimum detectable energy flux of ROSAT depends on the exposure time although it is in the range a few times 10^{-14} erg s^{-1} cm^{-2} for typical neutron star observations. Our model predictions are very different from those predicted by BT model when the minimum detectable energy flux is much less than 2×10^{-14} ergs cm^{-2} s^{-1}. The next generation of X-ray satellites, e.g. XMM, AXAF etc, can easily differentiate these two models.

Acknowledgements

We thank W. Becker for useful suggestions. This work is partially supported by a RGC grant of Hong Kong Government and Natural Science Foundation of Yunnan Province.

References

Becker, W. & Trümper, J. 1997, A&A, 326, 682
Becker, W. & Trümper, J. 1999, A&A, 341, 803
Cheng, K.S. & Zhang, L., 1998, ApJ, 498, 327
Cheng, K.S. & Zhang, L., 1999, ApJ, 515, 337
Emmering, R.T. & Chevalier, R.A. 1989, ApJ, 345, 931
Paczynski, B. 1990, ApJ, 348, 485
Press, W., Flannery, B., Teukolsky, S., Vetterling, W. 1992, Numerical Recipes: The Art of Scientific Computing 2nd ed., Cambridge Univ. Press, Cambridge
Zhang, L. & Cheng, K.S. 1999, ApJ, 526, 327

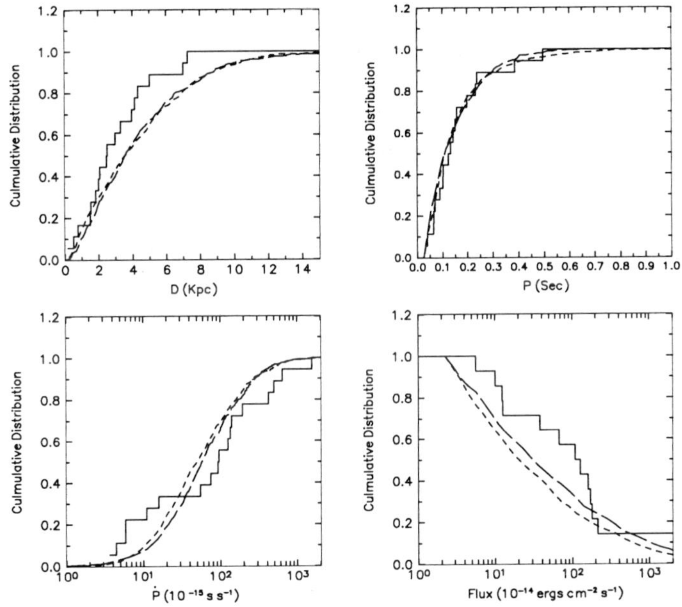

Figure 1 The normalized cumulative distributions of distance, period, period derivative and X-ray energy flux of Galactic young X-ray pulsars for both CZ and BT models. In our simulations, $S_{X\,min} = 2 \times 10^{-14}$ ergs cm^{-2} s^{-1}. The X-ray luminosity $L_X = 3.6 \times 10^{-5}\ L_{sd}^b$ for BT model. Long dashed curves are the simulated results for CZ model, and short dashed curves are those for BT model. Corresponding distributions of 18 X-ray pulsars observed by ROSAT are also shown (solid histograms).

A POLY-COMPONENT MODEL FOR GALACTIC HALO

Yuan-Zhong Zhang[1] and Ke-Jian Jin[2]
[1]*Institute of Theoretical Physics, The Chinese Academy of Sciences, P.O. Box 2735, Beijing, China*
The State Key Lab. of Scientific and Engineering Computing, The Chinese Academy of Sciences
[2]*Department of Physics, Northern Jiaotong University, Beijing, China*
[1]yzhang@itp.ac.cn

Abstract A polycomponent model for the dark halo of galaxy is proposed. For some two-component dark halo systems We give the rotation curves which are flat at large radii. We calculate deflection angles. In particular, for the two-component model as a gravitational lens we find four critical curves, three of them radial and one tangential. It implies that polycomponent dark matter system might give more critical curves.

Keywords: Dark matter, galactic halo, rotation curves, gravitational lensing

1. INTRODUCTION

The observed flat rotation curves at large radii for galaxies suggests a massive dark halo surrounding the galactic disk. It is shown that the rotation curves have different shapes and some of them have several peaks in both inner and outer regions. This non-smoothness might come from non-smooth mass distribution, i.e., polylaminate distribution. For the inner region of galaxy, a two-component model has been widely used (Miyamoto and Nagai, 1975). Recently, a three-component model for the visible matter has been proposed too (Sofue, 1996). It should be similar that the non-smoothness of the rotation curves at large radii might imply the polylaminate distribution of dark matter inside the halos. In this paper we shall study the poly-component model of dark halo consisting

*The project supported partially by National Natural Science Foundation of China under Grant Nos. 19745008 and 19835040

K.S. Cheng et al. (eds.), Stellar Astrophysics, 501–504.

of some polytropic spheres. The non-relativistic fermion-fermion star (Jin and Zhang, 1999) is an example of the poly-component model.

2. POLY-COMPONENT DARK MATTER SYSTEMS

Poly-component dark matter systems consisting of several types of polytropic spheres having the equations of state $p = K\rho^\gamma$ with different values of the parameters K and γ. Only gravitational interaction exists among the different spheres. The metric for these systems is the spherically symmetric and static. The dynamics is described by Einstein's field equations of the metric and the covariant conservation equations for the energy–momentum tensor of the polytropic fluids.

Deflection angle in the gravitational field is given by $\hat{\alpha}(r_0) = \Delta\phi(r_0) - \pi$, where $\Delta\phi(r_0) = 2\int_{r_0}^{\infty} \frac{e^{\mu/2}}{\sqrt{(r^4/b^2)e^{-\nu} - r^2}} dr$ where r_0 is the closest distance between the light ray and the center of the gravitational force, the impact parameter b is defined by $b = r_0 exp[-\nu(r_0)/2]$, and $exp(\mu)$ is the time-time component of the metric temsor. An observer O and a point source S are assumed to be located in an asymptotically flat spacetime far away from the dark matter star (as a lens L). Let D_{ol} denote the distance from the observer to the center of the lens, D_{ls} the distance between the lens and the source, and D_{os} the distance between the observer and the source. An image position is specified by the angle θ between OL line and the tangent to the null geodesic at the observer. β stands for the true angular position of the source. The lens equation may be expressed as (Virbhadra et al., 1998; Dabrowski and Schunck, 1998) $\sin(\theta - \beta) = (D_{ls}/D_{os}) \sin\hat{\alpha}$. The geometry gives $\sin\theta = b/D_{ol} = (r_0/D_{ol}) epx(-\nu(r_0)/2)$.

3. RESULTS AND DISCUSSION

In the present paper we consider the two-component dark matter stars with some values of γ_1 and γ_2. We calculated numerically the velocities of test particles $v(r)$ and the deflection angles $\hat{\alpha}$. Figures 1 and 2 show two rotation curves and two gravitational lensing effects for the two two-component polytropic stars with the values of parameters: $K_1 = 2.75 \times 10^4$, $K_2 = 0.223$, $\gamma_1 = 2.17$, $\gamma_2 = 1.8$, $\rho_{20} = 3.50 \times 10^{-8}$, $\rho_{10} = 8.11 \times 10^{-10}$ for the solid lines, and $\rho_{20} = 9.673 \times 10^{-10}$ for the dashed lines. We can see in Fig. 1 that the curves have two peaks which imply the number of images being from 1 to 5. Thus we can find four critical curves, three of them radial and one tangential, similar to the case of the gravitational lensing for the fermion-fermion stars (Jin, Zhang, and Zhu, 1999).

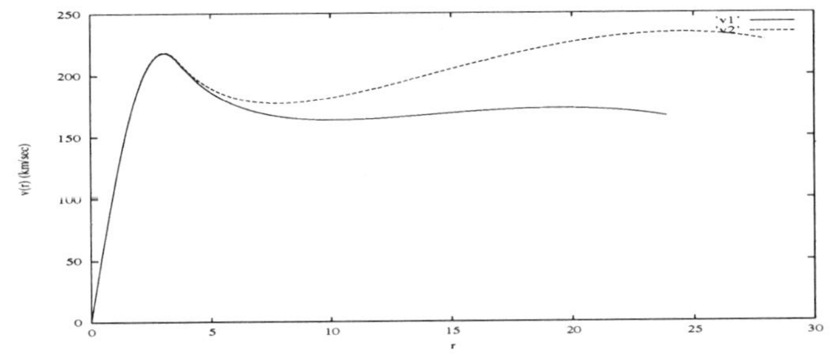

Figure 1 Rotation curves for the tow-component polytropic spheres

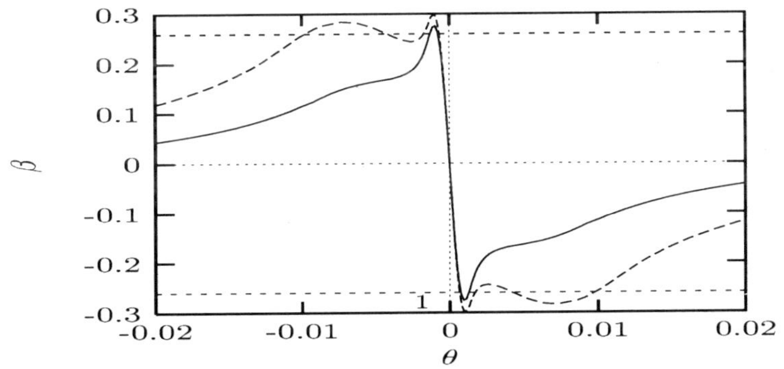

Figure 2 Gravitational lensing for the tow-component polytropic spheres

In calculations we did not use a specific unit which can be found by comparison with the that in the fermion-fermion stars (Jin, Zhang, and Zhu, 1999). It is well known (Weinberg, 1972) that a non-relativistic fermion star is a polytropic sphere with the parameters

$$\gamma = \frac{5}{3}, \qquad K = \frac{\hbar^2}{15 m_e \pi^2} \left(\frac{3\pi^2 \rho}{m_N \mu} \right)^{5/3}$$

, and an extremely relativistic fermion star is a polytropic sphere with the parameters

$$\gamma = \frac{4}{3}, \qquad K = \frac{\hbar}{12\pi^2} \left(\frac{3\pi^2 \rho}{m_N \mu} \right)^{4/3}$$

Therefore we know that both non-relativistic and extremely relativistic fermion-fermion stars are two-component polytropic spheres, and that the fermion stars are the special cases of fermion-fermion stars.

The shapes of rotation curves for two-component polytropic stars depend on the values of the parameters K_i and γ_i and the boundary conditions ρ_{i0}. By choosing appropriately the values for the parameters and boundary conditions we can get a flat rotation curve at large radii. This presents a striking contrast with the rotation curves of fermion stars, which have no flat part. At the same time the gravitational lensing effects of the two-component polytropic stars are similar to those of the fermion-fermion stars, there being four critical curves, three of them radial and one tangential. This again present a striking contrast with those of fermion stars. The results in this paper might suggest that the polycomponent dark mater stars could give different rotation curves of test particles and more critical curves of gravitational lensing.

References

Dabrowski M.P. and Schunck F., *Boson stars as gravitational lenses*, astro-ph/9807039.

Jin K.J., and Zhang Y.Z., *Fermion-Fermion Stars* Commun. Theor. Phys. bf 32, 457, 1999.

Jin K.J., Zhang Y.Z., and Zhu Z.H., *Gravitational Lensing Effects of Fermion-Fermion Stars: Strong Field Case*, Phys. Lett. A (in press), (gr-qc/9907035).

Liu D.H., *Gravitational Lensing of Fermion Stars*, Thesis (Institute of Theoretical Physics, the Chinese Academy of Sciences, Beijing, China), August 1999.

Miyamoto M., and Nagai R., PASJ, **46**, 533, 1975.

Sofue Y., Astrophys. J. **458**, 120, 1996.

Virbhadra K.S., Narasimha D., and Chitre S.M., *Role of the scalar field in gravitational lensing*, astro-ph/9801174;

Weinberg S., 1972, *Gravitation and cosmology: principles and applications of the general theory of relativity*, John Wiely & Sons, NY.

AUTHOR INDEX